MODELLING STEEL AND COMPOSITE STRUCTURES

MODELLING STEEL AND COMPOSITE STRUCTURES

PEDRO COLMAR GONÇALVES DA SILVA VELLASCO
Rio de Janeiro State University, Rio de Janeiro, Brazil

LUCIANO RODRIGUES ORNELAS DE LIMA
Rio de Janeiro State University, Rio de Janeiro, Brazil

SEBASTIÃO ARTHUR LOPES DE ANDRADE
Rio de Janeiro State University, Rio de Janeiro, Brazil
Pontifical Catholic University of Rio de Janeiro, Rio de Janeiro, Brazil

MARLEY MARIA BERNARDES REBUZZI VELLASCO
Pontifical Catholic University of Rio de Janeiro, Rio de Janeiro, Brazil

LUÍS ALBERTO PROENÇA SIMÕES DA SILVA
University of Coimbra, Coimbra, Portugal

ELSEVIER

Butterworth-Heinemann
An imprint of Elsevier

Butterworth-Heinemann is an imprint of Elsevier
The Boulevard, Langford Lane, Kidlington, Oxford OX5 1GB, United Kingdom
50 Hampshire Street, 5th Floor, Cambridge, MA 02139, United States

British Library Cataloguing-in-Publication Data
A catalogue record for this book is available from the British Library

Library of Congress Cataloging-in-Publication Data
A catalog record for this book is available from the Library of Congress

ISBN: 978-0-12-813526-6

For information on all Butterworth-Heinemann publications
visit our website at https://www.elsevier.com/books-and-journals

Publisher: Matthew Deans
Acquisition Editor: Ken McCombs
Editorial Project Manager Intern: Gabriela Capille
Editorial Project Manager: Mariana Kuhl
Production Project Manager: Anusha Sambamoorthy
Cover Designer: Mark Rogers

Typeset by SPi Global, India

CONTENTS

PREFACE

The main objective of the present investigation is to describe, with the aid of a series of examples, the methods and procedures used for the modelling of steel and composite structures. Initially the main models used to determine the structural behavior of these elements are introduced. This is followed by a detailed description of the experimental models, as well as their main requirements and precautions that have to be taken into account when performing steel and composite structural experiments in the laboratory.

The work proceeds focusing on numerical models based on the finite element method. Numerous simulations presenting a nonlinear response are illustrated, together with their main details and restrictions in terms of boundary conditions, main difficulties, solution strategies, and methods adopted to surpass convergence difficulties. Finally, examples of the use of computational intelligence methods to simulate steel and composite structures response are also presented. This is done with the aid of neural and neuro-fuzzy networks and genetic algorithmic focusing on their main advantages, scope, and limitations.

ACKNOWLEDGEMENTS

The author Pedro C.G. da S. Vellasco thanks his son and daughters: Murillo, Mariana, and Maria Clara for all the caring and incentive for the conception of this book.

The authors thank the support of CAPES, CNPq, and FAPERJ that financed these investigations. Thanks are also to all co-authors: Ricardo R. de Araújo, José G. S. da Silva, Juliana da C. Vianna, Flávio Rodrigues, Luiz Biondi Neto, Marco A. C. Pacheco, Fernando B. Ramires, Luis A. C. Borges, Alexandre A. Del Sávio, Luiz F. C. R. Martha, Aluísio J. R. Mergulhão, Ronaldo S. de Souza, Luiz C. V. de Carvalho, Alex V. D'Este, Leon T. S. Ferreira, Olavo F. Brito Jr, Tadeu H. Takey, Yuri R. de S. Rosa, Allyson J. do N. Beltrão, Luis F. da C. Neves, Alan da S. Sirqueira, David A. Nethercot, Mateus C. Bittencourt, João de J. dos Santos, Monique C. Rodrigues, André T. da Silva, Elaine T. Fonseca, and Bruno C. da Cruz.

CHAPTER ONE

Introduction

1.1 INITIAL CONSIDERATIONS

The current dynamics of scientific progress and technological innovations, in addition to the markets globalisation, imposed changes in the engineers training. On the other hand, an innovation of processes and integrated computer systems was necessary, in order to enable the construction industry to compete at international level. This change of attitude can also be seen in the adoption of more efficient and economical structural systems like the ones that use steel and composite solutions aiming to become cost-effective and viable alternatives.

One of the main objectives of this book is to enable and boost the use of steel and composite structures in buildings. For that, it is intended to improve the training of a new generation of engineers who are familiar with its behaviour to widespread its use in Brazil. The structural design development associated with new constructive techniques turns out to be a direct consequence of such ideas, but this will only be possible with a better understanding of the behaviour of the structural elements that form the global structure.

This understanding is based on the complete development of experimental and numerical models contemplating the behaviour of elements and structural systems. This strategy enables a better understanding of phenomena such as strength, structural stability, and stiffness; fabrication processes effects; erection aspects; and the steel and composite structural systems' dynamic response.

The structural behaviour requires the understanding of a series of physical phenomena related to the occurrence of ultimate limit states like flange local buckling, web local buckling, lateral torsional buckling, plastic hinge formation, or even, the crack distributions shown in Figs 1.1–1.5.

Naturally, over the last few years there has been a considerable evolution of procedures and models used in the steel and composite structures design. This evolution was the result of an increased knowledge level of structural

Modelling Steel and Composite Structures
http://dx.doi.org/10.1016/B978-0-12-813526-6.00001-5

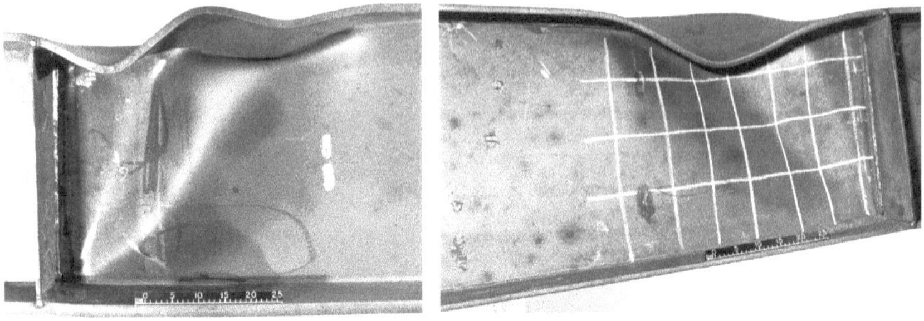

(A) (B)

Fig. 1.1 Flange local buckling.

(A) (B)

Fig. 1.2 Web local buckling.

Fig. 1.3 Lateral torsional buckling of a cantilever beam [8].

(A) (B)

Fig. 1.4 Plastic hinge formation.

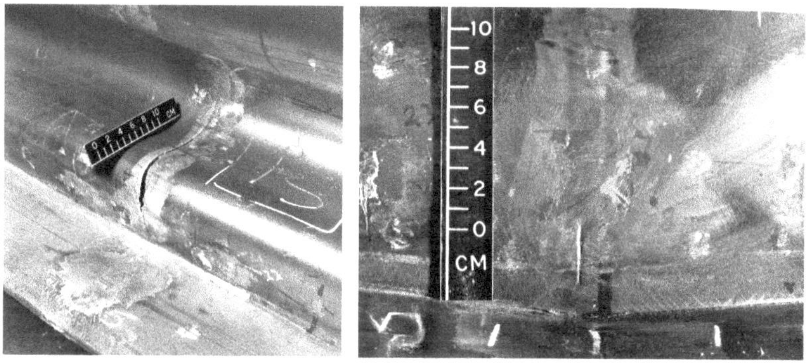

(A) (B)

Fig. 1.5 Crack formation.

behaviour obtained with the use of new numerical and experimental techniques. The consolidation of such knowledge has been made through structural design scientific publications present in journals and conferences. These publications generated discussions and ideas that were later consolidated by the technical committees of structural design standards, and scientific societies such as ECCS, Eurocodes, and ABNT.

However, the incorporation of the outcomes of experimental and numerical investigations in the design standards is not a simple process, giving rise to numerous discussions and to the development of more accurate models to be inserted in their design recommendations. These structural behaviour models can be associated with various natures, complexities, and formats where the most frequently used will be detailed in the next section of this chapter.

1.2 STRUCTURAL BEHAVIOUR MODELS

1.2.1 Analytical, Mathematical, and Hybrid Models

One of the simplest ways to understand these types of models comes from examples such as the structural joints. The joints classification in terms of moment versus rotation curves can be divided into three types: the analytical, mathematical, and combined models. In the analytical models, the moment versus rotation curve is based on their physical characteristics. In mathematical models, on the other hand, this curve is expressed by a mathematical function in which the parameters are determined by a curve adjusted to experimental results. Finally, the combined models use both analytical and mathematical models.

The analytical models can be used to predict the joint stiffness based on their geometric properties and components arrangement. The joint mechanical behaviour can be predicted by numerical methods such as finite elements based on the connection components deformation mechanism hypothesis. With this in hand, the components deformation and the connection moment capacity can be determined as well as their associated moment versus rotation curve.

Generally, parametric studies are conducted considering the effects of several geometric variables related to joint components. Practical values of these variables are then analysed to produce data for the analysis. However, the cost and time involved are usually unsatisfactory for practical applications, because each type of joint or joint component configuration requires a new formulation for obtaining the moment versus rotation curve, Chan [1]. In addition, the joint uncertainties can significantly affect their model predicted stiffness. There is still the fact that additional data-handling procedures are necessary to incorporate analytical results in the semirigid frame analysis. Another classic model for evaluating joints is characterised by the components method present in the Eurocode 3 pt. 1.8 [19], which is also known as a classical mechanical model. Further details on this model will be made in the examples present in Chapters 2–4.

Another method used to determine the joints moment versus rotation curve consists in approximating a curve to match experimental data by using simple expressions. These expressions are called mathematical models, which directly relate the connections moment and rotation by mathematical functions, using curve-fitting procedures. When these procedures are determined by experimental data adjustment, the moment versus rotation curve can be explicitly expressed and directly used in structural analysis.

The mathematical models are simpler than the analytical models previously mentioned. Examples of mathematical models include the model of Richard and Abbott [2], Lui and Chen [3,4], and Al-Bermani et al. [5]. Since numerous tests on various types of connections have been carried out in recent decades, many data, for various types of connections, are accessible for determining the parameters or constants required for the mathematical models.

Composite models are combinations of analytical and mathematical models. In the composite models formulation, the moment versus rotation curves are expressed in terms of both adjustment constants as well as geometric parameters. With the goal of developing a general expression for all connections with a similar component layout, usually these functions are, for convenience, standardised. The curves adjustment constants are determined by curves approximation techniques whilst the geometric parameters are based on the connection components geometry.

Composite models require few parameters when compared with the mathematical models and, in addition, similarly to analytical models, they keep the geometric parameters that establish a correlation with the joint physical characteristics, a feature that is not found in the curves adjustment mathematical models. In general, the hybrid models can be used to evaluate the initial stiffness of particular types of connections and to predict their nonlinear response. The polynomial model [6] and the Ramberg–Osgood [7] are the examples of hybrid models.

Another example of a mechanical/analytical model is shown in Fig. 1.6. In this figure, the web local buckling phenomenon is illustrated in the region where a direction change in the beam bottom flange occurs. This ultimate limit state is particularly significant in tapered beams. Chapters 2 and 3 will depict experimental and numerical models related to this phenomenon. These results enabled the development of a simple column model where the constants K_1 and K_2 were calibrated to resemble an effective width and a buckling length compatible with the already mentioned experimental and numerical results.

1.2.2 Experimental Models

The steel and composite structural models may also involve experiments carried out in the laboratory. These experiments are crucial to the understanding and calibration of the investigated structural behaviour and will be described in detail, as well as the arrangements, instrumentation, and techniques usually adopted in these tests. This chapter will also address

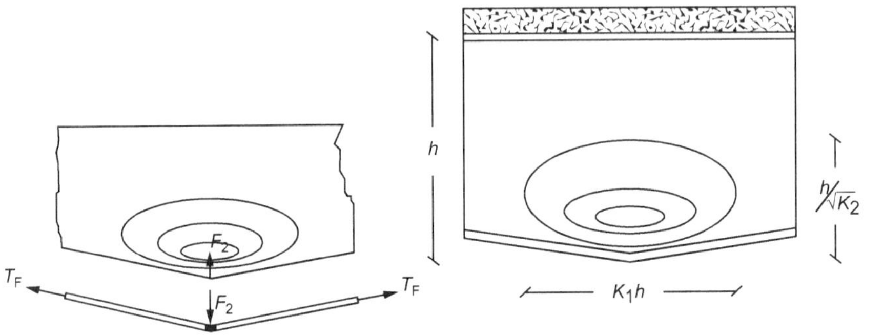

Fig. 1.6 Tapered beam's local web buckling analytical model.

some aspects and general questions to elucidate their key planning points in order to ensure the success of the demanded experimental programme.

1.2.3 Numerical Models

Structural behaviour models can also involve computer simulations. Many types of simulations are possible involving finite differences, finite strips, finite elements, computational intelligence techniques, amongst others. Chapters 3 and 4 of this book will focus on the latter two types of steel and composite structures computational modelling. First, the numerical models developed based on the finite element method will be discussed and presented. Several simulations showing nonlinear behaviours will be presented, as well as their main details and restrictions in terms of boundary conditions, major difficulties, solution strategies, and methods to overcome convergence difficulties. This will continue with the presentation of a series of examples of computational intelligence techniques application in the simulation of steel and composite structures response. Thus, neural networks, neuro-fuzzy networks, and genetic algorithms nebulae will be described and applied focusing on their main advantages, scope, and limitations.

1.3 STATISTICAL ANALYSIS OF INPUT DATA AND RESULTS

The correct planning of the investigated models must be carried out very carefully in order to consider the effects of the main variables that influence the studied structural behaviour. One of the classic forms of planning uses a number of techniques and procedures based on statistics and design of experiments.

The design of experiments aims at determining enough information with the least possible number of tests seeking to optimise the number of tests necessary to encompass all the possible variables. For that, it is necessary to follow some suggestions for a better planning: recognise, establish, and clearly delimit the problem; identify the possible factors that can affect the investigated problem; check which factors can be kept fixed, therefore, will not have their effects assessed in the experiments; identify, for each factor, the range and levels that will be studied; choose a suitable design of experiments (i.e. how the levels of the factors can be combined in order to solve the proposed problem with the lowest number of tests); and choose which variables properly measures the process results (quality, performance, etc.) for the tests data analysis planning.

This can be done through an orthogonal arrangement, in which the sequence and the number of experiments to be carried out can be determined. This type of arrangement includes the combination of variables and their adopted levels. Several methods were developed to represent the design of experiments such as 'Latin Square', Taguchi, 'D-Optimal', and others. The Taguchi method, for instance, works with equal levels for all variables of the project. Alternatively, the factorial method 'D-Optimal' allows the use of different levels for each variable through a quadratic optimisation of the complete factorial arrangement minimising the amount of experiments required to encompass the entire set.

The correct interpretation of the results, curve adjustment, and the main variables significance assessment that influence the problem can also be done with statistics-based methods such as those that use tables like the new Taguchi processes. These procedures, when used properly, can simplify the process of generating structural design equations. They can identify and quantify the terms that should be present and which terms may be discarded without an accuracy loss. Further details of these processes can be found in Lochner and Matar [9].

1.4 BOOK SCOPE

This chapter presented a description of the main models used in the simulation of the of steel and composite structures behaviour. Chapter 2 contains a detailed description of the experimental models, as well as the main requirements and precautions that should be taken for a steel or composite structure test in the laboratory.

Chapter 3 introduces a series of numerical models, developed with the aid of the finite element method. Various simulations depicting nonlinear responses are presented as well as their main details and restrictions in terms of boundary conditions, major difficulties, solution strategies, and methods to overcome convergence difficulties.

Chapter 4 summarises the use of computational intelligence techniques in the simulation steel and composite structures response. Neural networks, neuro-fuzzy networks, and genetic algorithms are described and applied focusing on their main advantages, scope, and limitations.

Finally, Chapter 5 presents a synthesis of what was presented in this book and establishes the main contributions, finalising with some considerations and thoughts about the present publication.

CHAPTER TWO

Experimental Modelling

2.1 INITIAL CONSIDERATIONS

This chapter aims to present the different aspects and techniques usually adopted in steel and composite structures tests carried out in the laboratory. The framework in which this book was conceived was based on a model in which the most important aspects of modelling are presented as tests case studies. This makes it easier for the general understanding throughout the use of actual practical situations without demanding the reader's a previsualisation capacities for the full understanding of the presented issues. Despite this fact, some general aspects and questions will be initially addressed to elucidate the tests planning key points in order to ensure the success of the experimental programme.

Normally, the dimensions of the structures to be tested must be adequate to the space available in the lab. This always raises questions related to which strategy would be most appropriate whether reducing the experiments' dimensions in order to adapt them with the available lab space or to execute them with their original dimensions, in an external experimental field. Even in nondestructive tests, as in the case of a structure proof load test, many questions remain and are not only related to the conditions of *in situ* temperature and humidity, but also to the actual availability of instrumentation systems to be used in the field.

Typically, in order to simplify the tests execution, the usual strategy is to modify the dimensions of the structure to the available lab space. Amongst other reasons such as convenience, ease access to apparatus, and measuring devices, another condition that proves to be extremely relevant is related to the possibility of repetition of the same test conditions. Although obvious, this condition of feasibility of repeatability is often of critical importance to ensure that doubts regarding the results causality or the checking of any parameter or even some type of instrumentation did not affect the test results.

Another fact that induces the use of the laboratory is related to the handling and transport of heavy or bulky loads or with unusual dimensions. The

Modelling Steel and Composite Structures
http://dx.doi.org/10.1016/B978-0-12-813526-6.00002-7

labs are usually provided with cranes and other similar lifting equipment for the transport of heavy loads that facilitate the assembly, disassembly, and the test execution. This can also be done on site, with cranes, munck trucks, or even derricks, however, that generally implies high additional costs, which often inhibit or at least reduce the use of such equipment, in order to balance the final test budget.

Nevertheless, not everything is an advantage in a test carried out in the laboratory. Whenever planning a test in such conditions, one should always bear in mind that the effect derived from the reduction of the experiment dimensions or even the cutting off of part of the structure being tested to fit the available lab space. This should be carefully planned in cases with reduced scale structural models or when part of the structure is simply suppressed. In the former, the various nonlinearities associated with possible buckling, yielding, or even manufacturing conditions are highly influenced by the plate's thickness that composes the structure. The reduced models often lead to the use of thicknesses that hinder the welding or even change the controlling of ultimate limit states that modifies the structural response. On the other hand, the suppression of part of the structure can also lead to a significant change of boundary conditions that modify the final test results. In these cases, well-planned and executed numerical simulations are often of fundamental importance for the test planning.

2.2 EXPERIMENTAL MODELS

The conception of the experiments must be carefully done, weighing the various factors that can influence their associated results and above all, keeping in mind that what is expected to be observed and measured during their execution. Typically, a useful technique can be adopted to assist in planning. It consists in the execution of a detailed description of the trajectories of the structure acting forces. Each acting force or moment has its behaviour within the structure individually identified focusing on the various steps required for their transmission within the structure. With this simple measure, the clear understanding where every force arises and how it is transmitted to its end can be achieved. This identifies the components and the structure parts that are subject to actions facilitating the instrumentation planning and their correct positioning within the tests. This identification also enables a preview and design of each structural component for a later comparison to the test results.

It is also necessary to emphasise the importance of the tests' repetition in an experimental programme. The repetition is crucial to elucidate questions relating the results of instrumentation, components behaviour and, most importantly, to support the fact that the tests performance was not accidentally consistent and by that, generating more reliable results. Despite this fact, it must be also taken into account that systematic errors of planning, instrumentation, or even of tests procedures cannot be detected through repetitions, since they are classified as systematic errors. Their real identification can only be made through a detailed analysis of the structural behaviour and can be facilitated through comparisons with the performance of analytical or numerical models of the same physical phenomenon.

2.3 LOAD APPLICATION SYSTEMS

The first step in the test planning is in the specification of the load application system to be used in the tested experimental model. The most simple and usual system is the one that, with hydraulic actuators, monotonically applies static loads to fixed points of the structure.

However, the system simplicity still requires some care to ensure that the planned loading trend is actually being applied to the structure. When more than one point of the structure receives these loads that can be done directly or indirectly. The direct method uses an actuator at each point where the load must be applied and generates the need for a careful control as the experiment evolves. This can be made by using pressure transducers in hydraulic system outputs and/or by using load cells located, if possible, at every point where loads are applied and at the structure's supports. This enables possible load imbalances to be detected at the test early stages ensuring their full development as initially planned. An indirect method consists, for example, in the use of beams distribution that divides the single point load applied by the actuator in more than one point along the tested structural model.

The control of the load application is a key feature to ensure the planned test programme success. It is usually done by using a load or a displacement control. The load control demands less complex application systems, but this strategy is inefficient in models that are subject to large deformation close to their ultimate limit state. In these cases, the displacement control strategy enables the clear detection of the ultimate load and the acquisition of the structure response in the unloading or postbuckling stage.

The models sometimes generate the need for the use of load application systems that follow the structure as it deforms. This must be carefully planned, with the use of universal hinges and other devices and apparatus that may be used as the test develops. The idea is to hold the structure as the load is removed, to increase the hydraulic actuators course to compensate for the displacements that are already present in the structure. More details on these aspects will be illustrated later at the examples present in this chapter. Cyclical nature loadings also demand caution in their application and control. In these cases, complex systems usually based on the adoption of sensors and servo-valves are used and can be indispensable in tests like push-outs (according to Eurocode 4 specifications), aiming to evaluate the model degradation with loading/time (based on their energy absorption capacity), tests involving fatigue model behaviour, amongst others. In these cases, one should always bear in mind that the load application system cannot be subject to overheating and to avoid that, an easy inspection and maintenance efficient cooling system must be adopted.

Finally, cases with dynamic loads, sudden, periodicals, or even simulating seismic functions can be cited. These particular load types require more care in their follow-up and monitoring, which are not part this book scope.

2.4 INSTRUMENTATION

One of the most important elements to ensure the success of an experimental programme consists in the correct specification of its instrumentation. Within this perspective, the adopted instrumentation design to be used in the tests shall, whenever possible, contemplate a series of measurements with redundancy. The reason for this measurement strategy comes from a large variability associated with the measured variables mainly used to calibrate and guarantee their trustworthiness. Additionally, some elements used in instrumentation, like the strain gauges, have a nonnegligible chance of damage or even measurement inconsistency, even before the test started. In these and in many other cases, the redundancy becomes essential for a correct instrumentation performance in the adopted experimental model.

A structural test normally uses sensors and transducers for measuring displacements, slips, rotations, forces, moments, pressures, natural frequencies and strains, amongst other variables. On the other hand, due to the use of hyperstatic and/or structural systems containing semi-rigid connections, many of these measurements can give rise to the use of indirect monitoring strategies to detect their behaviour since direct measurements are often not viable. One should also emphasise that support conditions correctly specified and adopted, as well as the use of bracing can decisively contribute to simplify the adopted instrumentation system. In addition, the instrumentation is normally conceived to monitor isolated effects in parts of the structure or its components and/or global effects in the structural model.

The measurement of displacements in structures is usually done with displacement transducers, LVDTs (Linear Variable Displcement Transducers), dial gauges, or similar instruments, etc. Its main objective is to determine values to be compared with service limit states to assess if these maximum limits have been exceeded. These values are also fundamental to measure the structure stiffness, and, consequently, the proximity of the structure ultimate load. Displacements may also be used for assessment of symmetry conditions, detection of rigid body movements, and even plate out of plane deflections indicating the onset of local lateral and/or torsional buckling. Slips at the concrete to steel interface usually use the same sensors, and their measures are generally obtained by the difference of displacements between the elements present in the interface.

The rotation measurement is usually not as trivial as the previous case. Systems using inclinometers, optical and laser measurements amongst others can be used, but the most usual ways can be done with LVDTs measurements positioned in structural elements. In the latter case, the rotations are acquired indirectly, with simple geometry concepts. More details about this process will be presented later in the examples of this chapter.

The structure's strain monitoring is usually done with electric strain gauges, but other processes such as photo-elasticity-based optical reflection, fibres optical, or even piezoelectric elements can be adopted. In steel, these measurements are simple and direct, but when the concrete needs to be monitored, elements immersed in this structural material must be used, otherwise only strain values on the material surface can be obtained. Immersion in concrete generates the need for protection that can be indirectly done with the use of thin steel bars embedded in concrete in order to measure equivalent deformations due to the compatibility between these two materials. Cracks and their development are also often tracked and monitored in the concrete where the visual inspection is one of the most effective instrumentation processes.

Temperature and quantities associated with characteristics related to the structure dynamic performance, like natural frequencies and vibration modes, are also often object of study that requires thermocouples and accelerometers for their monitoring. However, these processes are usually more specialised, and in the case of dynamic instrumentation measurements, specialised systems that compensate and/or reduce noises that usually comes associated with these measurements are necessary.

2.5 DATA PROCESSING AND FORMATTING

Whilst in a not-too-distant past, experimental measurements were performed using manually demanding readings and individual notes, the development and improvement of automated data acquisition systems facilitated the monitoring process of the structural models studied.

On the other hand, many of these systems still do not have an immediate response to all the measures, therefore encouraging the use of measuring instruments of immediate performance visualisation, such as dial gauges. During a test, many variables and operations are usually performed simultaneously, often demanding a quick decision, fact that could be crucial to the correct development of the test in course. Quick previews of the dial gauges can be the key to the success of a correct evaluation of the imminence of the structural collapse. In these gauges, very fast laps of its pointers indicate an increase of the displacements speed increment, signalising the load application system operators to immediately take steps to better overcome this test phase, like decreasing the load application rate, etc.

It must also be noted that either unnecessary or excessive measures can complicate the data interpretation and analysis. Most of the times, tests are extremely well described and treated, containing thousands of information and graphics of the studied models structural performance. However, a lack of analysis and the correct interpretation of these values have become increasingly common, generating a mass of raw data with little immediate use. In such cases, it is crucial to filter the relevant information in order to enable a correct interpretation of the structural model performance. The next section of this chapter will present 15 examples of model experiments involving steel and composite structures, as well as details regarding their design, instrumentation, and correct interpretation of their main results.

2.6 APPLICATION EXAMPLES

2.6.1 Local Web Buckling in Composite Tapered Beams

An ever-growing demand for very heavily serviced widespan office floors has led, as a natural development of the concept of hunched and tapered frames, to the utilisation of composite flooring systems supported by tapered (varying web depth) beams (Fig. 2.1). One of the aspects that deserve greater attention is the local buckling behaviour of the relatively deep slender web at changes of slope in the lower flange, in particular, at the slope change, which occurs in the centre of a beam whose depth increases linearly from each support. The provision of local stiffening is technically feasible, but very expensive, especially when automated fabrication systems involving cuts and welds are uses.

The principal aim of the present investigation was to identify the significance of the main variables that influence the phenomenon, such as angle of taper, web thickness, and others [10–12]. For practical reasons, the

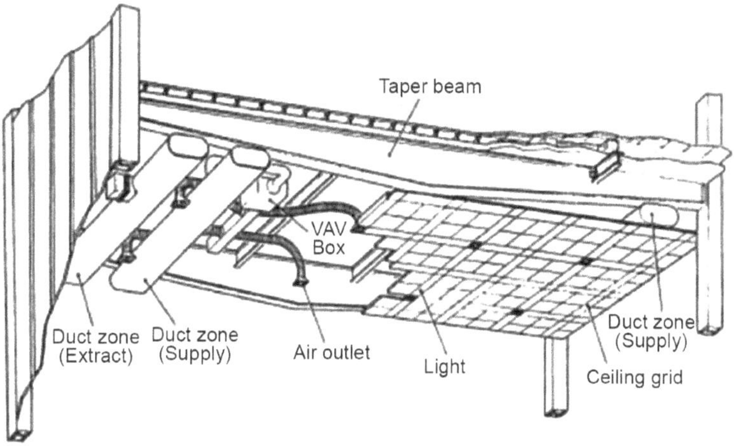

Fig. 2.1 Composite tapered beams.

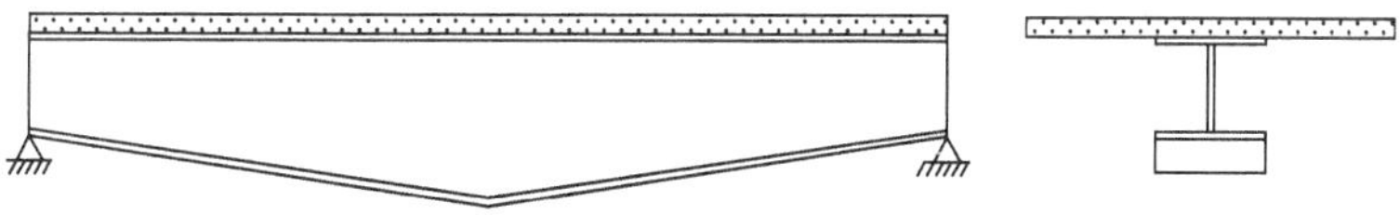

Fig. 2.2 Composite tapered beam model: scale 1/2.

experiments were executed at half-scale and the models simulated the composite girder concrete slab by a heavy steel top flange (Fig. 2.2). The only drawback to the use of the steel top flange was the reduced lateral torsional restraint, which was given for the same axial stiffness, but it was relatively easy to provide external lateral restraints, as it will be later described. The main objective of the experiments was to access whether vertical stiffeners were necessary in the web at the point where the bottom flange changes direction.

A first series of tests was performed to evaluate the influence of web thickness on the load-carrying capacity of the girders. The plastic neutral axes of these beams were in or close to the top flange. Three tapered models have been tested with different web thicknesses (i.e. 3, 5, and 7 mm). A second series was conducted, in which the plastic neutral axis was lowered progressively in order to assess the influence of neutral axis position, including the construction case where the steel girder has to carry the freshly cast slab. The test rig was made of a simply supported beam 11.46 m long subject to two symmetrically disposed concentrated loads (Fig. 2.3). The girder was composed of three distinct parts: a central 'specimen' and two 'loading arms'. The loading arms were reutilised in all of the tests with the purpose of just receiving the load from the hydraulic jacks and transmitting it effectively as a

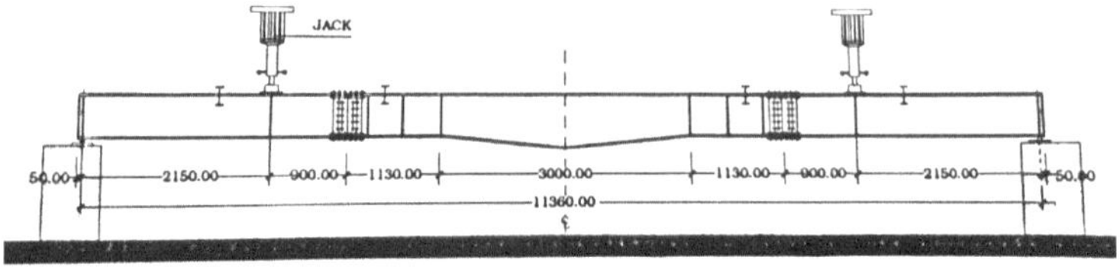

Fig. 2.3 Complete test layout.

Fig. 2.4 Adopted jack hinges and simple supports.

pure moment to the specimen. The specimen was changed in every test and was the main objective of the experiments.

Loading has been monotonically applied via two hydraulic actuators supported on loading frames. Steel rollers provided the simple supports. Everything was arranged on and firmly bolted to the laboratory strong floor structure (Fig. 2.4). The horizontal movement of the steel rollers did not exceed 15 mm during the tests and their performance was considered satisfactory.

Since the loading arms were shallower than the centre of the specimen, and still had to carry the same moment with no risk of failure, they were constructed in grade 55 steel, whilst the specimens were assembled from grade 43 plates. Each loading arm was 3.1 m long with an effective lever arm between the roller support and the jack centre line of 2.15 m. The total height of the loading arms was 500 mm, the top and bottom flanges were 25 mm thick and 300 mm wide, and the web was 12 mm thick measuring 3.1 m long generating an effective lever arm measured between the armrests and the load application point of 2.15 m. The specimens were 5.26 m long of which the central 3 M formed a tapered test panel with minimum and maximum heights of 500 and 650 mm, respectively. These dimensions correspond to a taper angle of 1:10 or 5°. The top and bottom flanges were 25, 12 mm thick and 300 mm wide. The outer ends of the specimens had to match the loading arms, and were thus 500 mm deep with 300 mm by 25 mm flanges and a 12-mm web thickness.

The moment was transmitted from the loading arms to the specimens by means of the full moment friction-grip connection. It used 20 M20 bolts in the web and 16 M24 bolts in each flange. The bolt torque was applied to ensure friction resistance, according to the part-turn method. In each flange connection, an outside plate 500 by 650 by 20 mm and two inside plates 120 by 650 by 20 mm were used. The difference in bottom flange thickness between the specimen and loading arms required the use of four extra 320 by 120 by 13 mm plates. In the web, two plates 400 by 375 by 10 mm were used. All plates were grade 55 steel.

As explained previously, four lateral restraint frames were utilised to provide the necessary support for the top flange to prevent a lateral buckling failure (Fig. 2.5). They were located 500 mm from the two jacks and 50 mm from the centres of the connections. The restraint was provided by guide rollers carried on the top flange of the beam acting on running surfaces mounted on robust portals bolted to the strong floor of the laboratory. During the tests, it was noted that the jacks could run out of stroke due to the large vertical deflections developed by the beams. Thick plates connected to the laboratory strong floor by long floor bolts were then introduced to the rig (Fig. 2.5).

The instrumentation used to monitor the trials was complex and motivated the development of a series of new apparatus for measurement of displacement. The value of the load in the jacks was obtained through a pressure transducer located at the return tube to the loading cabinet, and was frequently checked against the pressure indicated in the control cabinet. A plate carrying 36 displacement transducers was built to monitor the lateral deflections of the web at its critical central area. Their location and arrangement can be seen in detail in Fig. 2.6. The purpose of this transducer array

Fig. 2.5 Bolted splices designed for friction, lateral bracing, and adopted plates to compensate for jacks reaching their maximum strokes.

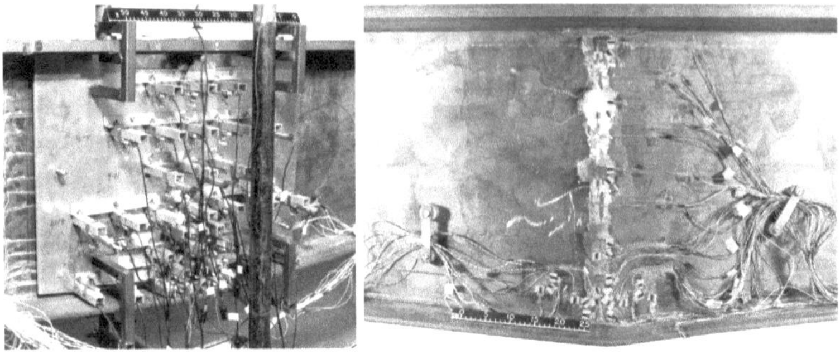

Fig. 2.6 Adopted test instrumentation.

was to map the vertical buckling pattern for later comparison with the finite element simulations. Vertical displacement transducers were positioned on the top flange at the ends and centre of the tapered beam section. The deformations were measured by electrical linear strain gauges positioned at the flanges and next to the extremes of the models and by rosettes located in the centre pane of the beam of variable inertia (Fig. 2.6).

The results of the first series of test confirmed that the web thickness is one of the most important variables in the vertical buckling problem. The 7 and 5 mm girders reached their ultimate load with no sign of web buckling (Fig. 2.7). They failed by the development of a plastic hinge at the edge of the tapered panel.

Fig. 2.7 Original and deformed beams with 7 and 5 mm webs.

Fig. 2.8 Deformed shape of 3 mm web beam and web vertical local buckling.

Fig. 2.9 Original and deformed 3 mm web beam.

On the other hand, the beam with the 3 mm web failed by vertical buckling (Figs 2.8 and 2.9). A graph that summarises the maximum moments (normalised by M_{pl}) achieved by the three specimens is shown in Fig. 2.10. It is very clear from that diagram that the reduction in resistance between the 5 and 3 mm web specimens was very significant. For beams with the present geometric and material characteristics, it is possible to say that the limit for the vertical buckling of the web lies between the values of 3 and 5 mm for the thickness of the web, which corresponds to h/t_w limits of 130 and 216.67.

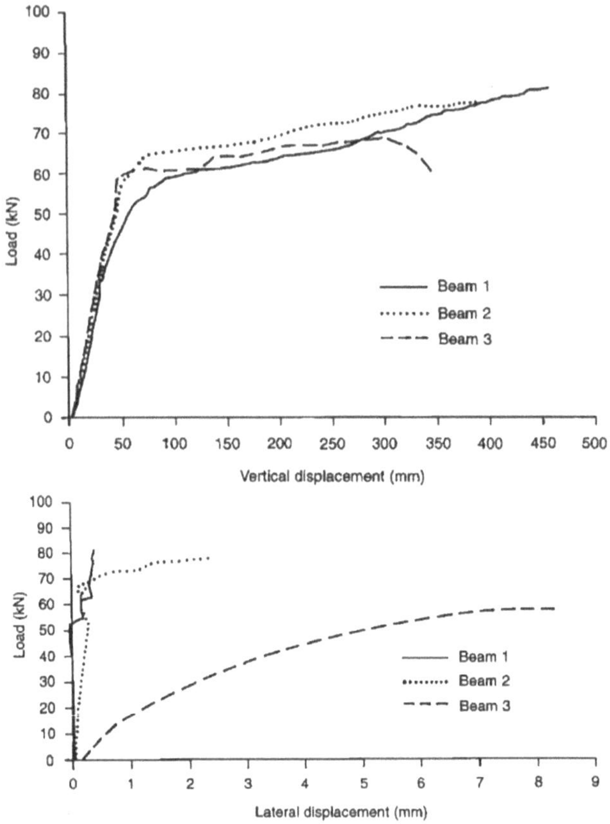

Fig. 2.10 Load versus vertical and lateral displacement curves: 1st series.

Alternatively, the plastic neutral axis position had no significant influence on the vertical buckling. This was demonstrated by the second series of tests summarised in Fig. 2.11. The ultimate moments (normalised by M_{pl}) for beam 2 and beams 4–6 are plotted against the plastic neutral axis position (Fig. 2.12). The significant decrease in resistance for beams with the neutral axis depth greater than 300 mm is explained by the sudden top flange buckling failure. This is the case of the fifth and sixth beams. Of course, in a real composite beam with the section properties of beam 5, the steel deck and the concrete slab might have prevented flange buckling so that local web buckling governed. However, the failure of beam 6, modelling uncropped construction, is not open for this objection. Even with its neutral axis in the lower flange, top flange buckling rather than web buckling governed the behaviour. This strongly suggests that plasticity rather than web buckling would govern the behaviour of a steel/concrete composite beam with the section properties of beam 5.

Fig. 2.11 Second series beams presenting flange local buckling.

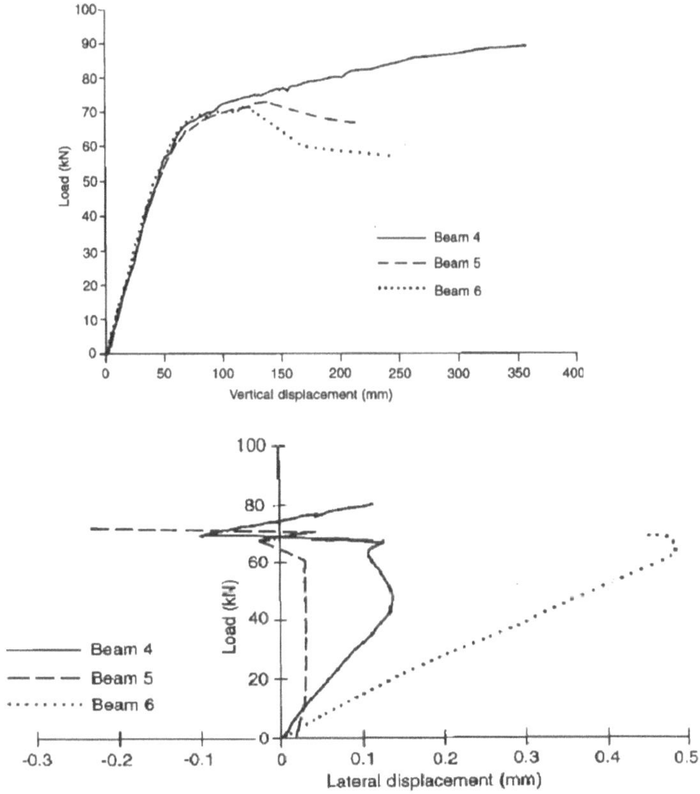

Fig. 2.12 Load versus vertical and lateral displacement curves: 2nd series.

The experiments indicated that vertical stiffeners in the web would only be needed in extreme situations. Most of the cases found in practice will not need stiffeners. Given the high cost of fitted stiffeners, the approach of slightly increasing the thickness of the web to avoid the vertical buckling seemed likely to be the most satisfactory design expedient in limiting cases.

One of the central issues of experimental modelling performed was whether the model conceived would really represent the worst case associated with the onset of local vertical buckling of the web. Some questions were raise regarding the proof that the maximum force developed in the lower flanges, and responsible for the local buckling, would really be achieved. The solution was quite simple and involved the use of linear strain gauges in the flanges. When these strain gauges reach the maximum load generated by the flanges (naturally neglecting the slight stress variation during strain hardening that occurs with the excessive strains leading to a phase of substantial beam deflections that are incompatible with the simply supported system service limits states), they ensure that the maximum vertical load imposed on the web by the flanges angle change, reaching its upper limit point.

Another relevant modelling issue in reduced scale models regards the problem of reducing the thickness in cases where buckling nonlinearities occur. In this case, there were no further complications because the scale reduction was not extreme but reductions of up to 2 or 3 can lead to very thin thicknesses, making the models difficult to weld and/or to the onset of local buckling limit states nonexistent in real cases.

A very significant aspect for any numerical or theoretical prediction modelling concerns the determination of the mechanical characteristics of the materials used in experiments. Often incorrect results are obtained when using incorrect loading speeds in the stress versus strain tests of the steels used in the experiments.

Often yield and rupture stresses present very close values particularly for steels without a defined yield plateau. This turns out to be even more problematic because in these cases, a straight line parallel to a linear segment passing through 0.2% or 0.1% strain values that could jeopardise the results determines the equivalent yield stress. This significantly complicates any subsequent attempt to use these values to model or calibrate numerical simulations. It is then essential that care must be taken on these aspects and preferably different operators to assess if any systematic error is still present due to the manual load application should perform repetitions.

The test instrumentation created the need for the development of a special measurement system for the beam web out of plane displacements. A plate board containing various LVDTs was then built to map this local buckling. In addition, pressure cells had to be combined with load cells for acquiring the structure-applied loads. Some additional LVDTs were also utilised in selected points of the structure. Although is usual to have redundant vertical deflection measurements to acquire the model displacements, care should be taken when detecting possible rotation or translation rigid body movements in the models. Without these data, it is not possible to correct the models behaviour or check for some possible inconsistencies. The use of dial gauges is also of fundamental importance to alert for the eminence of collapse in tests involving plastic, or buckling structural phenomena. Automatic acquiring systems often fail or may be stuck due to a data acquisition delay, or to the reading to viewing delay response. On the other hand, the dial gauges instantly alert for the eminent collapse and are not affected by the already commented problems, serving as an instant alert for some ultimate limit states that control some of the structural responses.

When more than one load application point is used, possible load asymmetries can occur; therefore, a redundancy in the measurements could be used to solve future questions that may appear in later checks of the experimental results. A systematic error source could occur when hinges are not used in the supports and actuator heads. Partial rotation restraints or changes in the load directions can be generated when these hinges and boxes with cylinders rollers (simulating simple supports) are not used. Another very common error is related to not forecasting the possibility that the actuators end of strokes can be reached due to excessive deflections. In this test, a plate was used to hold the beam and reposition the actuators head without fully removing the load. Many times, if this is not done, especially in tests involving composite structures, beam buckling or even fatigue can be induced in these elements.

A series of strain gauges was used to monitor the investigated beams stresses. A series of linear strain gauges was used to determine whether plastic hinges would occur near the minimum height beam sections. Other linear strain gauges were utilised on the flanges to determine the vertical component force that is imposed to the web in the region where the flanges change direction. A number of rosettes were also used on both sides of the web to determine the stress evolution during the tests.

The web local buckling was best identified with the use of a mesh painted in one of its sides. This mesh allowed a better visualisation of this buckling shape to enable comparisons to web lateral deflections obtained in numerical simulations. Another extremely relevant fact in this kind of test regards the importance of measuring the imperfections present in the models before performing the tests. These data are essential for a subsequent incorporation in numerical models, not only in terms of its global qualitative shape, but also in terms of their magnitude.

In this test the concrete slab was simulated by increasing the steel beam top flange thickness. This allowed the position of neutral axis to be at an equivalent position of the composite models plastic neutral axis. However, the concrete slab absence required the use of lateral bracing points positioned along the girder to prevent the lateral torsional buckling occurrence. This was made possible by the use of steel frames that have been specially designed and used along the girder to prevent the occurrence of this ultimate limit state.

Two aspects that should also be mentioned concerning the reuse of some test parts and the bolts friction design at the joints between the reused test parts adopted for introduction the loads and the models themselves. This reuse enabled a great speed and economy in tests, being fundamental for the success of the experimental series.

The bolts friction design at the joints enabled a vertical displacements reduction, but because of the adopted standard boltholes, slips were still present during the tests, but these were associated with fewer occurrences and with smaller load reductions when compared with shear designed bolts.

2.6.2 Spatial Trusses With Flattened Ends and a Single Bolt

The present study was concentrated on an experimental programme, consisting of a series of full-scale tests performed to clarify the structural behaviour of tubular space trusses containing end-flattened ends [13–17]. Its main objectives were to evaluate the effect of the stiffness variation due to the flattening of ends of the structure's bars. A study of the structural reinforcement, created to improve the structural load capacity, minimise deflections or reinforce structures with excessive deflections, is also presented.

In an attempt to reduce fabrication costs, several other systems were conceived. The most common is composed of tubular steel sections where chords end and diagonals end are flattened to be connected by a single large diameter bolt (Fig. 2.13). When the end-flattened bar systems are compared with standard spatial structural systems, it is clear that in addition to not requiring costly complex manufactured nodes these bars do not require any special gadget or modifications. Although it leads to an eccentric joint, numerous structural designers still neglect the bending moments that arise at member extremity.

A 20-m free span square structure was initially considered to be representative of real-sized structures results (Fig. 2.14). In the experimental model, six columns supported the spatial frame on the boundaries where four supports were located in the bottom chord plane whilst the other two were located at the top chord plane. The structure layout was a square-on-square diagonal configuration; 2-m structural module in both directions and a double-layer grid height of 1.5 m. The adopted materials were hot-dip galvanised steel deck (single sheet with 0.65 mm thickness); cold-formed steel tubular sections, made of continuously welded bent plate sheets; and ASTM A-325 high-strength bolts in all the joints. Due to the

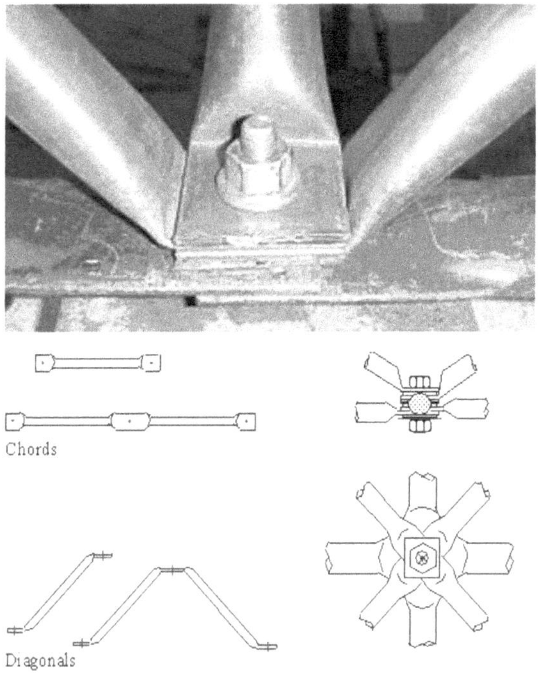

Fig. 2.13 Flattened ends node system and members and nodes utilised in the spatial system.

space limitations, the structures' length was reduced to a 12 by 12 m space truss with 416 bars, and five supports (Fig. 2.14).

The adopted tubular sections were SAC-41 steel similar to ASTM A-36, with mean measured yield and rupture stresses of 302 and 425 MPa. The first experimental series was made of four identical tests in terms of geometry, bars sizes, and thickness. In the first test, all the grid nodes utilised identical eccentric one-bolt joints (Fig. 2.15) reaching an ultimate of 99.5 kN. The failure was associated with the formation of a plastic hinge mechanism located at the structure most loaded support position, that is, the first diagonal top end (Fig. 2.15).

In order to improve the load-bearing capacity of the structure, nodes without eccentricities were utilised in the supports. This simple measure reduced by half the eccentricity present in the supports. The centred nodes consisted of two gusset plates to sustain the diagonal bars flattened ends, creating two shear plans in the connection bolts (Fig. 2.16). The second test ultimate load was 129.23 kN whilst the observed failure mechanism is shown in Fig. 2.16.

The third test had similar characteristics to the previous test. The main difference was that the support nodes only had one gusset plate to support

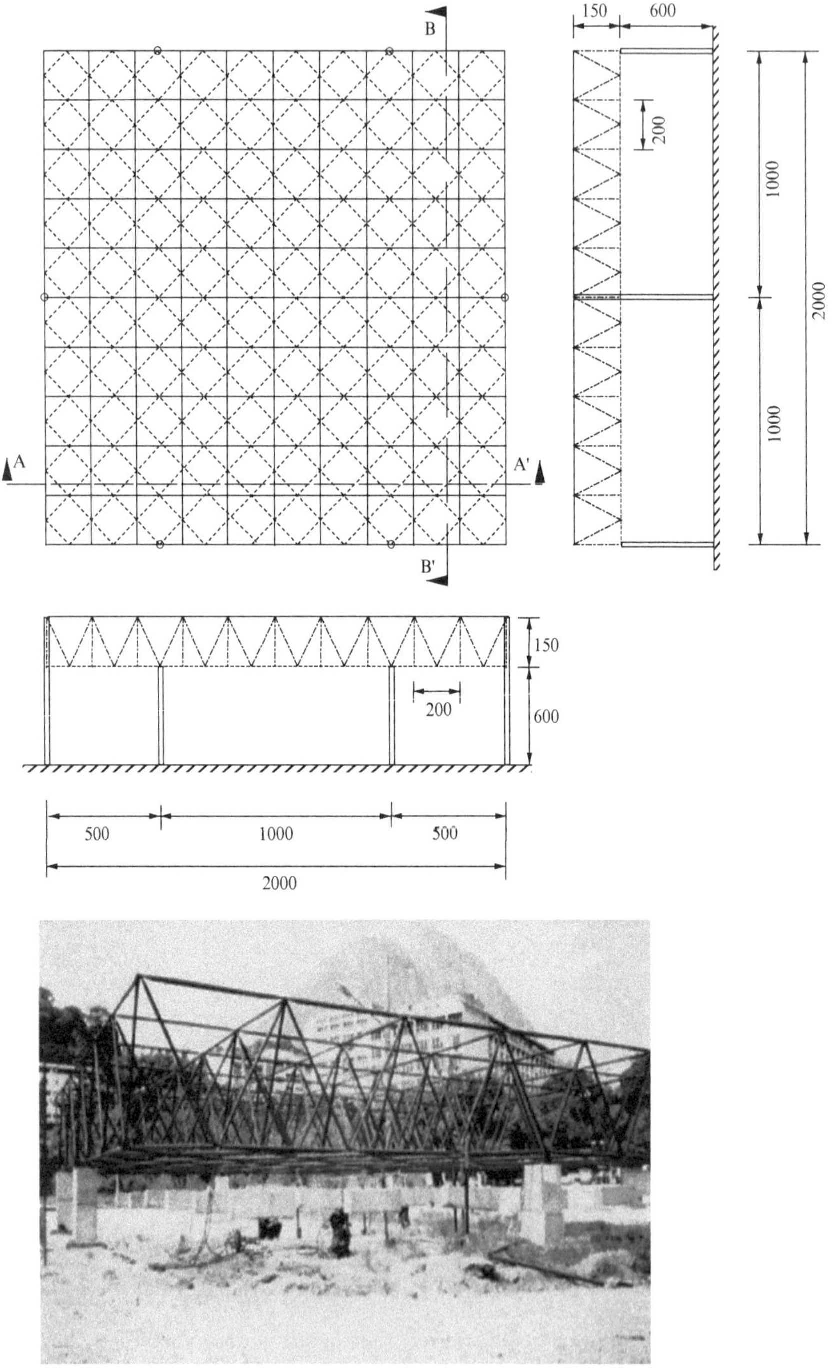

Fig. 2.14 Space structure initial layout & tested structure, 1st series.

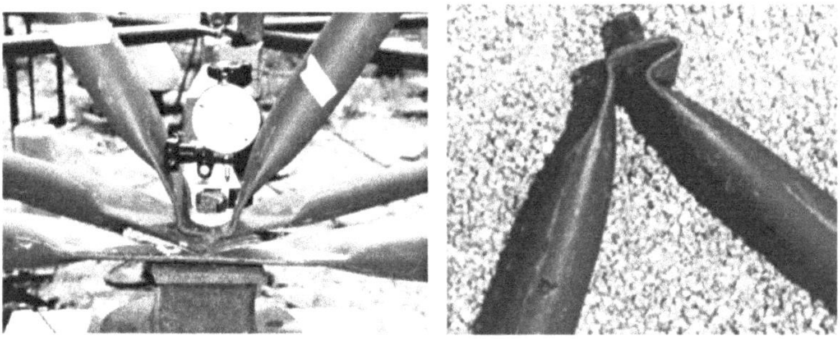

Fig. 2.15 Structural node at support position & plastic hinge mechanism, 1st test.

Fig. 2.16 Second test centred support nodes & diagonal detail before and after collapse.

Fig. 2.17 Third test centred support nodes & support node collapse, 1st series.

the flattened end bars reducing the number of shear plans to one, as shown in Fig. 2.17. The ultimate load of the structure was reduced to 72.32 kN where the gusset plate of the first support node failed by bending in a transverse direction to the support diagonal plan (Fig. 2.17). The failure occurred in a sudden and violent way making it impossible to distinguish whether the failure happened in the flattened end support bars or in the gusset plate of the support node. It is important to note that the gusset plate support node thickness was lower than its design value. This structural node proved to be inefficient due to its reduced compression resistance, as well as the lack of any kind of bracing in the bending plan.

The fourth test was performed using the same nodes without eccentricities used in the second test. In this test, in addition to the support nodes, some of

Fig. 2.18 Fourth test nodes used near the support regions, 1st series.

the nearby support nodes also utilised centred solutions. These nodes adopted the same characteristics used in the third test (one gusset plate reinforced by a stiffener) (Fig. 2.18). The ultimate load value reached by the structure was 99.18 kN associated with a similar plastic hinge mechanism present in the first and second tests, this time in the second diagonal bar. This failure occurred due to the use of bars with flattened lengths larger than their design values and structural nodes with thickness less than their specified values.

The second test series assessed the efficiency of a set of proposed nodal reinforcements and the influence of the diagonals flattened lengths. The structures were assembled with the maximum geometry that could be fitted in the laboratory, differing from the first series that was tested in the outdoors experimental camp. Another difference was associated with the applied force direction. In the second series, to simulate gravitational loads, the jacks were utilised in the opposite direction of the first series to simulate the wind load action over the studied structure. Fig. 2.19 shows the experimental layout (80 m^2) utilised in the first and second tests: the load application and the instrumentation: three load cells, six vertical displacement transducers, and six linear strain gauges. The final loading sequence used two hydraulic jacks (A1 and A3) in the first test and three hydraulic jacks (A1, A2, and A3) in the second test.

Load versus strain curves, for the instrumented diagonals, are shown in Fig. 2.20, for the first and second test. It is possible to note from these curves that the maximum strain attained on the first and second tests were 550 and 850 $\mu\varepsilon$; both values lower than 1200 $\mu\varepsilon$, the test yielding strain.

Fig. 2.21 shows the plastic hinge failure mechanism related to the first test whilst Fig. 2.22 shows details of the diagonal settling mechanism close to the support and the double diagonal plastic hinge mechanism encountered in the second test. The main difference between the first and second test

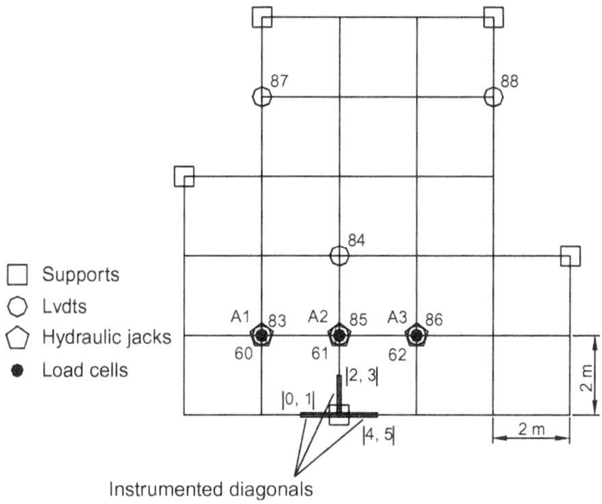

Fig. 2.19 Second series and first and second test layout (bottom chord plan).

was the load configuration. Despite this change in both cases the flattened ends length significantly influenced the structure load-carrying capacity.

The structures were designed and assembled with a smaller geometry (36 m^2) when compared with the first and second tests. Fig. 2.23 shows the experimental layout utilised in the third test: the load application point and the adopted instrumentation. The final loading sequence used only one jack to apply the load to the tested structure. Fig. 2.24 shows the details of the plastic hinge failure mechanism formed on a double compressive diagonal and excessive deformation on a bottom chord inferior node.

The main difference between this test and the previous one was the adoption of the nodal reinforcements in the support nodes and at the load application point. Three different types of nodal reinforcements were conceived using simple steel plates (Fig. 2.25). All the plates were made by an SAE 1020 grade with a 76.2 mm width, and 8.0 mm thick. The failure was related to a diagonal buckling when the load reached 117.13 kN. The nodal reinforcements increased the ultimate load by approximately 10% preventing the failure mode present previously. The nodes with reinforcements were highly stressed (Fig. 2.26), but were still able to sustain the imposed strain and stresses.

Fig. 2.27 shows load versus vertical displacement curves for the third and fourth tests of the second experimental series. These tests clarified the beneficial effects of the nodal reinforcements over the structure load-carrying capacity. The advantages of this solution were not restricted to a 10% increase in the structure ultimate load, it also contributed to increase the structural stiffness and to substantially reduce the deflections.

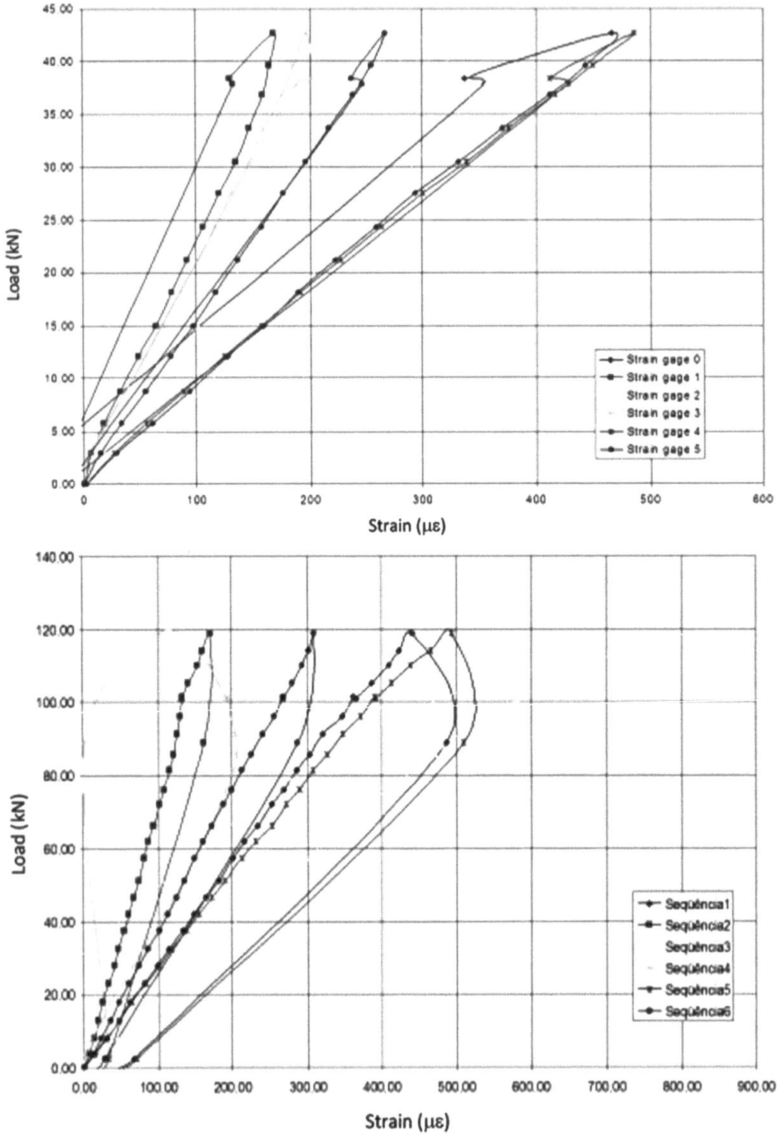

Fig. 2.20 Load versus deformation curves and second test, 2nd series.

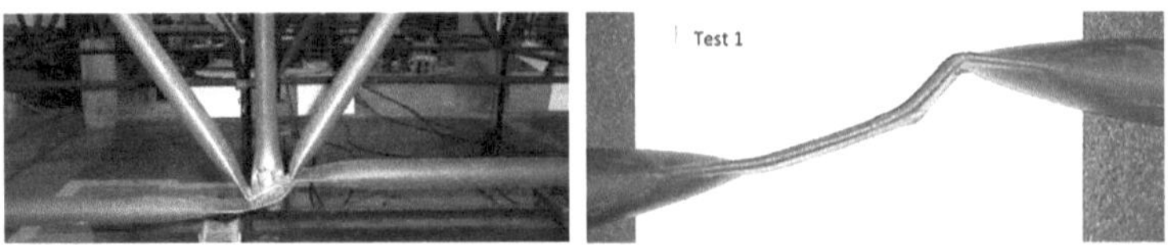

Fig. 2.21 Bottom chord plastic hinge mechanism, 1st test, and 2nd series.

Fig. 2.22 Diagonal plastic hinge mechanism at the supports, 2nd test, and 2nd series.

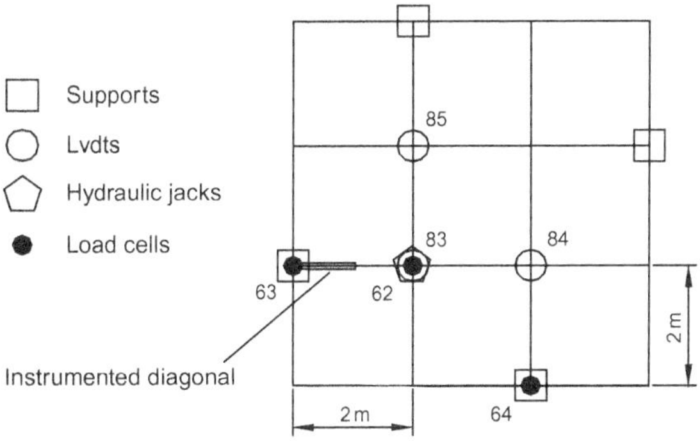

Fig. 2.23 Second series third and fourth test layout (Bottom Chord Plan).

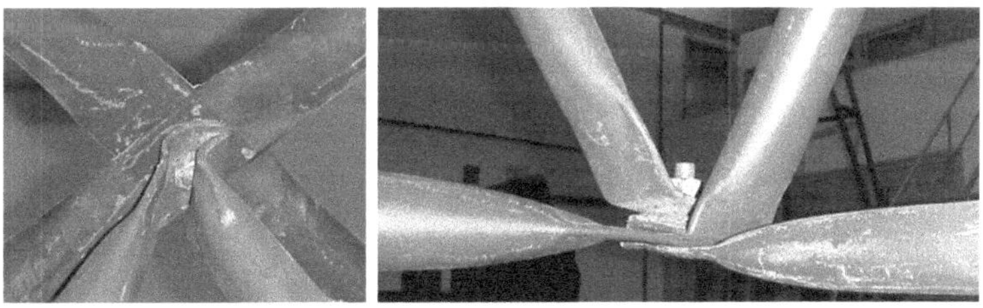

Fig. 2.24 Double diagonal plastic mechanism and bottom node deformation, 3rd test, and 2nd series.

One of the tests' main objectives was to assess the end-flattened length effect over the structure load-carrying capacity. To fully investigate this effect separately from all others, only the structure diagonal bars were sent back to be remanufactured. This process changed the diagonal bars angle to the horizontal plane by reducing the standard module height to 1.1 m.

Fig. 2.25 Reinforcement plates for two, three, and four diagonal nodes.

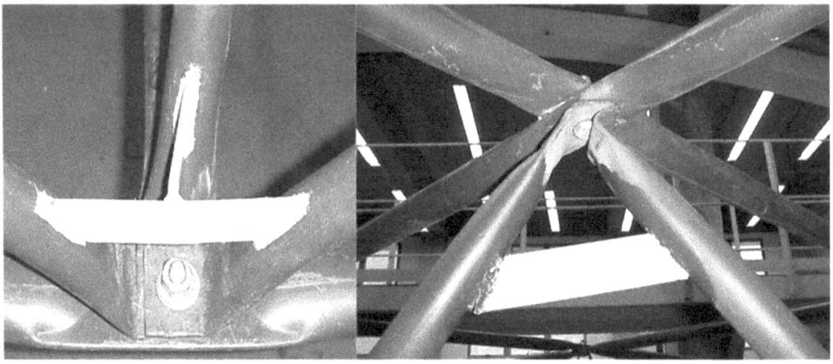

Fig. 2.26 Triple- and double-reinforced node deformations, 4th test, and 2nd series.

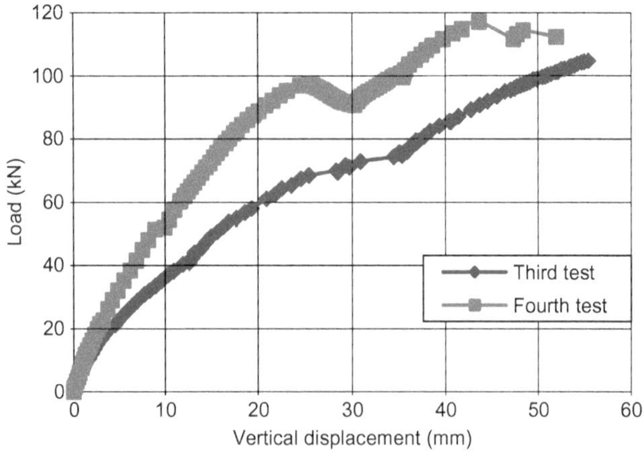

Fig. 2.27 Load versus vertical displacement curve, 3rd and 4th tests, and 2nd series.

Two basic issues were considered in the remanufacture process: minimising the flattened ends length and eliminating the semi-circular bulb-type fold left near the bar-flattened ends. Three tests were performed utilising the same adopted geometry of the third and fourth tests only using the remanufactured diagonal bars. The fifth and sixth tests did not utilise any type of welded nodal reinforcement, they only adopted extra washer plates

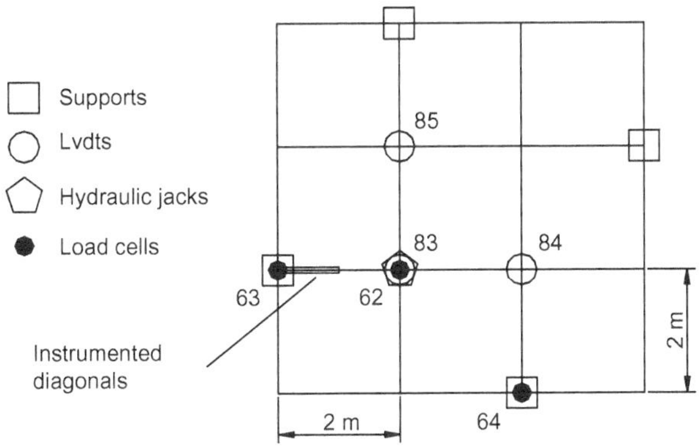

Fig. 2.28 Second series and fourth, fifth, and sixth test layout (Bottom Chord Plan).

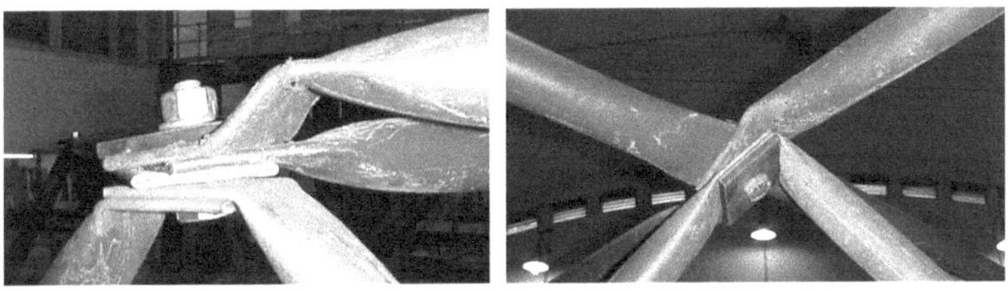

Fig. 2.29 Washer reinforced upper chord collapse, 6th test, and 2nd series.

located in the most stressed nodes. The seventh test used these washer plates and two types of welded reinforcement plates. Fig. 2.28 shows the experimental layout used in the fifth, sixth, and seventh tests: the load application point and instrumentation. These tests also used the same configuration of the two previous tests load with just a single hydraulic actuator.

The fifth test was performed after the diagonal bars remanufacture. These double diagonals were transformed into two single diagonals with the flattened ends length reduced to the minimum required, to maximise the structure load-carrying capacity. Extra washer plates were used in key points to minimise the nodal deformations.

The sixth test used a large number of extra washer plates located on the structure top chord nodes, to try to inhibit the development of a plastic hinge mechanism. However, the use of plastic hinge mechanisms was formed on the top chord (Fig. 2.29), making it clear that only their use was not enough to inhibit this ultimate limit state. Alternatively, the load magnitude imposed to the structure substantially induced large strains even in washer reinforced nodes, indicating that a serviceability limit state was close to be reached. A direct tensile rupture of the loading bolt where the hydraulic jack was connected determined the maximum load.

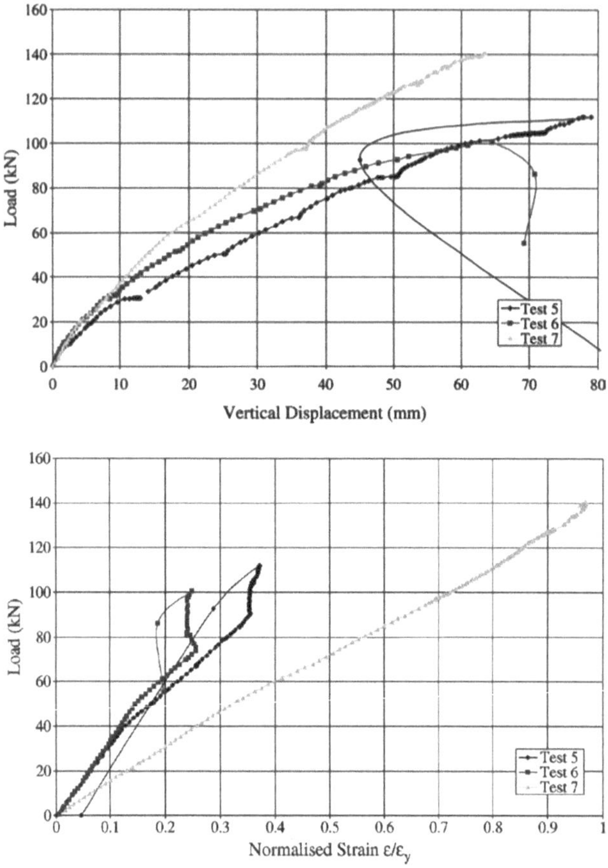

Fig. 2.30 Load versus vertical displacement and load versus normalised strain ($\varepsilon/\varepsilon y$) curves, 4th, 5th, and 6th tests, and 2nd series.

Fig. 2.30 shows interesting comparisons of load deflection behaviour for the three last tests in terms of vertical displacements and key diagonal bars attained strains. The welded plate reinforcements were responsible for at least a 25% increase in the structure ultimate load, as well as a stiffness increase. The maximum vertical displacement at the ultimate load was less than 80 mm. The seventh test depicted two interesting phenomena: a linear response was reached for loads greater than 60 kN showing that the loading bolt rupture prevented the structure from reaching higher loads.

The experiments confirmed the reduction in the structure global resistance. The main reason for this loss of resistance was the interaction between the bar flattening ends and the type of structural joint utilised in the structural system. This response could be substantially improved if, during fabrication, more care could be taken in the stamping and folding of the bar ends, as it was confirmed by the second experimental series. In order to improve the excessive loss of structural compression resistance noted in the first test, the support node configuration was changed to the use of a centred solution. The use of a stiffness welded to a thicker node plate significantly raised the load-carrying capacity of the third test. The loss in compressive resistance found in the second diagonal bar noted in the fourth test was very close to the value reached in the first test. This decrease can be related to the excessive length of the flattened ends of the second diagonal bar.

A relevant conclusion drawn from this study is that the reduction in the node eccentricities of the structure's supports not always leads to an improvement in the structure load-carrying capacity. This was confirmed by the third test, first series, in which a structural system using centred support nodes led to worse results than the eccentric solution present in the second test.

The experimental results confirmed and highlighted the detrimental effect of using an excessive length in the bars flattened ends extremes, and consequently reducing the structure load-carrying capacity and global stiffness. The nodal reinforcements proved to be capable of improving the ultimate load and global stiffness by reducing displacements and node rotations. The test without nodal reinforcements presented a failure associated with high deformations in the flattened ends lengths related to large rotations in the nodal joints.

Alternatively, the three last tests of the second series proved that the remanufactured diagonals, utilising the minimum possible flattened ends lengths, had a substantially improved structural response. The use of the minimum flattened ends length increased the node stiffness leading to less rotation. The minimum flattened ends length utilised in the present study was equal to the width of the flattened tube, approximately equal to the diameter, plus a minimum clearance proportional to the tubular bar thickness.

The use of the welded plate reinforcement improved the structure response but the fully use deserve further considerations. As in any welding process the correct type of electrode and the right current intensity, according to the plate and tube thickness utilised, are fundamental issues. The thin plates utilised in the tubular bars only allowed a single welding to be made. If the welds were to be removed, the tubular bars would be damaged, reducing the structure versatility associated with the disassembling process, and enabling a future structure reassembling in a new location.

The space trusses' experimental programme was developed in two stages comprising a first phase performed onsite followed by another carried out in the lab. In the first phase, it could be concluded that if the supports are not very well designed and detailed, the structural system can be associated with a premature collapse. The load application must be carefully done, because in the majority of cases, the collapse occurs without warning. In the present investigation, a number of possibilities were studied to determine the possible collapse modes, that is, buckling of the bars, nodes excessive rotations, collapse of support, amongst others.

The system was instrumented with dial gauges, LVDTs, and strain gauges arranged in key points of the structure. The deflections allowed the measurement of the structure stiffness, as well as the eminence of the structural collapse. Special care has been taken with the use of two or four strain gauges arranged every 90° or 180° on the tube circular section to detect the actual axial forces by eliminating the residual bending effects.

The system was continually modified to improve its performance by making changes in the nodes with use of reinforcement plates, washers, and stiffeners and, mainly, by reducing the bars end-flattened length. As the structure was changed, it could be noticed that structural response was controlled by the bars global buckling failure modes initially forecasted.

2.6.3 Tension Capacity of Carbon and Stainless Steel Plates With Staggered Holes

The current stainless steel design standards are largely based on analogies with the response of carbon steel structures. However, the stainless steel when subjected to axial tensile and compression forces presents nonlinear stress versus strain curves without a defined yield plateau, unlike the equivalent carbon steel curves, as shown in Fig. 2.31.

Bolted joints, unlike welded connections, tend to be associated with a reduced sections tensile strength, since, for the bolts use; it becomes necessary to drill holes in the connected elements, reducing their cross-sections. Due to the presence of these holes, various modes of rupture can control their design. Fig. 2.32 shows the joint main geometric parameters: t and L, respectively, represent, the plate thickness and width; s, p, e_1, and e_2 are distances between holes and from hole to the plate edge in directions parallel and perpendicular to the load direction prescribed by the Eurocode 3 Part 1.8 [20].

The Eurocode 3 Part 1.8 [20], for the design of carbon steel joints, defines the criteria for the resistance assessment of bolted connections under to tension. These criteria cover two ultimate limit states: gross section yield and net section rupture of the bolted joint base plates, that is:

$$N_{pl,rd} = \frac{(A \times f_y)}{\gamma_{M0}} \tag{2.1}$$

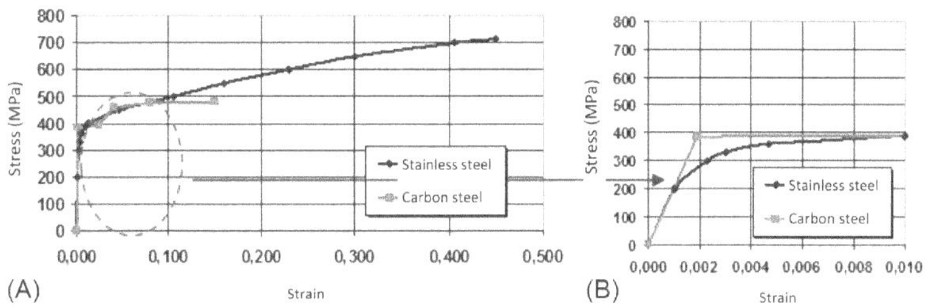

Fig. 2.31 Carbon & stainless steel stress versus strain curves. (A) Full range curves. (B) Initial segment.

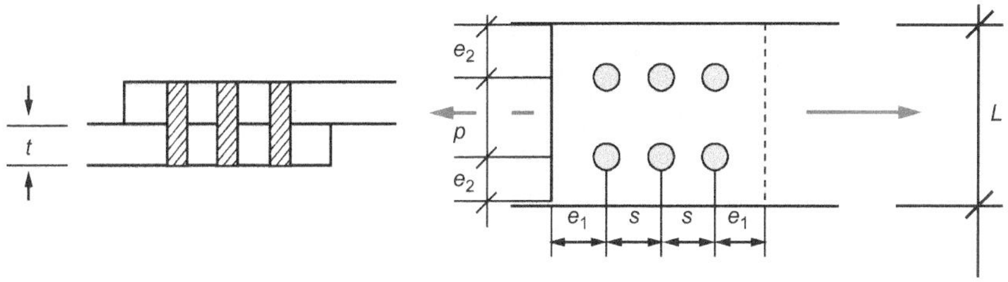

Fig. 2.32 Variables definition.

$$N_{u,rd} = 1.0\frac{(0.9 \times A_{\text{net}} \times f_u)}{\gamma_{M2}} \quad (2.2)$$

$$\frac{N_{u,rd}}{N_{t,rd}} \geq 1.0\frac{N_{pl,rd}}{N_{t,rd}} \geq 1.0 \quad (2.3)$$

where $N_{t,rd}$ is the tension force acting on the joint which must be less than or equal to the maximum value between $N_{pl,rd}$ and $N_{u,rd}$; A and A_{NET} are the joint gross and net areas, respectively; 0.9 is the net area reduction coefficient; F_Y and F_u read the yield and rupture stresses, respectively; γ_{M0} and γ_{M2} are partial strength coefficients that were taken as 1 to calibrate the actual behaviour of the investigated bolted connections. The 0.9 coefficient in Eq. (2.2) is obtained as a function of the element type (plate or profile) and the connection type (welded or bolted). Connections with staggered bolts use an equation developed by Cochrane in 1922 and still present in the Eurocode 3 [20].

$$A_{\text{net}} = A - t\left[n \times d - \sum\left(\frac{s^2}{4 \times p}\right)\right] \quad (2.4)$$

where n is the number of bolts. Stainless steel joints are designed according to Eurocode 3, PT. 1.4 [21] that again prescribes that tensile strength should be the smallest value between the gross section plastic resistance $N_{pl,rd}$ and the net section rupture $N_{u,rd}$, that is:

$$N_{pl,rd} = \left(\frac{A \times f_y}{\gamma_{M0}}\right) \tag{2.5}$$

$$N_{u,rd} = \left(\frac{k_r \times A_{\text{net}} \times f_u}{\gamma_{M2}}\right) \tag{2.6}$$

where

$$k_r = \left(1 + 3r\left(\frac{d}{u} - 0,3\right)\right) \leq 1.0$$

where r is the number of bolts on a section divided by the total number of connecting bolts and u is twice the value of the edge to hole distance perpendicular to the load direction, but less than or equal to p_2 which is the distance between the hole centres perpendicular to the load direction. Although the bearing resistance of a bolted connection in stainless steel is usually governed by the need to limit the hole deformations at service loads, a separate check must be made to prevent this effect at the ultimate limit state by replacing the f_u by $f_{u,red}$ given by:

$$f_{u,red} = 0.5f_y + 0.6f_u \tag{2.7}$$

The connections' ultimate limit states are shown in Fig. 2.33.

The present investigation was centred on bolted connections' tests without eccentricities in relation to the cross-section gravity centre, seeking to evaluate, without their influence, their limits and structural responses [22–24]. The first step was to evaluate the bolted connection load-carrying capacity. The following ultimate limit states were verified: gross section yield and net rupture on sections passing through two and three holes, the plate bearing and bolts shear strengths.

Thus it was possible to choose the best geometric configuration, which could provide a better use of the collected data for the staggered bolted joint evaluation. For that, the variable p was taken as 55 mm. Eurocode 3 equations enable the production of the curves shown in Fig. 2.34 in such a way that the net rupture limit net rupture on sections passing through two and three holes could be determined. A similar analysis was made for staggered stainless steel joints generating the limits identified by the curves shown in Fig. 2.34.

The characterisation of the staggered bolted joints strength under tension is represented by the stress versus strain curve or by the stress versus displacement curve that were important data for the joints design and analysis. The best way to obtain these curves is through experiments performed in the laboratory. With these tests, it becomes possible to evaluate the results to determine the joints stiffness and determine their ultimate load capacity.

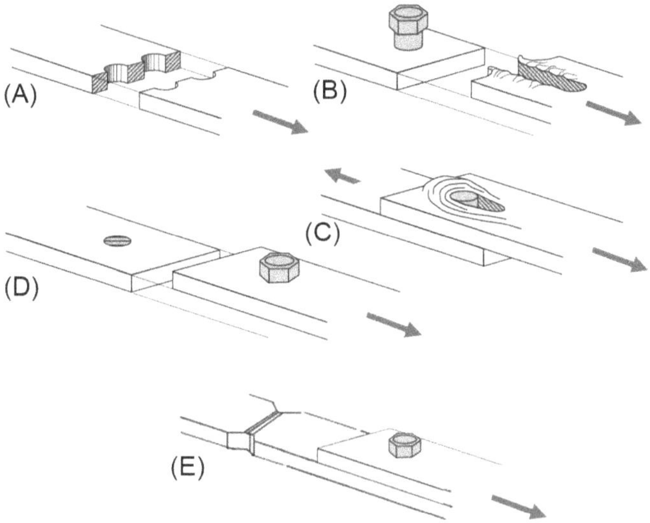

Fig. 2.33 Bolted joints ultimate limit states: (A) net section failure, (B) bearing with small edge distance, (C) bearing with large edge distance, (D) bolts in shear, and (E) gross section yielding.

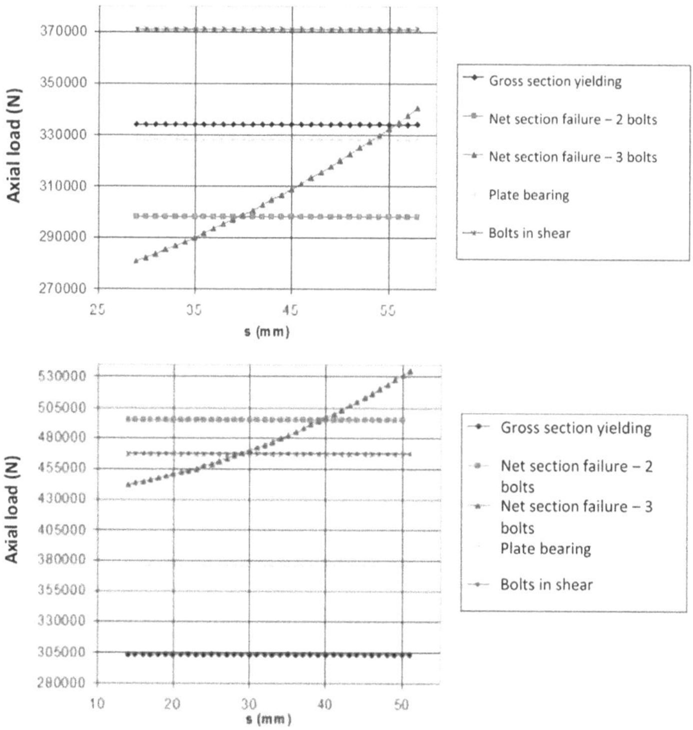

Fig. 2.34 Carbon and stainless steels ultimate limit states strengths.

Carbon steel plates made of a USI 300 steel grade, donated by USIMINAS and austenitic stainless steel sheets made of an A304 steel grade, donated by ACESITA, were utilised in the experimental programme. ASTM A325 bolts with a ½-inch diameter fully threaded were also used, ensuring that their shear strengths would not be exceeded in order to effectively assess the effects of the different possible bolt configurations. The main idea was to vary the *s* and p values to generate different cross-section net areas, that is, the possible rupture lines A, B, and C, enabling their evaluation as shown in Fig. 2.35. An over designed loading plate with a 15-mm thickness was adopted to ensure an elastic response as shown in Fig. 2.36.

Five strain gauges per plate were installed to monitor the strains in specific locations, as depicted in a photo of the E3_CARB_S50 test, shown in Fig. 2.36. Table 2.1 summarises the performed experiments. The letter a, in the first and second tests, represents the first adopted bolt layout, as shown in Fig. 2.37, whilst the other tests used the second bolt layout (Fig. 2.37).

Table 2.2 presents the performed experiments in terms of failure modes and associated ultimate loads. This table also contains their equivalent

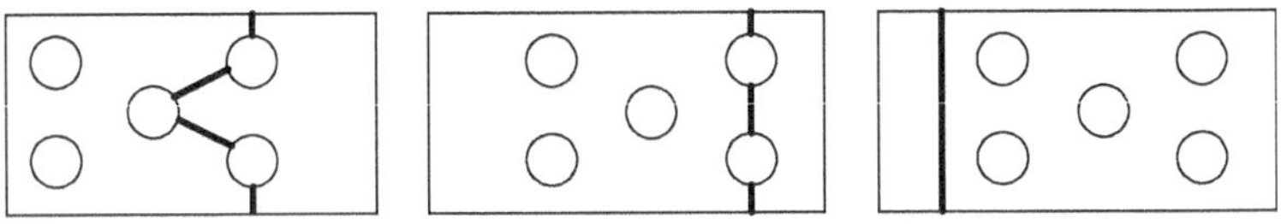

(A) (B) (C)

Fig. 2.35 Possible net rupture paths for the investigated connections.

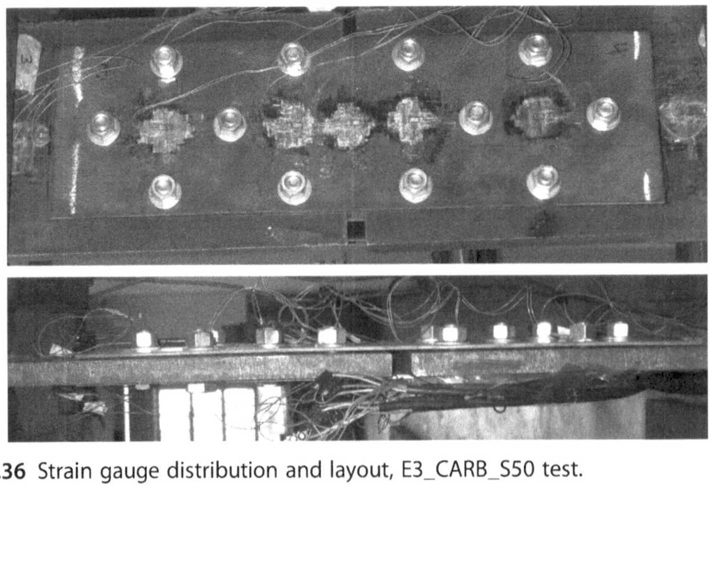

Fig. 2.36 Strain gauge distribution and layout, E3_CARB_S50 test.

Table 2.1 Tested Geometry Parameters

Test	Identification	s (mm)	p (mm)	e_1 (mm)	e_2 (mm)	d_0 (mm)	Steel Grade	Base Plate Thickness (mm)	Number of Bolts
1	E1_CARB_S50_a	50	55	40	17.6	14.7	Carbon	15	6
2	E2_CARB_S70_a	70	55	40	17.6	14.7	Carbon	15	6
3	E3_CARB_S50	50	55	40	17.6	14.7	Carbon	15	6
4	E4_CARB_S30	30	55	40	17.6	14.7	Carbon	15	6
5	E5_INOX_S50	50	55	40	17.6	14.7	Stainless	15	6
6	E6_CARB_S30_P10	30	55	40	17.6	14.7	Carbon	10	6
7	E7_INOX_S30	30	55	40	17.6	14.7	Stainless	15	6
8	E8_CARB_S50_P8	50	55	40	17.6	14.7	Carbon	8	6
9	E9_INOX_S23	23	55	40	17.6	14.7	Stainless	15	6

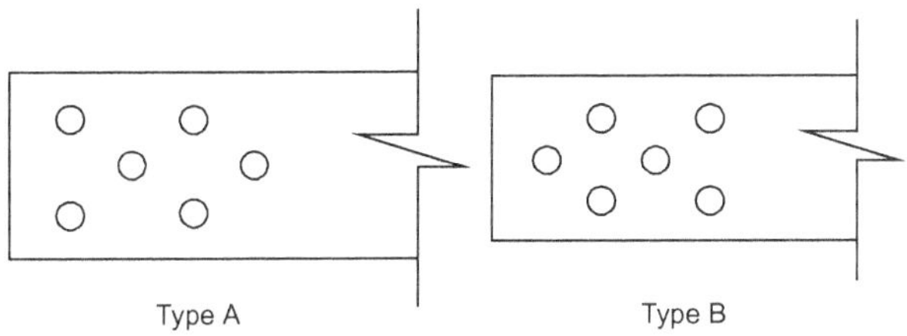

Fig. 2.37 Different bolt arrangement layouts.

Table 2.2 Summary of Experiments

Test	Failure Type	Ultimate Load (kN)	Failure Type EC-3 (kN)	Ultimate Load EC-3 (kN)	Difference (%) Exp—EC3
E1_CARB_S50_a	3 F	312.0	2 F	298.3	4.6
E2_CARB_S70_a	1 F	349.5	2 F	298.3	17.2
E3_CARB_S50	2 F	310.0	2 F	298.3	3.9
E4_CARB S_30	2 F	296.0	3 F	282.5	4.8
E5_INOX_S50	2 F	480.0	AB	302.9	58.5
E6_CARB_S30_P10	3 F	309.5	3 F	282.5	9.6
E7_INOX_S30	2 F	459.0	AB	302.9	51.5
E8_CARB_S50_P8	2 F	326.0	2 F	298.3	9.3
E9_INOX_S23	3 F	436.0	AB	302.9	43.9

predicted values by the Eurocode 3 and their differences in the experimental findings where AB is the gross area yield and NF is the net area rupture passing through N holes.

In a preliminary comparison of the load versus displacement curves, it is possible to note that the largest value of the ultimate load is associated with stainless steel bolted joints (Fig. 2.38). It is also observed that the stainless steel joints feature a larger displacement than the carbon steel bolted connections, cohobating their substantially more ductile behaviour.

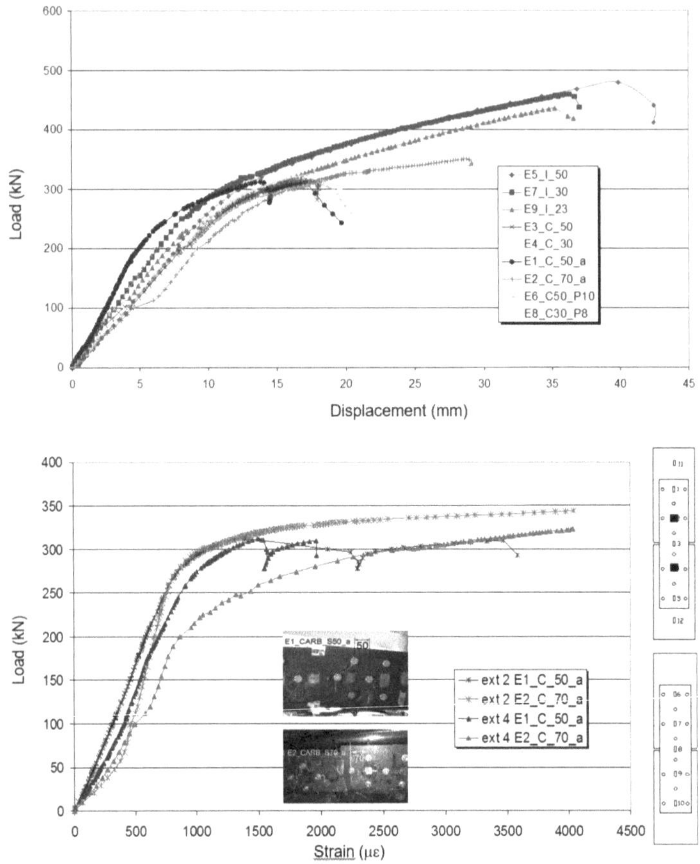

Fig. 2.38 Carbon and stainless load versus displacement, and load versus strain curves, E1 and E2 (SG 2 and 4).

The E1_CARB_S50_a and E2_CARB_S70_a tests had the same bolted layout but different values of the parameter s, allowing the assessment of its effect. This can be observed in the load versus deformation curves (Fig. 2.38), constructed with strain gauges 2 and 4 of the two mentioned experiments. It is possible to note that the E1_CARB_S50_a test presented a higher stiffness than its counterpart did by comparing the different deformations for the same applied load. This effect was caused by the lower value of the s parameter immediately mobilising the three bolts, as well as for the stresses concentration in the first holes caused by 15-mm thick loading plate. The highest ultimate load was associated with the E2_CARB_S70_a test, due to the need to achieve a higher load to surpass the higher resistance associated with the net rupture on an area passing through a single hole.

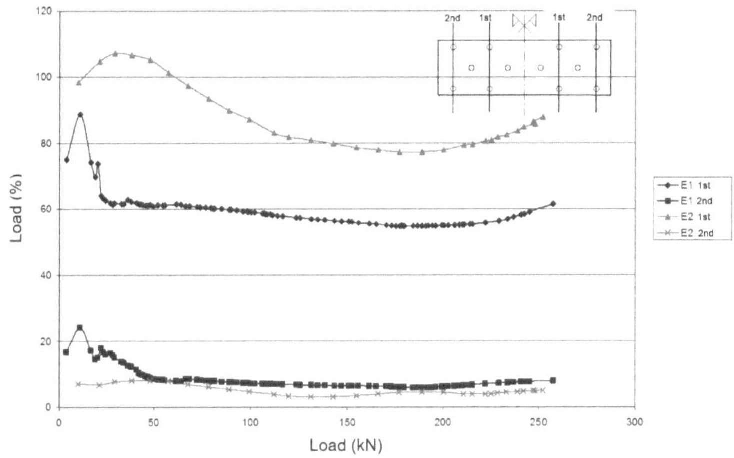

Fig. 2.39 Load percentage per bolt line versus total applied load curves, tests E1 and E2.

This behaviour occurred due to the large difference in load transmitting capacity of bolts present in the same line, promoted by the different bolt layouts, as can be seen from graph percentage of load per bolt line versus total applied load shown in Fig. 2.39. Despite this fact, the Eurocode 3 [21] limits can be interpreted as lower bounds for the problem. These data were evaluated by the average strains of the 3 mm test strain gauges.

The first cross-section in the load application direction is characterised by the first section closer to the plate centre, where strain gauges were positioned, 2, 4, 7, and 9. Additionally, strain gauges 3 and 8 were installed at the mid plate cross-section, considered as the most loaded plate section. Finally, the second cross-section in the load application direction is represented by the second section further away from the plate centre, where there were strain gauges 1, 5, 6, and 10 were adopted. These sections are for both bolt layouts and they are well defined in Fig. 2.40.

The difference of E1_CARB_S50_a and E2_CARB_S70_a experiments in relation to E3_CARB_S50 and E4_CARB_S30 tests was the bolt layout (i.e. type A and type B). It can be noted that in the first two tests, the

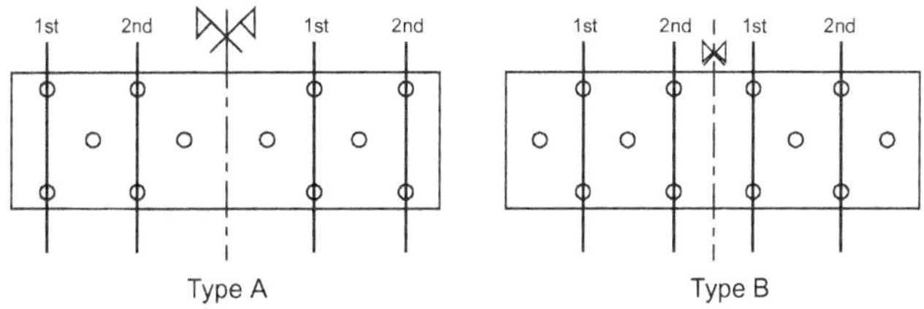

Fig. 2.40 Location of sections in the tested plates.

first section close to the symmetry axis contains one single bolt, whilst E3_CARB_S50 and E4_CARB_S30 tests had two bolts the same section. A comparison of load versus deformation curves in which the investigated experiments are grouped according to their structural responses is shown in Fig. 2.41. The highest initial stiffness indicated that tests E1_CARB_S50_a and E2_CARB_S70_a presented a better stress distribution at the joint end, since it was more balanced than the one depicted in tests E3_CARB_S50 and E4_CARB_S30.

The E1_CARB_S50_a, E3_CARB_S50, and E8_CARB_S50_P8 tests were also compared, since they are carbon steel staggered bolted joints with the same value of the s parameter. However, the inner plate in E8_CARB_S50_P8 test has 8-mm thickness, unlike the other two tests, where a 15-mm thick plate was used. Analysing the load versus deformation

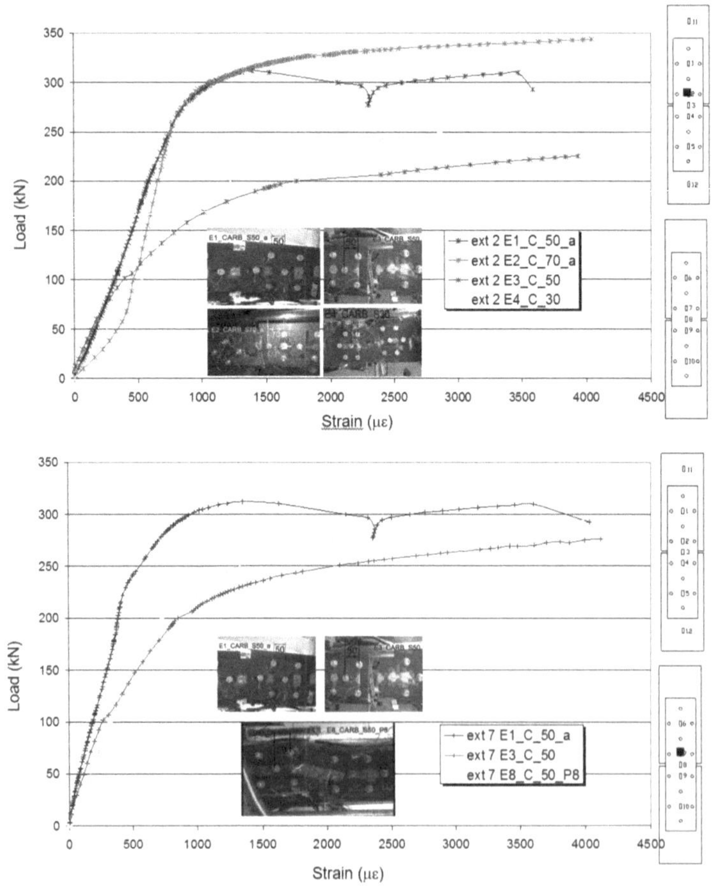

Fig. 2.41 Load versus deformation curves, tests: E1, E2, E3, and E4; and E1, E3, and E8.

curves shown in Fig. 2.41 a higher initial stiffness of the test E1_CARB_S50_a, in which the rupture occurred on a net area passing through three holes showing the influence of the bolt configuration in the connection stiffness.

An analysis was also made with the three stainless steel tests of identical dimensions, just varying the value of the parameter s. The load versus displacement curves shown in Fig. 2.42 indicated that the value of the s parameter governed the test ultimate strength. A higher initial stiffness was presented in tests, E7_INOX_S30 and E9_INOX_S23, but the E5_INOX_S50 test had the highest ultimate load. These curves also enable the grouping of the E7_INOX_S30 and E9_INOX_S23 tests' behaviour, due to the small difference between the values of the parameter s, compared with the E5_INOX_S50 test, showing an evolution of the joints' behaviour associated with the variation of the s parameter.

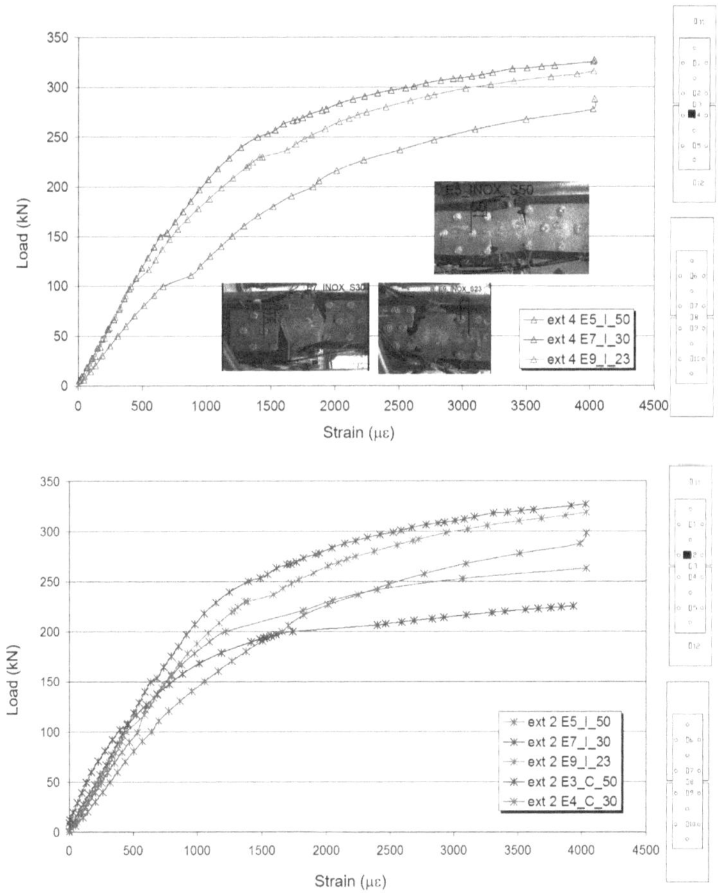

Fig. 2.42 Load versus deformation curves, tests: E5, E7, and E9; and E3, E4, E5, E7, and E9.

Comparing the E5_INOX_S50, E7_INOX_S30, and E9_INOX_S23 tests with the E3_CARB_S50 and E4_CARB_S30 ones, performed with identical main plates, via their associated load versus deformation curves (Fig. 2.42), it is possible to note that the initial stiffness of the carbon steel plate connection decreased faster than their equivalent stainless steel plates. A similar trend in terms of the plate ultimate load and associated ductility could be observed in the investigated tests.

A 12% mean difference between the $A_g \times f_y$ and $A_{net} \times f_u$ values was observed in carbon steel connections, as shown in Table 2.3. This difference confirmed the small strain hardening range of carbon steel plates, as well as their low stiffness and plastic strength capacities when compared with stainless steel plates. A 62% mean difference between the $A_g \times f_y$ and $A_{net} \times f_u$ values was observed in stainless steel joints, as presented in Table 2.3. This difference confirms the high capacity of hardening of stainless steel, showing their high strain hardening range, as well as their high stiffness and plastic strength capacities.

The carbon steel ultimate loads obtained through the Eurocode 3 [21], presented coherent values when compared with the experiments, when the design values were always safe. The bolt layout A showed the highest stiffness and a more balanced stress distribution at the joint ends than bolt layout B. It could also be concluded that the lowest value of the s parameter was associated with a better bolt force distribution due to their closer proximity. Whilst the maximum value of the s parameter provided a higher resistance, as it provides the largest net area.

The decrease in the loading plate thickness improved the load distribution on the carbon steel bolted connections, by its associated deformation to the tested plates, increasing the joint's resistance. The stainless steel tests showed large deformations and energy absorption capacities due to their strain hardening responses, as function of the higher rupture to yield stress ratios. The

Table 2.3 $A_g \times f_y$ and $A_{net} \times f_u$ Percentage Difference for Carbon and Stainless Steels

CAB (kN)	2 F (kN)	3 F (kN)	AB/2 F (%)	AB/3 F (%)	Mean (%)
Carbon					
334.1	298.3	282.5	112.0	118.3	115.1
334.1	298.3	320.1	112.0	104.4	108.2
Mean (%)					111.7
Stainless					
302.9	494.6	455.5	163.3	150.4	156.8
302.9	494.6	469.8	163.3	155.1	159.2
302.9	494.6	531.8	163.3	175.6	169.4
Mean (%)					161.8

expected collapse modes and the ultimate loads according to Eurocode 3 [18–21] for bolted stainless steel tests have not occurred, presenting experimental load values substantially higher than the standard's forecasting, making a very underestimated design. This fact is in line with the idea of avoiding large deformations, feature presented in stainless steel but not contemplating, in an intermediate form, the stainless steel higher load-carrying capacity.

A comparison of carbon and stainless steel staggered bolted joints, where ductility in the latter is characterised by the higher of f_u/f_y ratio, indicated that stainless steels had an energy absorption capacity around 60% higher than the carbon steels.

The high difference of the values of $A_g \times f_y$ and $A_n \times f_u$ of stainless steel when compared with carbon steel is due to their high strain hardening responses leading to higher ultimate loads. This is confirmed from the correlation of some carbon steels with some other stainless steel groups presented in Table 2.4. In this table, it is evident that the $f_{u,red}$ or the 0.9 factor is inadequate to represent the actual stainless steel behaviour. Alternatively, this value is suitable for carbon steels where the f_u/f_y ratio is much lower and where the use of this reduction factor reduces excessive joint deformations.

This experimental programme created the need for the design and the construction of a load cell for monitoring of loads in the investigated steel plates under tension. This load cell was calibrated and used in conjunction with the readings of the hydraulic system depicting a suitable response. The load cell was built with a series of linear strain gauges parallel and perpendicular to the load direction and had suitable precision because the instrumented load application plate has always remained in linear elastic regime.

The experiments proved the real need to measure, as accurately as possible, the stainless steel stress versus deformation curves in the perpendicular and parallel directions to rolling. Different results were found for these two types of specimens and indicated a large difference between the deformation limits associated with carbon and stainless steels. As expected, the stainless steel yield was not well defined, being determined by the default procedure specified in Eurocode 3 pt. 1.4 [21].

Table 2.4 Comparative Values Between Carbon and Stainless Steels

Steel Grade	f_y	f_u	0.9 $F_{u,red}$	Deformation (%)
MR 250	250	400	360 365	11
AR 350	350	490	441 461	9
Austenitic	350	700	630 595	40
Duplex	510	780	720 723	20
Ferritic	410	580	522 553	20

The results indicated that the loading plate thickness could affect the results as well as the staggered bolt geometric configuration. In one of the tests, unlike expected, the tensile rupture line occurred in a section that contained only a single hole, instead of the one containing two holes. This was due to the non-uniform stresses distribution in the bolts that had high concentrated stresses next to the plate section with a single hole generating unexpected failure mode.

It also became apparent that the Eurocode 3 pt. 1.4 formula must be revaluated. The constants used in the evaluation of the ultimate limit state of net effective section rupture under tension is based on serviceability limits to avoid very deformed holes are not properly calibrated. They should be made according to the different types of stainless steels (austenitic, Ferritic, duplex, etc.) reflecting the difference of the deformations associated with the rupture and yield stresses for these types of steels.

2.6.4 Prestressed Stayed Steel Columns Resistance

This section describes an experimental programme involving full-scale tests in a prestressed stayed steel column. The tests were performed at the Structures and Materials Laboratory of the Pontifical Catholic University of Rio de Janeiro [25–30]. The fabrication and assembly of the prestressed stayed steel column occurred in two stages. The first part was performed on the Metalfenas steel structures, where the tube was split in two parts, since the tubes generally come from steel manufactures with a standard 6 m length. Due to the complex welding of the perpendicular bars at the tube ends, they were divided in segments with 6.5 and 5.5 m. This strategy enabled the perpendicular bars to be welded at the 6.0 m segment. The tubes were later safely transported to the lab, where they were welded to for the final configuration of the stayed steel column structural system. The column top dimensions are shown in Fig. 2.43. After welding, the tubes were measured to determine their initial imperfection. Thin tubes were then welded to the perpendicular bars free end to enable the cable passage. The tubes were used to ensure that the cables would always pass through the perpendicular bars extremities, as shown in Fig. 2.43.

The cables were anchored to the column 'head' passing through hooks and locked with three steel cable clamps (Fig. 2.44). The cables were also locked in the middle of the column to maintain the prestress forces. One of the innovative contributions of this work consisted in designing a system for the acquisition of the cables forces. This was acomplished by using a load cell attached to a steel plate, thus simultaneously obtaining the forces in the four adopted cables.

A load cell would be externally supported in the column whilst the other load cell positioned close to the hydraulic actuator measured the structure's

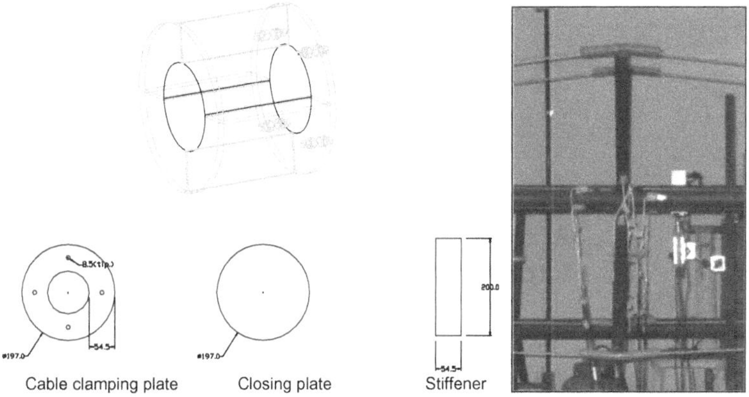

Fig. 2.43 Column head and cable deviators.

Fig. 2.44 Cable locks and load cells.

applied load. One of the problems of this load cell is that possible load unbalances on the cable cannot be detected, but this issue was later solved, as it will be later explained. On the other hand, the adopted system enables the assessment of global force acting on the stayed column by a difference of the readings of the two adopted (Fig. 2.44).

One of the major contributions of this investigation was the development of this test programme due to its novelty of being the first full scale three-dimensional experimental model. A reaction frame was designed to cope with horizontal loads to be applied to the column, since the lab did not have a clear height to vertically perform the tests (Fig. 2.45).

The instrumentation used was quite complex due to restrictions, geometry, flexibility of the tested parts, and adopted dimensions. Sixteen strain gauges were used, four measuring the strains in parallel to the stay axis, four

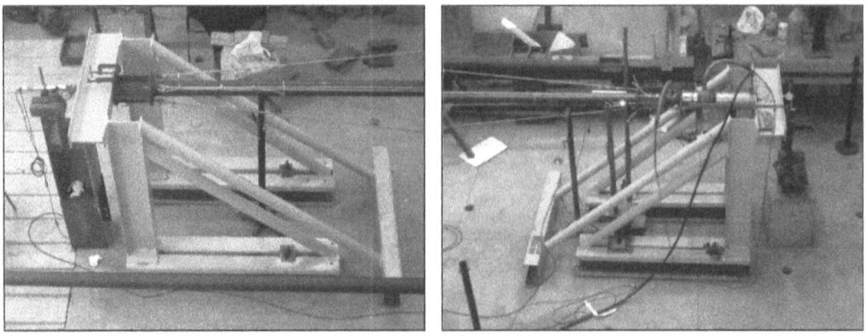

Fig. 2.45 Self-balanced loading frame for applying and reacting the horizontal loads.

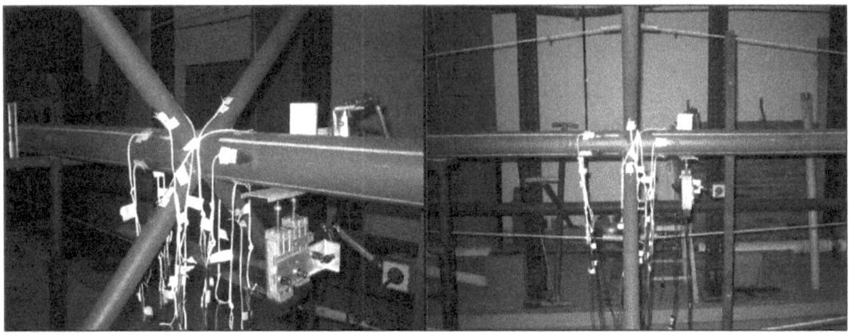

Fig. 2.46 Typically used strain gauges configuration.

positioned at a 45° angle to the stay axis. Finally, two diametrically opposed at each bar and perpendicular to the main column (Fig. 2.46) were used to acquire the loads acting on the primary and secondary tubes.

Ten LVDTs have also been used along the main column (eight) and at the reaction frames (two) (Fig. 2.47). The main column LVDTs were located at the column quarter spans, whilst four were adopted close to the column midspan. From the two remaining, one was installed at the column 'head' to measure the horizontal displacement whilst the other was positioned at the reaction frame opposite to the one where the hydraulic actuator was installed.

The column circular shape created a difficulty that was solved by the use of plates installed at the LVDTs points. At exactly the column midspan, it was not possible to measure the displacements due to the presence, at this exact point, of the cross-arm tubes. This was the main reason for locating the LVDTs at points 200 mm from the column midspan, whilst using an innovative measuring system with two LVDTs in each horizontal and vertical face of the plates. This system always enables the displacement measurements not being affected by the column rotation and/or translation movements.

Four dial gauges were utilised to provide an instant visual control of what was happening during the tests. This was done due to the previous knowledge

Fig. 2.47 LVDTs layout.

that the column will remain at the elastic range enabling its reuse. On the other hand, it was also known that, close to the ultimate load, the deflections quickly rise. They were important to avoid large deflections and to prevent large strains that could damage the prestressed, inhibiting their reuse or even compromising the test safety. Two load cells were adopted to measure the column forces. One was utilised for measuring the load applied by hydraulic actuator, whilst the other to measure the stays prestress forces. Fig. 2.48 shows the two load cells and the universal hinged used between them. The adopted steel properties were obtained by standard tension tests. The mean and standard deviation values were 403.8 and 29.3 MPa for the yield stress and 465.3 and 4.9 MPa for the tension rupture stress, respectively.

Due to the structural system dimensions, the self-weight induced a 115-mm vertical displacement at the column midspan, larger than the maximum imperfection value prescribed by NBR 8800 [31]. First compensation

Fig. 2.48 Load cells.

Fig. 2.49 Adopted system to compensate the column self-weight.

measures using the cables prestressing forces were tried out, but that created an imbalance and the introduction of a large eccentricity at the load application point due to a rotation at the universal hinge. An alternative solution was employed by using a cable that could relieve part of the system weight by anchoring it, at the column midspan, to a beam welded to the reaction frame (Fig. 2.49). This solution could, in principle, change the column buckling modes, but a nonlinear finite element of the investigated structural system proved that the use of this cable did not affect the overall system ultimate load or its associate failure mode.

Temporary supports were also used to support the column whilst a minimum axial load had not been applied (Fig. 2.49). Two G clamps were also utilised at the opposite end of the load application point (Fig. 2.50). Unfortunately, there was no room in the adopted configuration of the initial tests for using a second universal hinge at this point, issue that was corrected later.

Fig. 2.50 Adopted G clamps and cable stretchers.

These two facts contributed to this point behaving as a simple support with a partial rotation restriction.

The first series comprised two experiments on the stayed system where cable prestressing force magnitude varied. Initially, a structure preload was performed to evaluate and verify the correct operation of the measuring instruments. This was followed by a test where the cable slack was removed reaching an ultimate load of 900 N. In the second test, the cables were prestressed up to a maximum value achieved by tightening the anchoring hooks, totalling a 7.12-kN load in the column. The aim of this test was to achieve the minimum cables prestressing possible, where they would be only lightly tightened. Two steps were performed: First using stretches in series the cables were initially stretched (Fig. 2.50). The anchor cable hook tightening up to the desired prestressing magnitude followed this. With the cables positioned and stretched, the load was applied up to approximately the structural system critical load.

The load applied on column presented in the following graphs is the total load, that is, the sum of the hydraulic actuator load and the cable prestressing force reaction on the column. Fig. 2.53 shows the first tests' results in terms of the response of strain gauges at the column midspan in the direction parallel to the cables axis. The arrangement and identification of these strain gauges are shown in Figs 2.51 and 2.52. A linear behaviour of the four strain

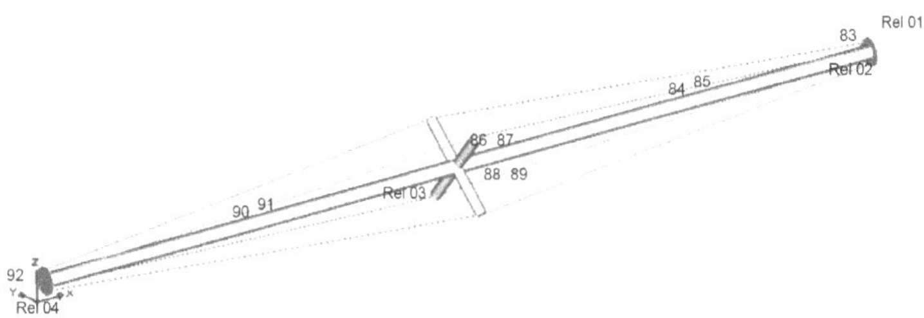

Fig. 2.51 Distribution and layout of LVDTs and dial gauges.

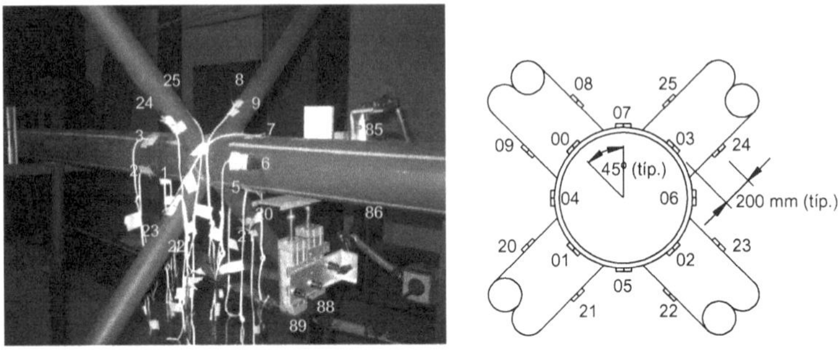

Fig. 2.52 Strain gauges distribution and layout.

gauges was observed, as well as the occurrence of bending in the tube. A comparative analysis between the column total load, measured by the internal load cell, and the load measured by strain gauges located in the column centre is shown in Fig. 2.53. A similar behaviour occurred in the second test of this series. To evaluate the forces from the strain gauge reading the column young modulus was assumed as 205,000 MPa, whilst the adopted column diameter and thickness were 89 and 3 mm.

Due to low prestressing force magnitude the perpendicular bars strain gauges were not significant. The diameter and thickness of the cross-arm tubes were 42.6 and 3 mm. Fig. 2.54 shows the presence of two plateaus at 5 and 10 kN. The first plateau represents the point where the column supports were removed, whilst the second occurred due to a slight loss of pressure in the pump that injects oil into the hydraulic actuator during the 1-hour period in which the column load was maintained to determine if there would be some loss of prestressing visible, which did not occur. Fig. 2.54 shows the column buckled configuration at this load level. The positioning of these parallel LVDTs was deliberate in order to assess if there was some sort of rotation at this point. As both deflections were very similar, no twist was detected at this point. From Fig. 2.54 it was possible to observe that the horizontal displacements were very small when compared to the column 12 m length, proving that the temporary support was able to prevent occurrence of an additional imperfection due to the column self-weight. It could also be noted that no column rotations were detected at this point. The use of this cable proved to be a suitable solution to minimise the self-weight effects (not present in stayed columns vertically tested).

The second test aimed to simulate a column with prestressing cable forces. Two steps were also made. First, using stretches in series the cables were stretched up to the maximum value that could be manually introduced, whilst in the second step, the anchor cable hooks were tightened to up their

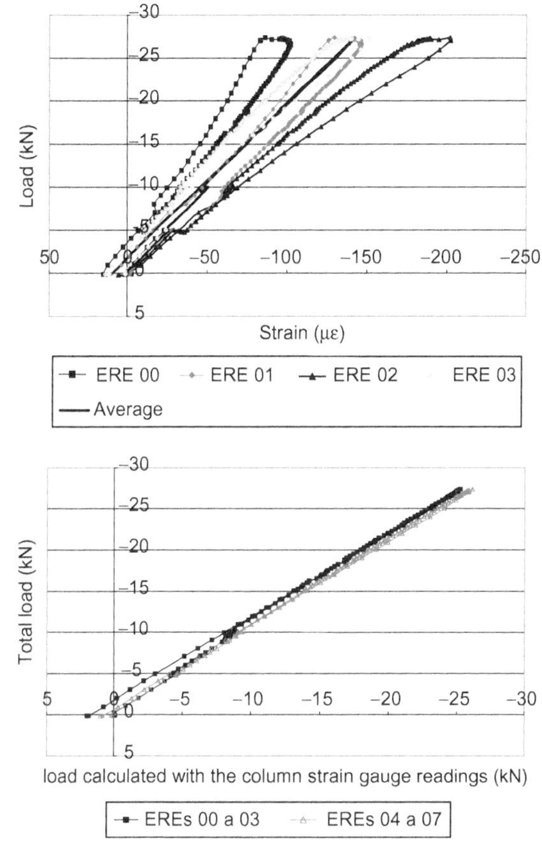

Fig. 2.53 Applied load versus central column strain gauges strains parallel to the cables and total applied load versus load calculated with the column strain gauge readings, 1st test.

maximum possible value (i.e. producing a cable force magnitude of 7.12 kN). This was followed by the hydraulic jack load application up to the column critical load. One more time, the load applied on column presented in the following graphs represents the total load, that is, the sum of the hydraulic actuator load and the cable prestressing force reaction on the column. During the test, close to 13 kN load level, G clamps were remove at the opposite location of the load application point to determine if these caused some kind of partial rotation restraint at the support influencing on the experimental results.

Fig. 2.55 shows the results for strain gauges located in the column centre parallel to the cable axis. In the first test, a linear response of the four strain gauges was expected, but it changed to a nonlinear response near the critical load. A similar explanation for the first test can be applied to this test, where some bending effects were detected in the main column tubular section. In this chart, two heights are

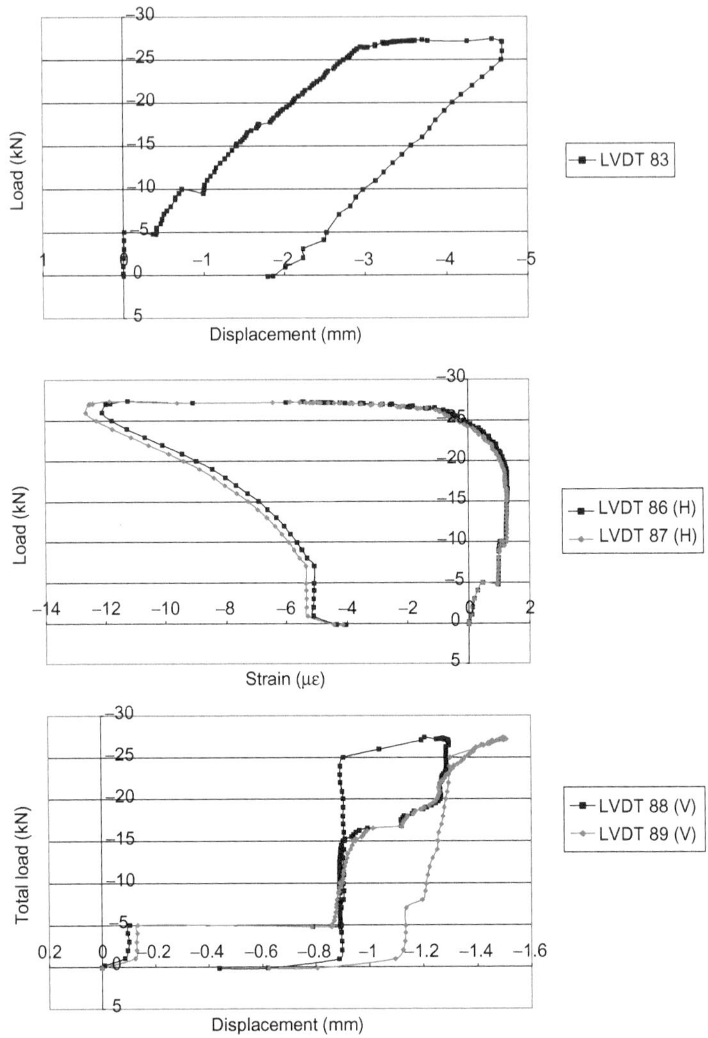

Fig. 2.54 Applied load versus LVDT 83 axial displacements and applied load versus horizontal and vertical midspan displacements curves, 1st test.

also visible, when the first occurred for the same reason in the first test (i.e. the withdrawal of support). Fig. 2.55 shows the presence of two plateaus, when the first one represents again the point where the column supports were removed around 13 kN was due to previously mentioned G clamps removal. Due to the higher prestressing load magnitudes on the cables, the cross-arm forces were not negligible as in the first test (Fig. 2.55).

Fig. 2.56 shows a rotation in the column axis that may have occurred when the supports were removed around 7 kN applied load magnitude. Due to the

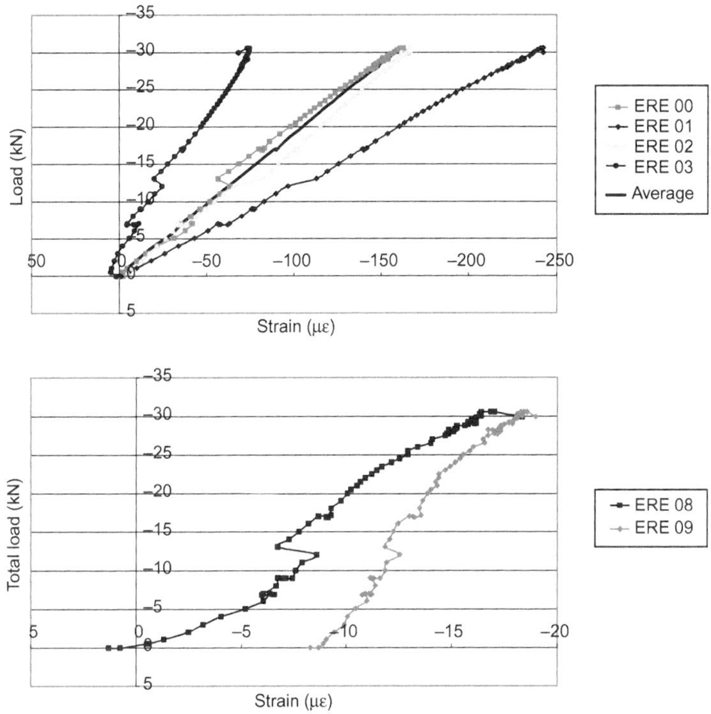

Fig. 2.55 Applied load versus central column strain gauges' deformations parallel to the cables and total applied load versus deformation in gauges 8 and 9 curves, 2nd test.

difficulties that have occurred during the lab tests assembly, it became necessary to perform a finite element analysis to better compare and assess the experimental results. As already mentioned, this problem does not occur in actual stayed columns, since most of these columns are set upright whilst their own weight acts in the direction parallel to the column axis. During the tests, due to adopted horizontal configuration, the self-weight acted perpendicular to the column axis, generating 115 mm at the column centre.

Two finite element analyses were performed using a cable to support the column at its centre for both, the first and second experiments. The first analysis was performed assuming a fixed support at one of the column ends to simulate the clips and the lack of a universal hinge at this point, thus creating an upper limit for the experiments. The second considered the two column ends as simple supported aiming to generate a lower bound for the experiments. Fig. 2.57 shows that the experimental column is stiffer than numerical model using the ANSYS software package [32]. Despite this fact, the tests ultimate loads were situated in between the two obtained modelling values.

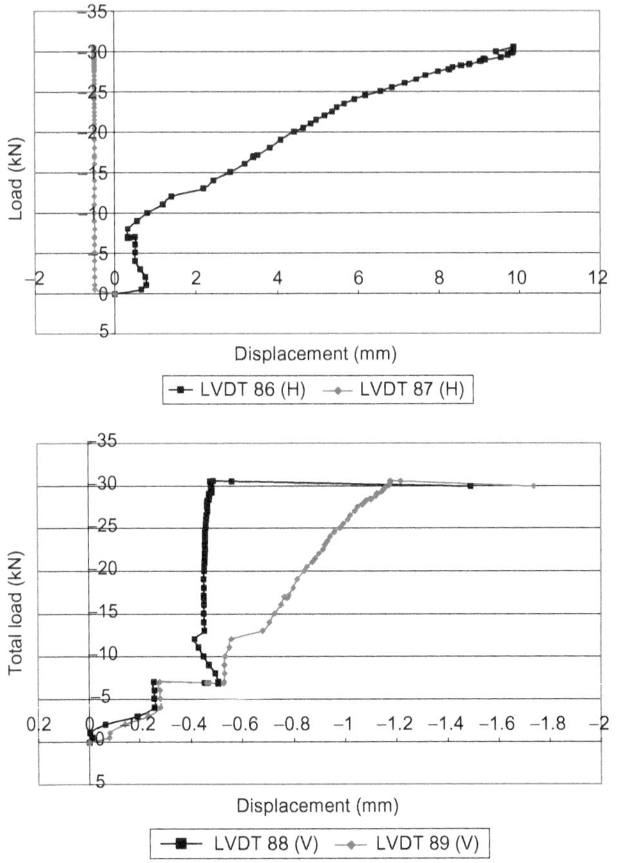

Fig. 2.56 Applied load versus horizontal and vertical midspan lateral displacement curves, 2nd test.

From Fig. 2.57, it is possible to observe that the modelling of experimental column was within the expected response. This was because the experimental model presented a partially fixed response being in between the simply supported and fixed limits. Fig. 2.57 also contains a graph where the value of the axial displacement of the experimental column was corrected by subtracting the value from the LVDT 83 (located at the column head) and from the value of LVDT 92 (located at the column opposite end). Fig. 2.58 shows the deformed configuration during the experiment where it is not easy to observe the column curvature due to the small deflections values when compared with the column length.

In the second test, as the cables forces are higher, it was expected that the critical load value would be higher than the value obtained in the first test, fact that was not observed in Fig. 2.59. These graphs indicate that when the

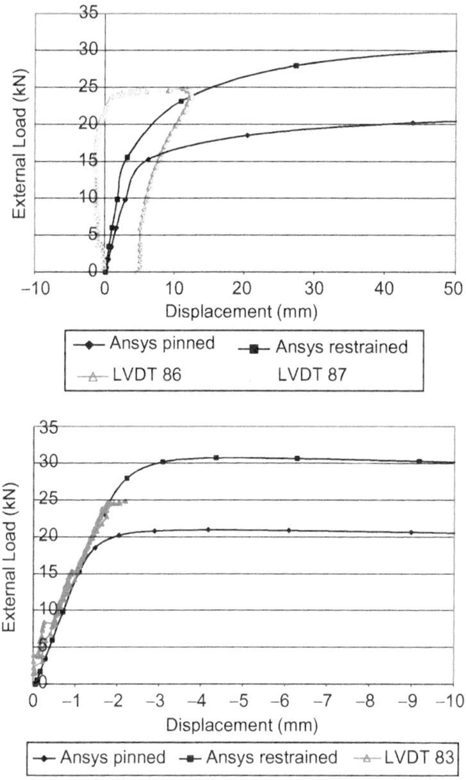

Fig. 2.57 External load versus horizontal lateral column displacements and external load versus axial column displacement curves, 1st test.

Fig. 2.58 Column deformed configuration, 1st test.

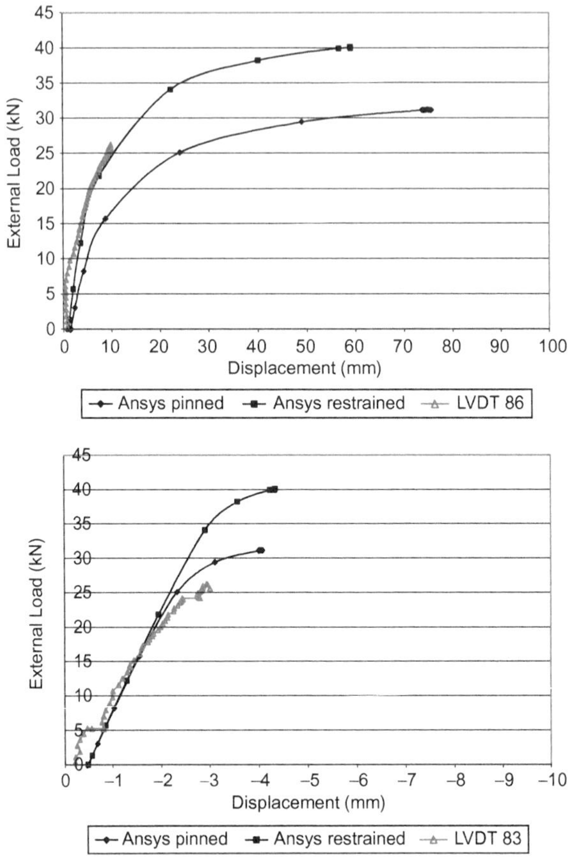

Fig. 2.59 External load versus horizontal lateral column displacements and external load versus axial column displacement curves, 2nd test.

cables prestressing force magnitude is increased, the experimental column, after approximately an applied load magnitude of 7 kN, had a similar response to the finite element model with a fixed end.

Fig. 2.59 shows a comparison between the ANSYS [32] axial displacements and the equivalent experimental values. The experimental load was lower than both values obtained from the finite element analysis and could be related to the fact that, in the proximity to the ultimate experimental load level, the deflections were increasing fast, and extreme caution was exerted to avoid the cables anchoring hooks rupture that would cause a sudden and dangerous failure mode. This fact ended the test at an early stage without a precise indication that the ultimate load was reached, fact that occurred in the first experiment. Fig. 2.59 also contains the axial displacement correction previously mentioned whilst Fig. 2.60 shows a clear deformed shape of the second test associated with the structural system controlling buckling mode.

Fig. 2.60 Column deformed configuration, 2nd test.

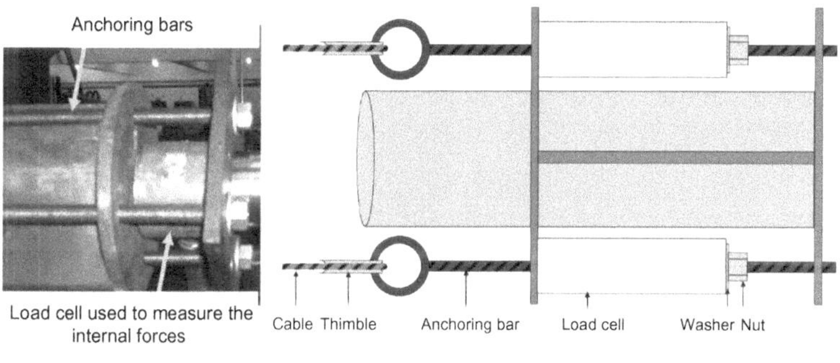

Fig. 2.61 Load cell-anchoring layout.

In the second series of tests, the assembly procedure was improved. The bars situated close to the external load application area have been fixed at the top of the steel plate used to hold the load cell used to measure the internal forces (Fig. 2.61). At the column, other end load cells were installed for measuring the stay forces. The anchoring bars had to pass through the stay load cells to be fixed to them enabling a direct transfer of the cable forces to the load cell. The anchoring bars must pass through the second plate to avoid column eccentricities and their consequent movement during the tests that could cause an imbalance. Fig. 2.62 shows a schematic procedure for installing the anchor bars at the column end whilst Fig. 2.63 shows some

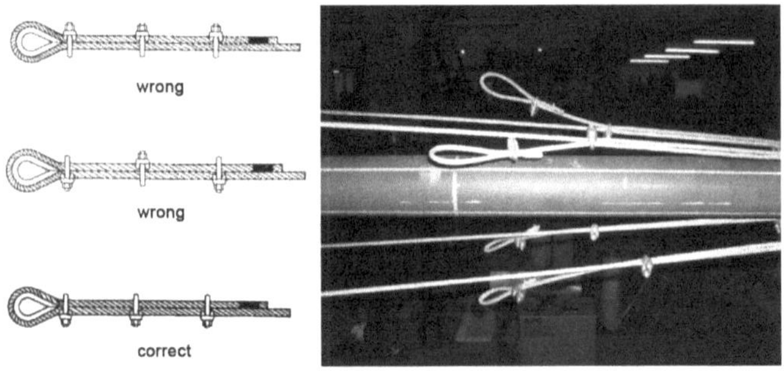

Fig. 2.62 Location of washers and wire ties for cable locking.

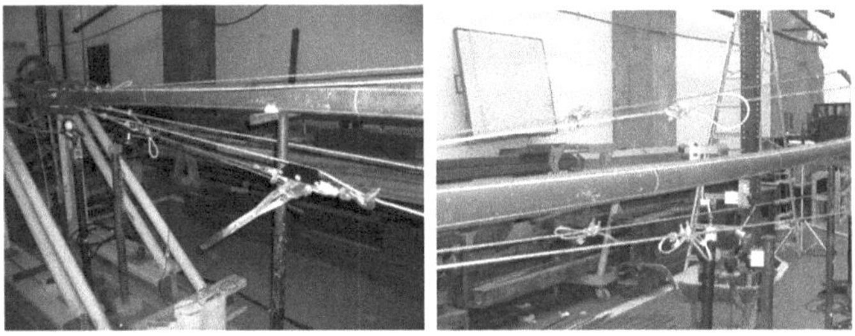

Fig. 2.63 First force application system (little grip at high stress levels) and adopted solution to improve the force application system.

of the details of the first force application system and the adopted solution to improve the original force application system.

During the tests, the importance of measuring the stay forces whilst applying the prestress was evident and motivated the development of an alternative solution to perform it. Initially, an S type load cell was thought to be used but this type of cell was not readily available in the lab or in the market. This particular load cell measures tension and compression and would be located at the column head (Fig. 2.64) between two plates. Unfortunately, their size was not compatible and could not be used in tests.

When the experimental programme was started, the initial idea was to use a single load cell to monitor the column-applied force whilst the cable stays were not registered. Since this was not a good solution, another load cell was used, attached to a plate, to simultaneously monitor the sum of the forces on the four cables. This new load cell would be externally supported at the column whilst the first cell located at the jack head acquired the column-applied load. One of the setbacks regarding this load cell was that it was not capable of individually acquiring the cable forces, inhibiting the detection of any cable force imbalance that could occur during the tests. The main gain of this second load cell was to obtain the value of the column global applied force, whilst the sum of the cable forces could be obtained by the difference of the two load cell results. Fig. 2.64 shows the system used to monitor the stayed steel column forces.

Therefore, it was necessary to develop an alternative solution to individually determine the cable forces in a size that could be fitted within the column head dimensions. This strategy has generated the need to conceive and design a load cell starting from a steel tube, with eight strain gauges located at diametrically opposed points in the tube. Its design was made in the linear elastic range, where the maximum allowed stress was equal to 50% of the steel tube compressive elastic capacity. The designed load cell consisted of a steel tube with a steel plate welded at one end, 10 mm thick, through which the loaded stays will pass (Fig. 2.65).

The adopted tube had a diameter of 38.1, 3.05 mm thick with an area of approximately 335 mm^2. The load cell was modelled on the ANSYS programme [32] with SHELL63 shell elements with a length of 150 mm along the z-axis, a 38.1-mm tube diameter for and a 50-mm end plate. The mesh, automatically generated by the programme, can be seen in Fig. 2.66. In order to have a better accuracy on the load application, a circle was defined between the tube diameter and the end plate hole forcing the node generation along the

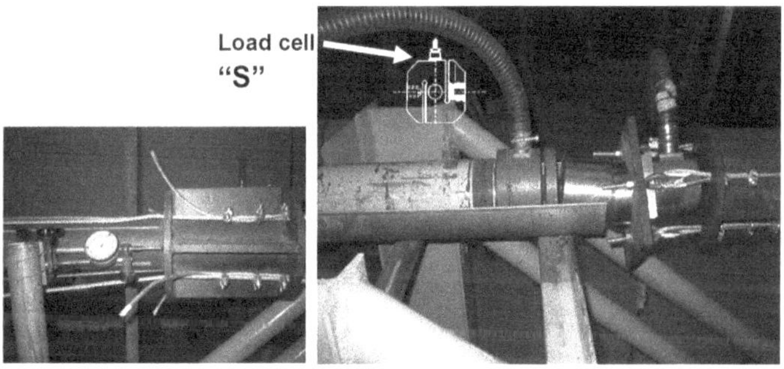

Fig. 2.64 First load cell positioning on the column top and second load to acquire column and stays forces.

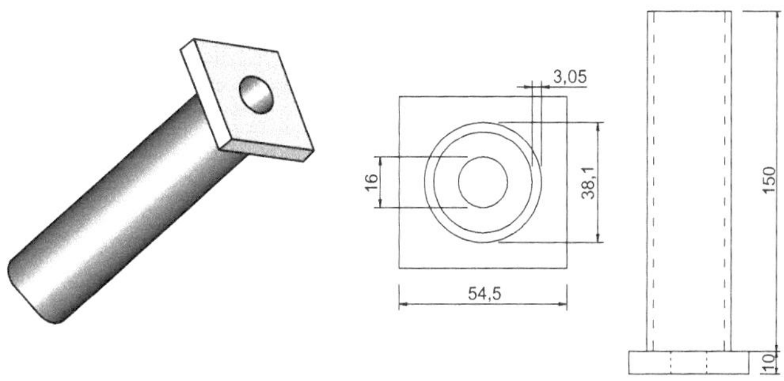

Fig. 2.65 Stays load cell outline.

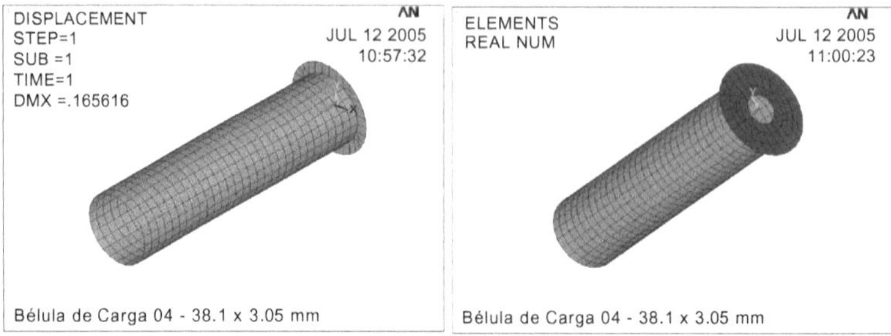

Fig. 2.66 Adopted load cell mesh.

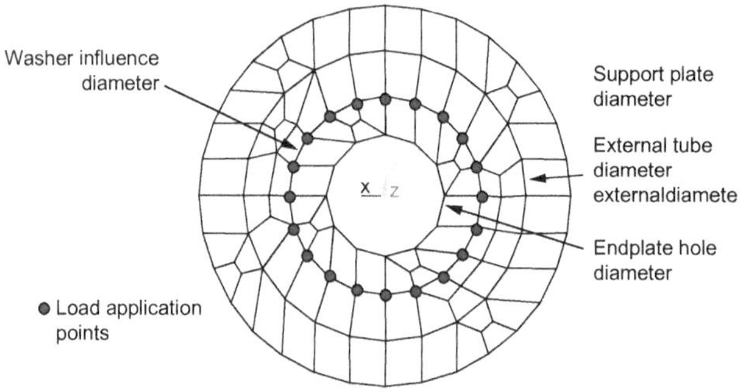

Fig. 2.67 Load application points.

this perimeter. This circle represents the washer influence on the end plate and the loads were applied in discrete points along this circle perimeter. Analyses were the performed on developed finite element model.

Fig. 2.67 shows the end plate circle containing the 20 load application points. The performed analysis involved the following load per node cases: M_1—Equal loads $P = 2700$ N (≈ 54 kN); M_2—1st Imbalance: $0.5P_1 = 3300$ N and $0.5P_2 = 1660$ N; M_3—2nd Imbalance: $0.25P_1 = 3840$ N and $0.75P_2 = 1920$ N; M_4—3rd Imbalance: $0.125P_1 = 4340$ N and $0.875P_2 = 21700$ N.

The imbalance load cases represent the possible problems that may occur during the tests, as for example, a small anchor bar eccentricity at the end plate, or an uneven cell surface. In the first model, all loads had the same magnitude (i.e. $P = 2700$ N). Figs 2.68–2.70 depict the numerical analysis results for the studied load cases. An inspection of Fig. 2.68 confirmed that the normal stress values corresponded to 50% of the maximum elastic stress design (i.e. 178.5 MPa). The distribution of stresses and displacements along the load cell confirmed that ideal point for positioning the strain gauges was at the steel tube centre.

The second model representing the first case of imbalance loads is shown in Fig. 2.69, where half the points were loaded with a load magnitude $P_1 = 3330$ N, whilst the remaining were associated with a load magnitude

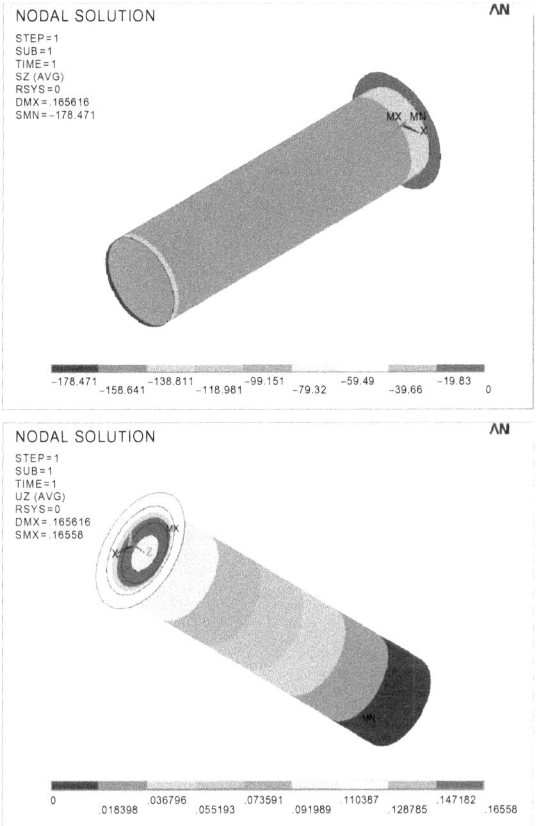

Fig. 2.68 *Z*-axis normal stresses (MPa) and deflections (mm), M_1.

$P_2 = 1660$ N. Despite being an eccentric load case, the maximum normal stress value was no more than 186 MPa. The stress remained, as predicted, confirming the tube centre as the best strain gauge location.

The third model representing the second case of imbalance loads is shown in Fig. 2.70, where quarter the points were loaded with a load magnitude $P_1 = 3840$ N, whilst the remaining were associated with a load magnitude $P_2 = 1920$ N. This is an eccentric load case where all the load is concentrated in a quarter of the area and produced a maximum normal stress value was equal to 190 MPa whilst the remaining tube points were under 167 MPa. This stress distribution confirmed the tube centre as the best strain gauge location.

The last model representing the third case of imbalance loads is shown in Fig. 2.71, where 1/8 of the points were loaded with a load magnitude $P_1 = 4340$ N, whilst the remaining were associated with a load magnitude $P_2 = 2170$ N. This is an extreme eccentric load case where the entire load

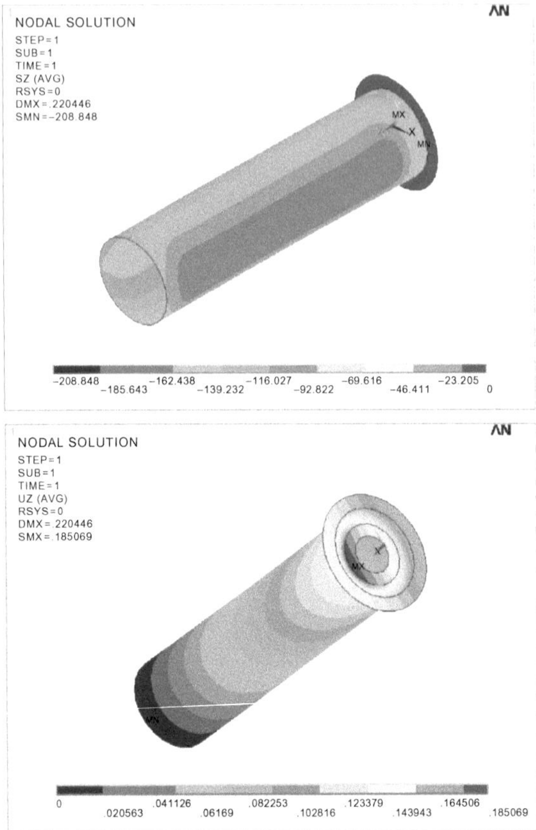

Fig. 2.69 *Z*-axis normal stresses (MPa) and deflections (mm), M_2.

is concentrated in 1/8 of the area but still confirmed the tube centre as the best location for positioning the strain gauges.

The numerical analyses results suggested a review of the initial load cell design. The end plate had its external part removed to fit the tube diameter, since the removed part could be an obstacle for installing the load cell at the stayed column head. The new load cell can be seen in Fig. 2.72.

During the test assembly, some difficulties appeared concerning the self-weight, rotation at one of the column extremities and the lack of space for using an additional universal hinge. They were solved by increasing the length of two external tubes of the reaction frames. At this stage, tests were carried out by varying the prestressing force magnitudes from 0 kN (no stays) to 3.00 kN per stay. The following results refer to cases with the prestressing force magnitudes: first a solution without using stays was assessed and it was followed by another test with a 0.23 kN magnitude applied in each stay. Fig. 2.73 shows the adopted column LVDTs location.

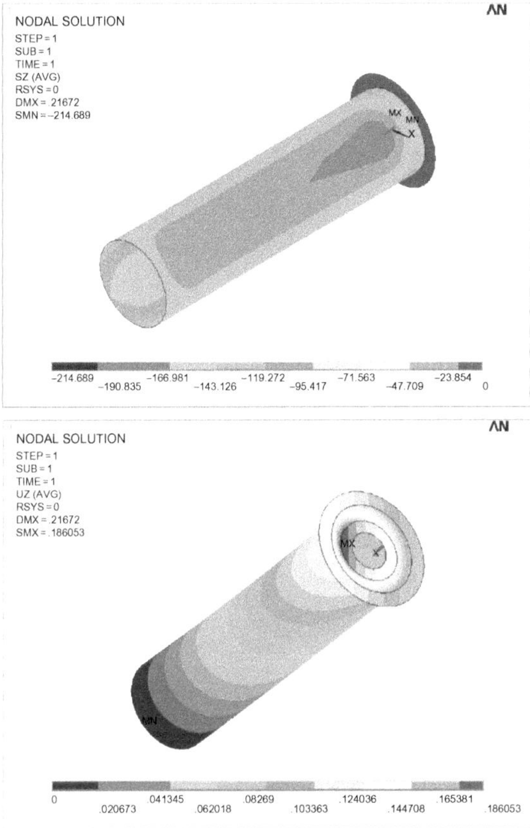

Fig. 2.70 *Z*-axis normal stresses (MPa) and deflections (mm), M_3.

The first column maximum load was 10.9 kN, close to the numerical value of 11.3 kN evaluated with the ANSYS programme for a column with 8 mm initial bow imperfection. Fig. 2.74 shows the test jack load versus horizontal displacement curves. Two parallel LVDTs were used at the column centre to check if any twisting occurred during loading, a fact that was not noted. A slip was observed close to the 8 kN load level due to the removal of the temporary supports at the column ends. The vertical deflection has not been significant, with values of approximately 2 mm.

In the column with 0.23 kN per cable, the force used to remove the cables slack with the same maximum load achieved during the test was 15.64 kN, that is, an approximately 40% increase in relation to the stayless column test. Fig. 2.74 also shows the test jack load versus horizontal displacement curves. In this second test, the column had larger displacements, up to its critical load value, but as in the previous test, no twist occurred. Once again, a slip was noted near the 8 kN load level due to the removal

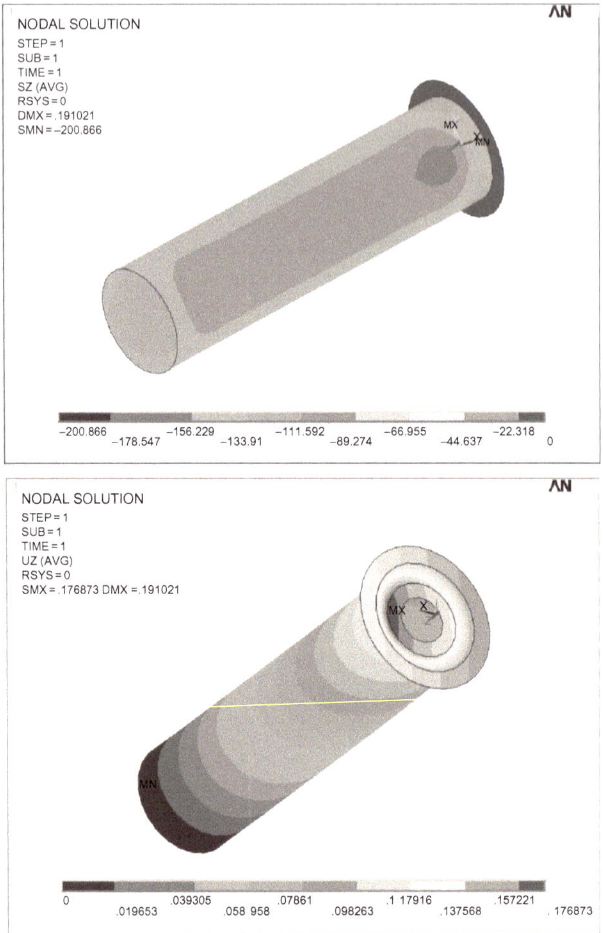

Fig. 2.71 *Z*-axis normal stresses (MPa) and deflections (mm), M_4.

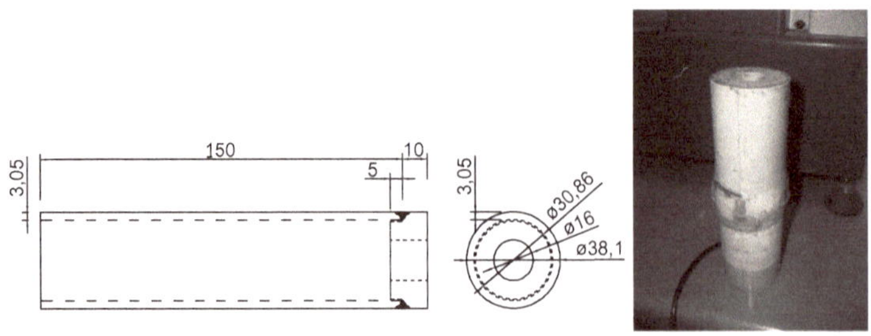

Fig. 2.72 Stays load cells.

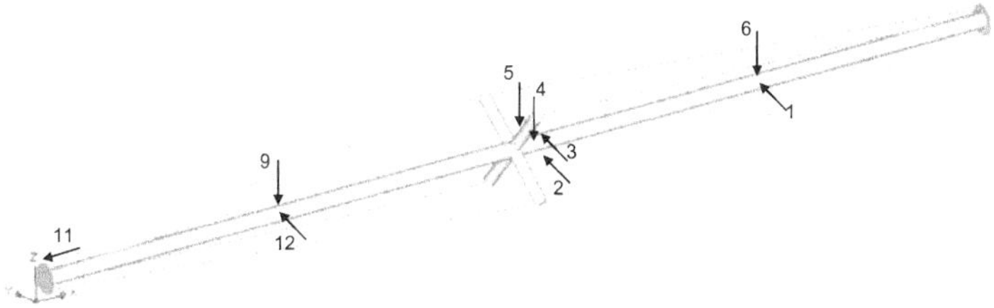

Fig. 2.73 LVDTs location layout.

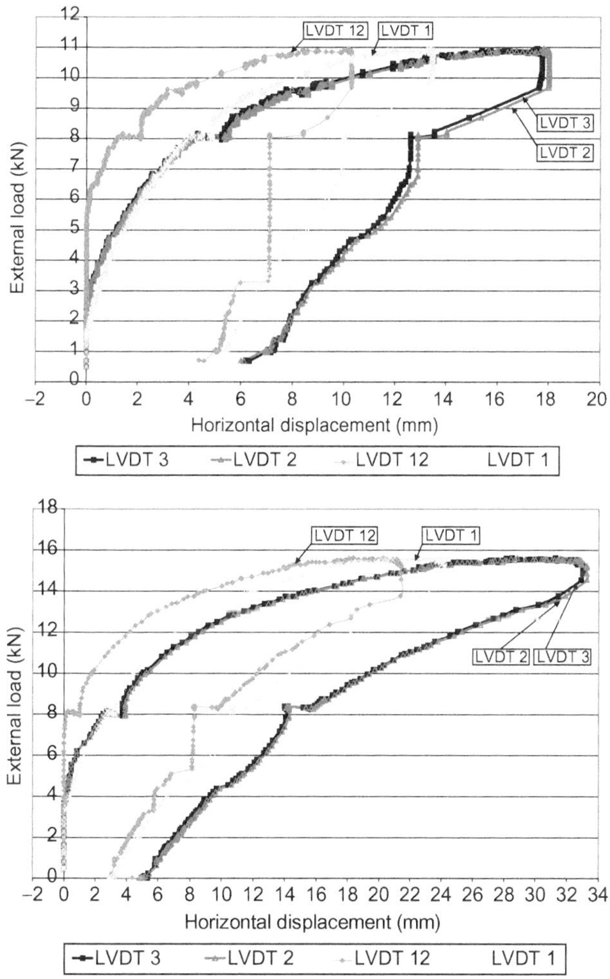

Fig. 2.74 Jack load versus horizontal deflection curves: column without stays and column with 0.23 kN per stay, 1st tested column.

of the temporary supports at the column ends. The vertical deflection was not large either, with values approximately equal to 2 mm.

The first test on the second column was performed on a stayless configuration with the aim at determining the structural system minimum capacity. The load supported by the column was 9.60 kN, value below the numerical estimation of 11.28 kN. This was due to a higher initial imperfection present in the main column. Fig. 2.75 shows the jack load versus horizontal displacement curves, where the typical elastic buckling curve can be

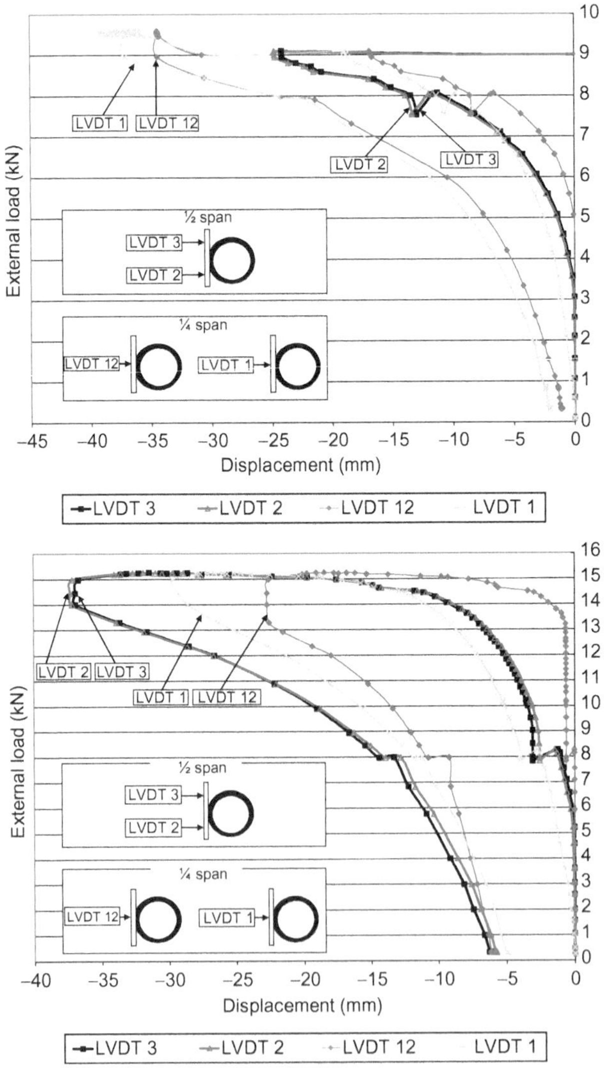

Fig. 2.75 External load versus horizontal deflection curves: column without stays and column with 0.31 kN per stay, 2nd tested column.

observed. The vertical displacements are not presented since they were not significant, for the column only moved horizontally.

The two LVDTs positioned in parallel were installed at the column centre to monitor the horizontal displacements and a possible twist, a fact that has not occurred, since the LVDTs 2 and 3 curves are very close. The larger displacement occurred at the column centre, characterising the structure first global buckling mode. As predicted by the first column tests, a small reduction of the applied load occurs around age 8 kN, due to the temporary supports removal near the main column ends.

The second test of this same column was carried out with a prestressing force of 0.31 kN per cable to remove their slack. The test ultimate buckling load was 15.26 kN, that is, 59% higher than in the first test, which was 9.60 kN, and 35% higher than the numerical buckling load, that is, 11.28 kN. Fig. 2.76 shows the jack load versus horizontal displacement curves, where a typical elastic buckling curve can be observed.

As well as in the stayless test, the vertical displacement results will not be presented since no significant values were recorded, because the column only moved horizontally. Fig. 2.76 shows the cables' responses during the test. The prestressing force increase is clearly noticed in stays 2 and 4, located on the side where the buckling occurred, whilst restricting the structure lateral displacement and thus increasing its load-carrying capacity. The individual cable force measurement was possible due to the smooth operation of the designed load cells.

The third stayed column test series were carried out to compare the results in terms of stay stiffness (cable or reinforcing steel bars). This column used

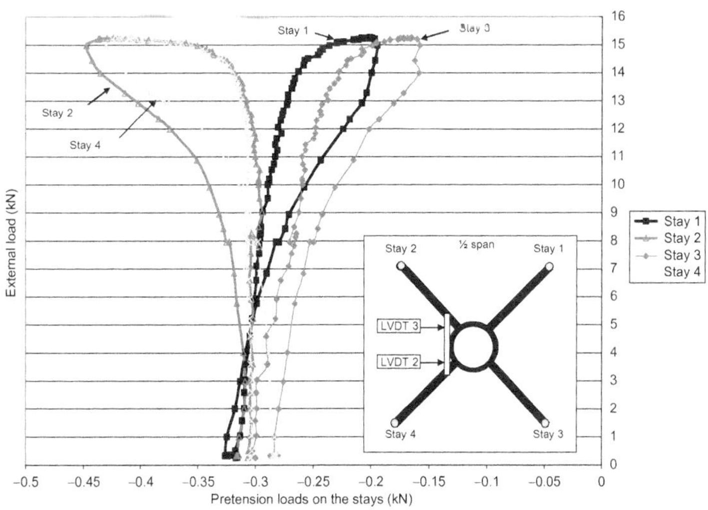

Fig. 2.76 External load curve versus forces acting on the columns' stays curves.

reinforcing bars instead of the cables used in the first and second columns. The tests were performed varying the level of prestress in the stays from 0 to 0.75 kN. The first test was conducted with the main column without the stays reaching an ultimate load of 11.43 kN, approximately equal to 11.28 kN of the numerical value. Fig. 2.77 shows the jack load versus lateral displacement curves for the investigated column where a typical elastic buckling response was again observed. One more time, vertical displacements were not presented

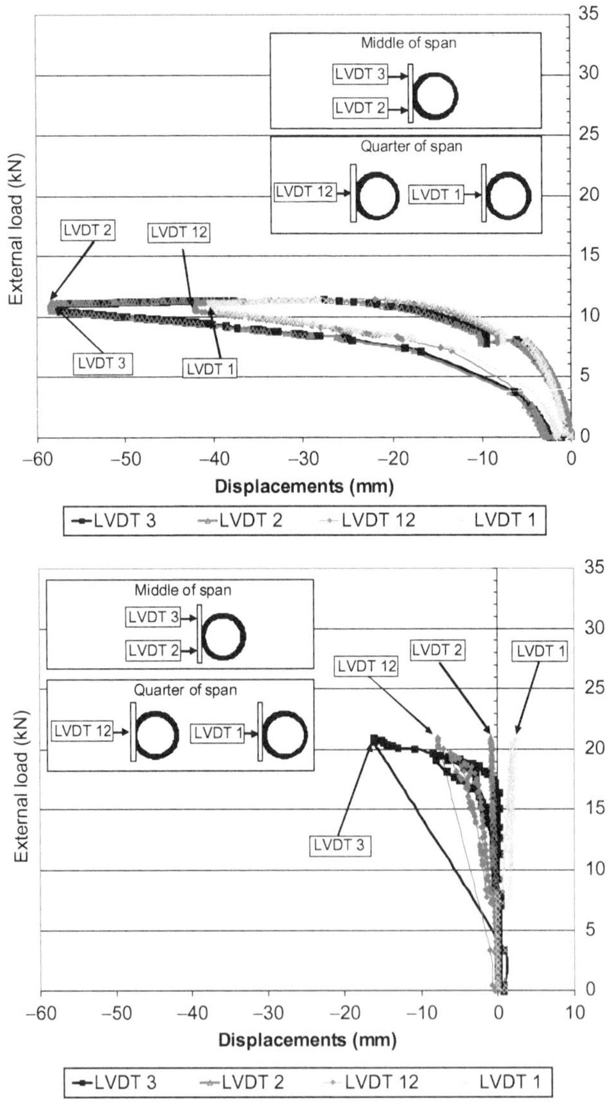

Fig. 2.77 Jack load versus horizontal deflection curves: column without stays and column with 0.31 kN per stay, 3rd tested column.

because they were not significant corresponding to values of approximately 5 mm. The two parallel LVDTs located at the column centre span were used to monitor if any torsion occurred during the test, fact that did not occurred. During the tests, the lateral displacement configuration had its maximum at midspan, typical of the first elastic buckling mode configuration.

The following test was performed with a 0.30 kN prestress level per stay aiming to remove the reinforcing steel bar slack. The ultimate load for this test was equal to 20.94 kN, approximately 83% larger than test result without prestress and/or the numerical buckling load. Fig. 2.78 shows the jack load versus lateral displacement curves for this test where a typical elastic buckling behaviour was one more time noted.

Once again, the vertical displacements in this test were insignificant. During the test, the first buckling mode configuration occurred as can be seen with the results of the middle span LVDTs shown in Fig. 2.77. The LVDTs' results also confirmed that no torsion occurred during the tests. Similar to the test without prestress, a small load drop occurred around 8 kN, due to the temporary supports removal located at the column ends.

The ultimate load for the test with 0.75 kN on each stays was 30.96 kN, approximately 170% larger than test without prestress and/or the numerical buckling load. Figs 2.78 and 2.79 show the jack load versus lateral displacement curves for this test where a typical elastic buckling response was again visible. Different from the previous tests this test was associated with the second buckling mode configuration.

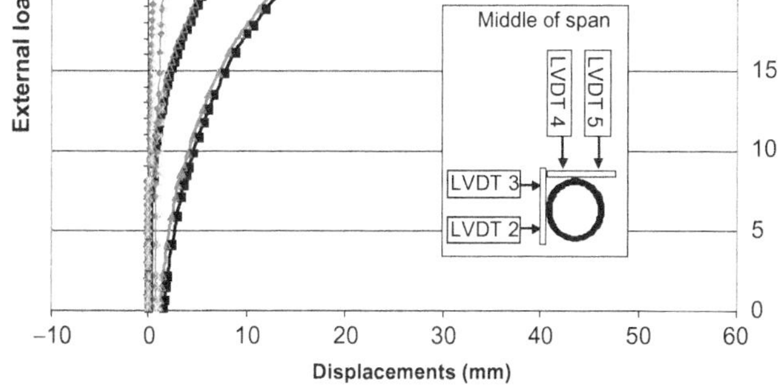

Fig. 2.78 Jack load versus lateral displacement curves, midspan, 0.75 kN prestress force magnitude.

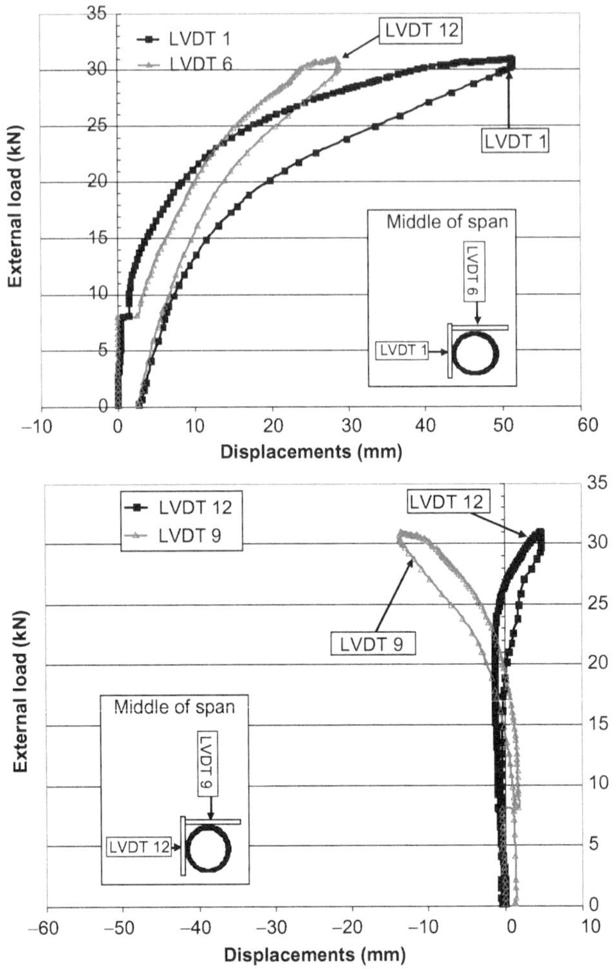

Fig. 2.79 Jack load versus lateral displacement curves, quarter span, 0.75 kN prestress force magnitude.

At midspan the stayed column vertical displacement was again not significant. The LVDTs' results also showed that no torsion was observed during the tests. Alternatively, the LVDTs located at quarter span recorded significant vertical displacement indicating the occurrence of the second buckling mode configuration. A significant loss of prestress in the stays is shown in Fig. 2.80, where the reinforced bars reached the slack mode.

The prestressed stayed steel column structural systems are effective for situations where it is needed to withstand substantial loads at great heights, enabling a fast assembly. The load-carrying capacity of a 12-m slender column was significantly increased with the use of prestressed stays. This

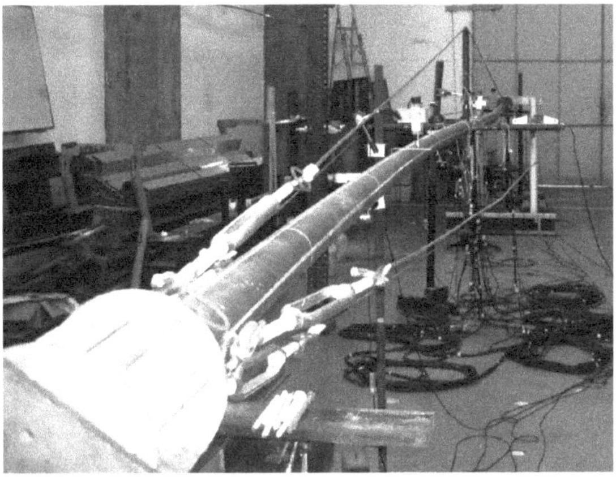

Fig. 2.80 Second buckling mode, 0.75 kN prestress force magnitude.

strategy effectively reduces the column effective buckling length in at least 50%, substantially increasing its strength. This resistance increase is because the stay forces are transferred to the main column, through bars welded at right angles in the central span, creating, at this point, an additional restriction to displacements and rotation.

The simple use of the stays with enough prestress to remove the slack increases in approximately 40% of its load-carrying capacity, where its resistance without stays was approximately 10 kN (slenderness $\lambda = 400$). The use of individual load cells for measuring the stay forces was essential, since they enable a better control over the application of the prestress forces and are able to record any load imbalance during the test. The computational analysis confirmed the maximum stresses within 50% of the elastic design so that the behaviour of the load cell was always linear. The load unbalance consideration was also important, avoiding that these problems could affect the load cell measurements due to unavoidable eccentricities.

The full-scale test results qualified the use of stays in slender columns. The ultimate load obtained in the second column test without stays was 9.60 kN. The numerical load for the column without stays was 11.28 kN, indicating that the column imperfections were significant. It is interesting to observe that when stays were used with enough prestress force to remove the cable slack (i.e. 0.30 kN), the column load-carrying capacity has been increased by 35%, when compared with the 15.26 kN numerical ultimate loads. When compared with the ultimate load for columns without stays, a significant 57% increase was obtained showing the structural system effectiveness.

A substantial increase of ultimate load occurred in the third column. The final test load without stays was of 11.43 kN, value close to the numerical load. The ultimate load of the test with prestress force of 0.31 kN per stay was 20.94 kN, approximately 83% higher than the ultimate load of the test without stays. Finally, the ultimate load of the test associated with a prestress force 0.75 kN per stay was equal 30.96 kN, approximately 170% higher than the ultimate load of the test without stays.

The developed experimental programme had to be conducted horizontally, due to the column dimensions (12 m) and the lab restrictions. This configuration generated excessive displacements on the columns due to self-weight that do not occur when these systems are used in vertical position. As previously shown a cable was used to compensate the cable weight and showed, in numeric equivalent models, no influence over the ultimate loads or collapse modes. To further investigate this issue, experimental models were tested with two settings: first, with cables parallel to the two cross arms named + configuration, and then, with cables of a 45° angle between two cross arms named × configuration. Very similar results were found in these two cases.

Greater care was taken in the final test series where hinges were used on both ends of the columns, since in the previous tests, when only one hinge was used; the obtained results were influence by a partial rotation restriction that occurred at one of the column ends. The application of prestressing force demanded the development of a procedure to balance the four stay loads whilst enabling a fine-tuning to determine their influence over the column ultimate load. For that, cable turnbuckles and threaded bars were employed, showing that the prestress application must be carried out in stages in diametrically opposed cables so as not to induce premature collapses only caused by the prestress procedure and associated imperfections.

The test required the development of a nontrivial instrumentation. As the tubes had circular sections, it became necessary to use two LVDTs arranged at a 90° angle per point, where the deflections should be monitored. A combination of these two LVDTs' results enables the determination of the displacements of this point in any direction. The cross bar tubes and the main tube were also monitored with four linear strain gauges located in the tube sections to measure the axial forces and assess possible bending effects.

The monitoring of the column loads was initially made with two load cells. The first was located between the prestress plate and the jack, which was used to measure the applied load, whilst the second one was located between the prestress plate and the column endplate measuring by difference in the first cell the applied global prestress forces. The test result indicated that, during the test, the cables forces could vary substantially. This fact created the need for the development of individual cable or bar load cells for their individual monitoring. During the tests, these cells had a suitable performance and enabled the detection of possible stay unbalances for the correct application of the prestress forces and to determine the individual performance of each cable or bar.

2.6.5 Structural Evaluation of Major Axis Semi-Rigid Joints With Angles

The experimental series consisted of three connection tests carried out to investigate their structural response and identify their associated moment versus rotation curves [33,34]. The joint test layout is shown in Fig. 2.81. The tested beams and columns were made of I $10'' \times 37.7 \times 1500$ mm,

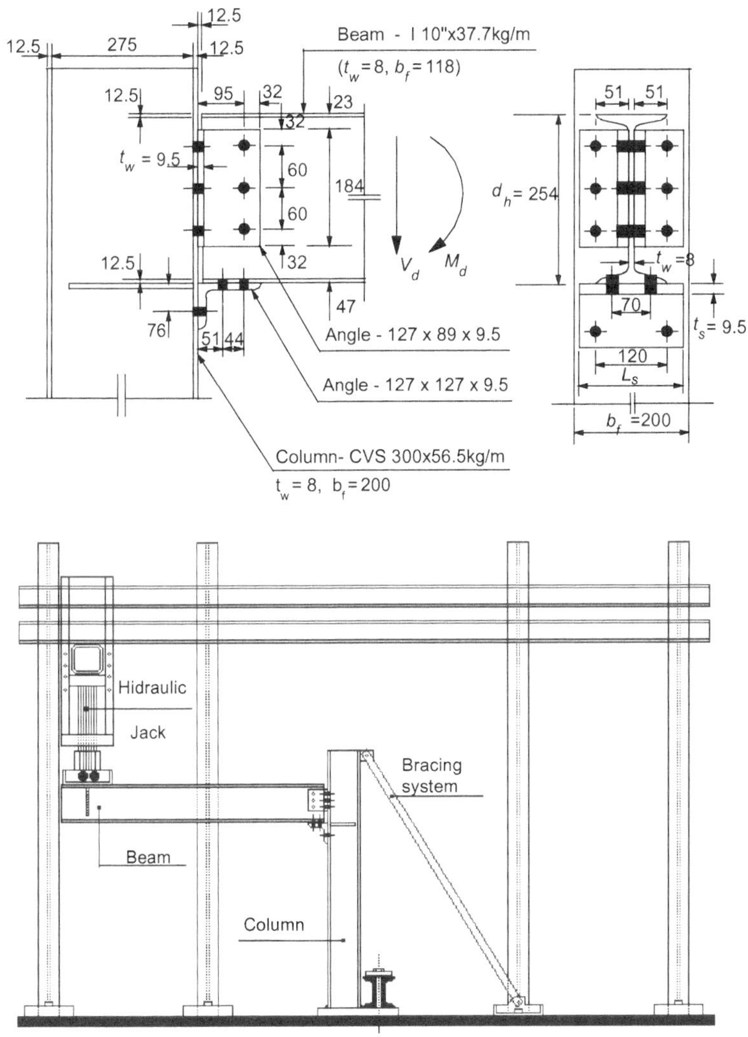

Fig. 2.81 Tests layout.

WWF-300 × 56.5 × 1700 mm. The first test utilised a 127 × 127 × 9.5 mm and a 127 × 89 × 9.5 mm angles, photo 2, whilst the second and third tests used 127 × 127 × 12.7 mm angles. The columns (F_y = 371 MPa, F_u = 494 MPa), beams (F_y = 309 MPa, F_u = 419 MPa), and angles (F_y = 327 MPa, F_u = 506 MPa), where all made with ASTM A36 steel grade. All the investigated joints utilised 19.05 mm, ASTM A325, high strength friction bolts. The first joint tested utilised 15 bolts, whilst the second and third tests utilised only 10 bolts. A 6.35-mm reinforcement plate was utilised in the third joint, when it was welded at the web angles top. The joints were tested up to their ultimate or serviceability condition under monotonic static loads.

The collapse mode was predicted for the first and third tests and was associated with large rotation/displacements and strains. These modes confirmed the structural behaviour observed in the tests. The first test had its failure associated with the formation of a plastic mechanism in the web angles, whilst the third test was stopped due to significantly large displacements. A bolt failure at the bottom flange to angle joints associated with the compressive force path was forecasted for the second test. This forecast proved to be correct, since soon after the applied load reached the predicted load value, a sudden shear bolt rupture occurred (Fig. 2.82).

The values of the displacements were measured with LVDTs and dial gauges (Fig. 2.83). To compare and access the error involved in evaluating the relative rotations, an optic measuring system was utilised. For the first test, a set of laser devices was used, whilst for the second, two topographical devices were used (Fig. 2.83).

Load displacement curves for the tested specimens are shown in Fig. 2.84 where it is possible to note that the ultimate load increased from the first to the last tested joint. Fig. 2.84 shows the moment versus relative rotation (beam minus column) curves for the tested joints. The rotations present in these curves were evaluated by two methods. The first method utilised the displacements, results obtained with the electrical transducers and dial gauges combined with the use of simple geometry and structural mechanics

Fig. 2.82 Deformed configuration, 1st test; bolt rupture, 2nd test; and deformed configuration, 3rd test.

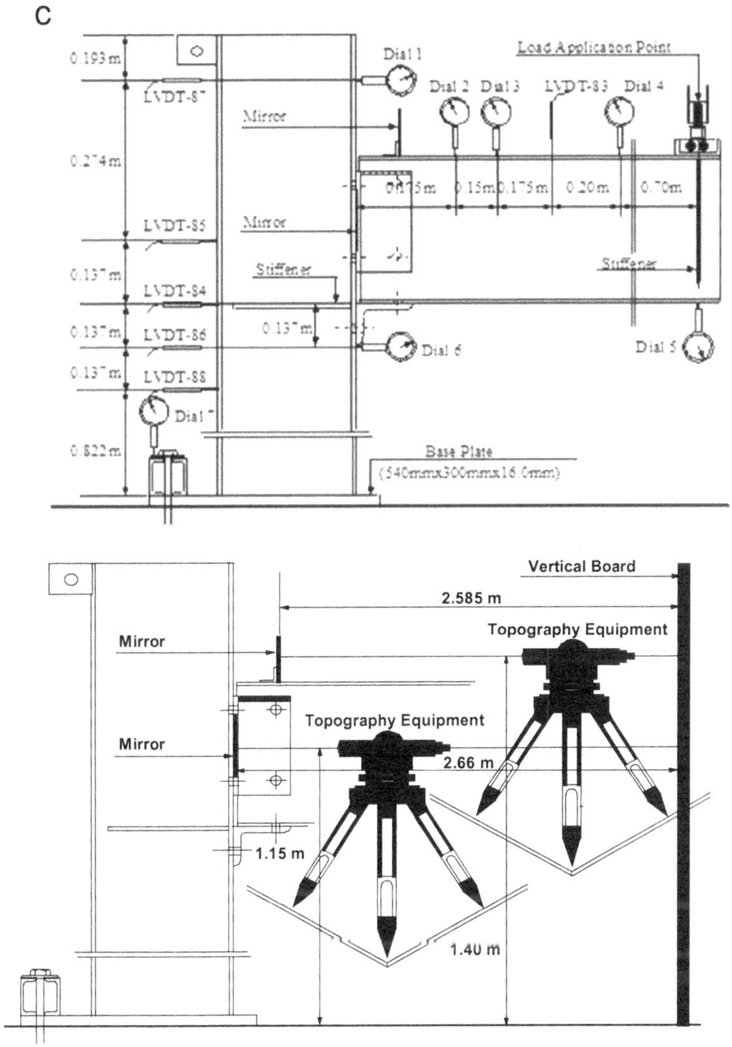

Fig. 2.83 Final test layout and rotation measurement system.

techniques. The second method utilised the optical system previously described. A substantial part of the difference observed by utilising the two systems was related to the fact that the optical measurement system was manually operated, whilst the LVDTs data were acquired through a computerised data acquisition system.

Connections are classified according to their moment transmission capacity (i.e. rigid, semi-rigid, or flexible). In the present study, two classification systems were used. The first system is recommended in the Eurocode 3 [19], whilst the second system was proposed by Bjorhovde.

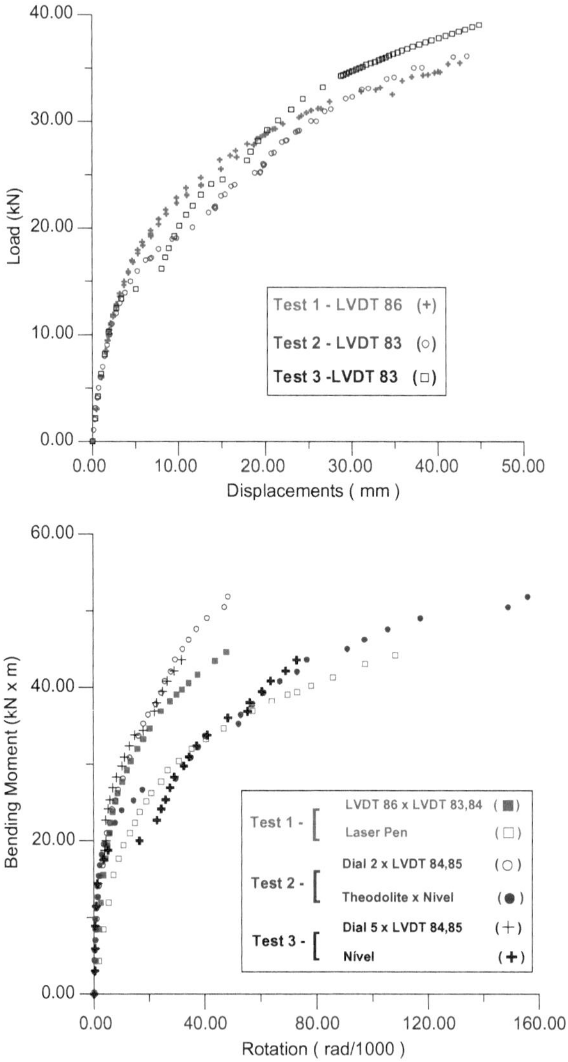

Fig. 2.84 Load versus displacement and moment versus rotation curves.

From the graphs shown in Fig. 2.85, it was possible to conclude that the three joints tested are semi-rigid.

Strain measurement was made with six linear strain gauges located at the seat angle and eight linear strain gauges in each web angle (Fig. 2.86).

Fig. 2.87 shows load versus normalised strain ($\varepsilon/\varepsilon_y$) curves for strain gauges numbers 0–3. All strain gauges yielded at an approximate load of 25 kN, signalising the start of the plastic mechanism formation.

The left and right web angle linear strain gauges are shown in Fig. 2.88. Strain gauges 6 and 20, located on the left web angle, almost reached yield at the test

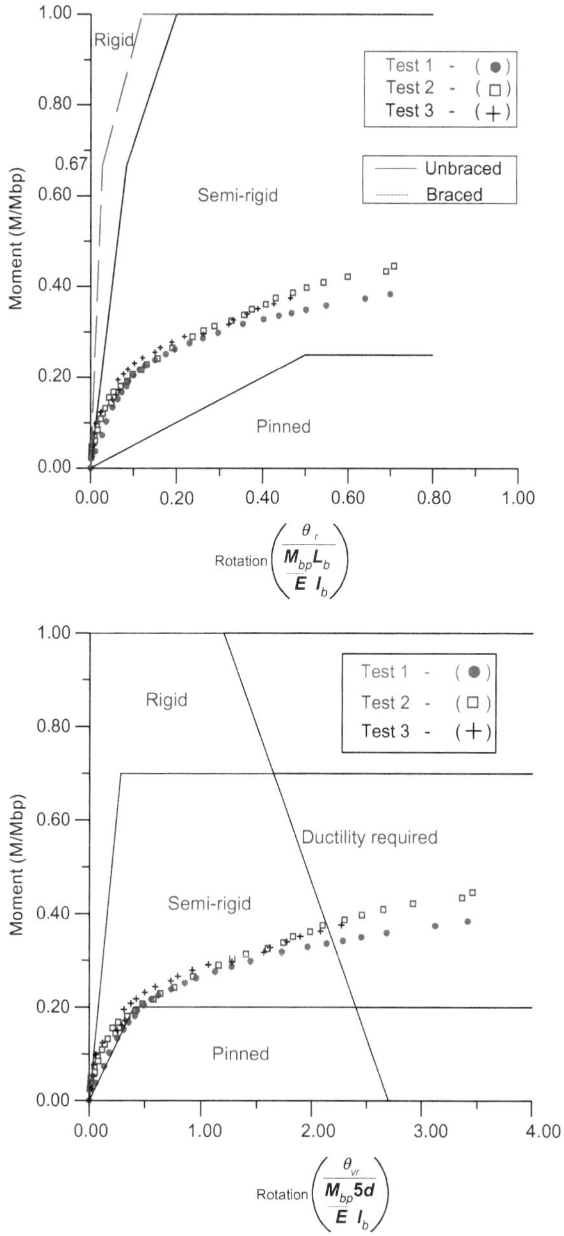

Fig. 2.85 Eurocode 3 and Bjorhovde connection classification systems.

ultimate load suggesting that the plate welded to angle significantly reduced their bending strains. All the other strain gauges presented insignificant results.

Experiments were undertaken to assess the moment versus rotation curves and the flexural capacity of bolted semi-rigid steel joints using web and

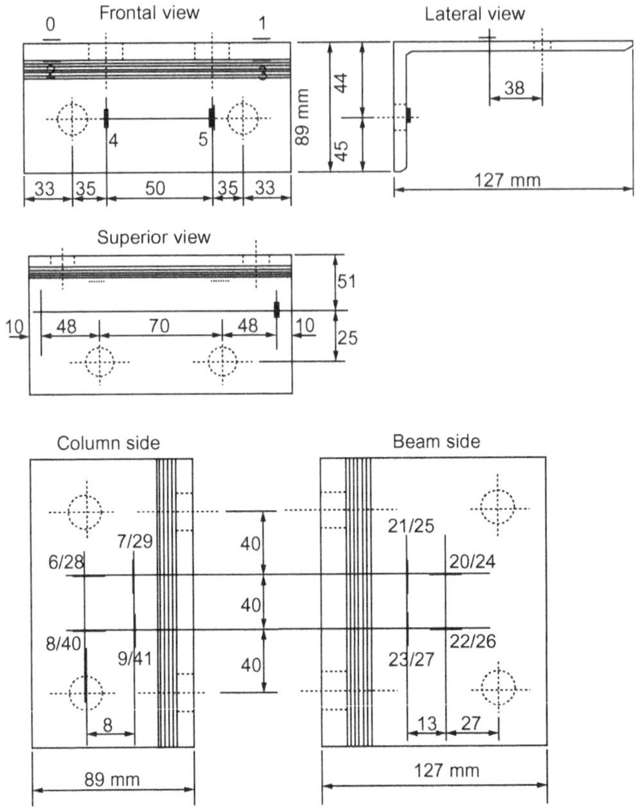

Fig. 2.86 Linear strain gauges used in seating and web angles.

support angles. The relative rotation measurement systems had a satisfactory response. The obtained moment versus rotation curve depicted a nonlinear behaviour for all three tests. The joints bending capacity was consistent to use in low-rise residential buildings with the extra benefit of being an economic and easy to assemble solution. The tested joints ultimate load was 35.5 kN for the first test. The second test ultimate load reached 38.3 kN, depicting a 7.7% increase in relation to the first test, but in relation to the initial stiffness, a significant increase of 297% was obtained. This was mainly due to the reduction of the bolt distances in relation to the column web and to the increase of the angle thickness.

The third test presented gains of strength and initial stiffness of 9.7% and 16.2%, when compared with the second test results. This occurred due to the use of a reinforcement plate on top of the web angles. The adopted plate shifted the gravity centre of the forces transmitted by the beam at the web angle and reduced the shear force acting on the seat angle. This led to smaller forces on these bolts and changed the ultimate joint behaviour associated with excessive beam

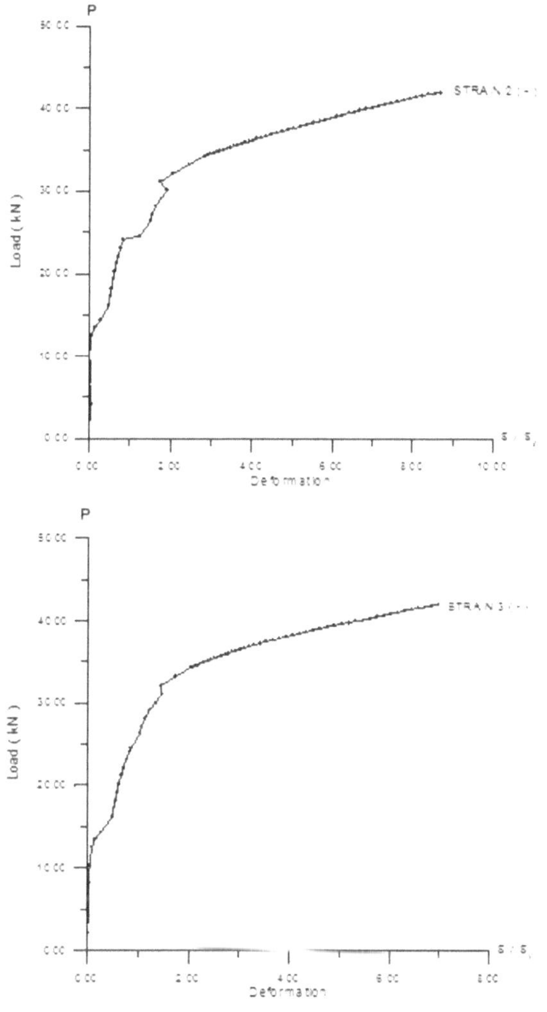

Fig. 2.87 Load versus normalised strain curve, seating angle, 3rd test.

deformations, bearing at the web and seat angle plates and angles strains induced by the joint large displacements.

The studied joint required the development of an instrumentation system for measuring the beam-to-column relative rotation and their main element strains. The adopted experimental model consisted of a column fixed on the base and anchored at the top, where it was connected to beam by two web angles and a seat angle. It should be noted that in the third test, the seat angle was stiffened to increase the rotation stiffness. A hydraulic jack, rollers, and a load cell were used at the beam-free end for the application and monitoring of a monotonic load.

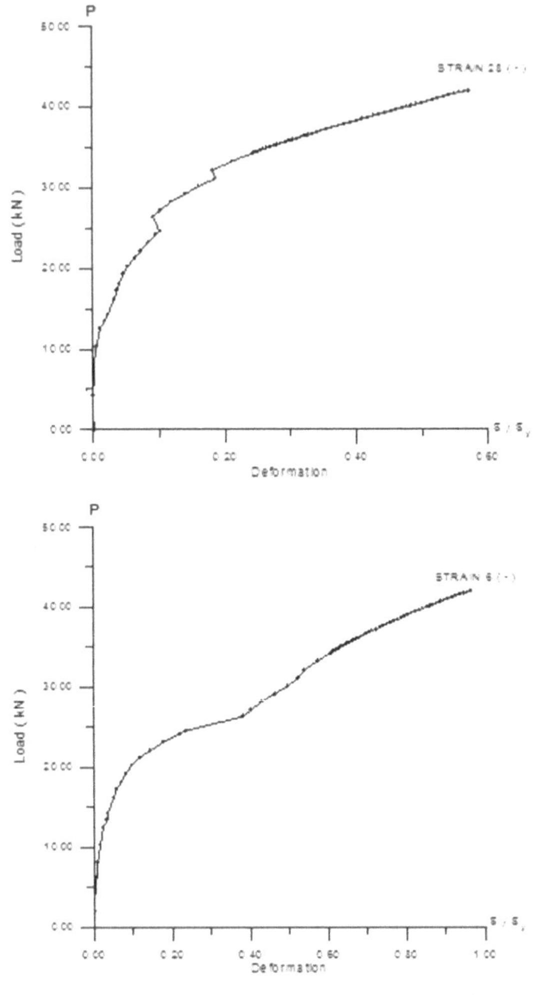

Fig. 2.88 Load versus normalised strain curve, web angles, 3rd test.

Strain measurement was performed with rosettes and linear strain gauges in the beam and angles enabling an assessment of the stress distributions evolution in these. On the other hand, the beam-to-column relative rotation measurement generated the need for redundancies to its correct assessment. The joint rotation was evaluated by the difference between the beam and column rotations. The column rotation was determined with LVDTs and dial gauges positioned in these elements, whilst the beam rotation was acquired by three methods. The first, and more conventional, used LVDTs and dial gauges installed in these elements, always keeping in mind that the beam displacement, at these points, should be deducted. The second and third methods used mirrors, graduated rules systems and laser pens (second), and theodolites (third) to evaluate the rotation. Although, qualitatively, the three methods led to consistent results, the accuracy of the latter two methods was jeopardised by the manual reading that failed to follow the load application in advanced loading stages.

This fact only came to confirm one of the key requirements of a good experimental analysis, that is, redundancy measures. Additionally, the use of seat angle stiffening proved to be quite effective in increasing stiffness and changing the ultimate limit state, which controls the connection design from the angle deformations to the shearing of the bolts present at the interface between the beam lower flange and the seat angle. It was also evident that this could be substantially improved with the use of a spacer positioned in the gap between the beam and the column flange or by using larger diameter bolts.

2.6.6 Structural Evaluation of Minor Axis Semi-Rigid Joints With Angles

When beam-to-column joints to the column minor axis are designed, this process usually assumes them as pinned, but this is not strictly true for a great majority of structural joints. The response of beam-to-column connections can be assessed with the Eurocode 3 component method [19]. From the theoretical point of view, it can be used in any kind of joint provided that the basic sources of strength and deformation are properly identified and modelled [8]. A usual component model for a minor-axis connection made of double web angles and a seat angle to the column minor axis is shown in Fig. 2.89.

The lack of a central stiffening web in minor axis joints suggests that the column web must resist the tensile and compressive forces coming from the beam flanges in bending, similar to a plate supported on its vertical sides. This additional source of deformation defined by the new component 21 [35–39]. The evaluation of this component is directly related to the column web effective width ($b_{eff,1}$), as shown in Fig. 2.90. The loaded region is

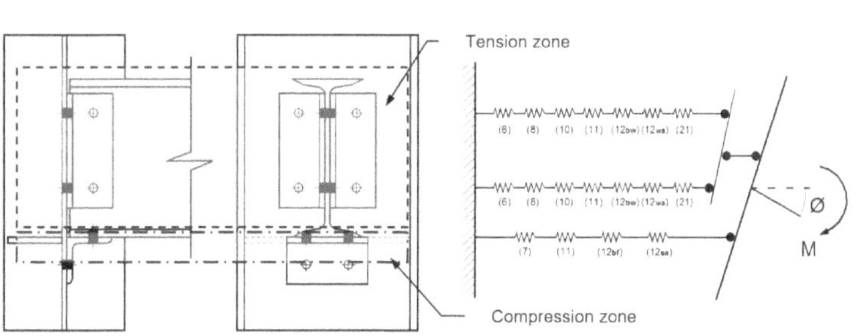

Fig. 2.89 Characterisation of the beam-to-column joints components.

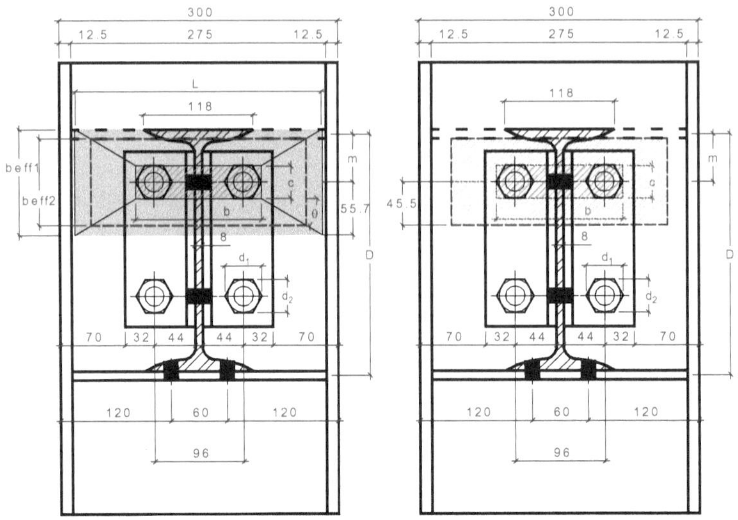

Fig. 2.90 Effective width characterisation.

limited by the dimensions b and c shown in Fig. 2.90, which contains the effective width used in column web in bending component. However, due to the stiffener, this effective width ($b_{\mathrm{eff},2}$) should be re-calculated. It is important to note that this component response is not considered in the Eurocode 3. This study involved reflection photoelasticity experiments that were later compared to numerical simulations. With these results, it was possible to assess the correct value of the effective width to be considered in the component response. The experiments were performed at the Structures and Materials Laboratory of the Pontifical Catholic University of Rio de Janeiro, Brazil. One of the most relevant outcomes of the test was the use of a reflection photoelasticity plate at the column web.

Many transparent noncrystalline materials that are optically isotropic when free of stress become optically anisotropic and display similar characteristics to stressed crystals. These characteristics persist whilst the loads on the material are maintained, but tend to fade when they are removed. This is the temporary double refraction response or birefringence (different refraction indices in the planes passing at any model point). The photoelasticity method is fundamental on this specific physical property of transparent noncrystalline materials [40–42]. For an optically isotropic material, the three principal refraction indices are equal in all the directions of light that propagates through the material.

The photoelasticity technique is utilised for the qualitative determination of stresses in structural components, determining the potential key points (where stresses are concentrated) and the principal directions for posterior analysis with strain gauges. Two types of photoelasticity may be utilised: by

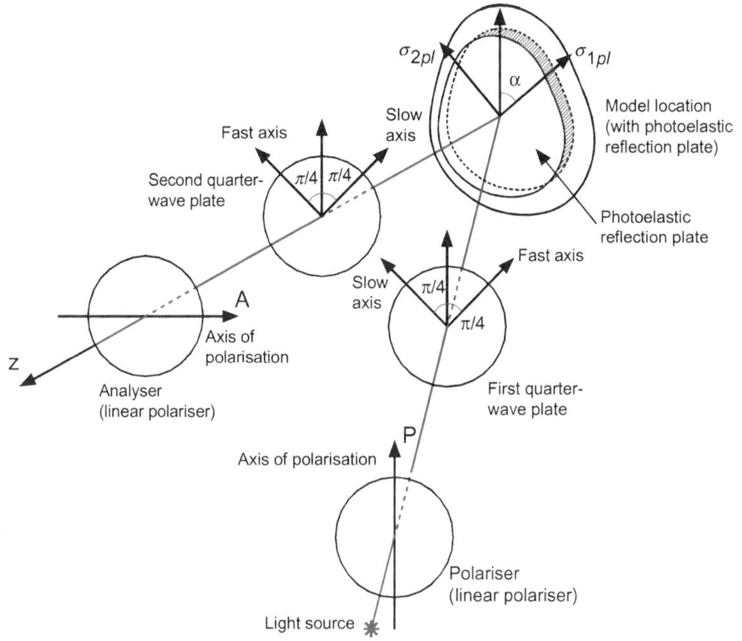

Fig. 2.91 Arrangement of the optical elements in a circular polariscope [41].

refraction or by reflection. The reflection photoelasticity technique was used in this study. This technique uses a polariscope that utilises the properties of a polarised light. Experimental stress analysis studies usually adopt plane and the circular polariscope. The polariscope arrangement and the wave plate, shown in Fig. 2.91, produce in the model dark fringe patterns.

The principal stress difference σ_1–σ_2 can be evaluated in a two-dimensional model if the fringe order N is acquired at each model point. Additionally, the model optical axes (a temporary wave plate) coincide with the principal-stress directions. These two facts can be effectively used to evaluate σ_1–σ_2 difference if a method to acquire the stressed model optical properties has been established.

When a monochromatic light is utilised, the fringe pattern appears in the model as a series of dark fringes, since the intensity of light is zero. However, when a model is observed with white light (all wavelengths of the visible spectrum present), the isochromatic fringe pattern appears as a series of colour bands. The intensity of light is zero, and a black fringe only appears when the principal stress difference is zero. Since a fringe order at any point on the model has been established, it is possible to evaluate σ_1–σ_2 and ε_1–ε_2. The values of the strain fringe $f\varepsilon$ are constant, independently of the plastic strains that are occurring in the model [40–42].

Fig. 2.92 Experimental test layout and connection's final configuration, first test.

Table 2.5 Experimental Test Geometric Characteristics

Test	Beam	Column	Support Angle	Double Web Angles
1	I 254 × 37.7	HEA 300	76 × 76 × 9.5	76 × 76 × 9.5
2	I 254 × 37.7	HEA 300	Plate 270 × 90 × 9.5	127 × 76 × 9.5
3	I 254 × 37.7	HEA 300	Plate 270 × 90 × 9.5	127 × 76 × 9.5

The experimental study encompassed three experiments utilising a cantilever-loading beam (Fig. 2.92). This figure also depicts a top view of the residual deformation shape in the column web of the first test, as well as the column web yield line pattern. The beams (S shapes) and angles were made of rolled steel sections whilst a welded steel section was utilised in the column.

The adopted steel profiles in the three experiments are presented in Table 2.5. The bolts utilised in the joints were of high strength type of ASTM A325 with a 19.05 mm diameter. The columns steel grade utilised 250 MPa yield strength whilst for the beams and angles the ASTM A36 steel grade was used. The steel yield and ultimate stress measured in the beam, column, first test angles, and other angles, acquired in accordance with ASTM E8-69, were 364, 497, 309, 419, 325, 472, 417, and 528 MPa, respectively. In the first test large, deformations occurred due to the web plate flexibility. The ultimate load reached 25.3 kN (corresponding to a 38.0 kN m joint moment) when the load cell slipped out due to the large rotation present in the cantilever beam end.

The first joint detail of the test is shown in Fig. 2.93. In the second and third tests, a stiffener replacing the seat angle and the web angle was used with unequal legs (127 × 76 × 9.5 mm) to ease the assembly (Fig. 2.94). In the third test, a stiffener was utilised in the column web close to the beam top flange.

Fig. 2.95 shows the maximum web centreline displacements (in mm) in the column web and the web angle strains. The web angle reached the yield deformation for a load of approximately 15.0 kN indicating the start of the yield line formation. The column web displacement (in mm) and the web angle deformation for the second test are shown in Fig. 2.96 where, unlike

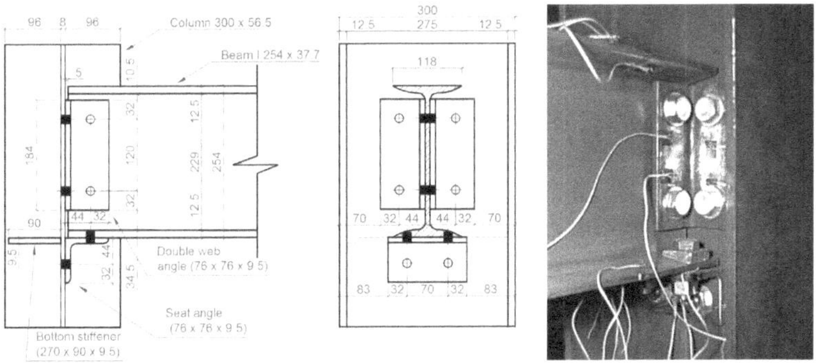

Fig. 2.93 Detail of the first connection tested (dimensions in mm).

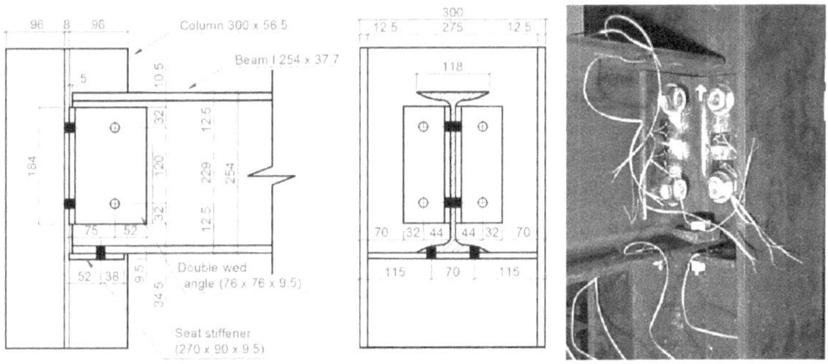

Fig. 2.94 Detail of the second connection tested (dimensions in mm).

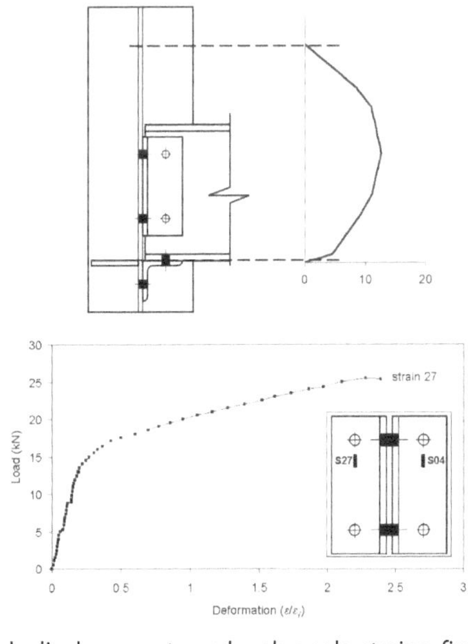

Fig. 2.95 Column web displacements and web angle strains, first test.

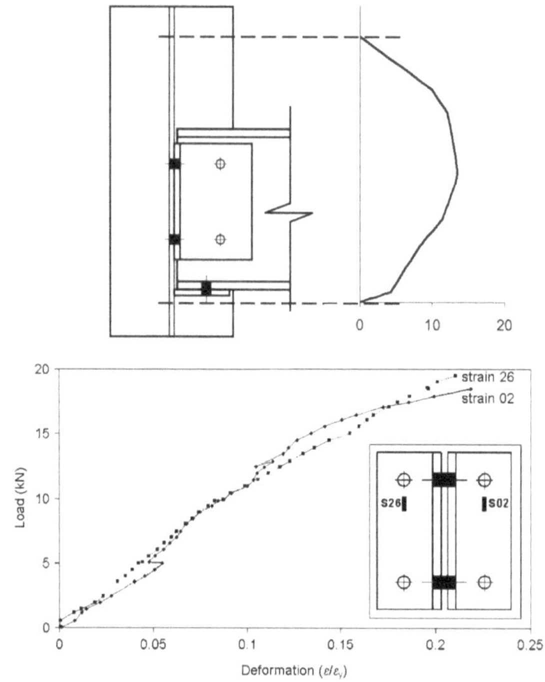

Fig. 2.96 Column web displacements and web angle strains, second test.

the first test, the yield deformation was not attained. Moment versus rotation curves for all the experiments are shown in Fig. 2.97.

The study continued with a test performed in column's minor axis using a cantilever loading beam (Figs 2.98 and 2.99). The beams and angles consisted of rolled steel sections (S shape 10″ × 37.7 and L 127 × 76 × 9.5). The utilised column was a three-plate welded steel section, CVS 300 × 56.5 (similar to W310 × 60). The stiffener thickness was 8.0 mm. All the test dimensions are shown in Fig. 2.98. A photoelastic plate PS-1B10″ × 20″ was cut and fixed on the free column web side and is shown in Fig. 2.100, where it may be noted that bolt holes were drilled on the photoelastic plate to accommodate the connection bolts.

The instruments utilised in the test are shown in Fig. 2.101 where the circular polariscope with wave plates can be noted. The load was applied by hydraulic jack, whilst the displacements were acquired with LVDTs located according to the test layout (Fig. 2.101). The joint reached an ultimate load of 25.0 kN producing a bending moment of 35.9 kN m. Both load versus displacement and moment versus rotation curves are shown in Fig. 2.102.

A simply supported beam test made of the photoelastic material was performed, due to the known linear fringe pattern distribution related to its

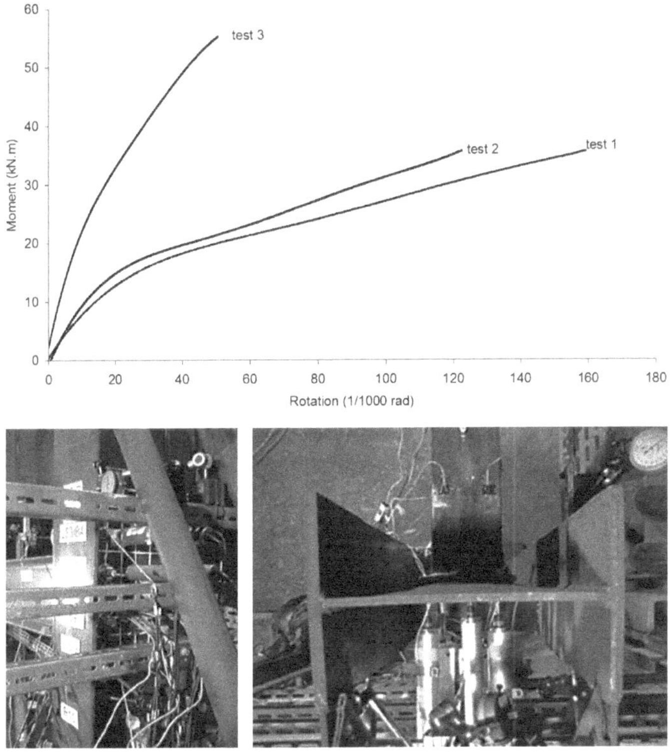

Fig. 2.97 Moment versus rotation curves of the experiments and LVDTs layout, first test.

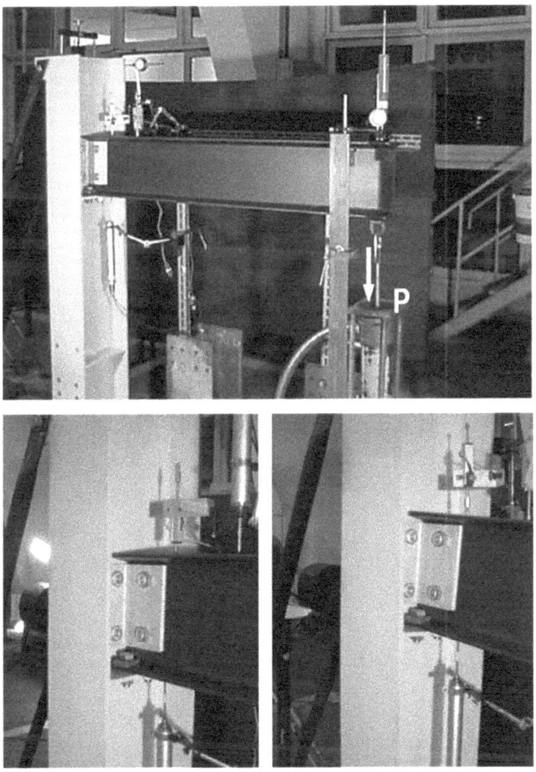

Fig. 2.98 Experimental test layout: original and deformed joint.

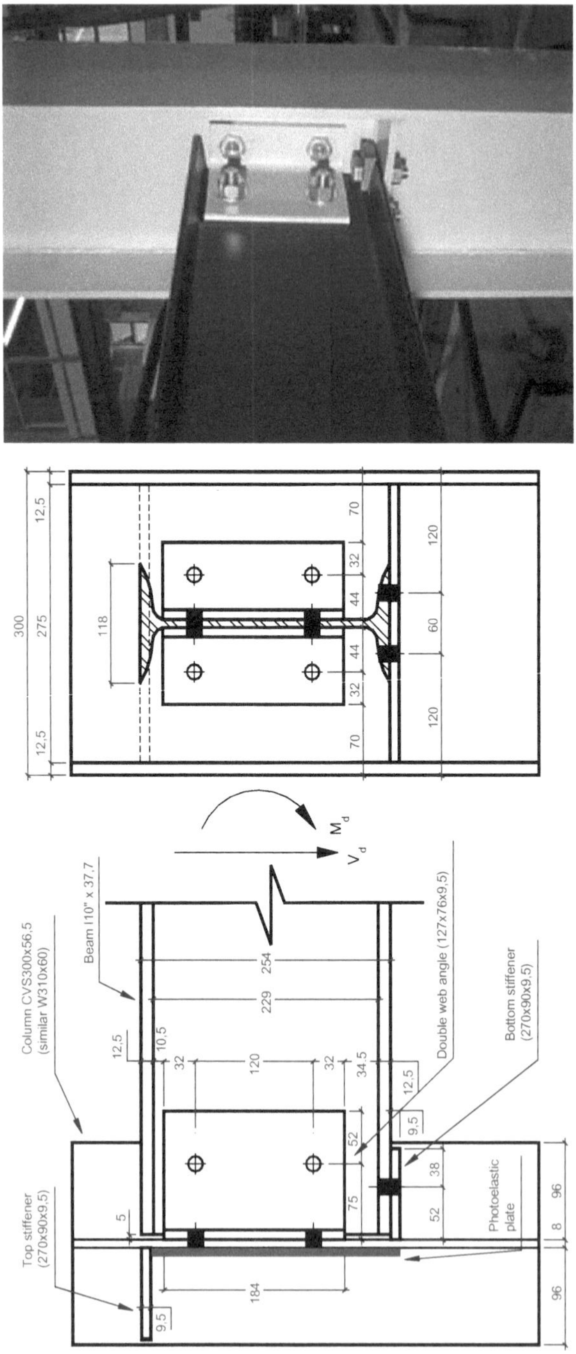

Fig. 2.99 Minor axis beam-to-column joint layout.

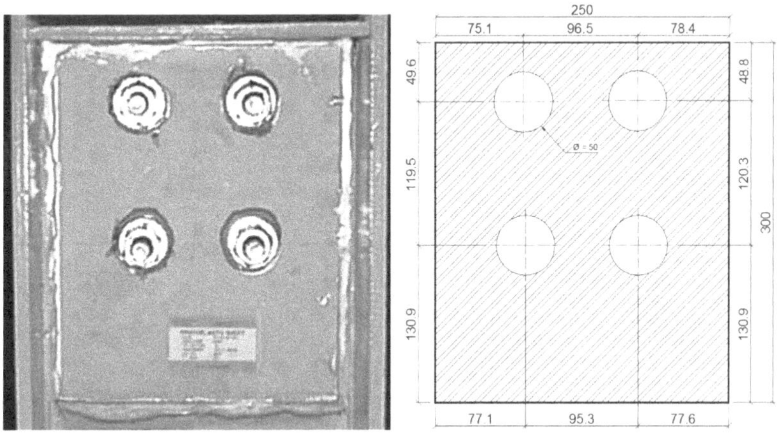

Fig. 2.100 Photoelastic plate dimensions.

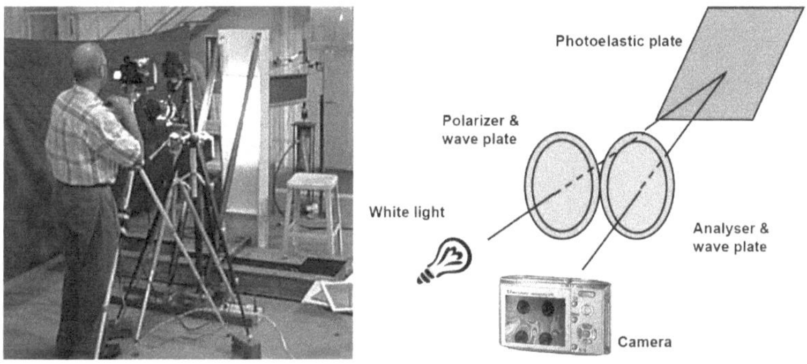

Fig. 2.101 Minor axis beam-to-column joint layout.

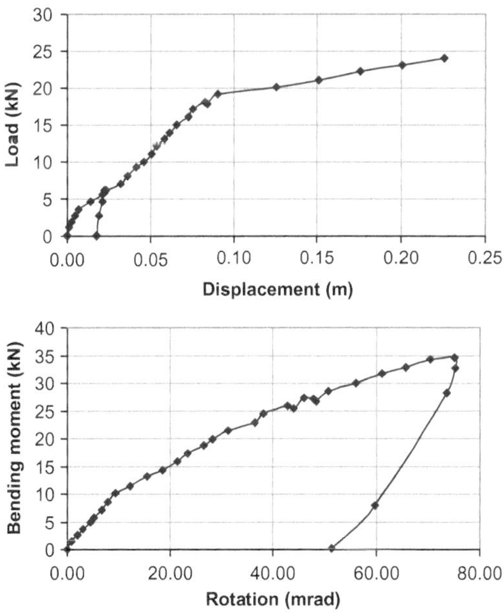

Fig. 2.102 Load versus displacement and moment versus rotation test curves.

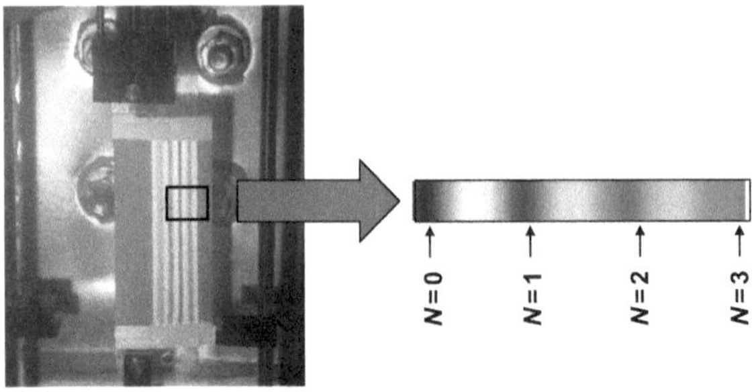

Fig. 2.103 Colour calibration for a known fringe pattern.

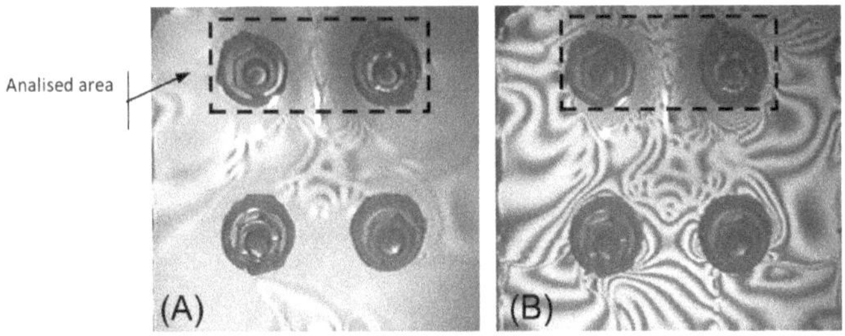

Fig. 2.104 Isochromatic fringes at the maximum load. (A) White light. (B) Monochromatic light.

structural response, to identify the colours that represent the fringe order change (Fig. 2.103). This calibration served as input data to identify the fractional isochromatic fringes distribution from the deformed model image according to the image pixel RGB colour pattern.

Fig. 2.104 shows two different alternatives for viewing the isochromatic fringes. The first case utilised a white light source producing a colour fringe pattern. The second alternative, with the adoption of a wave plate and a grey scale fringe pattern, produced a better picture of the fringe pattern.

Fig. 2.105 shows the fringe pattern evolution for three load levels, 3.0, 4.0, and 5.0 kN.

When the load reached 12.0 kN, the isochromatic fringe order identification was only possible by utilising the monochromatic light. At this load stage, the fringe order equals to six ($N=6$) was observed near the bolts (Fig. 2.106). Again the top bolt area was the column web most stressed region. When the load reached the value of 21.0 kN, the fringe order identification became difficult due to great isochromatic fringes proximity (Fig. 2.107). An analysis of the previous figures indicated that the out of plane column web response might be model as a rectangular plate limited by the bolts.

The tested connection had a configuration similar to the model previously presented only changing the column axis from major to minor. This joint also

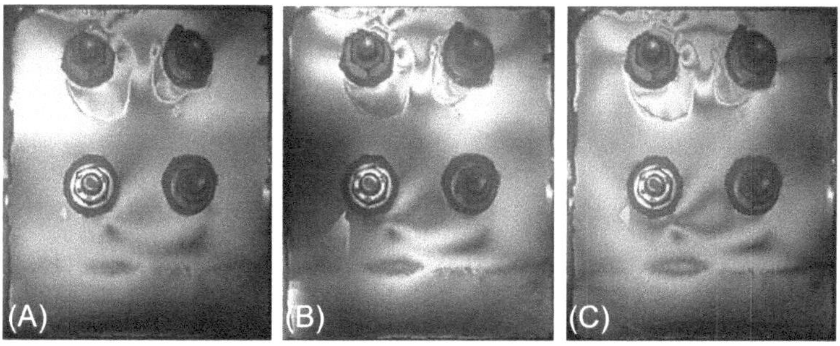

Fig. 2.105 Isochromatic fringes at 3, 4, and 5 kN load levels during the final test. (A) Load = 3 kN. (B) Load = 4 kN. (C) Load = 5 kN.

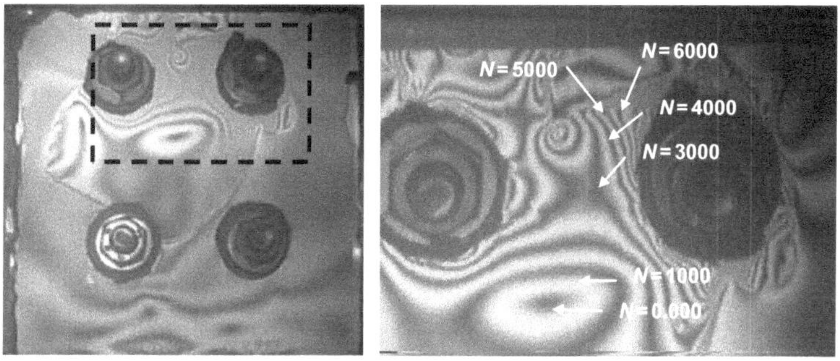

Fig. 2.106 Isochromatic fringes for an applied load of 12.0 kN.

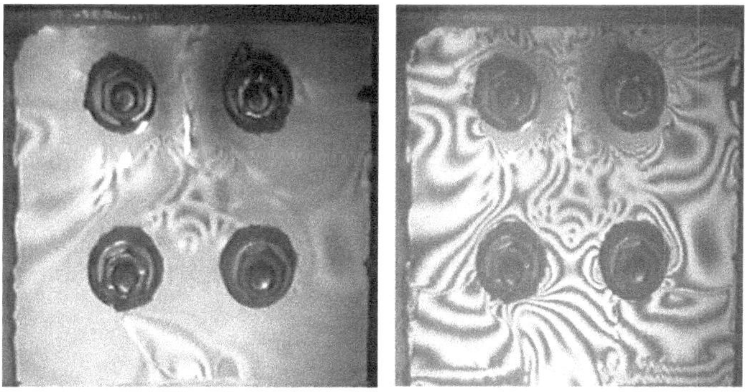

Fig. 2.107 Isochromatic fringes for an applied load of 24.0 kN.

required the development of an instrumentation system for measuring the beam-to-column relative rotation and the deformation in its main elements. The adopted experimental model was consisted of a column with a fixed base, anchored at the top, and connected to beam by two web and a seat angles. It should be stressed that in the second and third tests, a stiffener welded to the column web and flanges to increase the rotation stiffness has replaced the seat angle. A hydraulic jack, rollers, and a load cell were used at the beam-free end for the application and monitoring of a monotonic load.

Strain measurements were performed with rosettes and linear strain gauges in the beam and angles, enabling an assessment of the stress distributions evolution in these elements. An innovative system for monitoring the column web stress distribution was used in a subsequent test room based on reflection photoelasticity techniques. This system enables a qualitatively evaluation of the column web stresses evolution and, together with the strain gauges results, identified the points where the higher stresses were concentrated.

The measurement of the beam-to-column relative rotation was made by a difference between the beam and the column rotations. The beam and the column rotations were determined through LVDTs and dial gauges located in these elements. The use of a stiffener instead of the seat angle proved to be quite effective and increased the joint stiffness.

2.6.7 Structural Evaluation of Minor Axis Semi-Rigid End-Plate Joints

A series of static monotonic tests was performed in minor axis end-plate joints where an I section beam was connected to column web as shown in Fig. 2.108. The two bolt lines are present at top joint area [43–46].

The tests were divided in two series. The first series contemplated the five static monotonic tests, shown in Fig. 2.108, involving positive and negative moments. The second series again involved four static monotonic tests where positive and negative moments were applied, but utilised composite columns instead of the previous I sections. Table 2.6 summarises the test programme, indicating the beam and column profiles and the applied loading. The negative moment tests correspond to a bottom to top load shown in Fig. 2.109, where the beam top flange is under tension.

The materials specified in the design stage were the S275 steel grade for all I or H profiles, and plates. Hollow profiles were made from an S355 steel grade

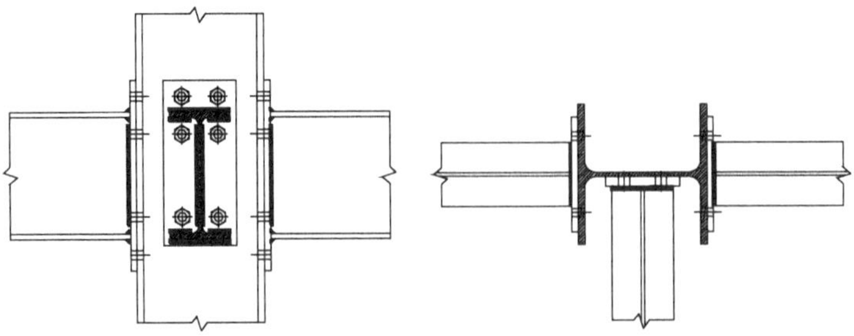

Fig. 2.108 First configuration: minor axis beam-to-column joints between rolled profiles.

Table 2.6 Experimental Programme

Test	Column	Beam	Type	Loading	Series
E 0	HEA 220	IPE 200	Steel	M −	1
E 1	HEA 220	IPE 200	Steel	M −	1
E 2				M +	1
E 4	IPE 330	IPE 240	Steel	M −	1
E 5				M +	1
E 7	HEA 220	IPE 200	Composite	M −	2
E 8				M +	2
E 10	IPE 330	IPE 240	Composite	M −	2
E 11				M +	2

whilst the bolts and threaded bars were specified as class 10.9. The bolts were prestressed with a torque wrench, aiming to induce a 150-Nm torque. The tests were conducted as shown in Fig. 2.109. A reaction frame and a reaction wall were adopted to support the top of the column and the load application system, respectively. IPE 200 beams where used at the column major axis to represent the actual flange rotation restriction present in a 3D joint. The asymmetry present on the wall axis in relation to the reaction frame centre was corrected by the introduction of a stiffer major axis beam segment (HEA240).

In the perpendicular direction (*X*-direction), the loaded beam is connected to the column web (1st series) or to one of the rectangular hollow profile sides (2nd series). The load was applied through a 200-kN hydraulic jack and 200-mm stroke. This stroke is a very important factor, since these tested joints can suffer large rotations in the postplastic range. In this study, it was possible to reach 200 mrad rotations in the monotonic test. The jack support was designed using a cantilever large stiffness beam, fixed by 12 prestressed ϕ 36 mm DYWIDAG bars to the lab reaction wall. This actuator had hinges at both ends around the axis of the *Y* axis to accommodate not only the beam vertical deflection but also the horizontal nonnegligible component of displacement in advanced stages of the test. The column was supported at its base by an articulated system, Fig. 2.110A fixed on a reinforced concrete base.

At the top end, the column is connected to the reaction frame, Fig. 2.110B designed to facilitate the assembly, allowing rotations of the column top section whilst still accommodating, if necessary, a load cell. The reaction frame, the reaction wall, and the base before the concrete was casted where the hinge anchoring details can be seen in Fig. 2.110C.

The load application was usually made in accordance to a procedure where the load or deflection was controlled. The first is used when it is desired to impose a load value previously known, assess the model ultimate load, or when the models always have positive stiffness behaviour, with little variation.

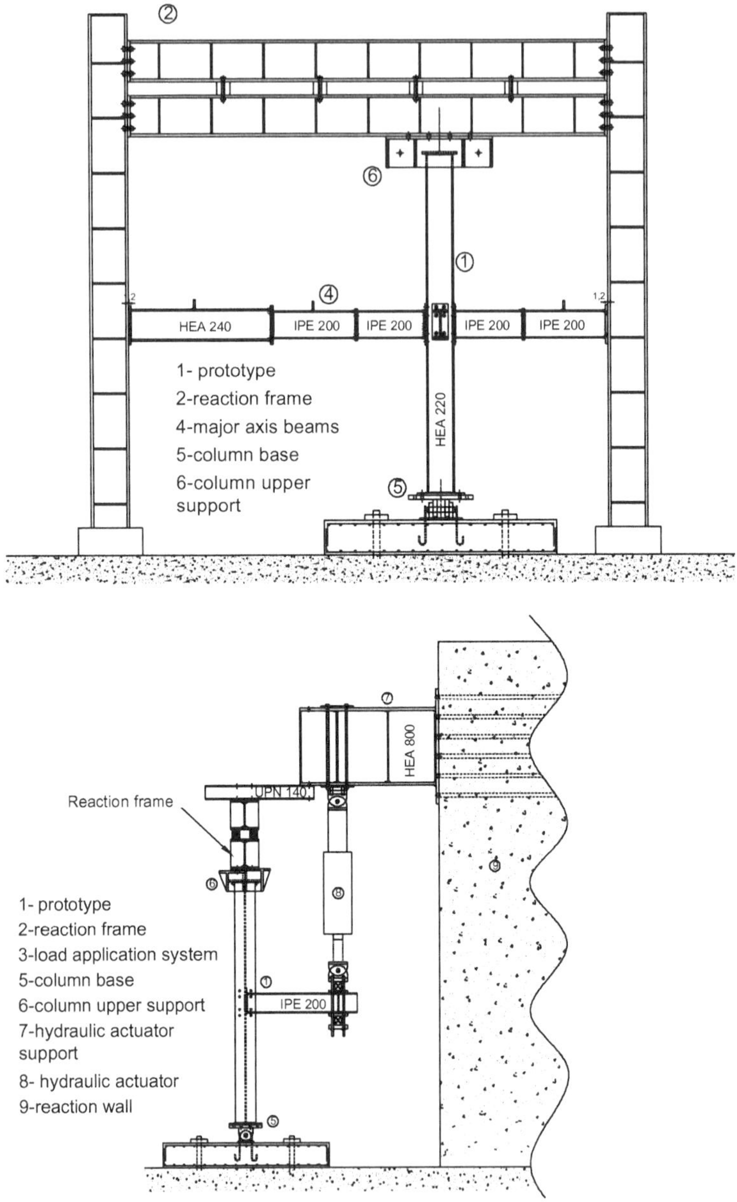

Fig. 2.109 Reaction frame and testing scheme in the *Y*- and *X*-directions.

The second procedure is used in material with little stiffness behaviour phases, null or even softening. This procedure is the best for structures depicting a plastic behaviour. With the available equipment, the loads could be imposed to the models by the two described procedures or by a combination of both.

Fig. 2.110 Reaction frame, wall, and floor.

Fig. 2.111 shows the loading history used in the tests. The entire test started with the application of a 5 kN load (elastic phase) followed by complete system unload to process all the models settings and adjustments, using a load control strategy. Then, a successive load was applied (always with displacement control) with unloads up to 5 kN (always with load control), until the ultimate collapse load or the jack maximum stroke was reached (corresponding to a 200 mrad joint rotation). The number of unloads varied with the model performance and was at least one. It is important to note that these loads and unloads enable the assessment of the joint initial stiffness.

The rotations' amplitudes corresponding to the considered unload were approximately equal to 90 mrad (for minor axis steel tests). The measurements aimed to evaluate the joints main mechanical properties: resistance, stiffness, stress magnitudes, and deformability of each component, in order to establish their contribution to the global joint response. Load cells and LVDTs were used to evaluate the beam and column deflections and rotations and the out of the plan deformation of the column plate. Strain gauges, BTM-6-C, were used in the bolts to acquire their forces and, consequently, the forces acting in the various bolt lines.

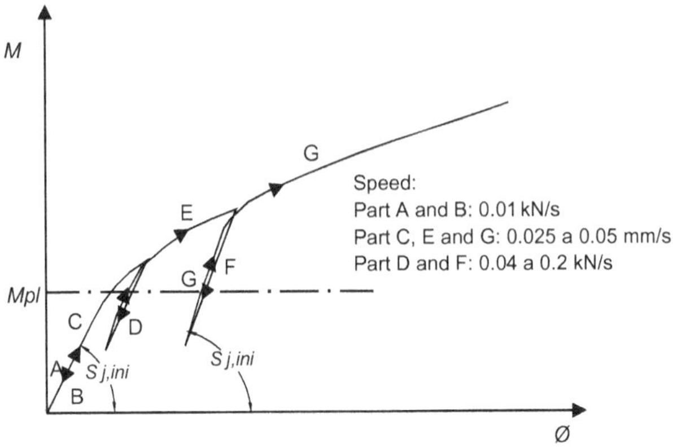

Fig. 2.111 Monotonic test load history.

Fig. 2.112 shows the locations of the load cells and the LVDTs. The LVDT 61 measures the displacement of the beam directly under the load point of application, and the LVDT 62 at an intermediate beam point. The LVDTs 63 and 64 measure (by difference of readings) the beam rotation in the section next to the joint. The same result can be obtained with LVDTs 61 or 62, when the beam displacement is compensated. The LVDTs 65 and 68 measure eventual column end displacements, thus allowing correction on the readings of the couple LVDTs 66/67. Through the difference between these readings, the column rotation at the joint section can be obtained. The joint rotation can be finally determined through the difference of the beam and column rotations.

Since it is essential to evaluate each one of these deformability sources (especially the column web deformation), this joint component was thoroughly monitored with linear and rosettes strain gauges and LVDTs. Fig. 2.112 shows the instrumentation used in the beam, minor and major axis endplates, bolts, and column flanges.

The loaded girder was instrumented in the section next to the joint with four linear strain gauges to detect the beam yield at higher stressed points (flanges). At the joint minor axis endplate, three strain gauges were used to monitor the force magnitude from which this component yields. A linear strain gage used in the major axis beam endplates will be used to detect whether this joint region always remained in the elastic range. The column flanges had two strain gauges to assess their strains.

All the bolts were drilled to create holes where strain gauges were inserted (Fig. 2.113). It is interesting to observe that according to the manufacturer's procedures the adhesive requires an oven cure at 140°C for 2 hours.

Fig. 2.114 shows some model instrumentation aspects, in particular the detail of the strain gauges and LVDTs located at the column web.

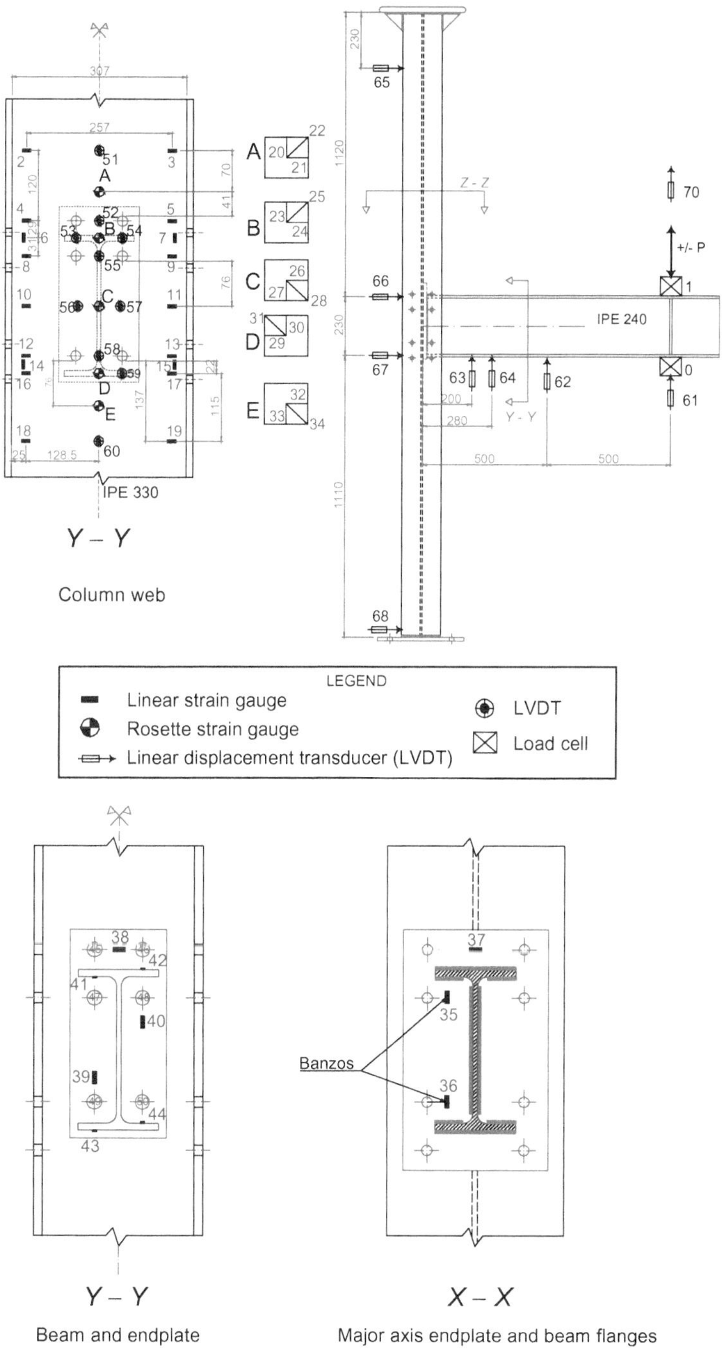

Fig. 2.112 LVDTs, load cells, and strain gauges used in column web and flanges, and beam and endplate.

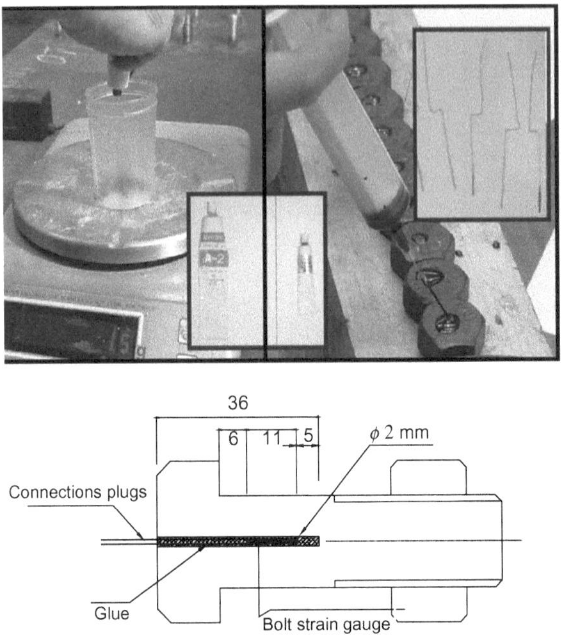

Fig. 2.113 Bolts instrumentation.

Fig. 2.114 Instrumented structure details.

Fig. 2.115 Views of the phases of the composite joints instrumentation.

In the composite tests, the instrumentation was similar to the previous, with some variations imposed by the new geometry. The joint region was instrumented with strain gauges located at the longitudinal reinforcing bars, stirrups, and on the four concrete face surfaces, at the tension and compression zones indicated and numbered in Fig. 2.115, to monitor the stresses in the reinforcing bars and on the concrete. In principle, the concrete prevented the use of LVDTs at the column web. This motivated the devel opment of an alternative to monitor the column web displacements that was made using small diameter bars immersed and isolated from the concrete by expanded polystyrene (Fig. 2.115). These bars were welded to the column web at the points to be measured whilst their outer edges were welded to plates where the LVDTs were positioned.

The strain gauges used inside the concrete column always required a special care for their protection. In addition to the protection used in the tests without concrete (TML-N1), the strain gauges used in the profiles and plates were also protected with a wax (TML-W1), whilst the strain gauges positioned in the reinforcing bars were involved in a tape (3 M VM). The instrumentation preparation can also be observed in Fig. 2.115, where the

Fig. 2.116 Global view of steel & composite joint models.

application of the strain gauges protection, the instrumented structure, and a global assembly layout before the simultaneous concrete casting of the six tests performed in the series is presented. An expansive foam that had the detrimental effect of causing some voids in the concrete surface was used to seal the joints between the forms. These imperfections were subsequently covered with mortar (Fig. 2.115). Finally, the six columns were painted with white lime to better reveal the cracking during the tests. Fig. 2.116 shows a general aspect of the steel and composite structures just before the tests.

Table 2.6 contains the performed tests in the first series: E1, E2, E4, and E5, corresponding to two different geometries, each under positive and negative moments. All moment versus rotation curves must be corrected to consider the column and reaction frame deformations. As a result of the joint flexibility, this correction was very small, but this was not the case of composite tests associated too much higher stiffness and strength.

Fig. 2.117 shows for the first test, M–ϕ curves obtained from readings of the pair 63-64, the column deflection induced by the portal frame deformation, and M–ϕ curves thus corrected, which are nearly coincident with the uncorrected curves. It is possible through this result to assess the relative importance of the joint deformation and the reaction frame and column, showing that in these tests the corrections' effects are negligible.

The analysis of the test results indicated that, in all cases, the column web is the dominant component, being the first to yield and contributing most to the joint deformability. It is interesting to note that there is an 'elbow' on the M–ϕ curves associated with the column web bending coincident with a

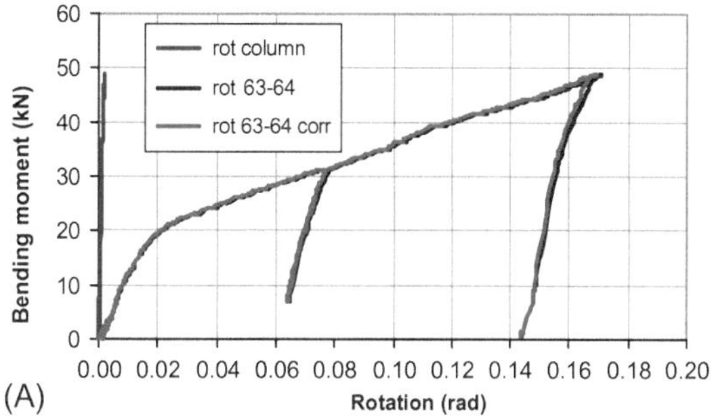

Fig. 2.117 (A) E1 test M–ϕ curve correction including the deformability of column and reaction frame and (B) deformed structure.

20 mrad joint rotation, value accepted by many authors as the joint ultimate limit state and, in particular, this component. The elastic stiffness (or initial) of the minor axis is almost entirely dependent on the column web bending stiffness. This slope should be measured in the unload segments with a subsequent load, because in the early part of M–ϕ curve the various adjustments and the premature plastification of the column web do not allow a correct assessment of this stiffness.

It becomes clear that the column web in bending works predominantly as a band between flanges. In fact, the x-direction stresses are far higher than the stresses in the y-direction. In this direction, due to the Poisson effect opposite signal, stresses appear contrary to the dominant perpendicular stresses. It is also possible to conclude from the stress analysis that once the advances in the y-direction are made and the force application quota (tension or compression) is observed, the horizontal stresses will decrease, which indicates the limited size of unidirectional operational zone of the column web.

The neutral axis does not suffer significant variations of position along the test, maintaining a distance of approximately 120 mm to the compression centre is located roughly at the beam compressed flange midline. The tests endplates were designed to remain elastic up loading stages where the column web yielded (plastic moment), which in the present tests, was always the critical component. The experimental studies performed in the minor axis varying the endplate indicated that this component had a very small effect on the joint overall behaviour, and the interaction of its behaviour with the bended web is also small.

In the present tests, the column web membrane effect can cause the moment supported by this component to be of an order three times higher than their plastic moment capacity. Although the endplate was instrumented with strain gauges, the adopted location has not detected the yielding in the most appropriate way. The horizontal positioning of the strain gauge in the extended part of the plate only enables the monitoring of strains in that direction, and not the mechanism formation observed in the positive moment tests. The strain gauge used on the major axis endplate confirmed the existence of stresses in this element, and the restriction imposed by it to the column flange rotation. The analysis of stresses on the major axis endplate for the various tests clearly showed that they remained in the elastic range in all tests.

All bolts were prestressed with an approximately 100 Nm torque, corresponding to measured bolt strains between 800 and 1100 με (approximately 200 MPa stresses). The measurement of the bolts strains enables the monitoring of the forces acting on them. Although the bolts have a relatively low-level force in relation to the magnitude associate to theoretical yield value of the critical component (we of the column web), the development, in this component of membrane stresses, causes the other components magnitude to be increased above their predicted values.

It was observed that the first two bolts lines were under significant tension forces, whilst in the third line, the influence of the web deformation in the compression zone led to a reduction in the bolts prestress forces. The stresses at the beams monitored points showed that, in reality, there were some flanges yielding but theoretically, the beams should always remain elastic. Table 2.7 compares the maximum moment values monitored during the tests as well as the adopted sections elastic and plastic moment capacities.

Table 2.7 Elastic and Plastic Bending Moment Capacities for the Adopted Steel Profiles

	M_{el} (kN m)	M_{pl} (kN m)	$M_{máx}$ (kN m)	$M_{máx}/M_{el}$
IPE 200	75.0	85.2	E1—48.8	65% (E1)
			E2—49.5	66% (E2)
IPE 240	120.3	136.0	E4—48.9	40.6% (E4)
			E5—53.7	44.6% (E5)

Fig. 2.118 Deformed joints, tests: E8 and E10.

In the noncomposite tests, the web stiffness and strength were constraints. In the composite tests, although the web can also be determinant, the concrete presence acts as a very effective reinforcement shifting the rupture to other components. The joint deformation may not be solely a result of the web deformations, since significant deformations were also detected on the beams and endplate, as well as extensive reinforced concrete cracking associated with the reinforcement bars yield that were recorded during the tests. In all cases, the failure was ductile, since it was accompanied by large deformations, as shown in Fig. 2.118.

The collapse occurred with the clear identification of an 'elbow' in the M–ϕ curves associated with a large rotation increase with a constant moment. This trend is not usual to the column web isolated behaviour that had extensive plastification and large rotations, but before the result of the involvement of other components. Strength and stiffness are much higher than those of the noncomposite joints, as well as the curve progress that was quite different, with a sudden stiffness loss happening at the referenced curve 'elbow'.

During the composite joint tests, the first yields appeared at bending moment magnitudes of approximately 80 kN m, corresponding to about 90% of the joint ultimate capacity. A generalised web plastification did not occurred at this load level whilst the joint failure was associated with other components. The low level of the web deformations equally justifies the absence of noticeable membrane effects. It can be noticed that in the noncomposite test counterpart, the web was subjected to high stresses and yield at a load level of just about 10 kN m, where there was widespread yielding accompanied by large deformations and consequent membrane effects.

Also in these composite joints, the web in bending is subjected to stresses in the x-direction far superior than the stresses in the y-direction, where due to the Poisson effect, the perpendicular stresses had an opposite sign. The web still works in a unidirectional way between flanges, and as one moves in the y-direction, from the force application point, the horizontal stresses are reduced. Even for joint maximum supported moment, the yield fails to be fully developed at the measured points along the vertical axis. Thus there is a strong limitation to the yield propagation, which does not exist in the noncomposite test.

The moment magnitude corresponding to the yield in the composite test is far higher than their noncomposite counterparts, and unlike those, there is not a general trend for the web to be under tension in both directions, since its deformations are not high enough to develop considerable membrane effects. The reinforced concrete behaved as effective web reinforcement where, at the compression zone, this deformation was practically zero, whilst the rotation centre was shifted to the compressed beam flange. For a better understanding of the composite structure effectiveness, Fig. 2.119 compares the longitudinal and transverse deformed profiles for the noncomposite (E1) and composite (E7) joints.

In this figure, two applied moment magnitudes are presented where the higher corresponds to the maximum value recorded in the noncomposite joint. For this magnitude, the web of the noncomposite joint presented maximum deformation magnitudes three times its thickness implying on a strong membrane effect. For the same magnitude in the composite structure, the maximum web deformation was very low (around 5% of the thickness). From the maximum moment onwards, there is an increase of maximum joint deformation at a constant applied moment, where the bottom of the web remained without considerable displacements. At the top, as the joint

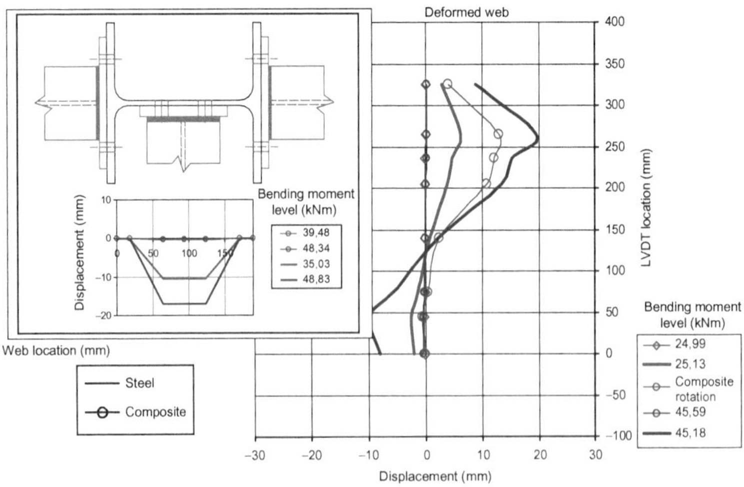

Fig. 2.119 Deformed web profiles, tests: E7 and E1.

deformation progressed at a constant applied moment, the web also deformation constantly increased. At this stage, the composite joint node already incorporated the column web membrane effects. Through the existence of concrete and its confinement, this component was submitted in the composite test, and for the same moment magnitude to substantially lower stresses than in the noncomposite tests. It can be observed that a wider mobilisation of the endplate and since the web does not yield immediately, and for low applied moment magnitudes, ceases to be the controlling joint component.

Certainly, in the composite tests, the forces transmission from the beam to the column will not be exclusively made by the bolts, but also by the reinforced concrete, where a strength mechanism will be formed. Consequently, it is not possible to establish any comparison between the M–ϕ global curves obtained through measurements of the load cells and the same curves constructed with the moment evaluated with the bolt forces. Naturally, the similarity observed in the remaining components, following the reinforced concrete collapse, induce the joint response to be reduced to the noncomposite joint response, and at this stage, almost all the joints forces will be transmitted by bolts.

In the composite tests for the same magnitude of applied moment on the noncomposite tests, the force absorbed by bolts is much smaller, since the forces transmission from the beam to the column does not solely depend on this component, but also on the resistant mechanism that forms in the reinforced concrete. As the reinforced concrete fails, the joint rotation increases whilst the moment transmission becomes increasingly more depend on the bolts action. At this point, it is observed a great structural response similarity between composite and noncomposite joints associated with a reduction of the bolt force difference in the two structures.

In the noncomposite tests, yielding was observed in the columns flanges close to the endplate, due to stress concentrations since, theoretically, the moment installed in these joints was lower than the beams elastic moment capacity. In the corresponding composite tests the achieved moments were far higher. Table 2.8 summarises the values of the beams elastic and the plastic moment capacities (calculated with nominal dimensions values and with the flange yield stress average values), and compares them with the maximum

Table 2.8 Elastic and Plastic Bending Moment Capacities for the Adopted Steel Profiles

	M_{el} (kN m)	M_{pl} (kN m)	$M_{máx}$ (kN m)	$M_{máx}/M_{el}$	$M_{máx}/M_{pl}$
IPE 200	71.5	81.2	E7—87.7	123% (E7)	108% (E7)
			E8—90.7	127% (E8)	117% (E8)
IPE 240	109.6	123.9	E10—109.4	100% (E10)	88% (E10)
			E11—94.5	86% (E11)	76% (E11)

values reached in the tests. Tests E7 and E8 clearly reached the beam plastic moment capacity, whilst tests E10 and E11 did not reach this value.

The theoretical web plastic moment was exceeded in Table 2.8, fact that could occur due to the steel strain hardening, but, what certainly occurred was a shift of the critical section of this interface to the reinforced concrete section face. The beam had a support at this point and, in the tension zone (top face) or in the compression zone (bottom face); the plastic deformations were clearly detected. In the compression zone, there is a flange plastic instability that forms the plastic hinge (a Eurocode 3 [19] class 1 profile).

Previous results have shown the reinforced concrete relevance in strengthening of all other components, completely changing the joints structural behaviour. The collapse of this component (reinforcing bars and concrete) will catalyse the joints collapse, since the remaining components start to be under very significant forces. After an extensive cracking, the stirrup next to the extended portion of the endplate was pulled out of the reinforcing bar arrangement, whilst the reinforcing bars were under significant stresses. Fig. 2.120 shows notable points of the seventh joint test.

Point (1) corresponds to the start of the concrete cracking, which passes from State I to State II, provoking the first joint stiffness loss, and a well-defined increase of the change rate of stress in the reinforcing bars. Point (2), around 30 kN m, can be associated with the beam first yields, and from this point, there is a growing divergence between the web and the joint curves, due to the yielding present on the beam. Point (3) corresponds to the column web first yields made possible by the reinforced concrete deformations. In point (4), the stirrups M–σ curves become horizontal,

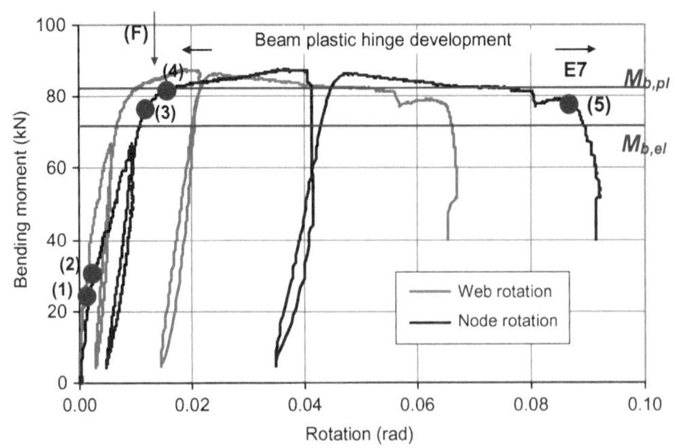

Fig. 2.120 Notables points of E7 test moment versus rotation curve.

corresponding to their yield, whilst allowing the onset of cracks in the concrete. The displacement marked with the letter 'F' roughly corresponds to the cracks opening, clearly visible, and a 15-mrad rotation. From this point, the concrete's contribution becomes less and less effective.

The beams elastic and plastic moments are also highlighted. After the plastic moment, the beam experiences large rotations in the plastic hinge, main responsible for the higher joint rotations when compared to the column web rotations. At the end of the curve, beam compressed flange plastic instability causes the decrease of the joint supported moment, point (5). The following paragraph will depict a similar description of the composite joint tests, as well as some comparisons to the already described noncomposite joint tests.

Despite the composite joint tests presented differences between the M–ϕ curves related to the joint and column web, the noncomposite tests presented identical curves for this two cases, the main differences occur particularly in plastic phase. Therefore, a comparison of the initial stiffness is also sensibly indifferent to the joint or column web curve, except in the case of test E7 where the web stiffness is roughly 50% higher than the joint stiffness, with measurements made in the reload on a stage before the curve elbow. After this elbow, the difference in the reloading stiffness tends to be attenuated. It can be noted that the beam component yielded at approximately the stage where the unload stage started.

It was also observed that negative moment tests were associated with a higher resistance and lower slips that the corresponding positive moment. This result is certainly due to the higher concrete confinement effect, in the case of negative moment, due to the presence of a stirrup directly over the endplate and to the presence of bolts located in its extended part, where a considerable part of the tension force is transmitted. All the composite tests were associated with a high ductility since their collapse involved several components, all with a ductile behaviour, which contributed to the joint deformability. The test E7 was associated with a smaller overall rotation, since its collapse mainly involved the beam yield.

An M–ϕ curve comparison for each composite joint test with the corresponding curve of noncomposite joint test counterparts is shown in Fig. 2.121, where the results previously presented for the components are clearly summarised. The global response of a composite joint test is characterised, and compared to the equivalent noncomposite tests, by a much higher elastic stiffness (in average, they are five times stiffer for tests with HEA columns and ten times for tests with IPE columns); higher moment capacity (in average, five for tests with HEA columns and eight times for tests with IPE columns); different failure modes, involving other joint components than the

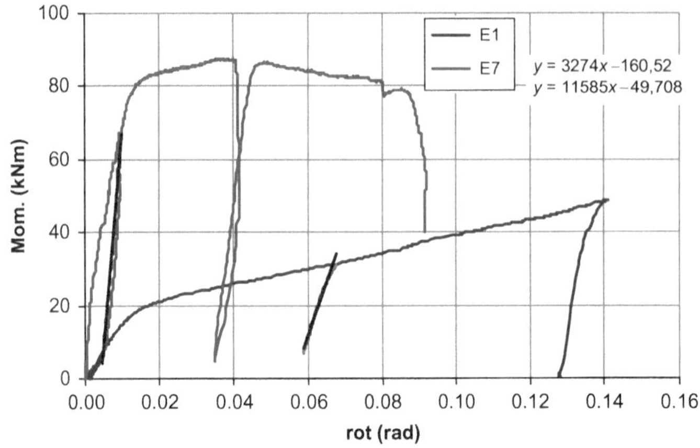

Fig. 2.121 *M–ϕ* curves comparison: noncomposite E1 and composite E7 tests.

column web, lower out of plane deformations; presence of a well-defined 'elbow' on the *M–ϕ* curve, making the transition from a stiffness segment close to the elastic value to a horizontal plateau tendency. This horizontal plateau appears in contrast to the membrane response observed in noncomposite joint tests associated with a predominance of the column web component, substantially deformed. In the noncomposite joint tests, the average M_{rot}/M_{pl} was equal to 3.3 whilst for the composite joint tests were equal to 0.97.

In the tested joints, the components that influenced the failure had a high ductility, but it is not legitimate to generalise the conclusion that a minor axis composite joint is necessarily ductile. As an example, if in a given joint the collapse is associated with the beam yield (that actually occurred in some of the tested joints), and if the beam was not class I, the joint will not have enough ductility to allow a structure plastic analysis to be made.

The results of the noncomposite tests showed that the column web component had a predominant effect over their structural response, being responsible for almost all the joints deformability. A stress analysis of the column web in the elastic domain experimentally indicated that there was a predominance of the component in one direction between the two flanges. The main stress directions at the web surface close to the flanges matched with the vertical and horizontal directions, whilst the stresses in a direction parallel to the flanges were much smaller than the perpendicular ones, and with an opposite sign due to the Poisson effect.

The remaining joint components, influenced by the column web component response, started to be subjected to more significant forces and stresses than those of a simple plastic analysis of joint could predict. The end-plate instrumentation, component in which a plastic resistance is always higher than the web, allowed the detection of yielding and even the formation of a collapse mechanism. It can be noted that if the additional force of a ductile component such as the endplate does not affect the joint ductility, this phenomenon in nonductile components can lead to a sudden fragile rupture.

Whilst in the noncomposite joint tests, the collapse mode usually involved only the column web, associated with large deformations and with the development of high postlimit strength due to the membrane effect, in the corresponding composite tests the behaviour radically changed. The reinforced concrete component proved to be of fundamental importance, since the collapse mode did not involve only the column web, but other components such as the endplate, reinforced concrete, or the beam. These components started to condition the global behaviour both in terms of stiffness and resistance.

In all cases, the M–ϕ curves were characterised by the existence of an 'elbow', around a 20-mrad rotation (value accepted by many authors as corresponding to joint failure), which was not present in the noncomposite tests. This elbow corresponds to the reinforced concrete degradation, and causes a relative abrupt stress increase in the remaining joint components. The joint component deformations inside the reinforced concrete (such as the column web or the endplate) were much more reduced in the composite test than in the noncomposite tests. Especially in the tension area, this deformation showed a strong dependence of the concrete reinforcement arrangement. Compared to the corresponding noncomposite tests, the stiffness and resistance increased by at least five times, whilst the ductility, now dependent on the component controls the joint collapse, proved to be satisfactory in the performed tests.

The designed instrumentation for this test programme contributed significantly to its success. The actual connections behaviour could only be deeply understood with the various stress distribution analyses, forces, displacements, and rotations. The use of bolts strain gauges was of fundamental importance for monitoring forces transmitted within the joint and served to compare and validate aspects like the moment versus rotation curve, prying effects, amongst others. The load application reversal in some tests also served to mobilise other connection components that may be triggered in the case where there is a potential load reversal such as in seismic events, progressive collapses, etc.

The individual assessment of each joint component was not a simple task and had to be calibrated by different means to ensure for their validation. In addition to contributing to the global understanding of the connection behaviour in terms of initial stiffness, transmitted bending moment, and rotation capacity, the individual behaviour of each component contributed to the assessment of collapse or failure sequence of each component as the external forces were applied to the investigated structure. This enables the studied joints to be optimally redesigned without excesses in some points which could, from a certain value, cease to contribute for improving the performance of the connection because, other components, less mobilised before, began to control their design.

The flanges confinement effect over the column web was also investigated through a test especially designed for this purpose generating, and clarifying key points to its understanding. Finally, the ingenious procedure for acquiring the web out of plane displacements in composite column profiles should be highlighted. Without this valuable experimental apparatus, performance comparisons between the steel and composite tests would not be made possible.

2.6.8 Structural Evaluation of Major Axis Semi-Rigid End-Plate Joints: Bending Moment to Axial Force Interaction

The beam-to-column joints test programme performed at the Civil Engineering Department of the University of Coimbra included 16 prototypes, 9 flush end-plate joints, and 7 extended end-plate joints, as shown in Fig. 2.122. In all tests, the columns were simply supported at both ends, consisting of an HEB240, the beams made of an IPE240 and the 15-mm thick end-plate, using an S275 steel. The bolts were M20, class 10.9. Nine experimental tests were performed for the flush end-plate configuration, involving numerous combinations of bending and axial forces. The tests started with the application of a

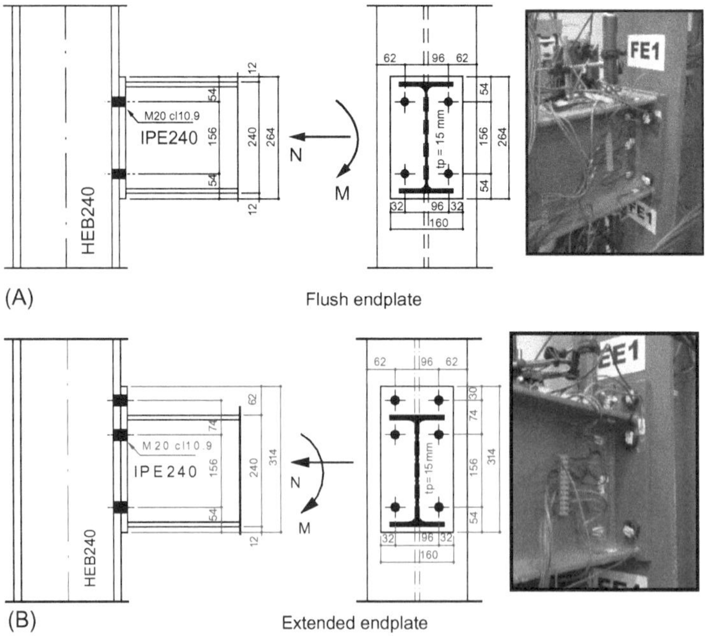

Fig. 2.122 Layout of joints. (A) Flush endplate. (B) Extended endplate.

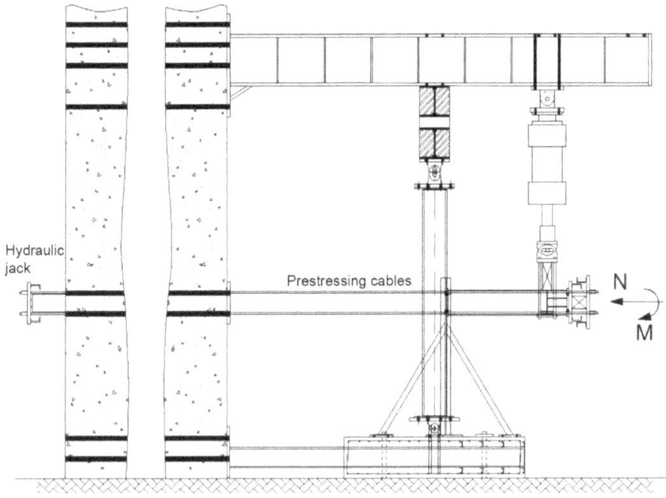

Fig. 2.123 Experimental test setup.

fixed level of axial tension or compression and continued with an application of a negative bending moment, incremented up to joint collapse.

In the first test, FE1, only bending moment was applied through a jack actuator, located a meter away from the face of the column flange. The next tests, FE3, FE4, FE5, FE6, FE7, FE8, and FE9, involved constant axial forces of −4%, −8%, −20%, −27%, −20%, +10%, and +20% of the beam plastic strength applied to the beam. The test layout is shown in Figs 2.123 and 2.124, where a hydraulic jack was used on the cantilever beam to apply the moment.

The system for the application of the compressive axial force adopted a hydraulic jack (Fig. 2.124) that applied a tensile force to four cables of diameter $\phi = 15.2$ mm, anchored to a reaction wall. The force transfer to the joint was made by a central load cell (Fig. 2.124). These cables passed by a deviator saddle (HEM100) to ensure that the axial force was always parallel to the beam axis. Load cells were also used at each cable to acquire their installed forces. The horizontal reaction at the column ends was transmitted to the reaction wall by (i) a steel beam, at the top, and (ii) a reinforced concrete footing prestressed to the strong floor using DYWIDAG bars and connected to the reaction using an HEB 200, at the bottom, Fig. 2.124.

The tensile force system is shown in Fig. 2.125, where four hydraulic jacks were installed in one of the ends of a circular hollow section profile to transmit the tensile force. These circular profiles were simply supported at the other end to enable free rotation and to ensure that the axial force was always parallel to the beam axis.

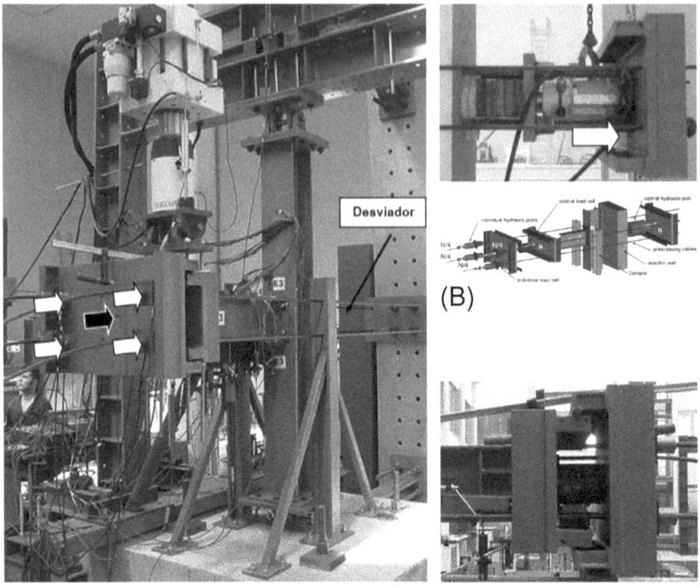

(A) (C)

Fig. 2.124 (A) Frame for applying compressive forces, (B) hydraulic jack, and (C) central load cell.

All tests were instrumented as shown in Fig. 2.126, with single strain gauges (FLK 6-11-TML), rosettes at 45° (FRA 5-11-TML), bolt axial strain gauges (BTM 6-C-TML), and LVDTs.

For all tests, a constant axial force was first applied, maintained constant during the test, with subsequent application of a bending moment incremented to collapse. Two unloading phases were made, the first for a bending moment of 25 kN m (down to 5 kN m, to eliminate possible slack in the joint) and the second for a rotation of 20 mrad. Force control was utilised in the first part of each test, and later changed to displacement control. In order to ensure an accurate control of the axial force application, a preliminary test was made (FE2) in the elastic range. During the test, it was observed that the axial load application with the cables was transmitted by the central load cell to the joint as shown in Fig. 2.127. It can be observed that the force applied to the joint, acquired either through (i) beam web and flange strain gages, (ii) central load cell, or (iii) by superposition of the individual load cells led to similar results.

For higher bending moment levels, it was noted that the beam rotation caused a reduction of the force in the bottom cables and an increment of load of the top cables. Consequently, hydraulic jacks were positioned at the bottom cables to correct the axial force as the test progressed. In Fig. 2.128, the

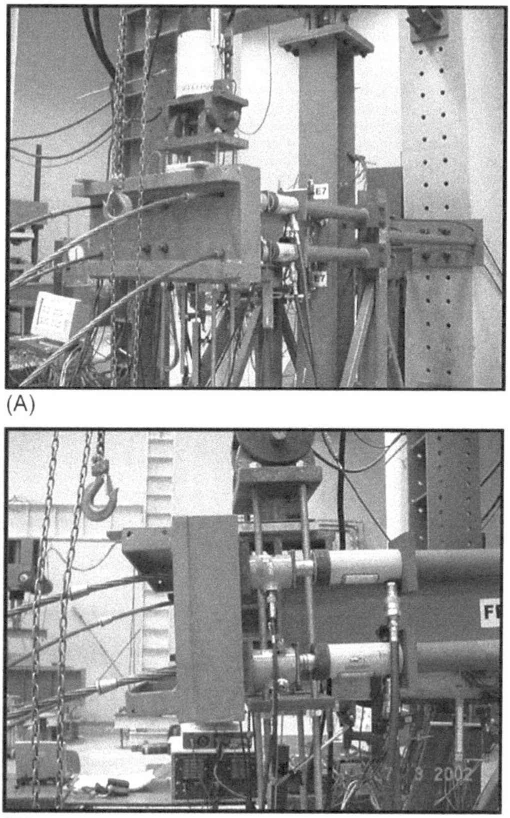

(A)

(B)

Fig. 2.125 (A) Application of tensile loading and (B) hydraulic jack layout.

variation of the axial force in each individual cable with applied moment is depicted. The clear difference between tests FE3, FE4 and FE5, FE6 reflects the fact that only for the latter four, individual hydraulic jacks were utilised to correct the load unbalance caused by deformation of the beam, thus leading to a virtually constant applied axial force.

The moment versus rotation curves for the eight tests are shown in Fig. 2.129, where it may be noted that even for a level of equivalent axial force of 27% of the beam plastic resistance, the bending moment still exceeds the pure bending result (FE1). In addition, the maximum bending moment was reached for an axial load level of 20% of the beam plastic resistance. Alternatively, the application of a tensile force in the beam results in a sharp reduction of the bending resistance of the connection. Fig. 2.130 shows the values of the moment resistance and initial stiffness of the tested connections. The theoretical values evaluated according to Eurocode 3 (FE1) were

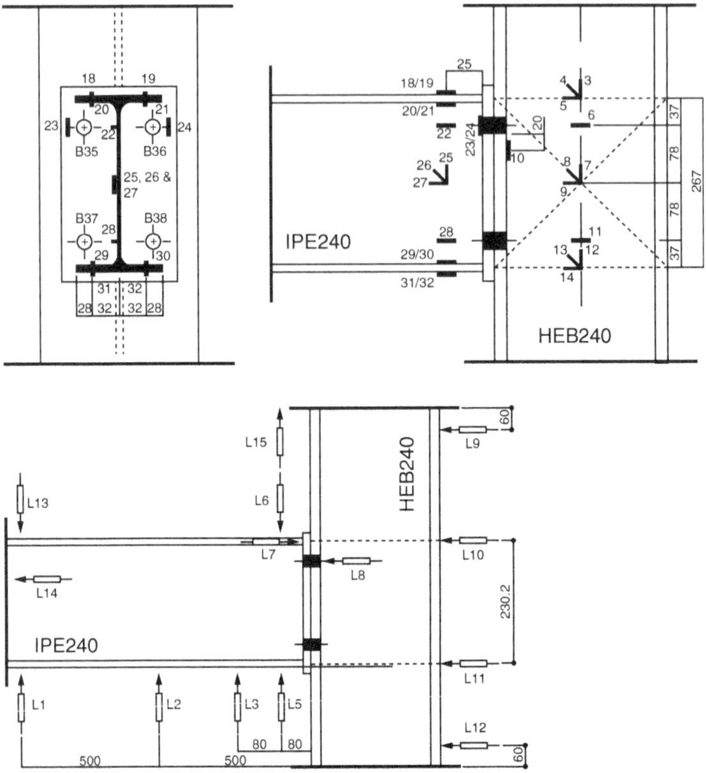

Fig. 2.126 Single strain gauges, rosettes to 45°, and bolts axial strain gauges layout.

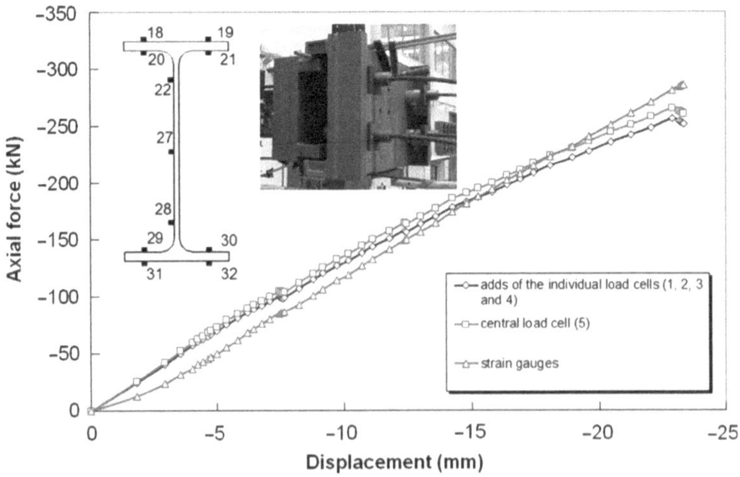

Fig. 2.127 Calibration of applied axial force.

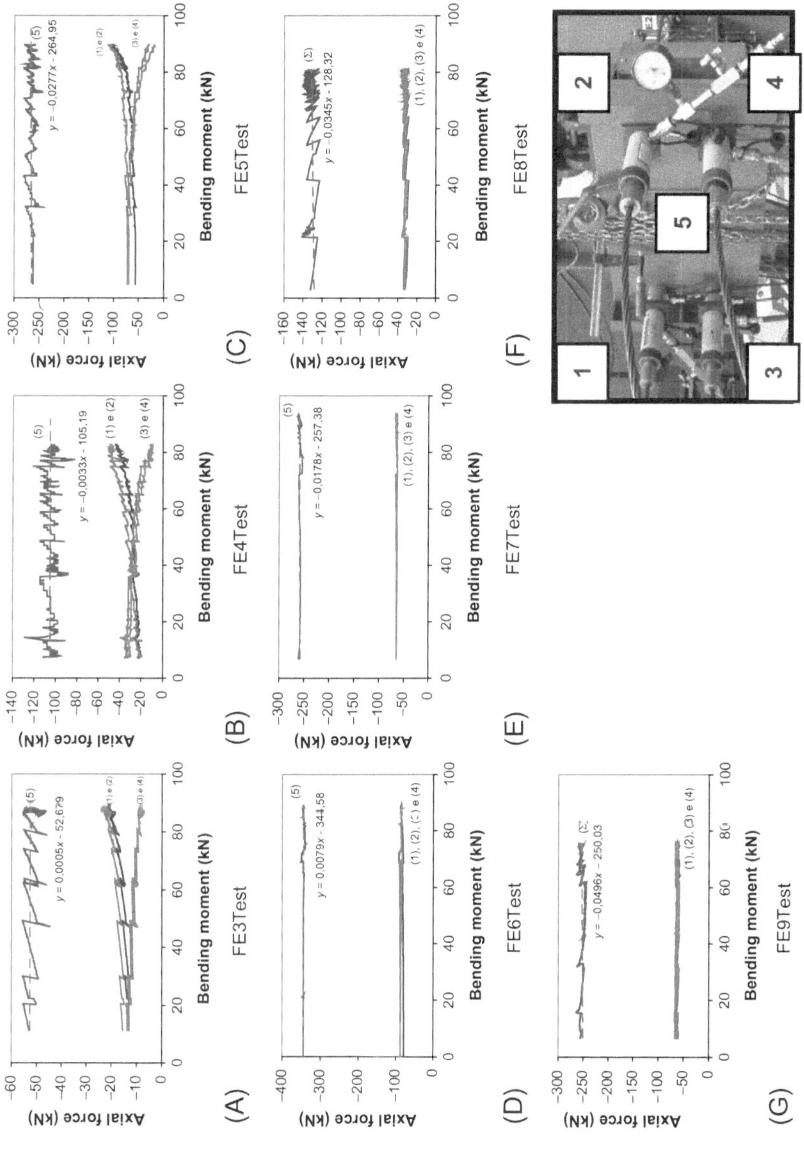

Fig. 2.128 Bending moment versus axial force curves. (A) FE3 Test. (B) FE4 Test. (C) FE5 Test. (D) FE6 Test. (E) FE7 Test. (F) FE8 Test. (G) FE9 Test.

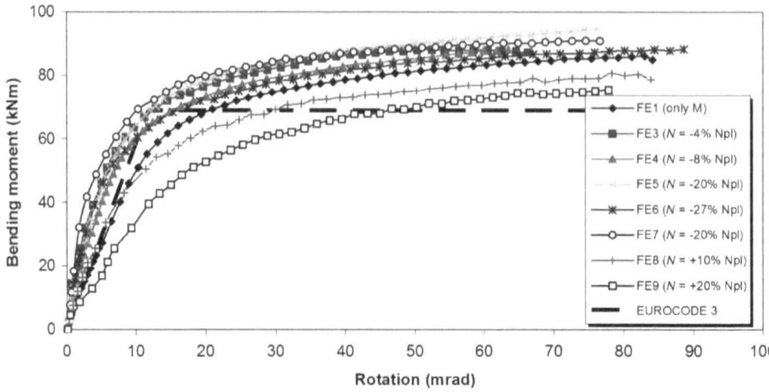

Fig. 2.129 Moment versus rotation curves of the experimental tests.

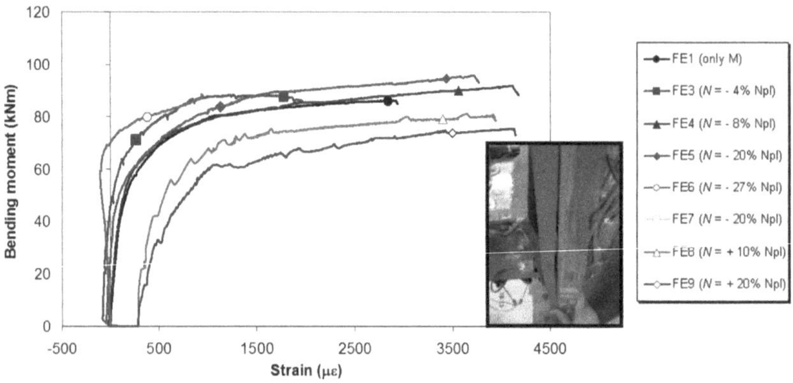

Fig. 2.130 Moment versus strain curves for component column flange in bending.

70 kN m and 11,865 kN m/rad. Fig. 2.130 shows that, for all the tests, the column flange in bending led to the flange full yield.

The influence of the axial force magnitude may be observed in Fig. 2.131, by comparing, for example, tests FE1 and FE5. For test FE1, the yield strain is reached for a bending moment of approximately 45 kN m. For this bending moment magnitude, the tensile load in the first bolt row, calculated using the strain gages located in the beam flange is 333.4 kN. However, for test FE5, due to the axial force contribution, the end-plate reached the yield strain at a higher moment magnitude. The moment versus strain curves for the component, bolts in tension are shown in Fig. 2.132, where it is noted that, with the compression axial force, this component is reduced, having initially negative strains for tests FE4, FE5, and FE6.

The test programme on extended end-plate beam-to-column joints was made of seven tests with several combinations of bending moments and axial forces. In all tests, the joint configuration was identical (Fig. 2.133), the column

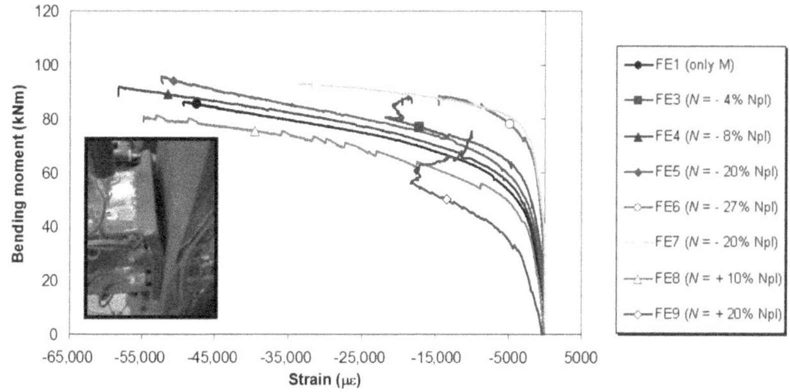

Fig. 2.131 Moment versus strain curves for component endplate in bending.

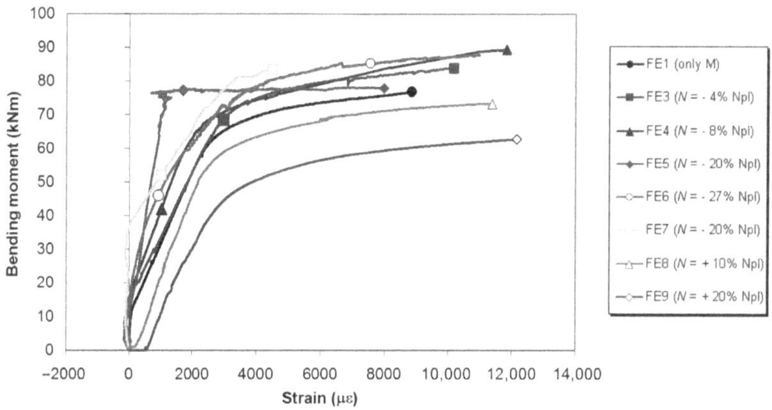

Fig. 2.132 Extended end-plate beam-to-column joint layout.

being simply supported at both ends and made of an HEB240, the beams' sections were IPE240 and the endplate was 15 mm thick, all made of an S275 steel grade. The bolts were M20, 10.9, prestressed with a 150-Nm torque.

The seven tests were made by first applying a fixed level of axial tension or compression and later imposing a negative bending moment up to the connection collapse failure. The first test, EE1, only involved bending moments. The next tests, EE2, EE3, EE4, EE5, EE6, and EE7, were under forces of −10%, −20%, −27%, −15%, +10%, and +20% of the beam plastic resistance applied to the beam.

The test setup is shown in 2.133 with the bending moment applied with a hydraulic jack, located 1 m away from the column flange face. The used compressive force application system was performed with a central hydraulic jack located behind the reaction wall, that applied a tensile force to four

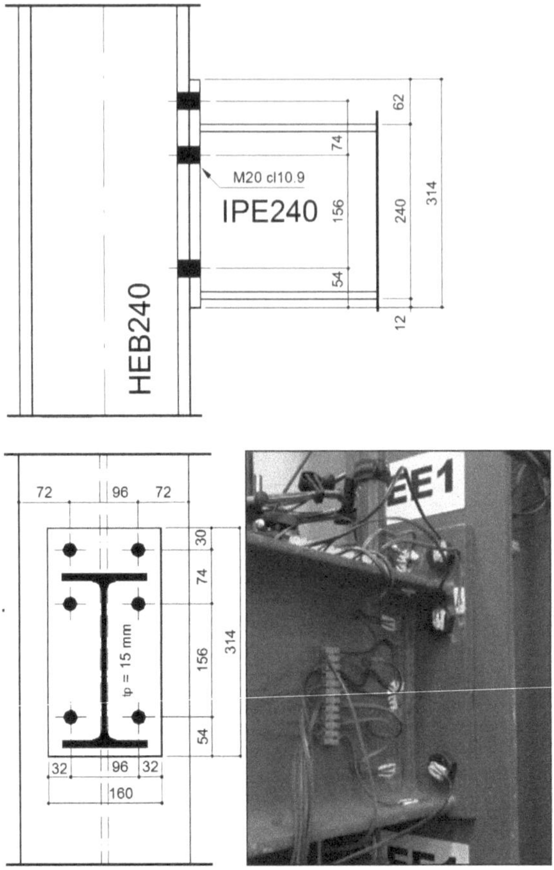

Fig. 2.133 Extended end-plate beam-to-column joint layout.

prestressing cables with diameter $\phi = 15.2$ mm. The transfer of this force to the joint was made through a 600 kN central load cell. These cables passed through a deviator saddle (HEM100) to ensure that the axial force was always parallel to the beam axis, whilst 200 kN load cells were utilised in each cable to acquire the cable forces (Fig. 2.134). The horizontal reaction forces at the column ends were transmitted to the reaction wall by (i) a steel beam, at the top and (ii) a reinforced concrete footing prestressed to the strong floor utilising DYWIDAG bars and connected to the reaction wall using an HEB 200, at the bottom.

The moment–rotation curves for all tests are shown in Fig. 2.135 where it can be noted that the connections axial force influenced the joint behaviour.

The maximum moment resistance was associated with test EE2, whose compressive axial force level corresponded to 10% of the beam plastic

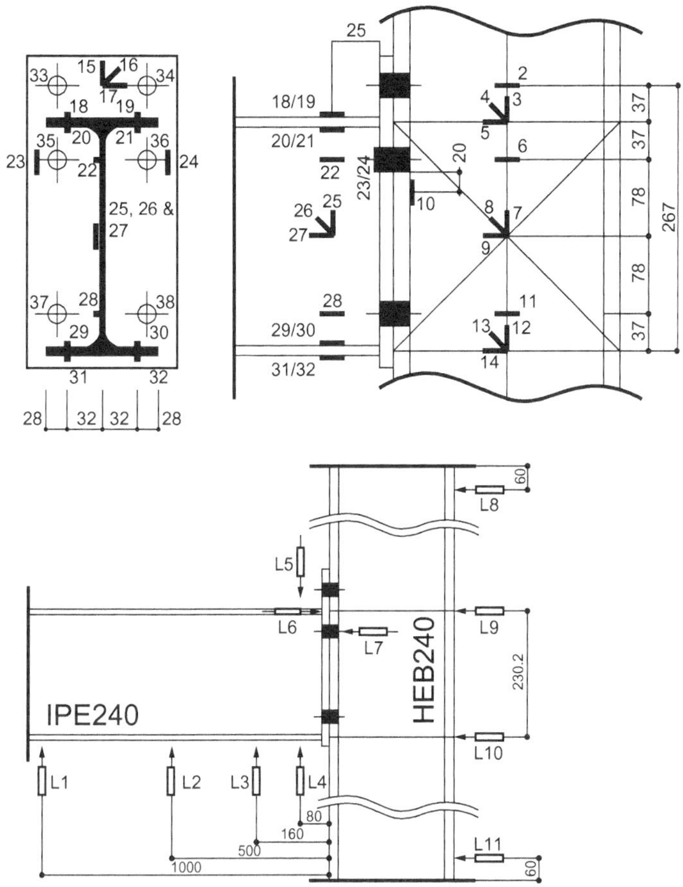

Fig. 2.134 Arrangement of strain gauges and LVDTs used in the tests.

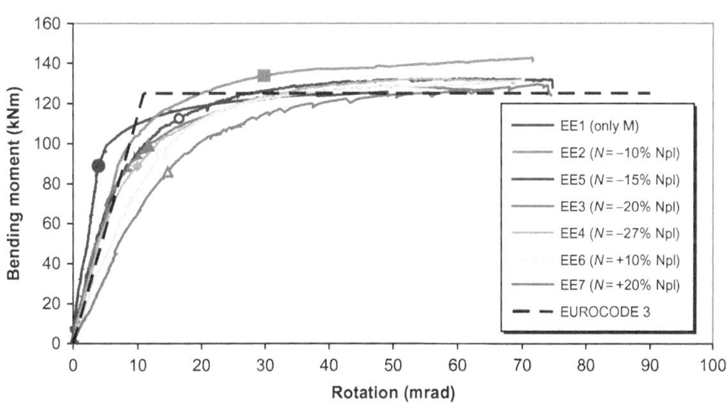

Fig. 2.135 Experimental tests: moment versus rotation curves.

resistance, leading to a value 6% higher than test EE1 without axial loads. For all other levels of compressive axial force, the maximum bending moments were smaller than in test EE2. For the two tests with tensile axial force, EE6 and EE7, the bending moment resistance was lower than in the EE1 test, due to premature mobilisation of the components in the connection tension zone.

Starting with the column web panel in shear, large deformations were expected for this component, since the connection corresponded to an external column. Fig. 2.136 shows the principal stresses acquired with aid of the rosette 'B', located at the column web panel in shear centre showing that this component reached the yield limit in all tests. Analysing the results of von Mises stresses calculated at the same point, it was noted that, for the first test, EE1, the yield limit was reached for a moment equal to 98 kN m, being the lowest bending magnitude where this component was yielded. Alternatively, the highest resistance was observed in test EE2, at 121 kN m, that is the test that presented the higher bending resistance.

Fig. 2.137 shows the moment–strain curves for the endplate in bending, acquired with two linear strain gauges. This component controlled the connection response in the tensile zone. For the tests under compression forces, an increase of the component resistance was noted, in contrast to the tests with tensile forces.

Fig. 2.138 synthesises, for each test, the sequence of yielding for all components. It shows that when the compressive forces are increased, the beam flange in compression component becomes progressively more critical. Analogously, when increasing tension forces, the endplate in bending turns out to be the critical component.

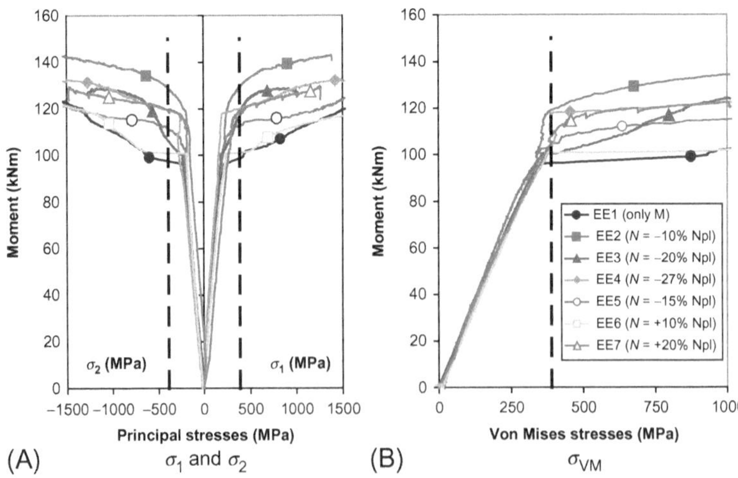

Fig. 2.136 Column web in shear (1): moment versus stress curves. (A) σ_1 and σ_2. (B) σ_{VM}.

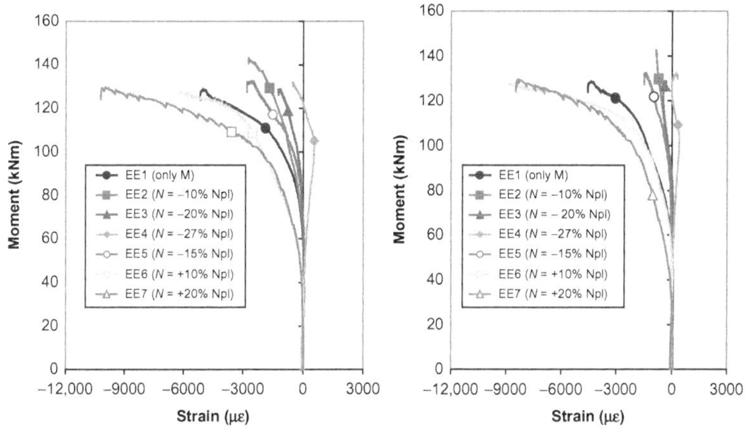

Fig. 2.137 Endplate in bending (5): moment versus strain curves.

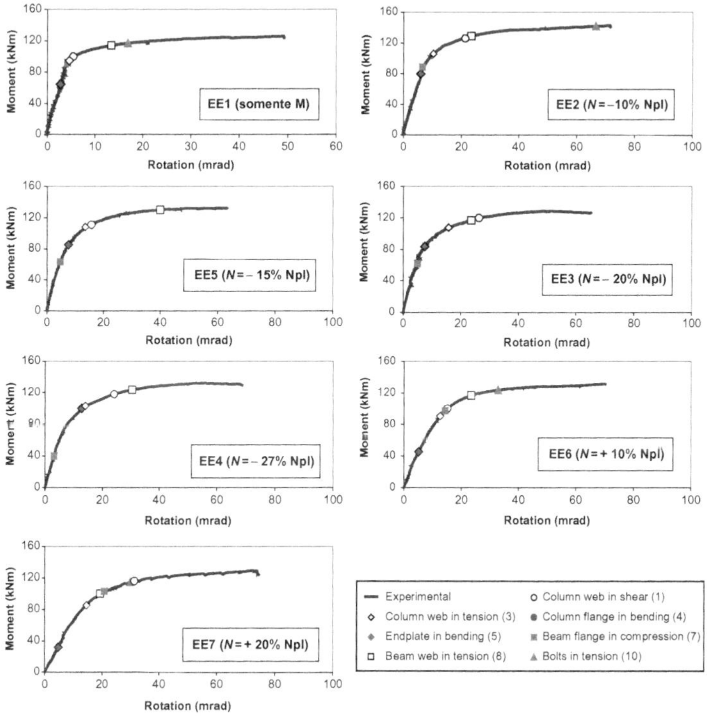

Fig. 2.138 Yielding sequence of the components.

The moment–rotation curves for all tests imply that the axial force presence considerably affects the connection structural response. The maximum moment resistance was obtained in test EE2, in which the compressive axial force level was associated with 10% of the beam plastic resistance, a value 6%

greater than in test EE1 (without axial load), alternatively, in the two tests with tensile forces, EE6 and EE7; the moment strength was lower than in test EE1 due to the premature mobilisation of the connection tension zone components. It was also noted that the compression zone critical component was the beam flange in compression, whilst for the tension zone, the critical component was the endplate in bending, results already predicted with EC3.

For the flush end-plate joints, an increase of the moment resistance was observed for compressive axial force below 20% of the beam plastic resistance, clearly indicating an asymmetry response, whilst the same connection loaded in tension exhibited a reduction of the moment capacity.

One of the major contributions in terms of development of these tests consisted in the axial load force application system design that follows the beam-free end, as the transverse force is monotonically incremented. Two system types had to be developed to maintain the balance of cables tension and compression forces. If the ingenious use of prestressing cables facilitated the movement of the beam-free end on one hand, on the other, it created a force unbalance in the cables as the beam deformed. This required the development of an individual control system of the cables forces, maintaining a uniform distribution forces despite the vertical displacement of the beam end.

The adopted system and supports were designed to simulate, as accurate as possible, the connections performance when subject to a bending versus axial force. The instrumentation designed for this test programme also included the use of strain gauges inside the bolts for not only monitoring the forces transmitted by joint, but also for comparing and validating aspects like the moment versus rotation curve, prying effect, amongst others. The individual assessment of each joint component was also made to contribute to the global understanding of the connection behaviour of terms of initial stiffness, transmitted bending moment, and rotation capacity. The individual response of each component served for determining the sequence of collapse or the collapse of each of these as the external forces were applied to the investigated structure.

2.6.9 Structural Evaluation of Major Axis Semi-Rigid Composite Joints With Angles

Although the use of composite action is becoming more popular in the semi-rigid joint design, its application to hogging moment areas in portal frame edge and corner columns is still not exploited. The solutions for semi-rigid composite joints under hogging moment action are based on two conditions. The first is that the hogging reinforcing, used in the slab's effective width, should have the required anchorage length in the positive moment regions. The second condition stresses the need of a minimum number of shear connectors to be utilised in the hogging moment region.

These joints are necessary, even though the concrete action is not effective, due to cracks. The shear connectors can progressively transfer the loading action from the reinforcing bars to the steel girder. This gradual load transfer would not happen if it were made only through the reinforcing bar's anchorage length. The usual composite solutions use studs and hogging reinforcing bars properly anchored in the positive moment regions.

These facts motivated the development of the present composite framing system that utilises Perfobond rib shear connectors [52–55]. This connector, initially used in composite girders by Leonhardt, was utilised for the first time in building structures. Its use was selected due to the fact that in addition to guarantee the concrete-steel bond, it also provides a better anchorage of the internal columns hogging moment reinforcing bars. These bars can be anchored passing by the Perfobond web holes or simply by being superimposed to the transverse reinforcing bars that are generally utilised on them.

The usual design solutions for external columns utilise reinforcing bars enlacing the column. This structural detail is not very well accepted by architects, due to the extra difficulty of adapting the architectural design solutions for the building facades. The need for enlacing columns with reinforcing necessarily leads to the elongation of the floor slab creating an overhang. This detail complicates the concrete slab construction and the use of vertically continuous window panels.

The adopted solution for external columns semi-rigid composite joints located at edge or corner introduces a new joint element, called T-Perfobond connector (Fig. 2.139). This connector has the objective of transferring the reinforcing bar forces, located in the hogging moment region, directly to the column flange. The other elements present in the internal and external joints were seated and double web angles.

In order to validate the proposed composite semi-rigid structural system, a three-phase experimental programme was performed. The first phase consisted of Perfobond push-out tests including an innovative push-out test configuration developed to measure the isolated shear strength contribution of the concrete dowels and the reinforcing bars present in the Perfobond holes.

The Perfobond shear connector is made from a rectangular plate with holes as shown in Fig. 2.140. Its mechanical response is dependent on the concrete slab in which it is embedded. During the concrete slab cure process, dowels are formed in a region near the Perfobond holes. The dowels contribute to the shear resistance and prevent slip and vertical split of the beam

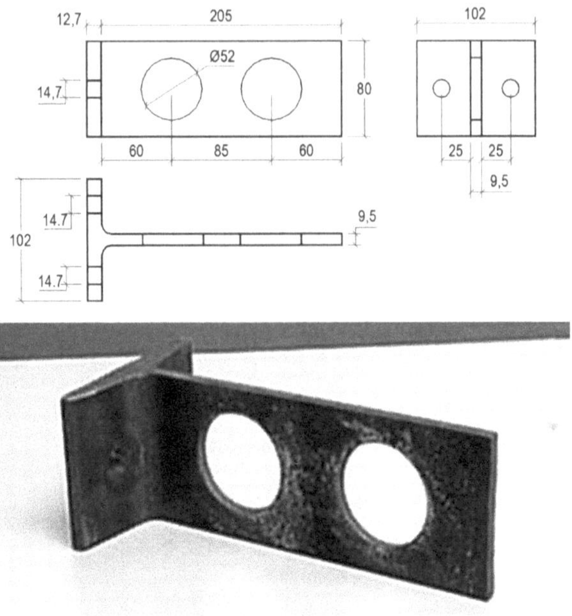

Fig. 2.139 'T-rib' connector geometry.

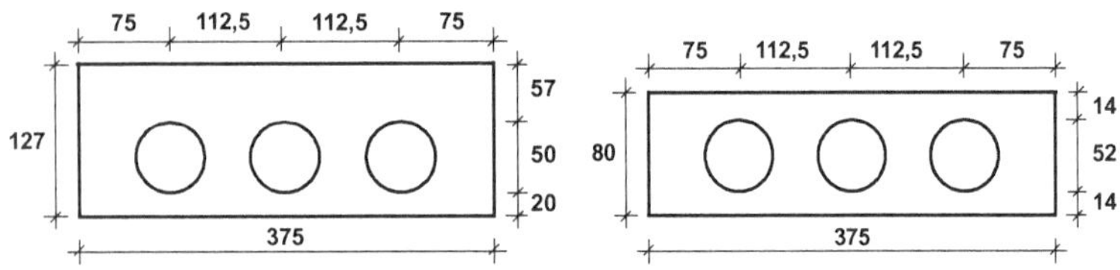

Fig. 2.140 Original and modified Perfobond rib shear connector.

from the concrete slab. The structural performance of this shear connector can be substantially enhanced by the use of reinforcing bars through its holes.

The Perfobond connector has an extra advantage when compared to shear studs. The Perfobond rib requires simple equipment for its assembly, fact confirmed by the full-scale tests. Push-out tests in which the number of holes and the plate's height varied in relation to the slab thickness serve as means to determine an optimal shape for the Perfobond. The adopted Perfobond connector had these optimal dimensions only substituting the plate's height according to the required slab thickness.

To compare the test results obtained by Oguejiofor [56] on 375 × 127 × 13 mm Perfobond, named EB and ED, another experimental series made of eight push-out tests (PB), was performed. The Perfobond specimens, 375 × 80 × 12.7 mm, were welded to I 203 mm × 27.4, 720 mm long sections (Fig. 2.141) and encased in two concrete slabs

Fig. 2.141 Perfobond rib shear connector geometry and push-out test layout.

720 × 720 × 100 mm. A bar square mesh of 4.76 mm ϕ @ 172 mm reinforcing bar square mesh was utilised in all the tests.

Tests PB1 and PB2 were performed in the same way as ED3 and EB3, whilst tests PB4, PB6, and PB7 were similar to ED7 and EB7. The only difference was the lack in EB3 and EB7 of a wire mesh layer near the slab outer face. The geometrical details of the concrete slab, Perfobond shear connectors, and reinforcing bars location utilised in the tests are presented in Ferreira [52,55].

Fig. 2.142 shows the load–slip curves for all the performed tests, where it is possible to note that the influence of the reinforcing mesh over the shear connector's strength is less significant for specimens with transverse reinforcing bars inside the Perfobond holes. Specimens without reinforcement bars inside the Perfobond holes, PB 01 and PB 02, had a 40% increase in the experimental ultimate load when compared to the forecasted values. This was due to the additional strength provided by a reinforcement mesh utilised in these specimens.

The T-Perfobond connector detail must minimise the prying action effect otherwise, a premature loss of stiffness in the joint can be expected.

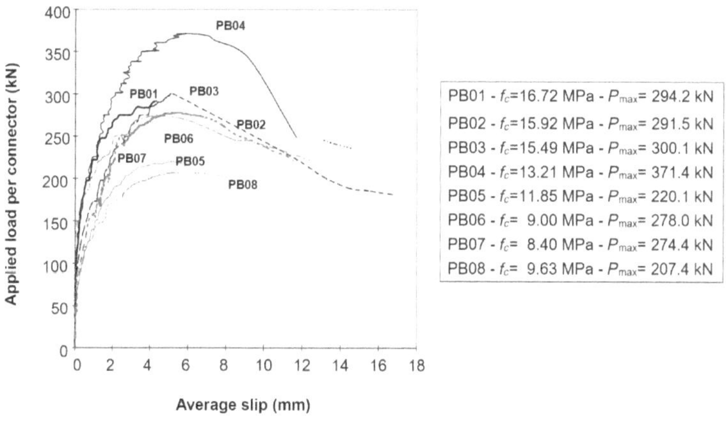

Fig. 2.142 Load–slip curves for all Ferreira et al. [3] tests.

Fig. 2.143 'T-rib' connector fabrication process.

The connector's height was set to 80 mm due to the slab thickness (100 mm) in which the connector is embedded, whilst thickness of 12.7 and 9.5 mm were utilised in the connector's flange and web. The T-Perfobond connectors are easily manufactured from leftover rolled sections avoiding extra welding work and reducing the costs. Fig. 2.143 shows the four steps of the T-Perfobond connector fabrication process: (a) initial profile, (b) web holes, (c) flange holes, and (d) opposite flange saw cut.

The two-storey portal frame consisted of three columns forming two spans of 4.5 and 6 m and a total height of 3.6 m. The first storey structure was a composite girder with a solid cast in place slab 120 mm thick and 1.5 m wide. This composite girder contained all the constructive system components, that is, the Perfobond rib shear connector, the T-Perfobond connector, web and seated angles, and reinforcing bars. The second storey beams were made by rolled steel sections, without the concrete slab. The three columns utilised welded steel sections WWF300 × 56.5 (b_f = 200 mm, t_f = 12.7 mm, t_w = 8 mm, h = 300 mm) with 5 mm fillet weld legs. All the girders were made by rolled steel sections I10″ × 37.7 (b_f = 118 mm, t = 12.7 mm, t_w = 8 mm, h = 254 mm). Figs 2.144–2.146 show the tested composite frame layout as well as the girder, slab, and reinforcing bars strain gauge locations.

Fig. 2.147 shows the first storey internal and external joint details. The first storey joints utilised 127 × 127 × 9.5 mm seated angle and 127 × 89 × 9.5 mm double web angles. The bolts were A325 with 19 mm diameter. The external column composite joints utilised the T-Perfobond connector. The second storey joints were made of 76 × 76 × 9.5 mm angles.

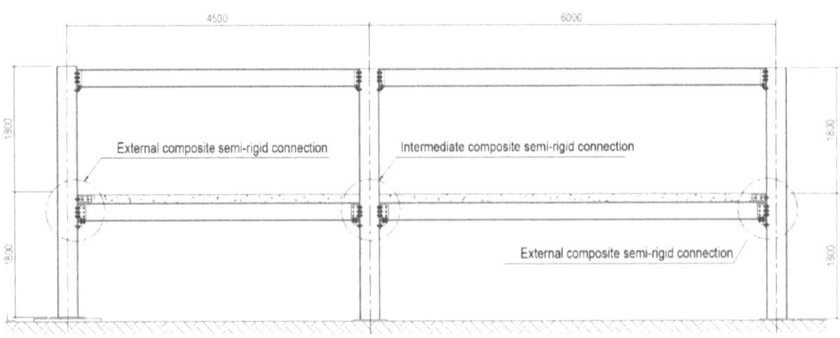

Fig. 2.144 Composite semi-rigid portal frame.

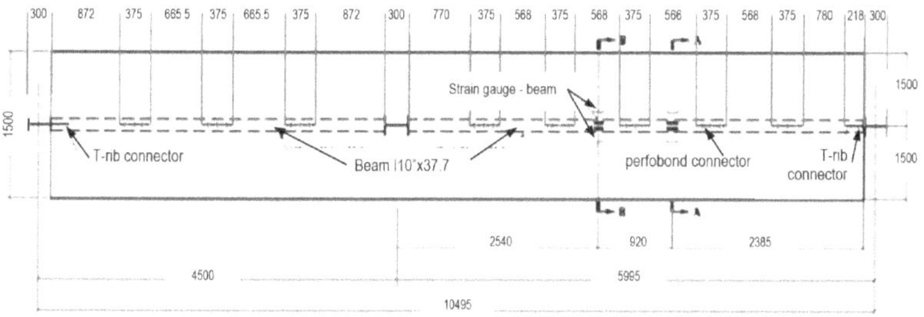

Fig. 2.145 Strain gauges used in the beam top flange and concrete slab.

The web angles were connected utilising A325 bolts of 19 and 16 mm diameter, in legs connected to the beam web and column flange, respectively. The seated angles utilised 16 mm A325 bolts. Figs 2.147 and 2.148 show the second storey joint details and the column base layout utilising 22 mm diameter anchoring bolts.

The used reinforcing bars were $\phi = 4.76$ mm in the longitudinal distribution bars and stirrups, $\phi = 10$ mm in the internal and external columns hogging moment bars, and $\phi = 12.7$ mm inside the Perfobond rib shear connector holes. Strain gauges were utilised in the reinforcing bars at two hogging moment regions to acquire the composite section strains. Fig. 2.149 shows the first storey joint details before concrete casting. The cast in place slab needed temporary framework and propping, as well as positioning the reinforcing bars. Only three and five Perfobond shear connectors were utilised on the small and large composite girder spans. When these values are compared to the equivalent numbers of 19 mm stud connector ($f_u = 415$ MPa), interesting conclusions can be drawn. Utilising the values usually provided for the Perfobond connector and the usual formulae for stud connector it was possible to determine that for the studied composite girder, each Perfobond connector is equivalent to 4.7 studs. This number, discarding the T-rib connector contribution at the external columns, totalises 15 and 24 studs to be utilised on the small and large composite girder spans.

The concrete nominal compressive strength, f_c', was estimated to be 20 MPa. A four-cycle vertical load test series was conducted. The load configuration utilised three 500 kN hydraulic jacks slaved to each other (350 mm stroke) and three load cells (two 200 and one 500 kN). One jack was positioned at the smaller span beam centre point whilst the others were located at the longer span beam (Fig. 2.150). Vertical and lateral deflections were recorded by 10 dial gauges and 10 LVDTs.

Fig. 2.151 shows load–displacement curves for points positioned at first storey smaller and larger span centre points. The maximum vertical displacements were 7.5 and 46 mm.

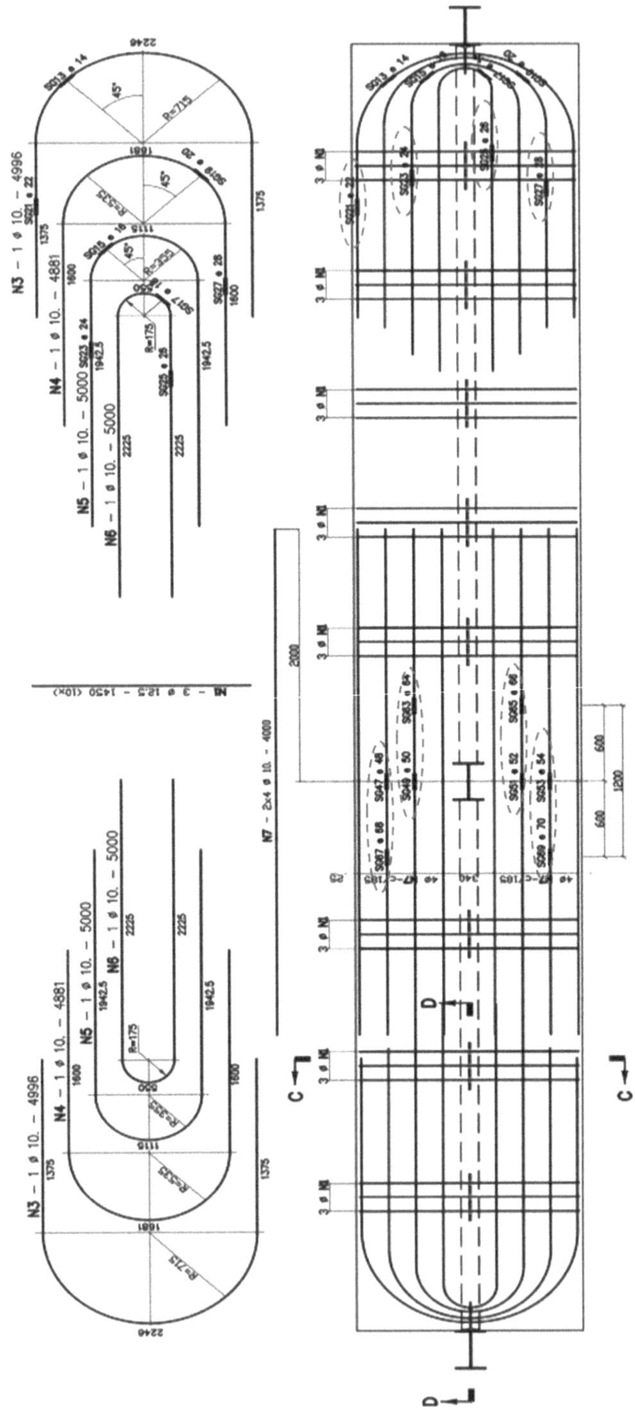

Fig. 2.146 Concrete slab reinforcing bars layout and strain gauge location.

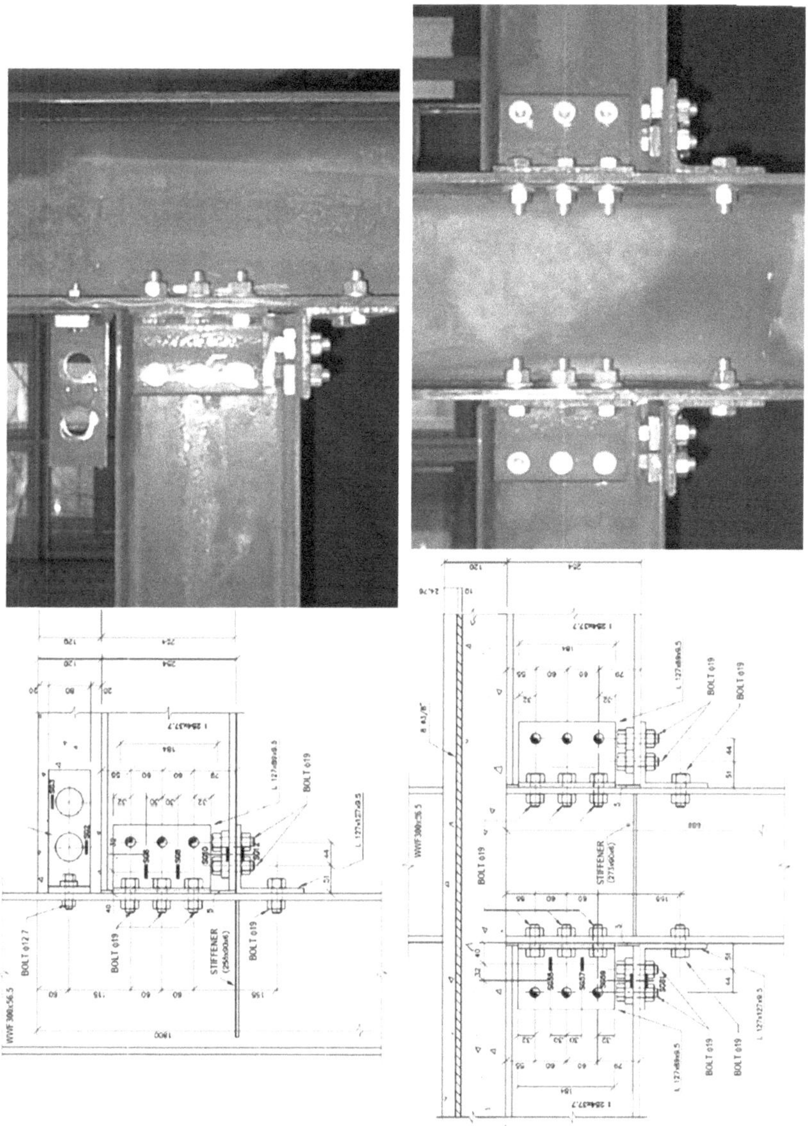

Fig. 2.147 Internal and external connections: strain gauge location layout 1st storey.

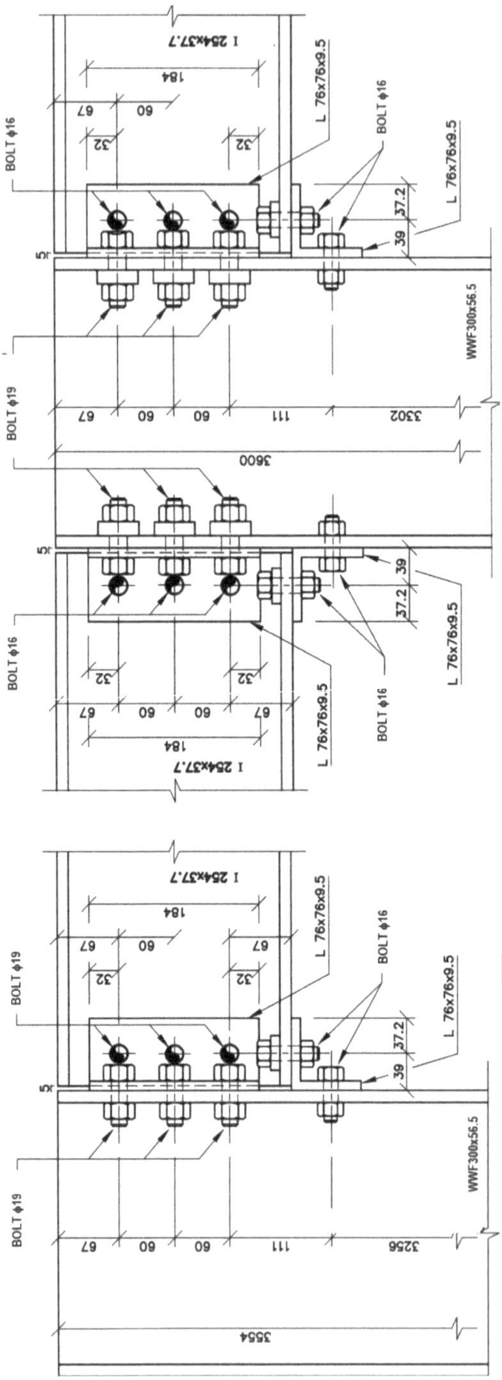

Fig. 2.148 Internal and external connections: strain gauge location layout 2nd storey.

Fig. 2.149 Details of the concrete slab and adopted reinforcing bars.

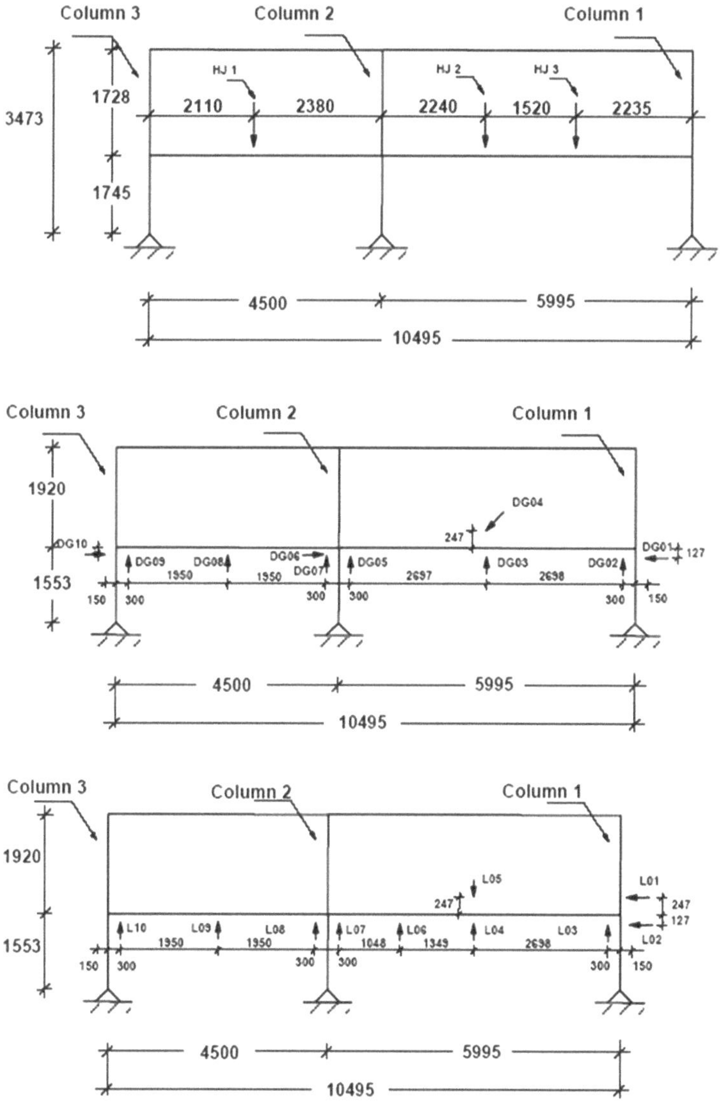

Fig. 2.150 Hydraulic jacks layout, LVDTs, and dial gauges locations.

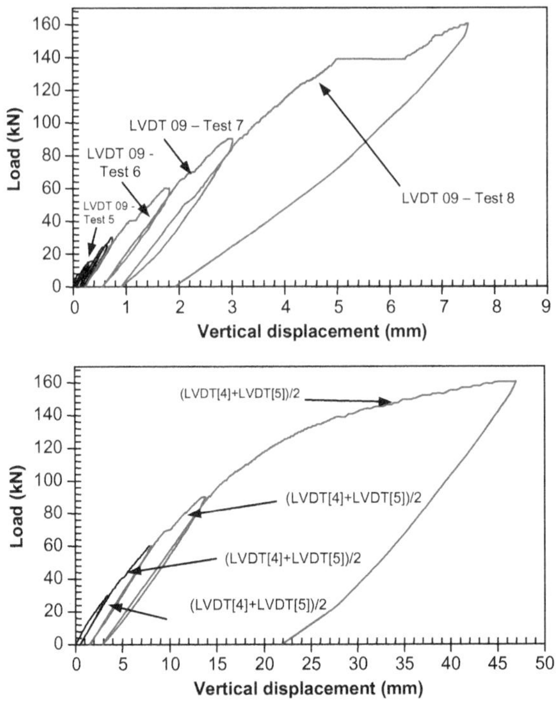

Fig. 2.151 Load–displacement curves at smaller and larger span centre points.

During the eighth test, when the load achieved 160 kN per jack, a significantly deformed portal frame configuration was reached surpassing by far its serviceability limit state (a 45 mm vertical deflection was measured at the largest span centre point). Around this load level, the concrete slab depicted a distributed crack configuration close to the T-Perfobond anchoring location on the external column. Despite this fact, no component or member failure was reached.

Since no collapse was reached two additional tests were performed on the composite frame to induce its failure. To achieve this goal, before conducting the ninth test, all seated angles were removed and the external column 'T-rib' bolts were untightened. The only active components left on these connections were the double web angles simulating standard simple joints. Alternatively, the internal connections could still mobilise the hogging moment reinforcing bars and the double web angles. The portal frame was loaded again to 160 kN per jack, but again, no collapse was reached.

Load versus rotation curves for internal and external composite joints were built through load–displacement curves. Although the portal frame was braced on its top and bottom by steel cables, the structure displaced

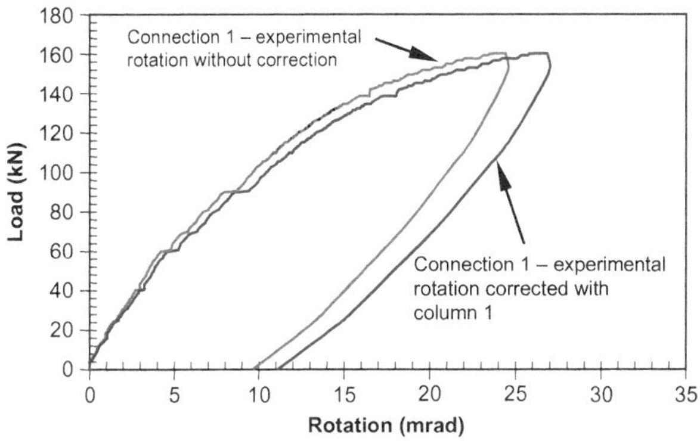

Fig. 2.152 External connection load–rotation curves.

longitudinally, affecting the connection rotations. The joint's relative rotations corrections were made by considering the column rotation to which the connection was connected.

The composite external connection rotation was evaluated in two steps: first, the global joint rotation was calculated neglecting the column lateral displacements. This angle was determined with the aid of an LVDT located on the beam bottom flange near the seated angle external leg. The column rotation angle was calculated using an LVDT located on the external flange of the first column. Fig. 2.152 shows the external joint load–rotation curves with and without the column lateral displacement correction where its influence could be easily noted.

An LVDT (L7) and a dial gauge (DG5) were utilised to measure the beam vertical displacements close to the internal composite connection. The rotation angle was evaluated similarly as in the case of the external connections. Two alternatives were evaluated to better represent this angle, the first used the dial gauge readings for the tests 5 and 6, whilst the LVDT readings were utilised in the subsequent test (Fig. 2.153). In the second alternative, the curve was built considering only the results from tests 7 and 8 acquired from the LVDT. When both curves are compared, it is possible to note that when only the LVDT readings are considered, the joint initial stiffness does not change, but the maximum recorded rotation is reduced from 33.4 to 31.4 mrad.

Regrettably, due to fixing problems, the internal column lateral displacement readings could not be utilised. Two options for evaluating the column rotation were considered by using the lateral displacements of the first and third external columns. Fig. 2.154 shows the two correction alternatives applied to the girder rotation calculated through LVDT (L7) and dial gauge (DG5).

One of the main objectives of this programme was to assess the transmission of forces process in the investigated composite semi-rigid. Two large

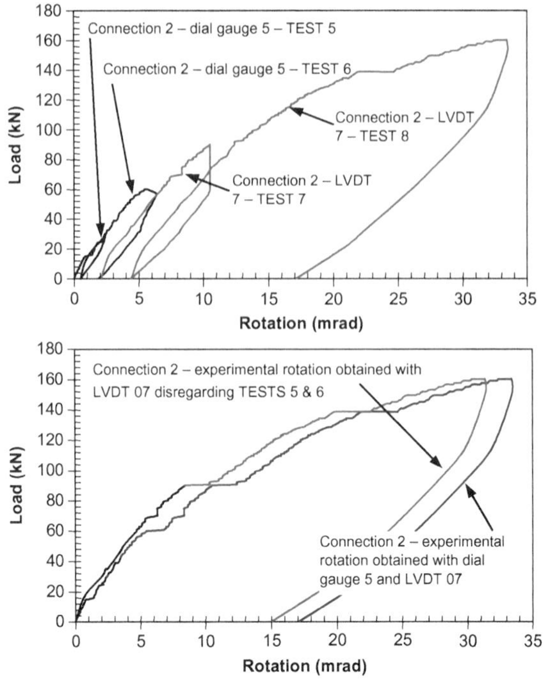

Fig. 2.153 Internal connection load–rotation curves without column correction: DG5, LVDT7, LVDT7 (7 and 8 tests only), and combined DG5 + LVDT7.

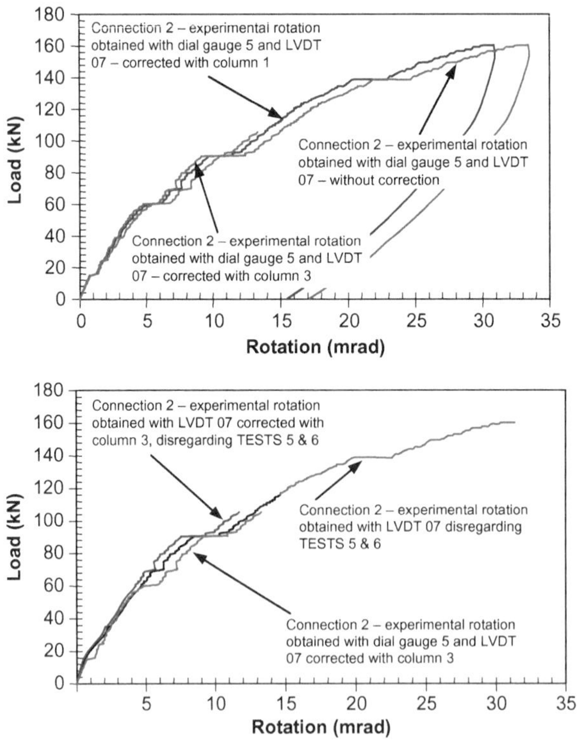

Fig. 2.154 Internal connection load–rotation curves presented with corrections or as measured.

joints groups, the externals and the internals, were studied by incorporating negative reinforcing bars in quantity enough to generate a full and, a partial rotation restriction optimising the compression and tension-resistant parts.

Perfobond shear connectors were used in positive moment segments to ensure the composite joint action and anchor part of the negative reinforcing bars in external joints. It was also designed and conceived a variation of the Perfobond connector, that is, the T-Perfobond connector for anchoring the negative reinforcing bars present in external column without the need to extend the slab beyond the columns external face.

The experimental programme generated a number of challenges in the instrumentation system since the value of the partial rotation restriction on the external and internal joints can only be measured indirectly. One of the alternatives for its determination used LVDTs and dial gauges positioned along the beam and column. This process proved to be efficient in the linear elastic range where classic formulas for the beams rotations and displacements can be obtained. However, this correlation was shown to be inefficient as the beam yield began. Another alternative used the stress distribution measured in points along the same beam cross-section, but once again, proved to be ineffective.

The experimental programme started from push-out and pull-out tests made on the already mentioned connectors and culminated in a full-scale test of a portal frame with two spans. The adopted instrumentation used in these tests included LVDTs, dial gauges, load cells, and strain gauges (linear and rosettes) distributed over key points of the structure. An innovative variation of the push-out test was designed to isolate the effect of the connector contact face with the concrete slab from the construction of slab prototypes slots before the connector face. The difference in behaviour and load between the tests with and without the slot provided the principle, the value of this resistant parcel. The pull-out test was also designed to determine the efficiency and strength of the T-Perfobond connector in transmitting and anchoring the concrete slab negative bars.

The experimental programme also enabled the investigation of efficiency of the rotation evaluation of joints through methods using LVDTs and dial gauges, evaluation of the effect of the column displacement and the girder under bending deformed shape on these rotations. The most important aspect was the perception that the most reliable behavioural curves for these joints were those related to the load versus displacement, where less approximations and assumptions had to be made. One more time, with this test programme, it was proven that the redundancy of the measures is the key to success, and even for future re-evaluations and comparisons with new tests.

2.6.10 Structural Evaluation of Major Axis Composite Semi-Rigid End-Plate Joints

The supports' region where negative moments occur, especially in continuous beams, is a topic of endless discussion due to the large number of variables involved in this force–transmission mechanism, which is constantly used in design. Aiming to fully investigate the actual structural behaviour of this region of hogging moments, a proposal for dividing this system into different components of greater influence is proposed.

These main components that can be observed in Fig. 2.155 are the concrete slab, the reinforcing bars, and the shear connectors, components still to be thoroughly studied in the hogging moment region. The transmission force path starts in the beam flange and continues through the shear connector weld and to the shear connector, and from this to the concrete, where interactions between these two components occur. This trajectory continues from concrete to the reinforcing bars, again leading to an interaction between these latter components. Finally, the force path ends at the reinforcing bars in tension component.

Each of the components, or variables that are represented in Fig. 2.155, has numerous possible variation levels that, directly or indirectly, influence their structural behaviour. The combination of all variables and their levels generates disproportionate amount of tests that urges for a reduction of these elements and/or levels that will lead to a consistent and feasible amount of experiments. Table 2.9 presents a new organisation for the required experiments focusing on the most significant components.

The length of the reinforcing bars is calculated by combining the spacing of connectors, represented by the distance from the column outside face to

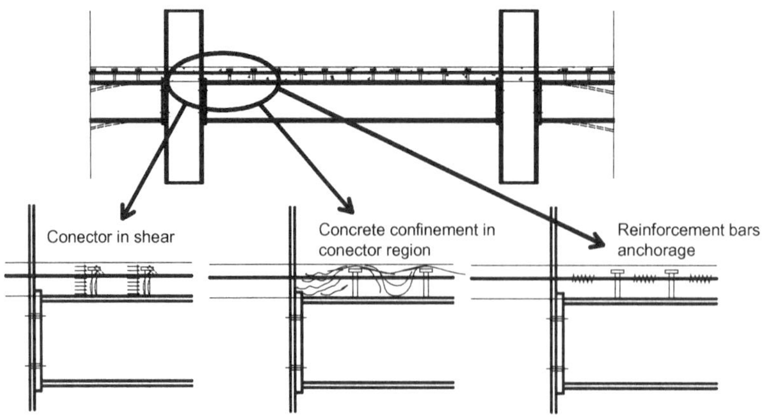

Fig. 2.155 Negative moment region components.

Table 2.9 Final Component Description and Characteristics

Reinforcing Bar			Concrete Strength	Connector		
Length (mm)	**Number (und)**	**Diameter (mm)**	**F_{ck} (MPa)**	**d_1 (mm)**	**d_2 (mm)**	**Number (und)**
$d_1 + d_2 + d_2$	2	16	20	100	100	3
$d_1 + d_2 + d_2$	4	20	30	200	250	2

the first shear connector, d_1, and the connectors spacing, d_2. A method known as robust design, developed by Taguchi [57], was used to create a series of experiments with a reduced amount of experiments. From the early 1950 onwards, Taguchi developed a methodology that used the application of statistical methods to improve the quality of manufactured products. With this method, it is possible to perform a coherent distribution of the variables within their levels, through combination arrangements represented by matrixes. Table 2.10 presents the conversion of the proposed methodology in terms of their actual variables and corresponding levels.

From this combination of variables, it is possible to make variations in this table, where some of slab confinement models were also proposed [58–60]. These confinement models aim to represent the continuity of the concrete slab usually present in actual structures, but applied in reduced scale elements, by means of restrictions, fully or partially, imposed their surfaces/limits.

Four confinement levels were proposed for these tests. The levels start with the full confinement and progresses to the partial confinement, without confinement and finally for the reinforcing bars anchor length capacity assessment, which will be explained in detail. These tests aimed to simulate, or reproduce, the hogging moment region, where cracks occur due to tension in the composite beam concrete slab. In this way, it becomes evident

Table 2.10 Combination Matrix With Variables and Levels

	Reinforcing Bar			Concrete	Connector		
Comb.	**Length (mm)**	**Number**	**Diameter (mm)**	**F_{ck} (MPa)**	**Sp. 1 (mm)**	**Sp. 2 (mm)**	**Number**
1	650	2	16	30	200	250	2
2	500	4	20	30	200	100	3
3	450	4	20	20	100	250	2
4	300	2	16	20	100	100	3
5	650	4	16	20	200	250	3
6	500	2	20	20	200	100	2
7	450	2	20	30	100	250	3
8	300	4	16	30	100	100	2

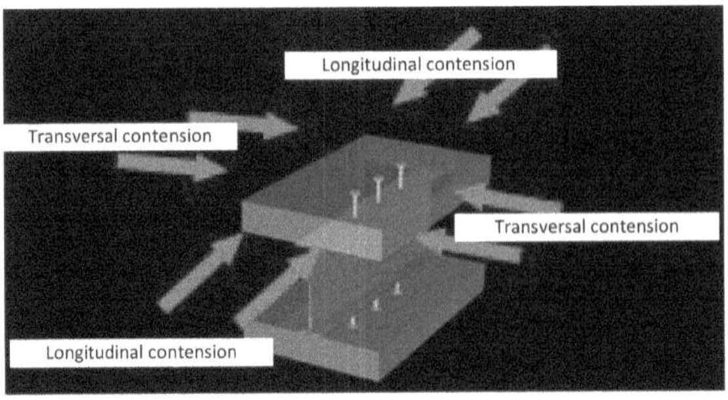

Fig. 2.156 Systems and types of confinement.

that the push-out test can inspire the newly proposed pull-out test, or a pull-out test variation using the reinforced bars present at the concrete slab under tension forces. The pull-out models subdivided by confinement modes will be discussed and presented according to the containment level (Fig. 2.156). The containment or confinement modes of concrete can be divided into two categories, in the lateral and longitudinal directions or by a combination of both directions.

Figs 2.157–2.160 visually present the confinement or containment, which is symbolised in Fig. 2.156, by arrows and texts. This representation is made through green plates, which represent the way to block the concrete expansion, searching through their combinations, and a closer representation of the continuous slab behaviour. The first test proposes the full confinement, where there is a restraint or confinement of the concrete slab in the acting tension force direction and transversely, as in an actual concrete slab, this confinement effect is much larger than the 1% in the proposed model (Fig. 2.157).

The full or complete confinement, made by the green-coloured plates around the concrete slabs, shown in Fig. 2.157, seeks to represent the continuity of the slab, which simulates the existence of a slab of width and length, in order to represent the continuity in all directions, situation found in actual structures condition. Fig. 2.158 focuses on the slab continuity in the element longitudinal direction (i.e. in the acting tension force direction) and, unlike the previous model, will not have the transverse or lateral containment, a generating partial confinement.

In the test without confinement, the model considers that the slab continuity does not influence this experimental test structural response (Fig. 2.159). Therefore, the model is opened in all lateral and longitudinal directions, free of any kind of confinement or restraint.

The following confinement model starts from the principle of non-confinement as described in the previous model. However, the test aims

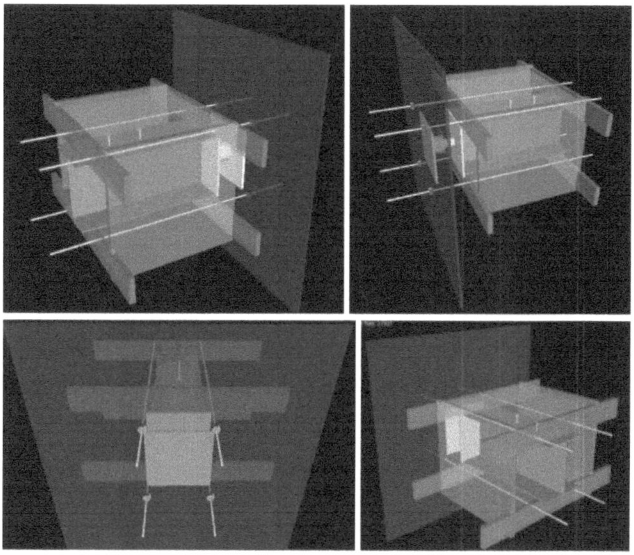

Fig. 2.157 Pull-out model: full or complete confinement.

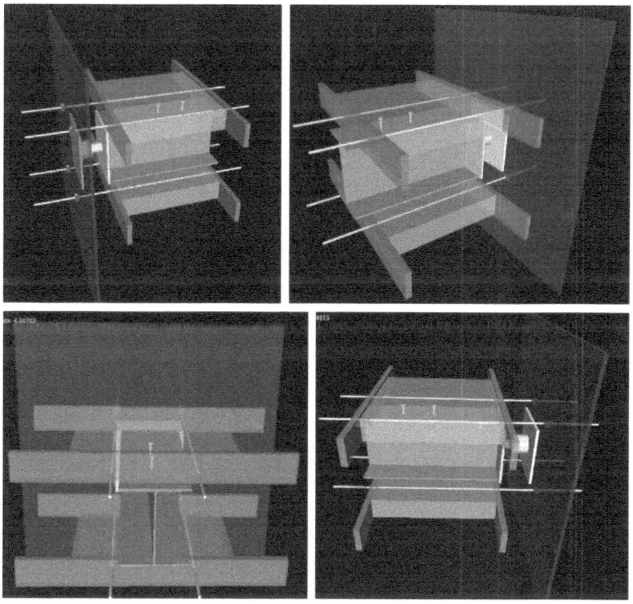

Fig. 2.158 Pull-out model: partial confinement.

to calibrate the formulation for the reinforcing bars anchor resistance without imposing loads or stresses to the shear connectors. Fig. 2.160 shows the reinforcing bars anchor test, where the load application is made through the hydraulic jack acting on the full extension of the steel profile extremity as

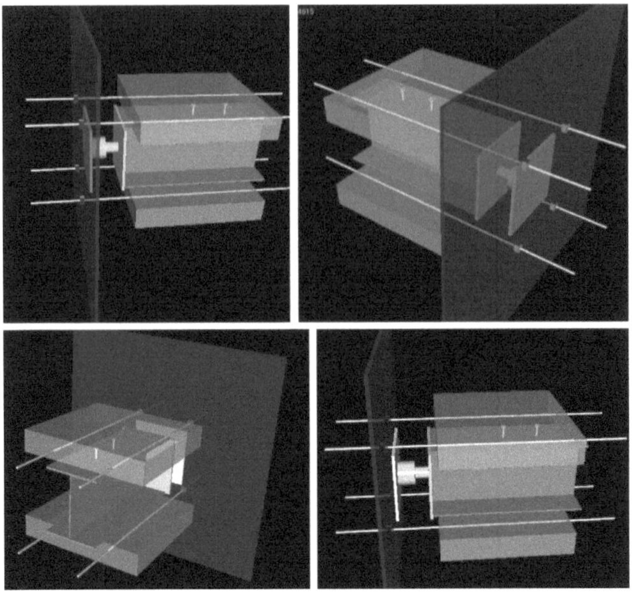

Fig. 2.159 Pull-out model: without confinement.

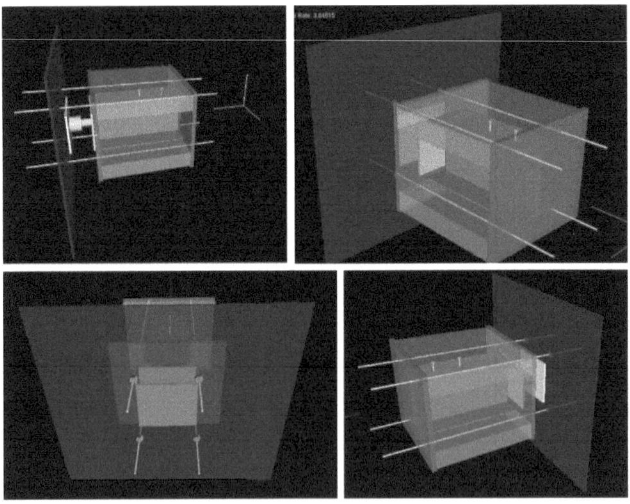

Fig. 2.160 Pull-out model: without confinement and anchoring.

well as the concrete slab, which are pushed, sand by this creating the reinforcing bar anchor and the shear connector resistance.

The last model configuration is intended to assess the concrete component confinement aiming to analyse their contribution within this set of components. Focusing on the contribution of the reinforcing bars anchor under tension forces along with the shear connector, their resistance to bending, shear, and tension, characterised by the point where the slip of the reinforcing bars occur, can be evaluated.

The preliminary tests intended to test and define the main test programme and also contributed to the definition of the quantity and quality of the results, PO 0.1 and PO 0.2. The preliminary tests used steel forms (Fig. 2.161) for a later reuse and assembling agility in the next planned test series.

A metal beam system was created to anchor the pull-out test to the reaction slab, so that the reinforcing bars were under tension generated by the load application by means of a hydraulic jack (Fig. 2.162).

Fig. 2.161 Steel formwork.

Fig. 2.162 Steel beams bracing system.

The initial test, PO 0.1, aimed to obtain the first data response of the structure, seeking to calibrate the applied load values and the reinforcing bars collapse mode. Its main objective was to prepare a test structure that sustained the applied loads order to produce reliable results mode, without compromising safety. In the second test, PO 0.2, no problems have occurred in the bracing structure; therefore, all the necessary procedures were carried out to start the test and finalise it with its collapse now associated with a different collapse mode. Initially, the predicted mode was a reinforcing bar tension rupture, but in this test, the collapse was associated with the shear connector failure.

The concrete was produced in the laboratory, and its composition was generated from an experimental dosing study. The reinforcing bars were divided into three different types, each with its respective function. The first reinforcing bars type had the function of measuring the concrete deformation by using strain gauges located along its length at positions close to the shear connectors line.

At this point, the hypothesis that this region, which will involve the concrete slab cracks lines, would have a deformation, in these 5 mm diameter bar was applied, similar to the concrete deformation on its surrounding. The second type of reinforcing bars was characterised by the stirrups, which followed the recommendations of EC4 [64]. These bars had the objective of controlling the slab cracking and were arranged in the longitudinal and transverse directions distributed along the slab length. The last type of reinforcing bars had a diameter of 16 mm. They were fixed in the other extremity by steel plates, through a cut on them, where the reinforcing bars were positioned and later welded to the plate. The reinforcing bars were positioned next to the point corresponding to one-third of the slab width. A pair of strain gauges was located at each point, where the measurement of these strains was required. Another pair of strain gauges was positioned on each bar, outside the concrete slab. The remaining strain gauges were installed in the vicinity of the shear connector's intervals (Fig. 2.163).

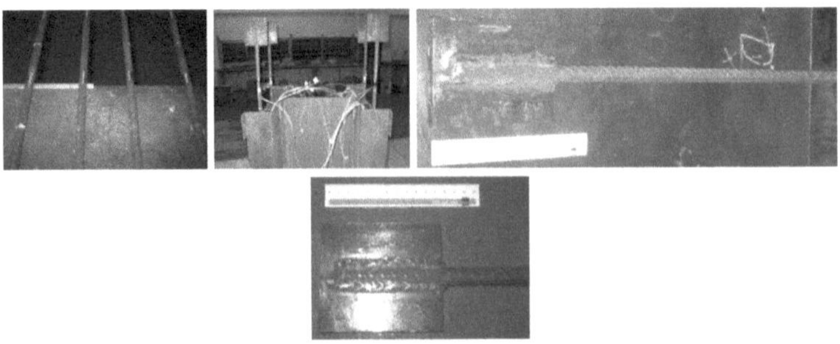

Fig. 2.163 Main reinforcing bars.

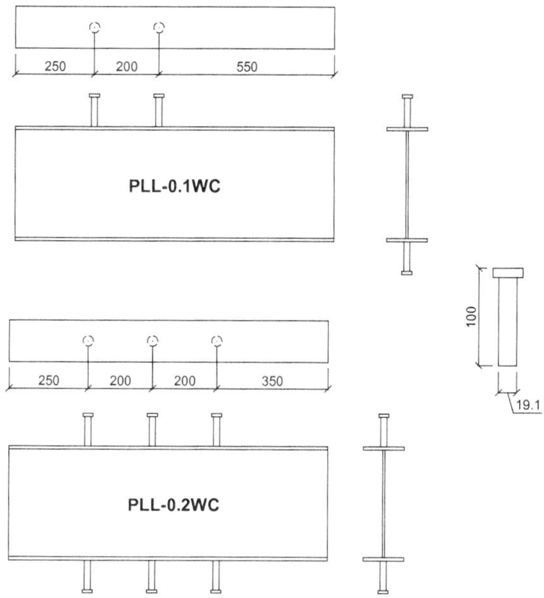

Fig. 2.164 Shear connectors' layout.

The beam of this test was constituted by an ASTM A-572 W410 × 46.1 produced by AÇOMINAS, with dimensions 410 × 140 × 6.7 × 11.2, that is, height, flange width, web, and flange thickness, respectively (Fig. 2.164). The profile length was of 1000 mm. This test also used composite structure longitudinal bracing profiles made of ASTM A-36 6 rolled U 6″ × 12.5 profiles, for the first bracing level. For the transverse bracing, also called second-level bracing, profile ASTM A-36 I3″ × 10 was utilised.

The shear connectors welded to the steel profile's flange had the function of preventing the slip of the steel to concrete slab was three connectors in the PO 0.1 pull-out test, and two in the PO 0.2 pull-out test with a 19.1 mm diameter and a length of 120 mm. Both preliminary tests used the same ASTM A-572 Gr. 50 W 410 × 46.1 steel profile.

The preliminary test PO 0.1 used three shear connectors, positioned at the profile flange width centre and was located 250 mm from the slab up to the first connector. The adopted distance between connectors was 200 mm. The preliminary test PO 0.2, used two shear connectors welded in each of the profile flanges, 250 mm spaced from the slab end and 200 mm between them.

The concrete slab was made up of two equal 120 × 600 × 100 mm parts, where each part was connected to the steel beam flanges. The tests main reinforcing bars used the same 650 mm length and 1050 mm total anchorage length (Fig. 2.165), where each slab contains two bars. The stirrups

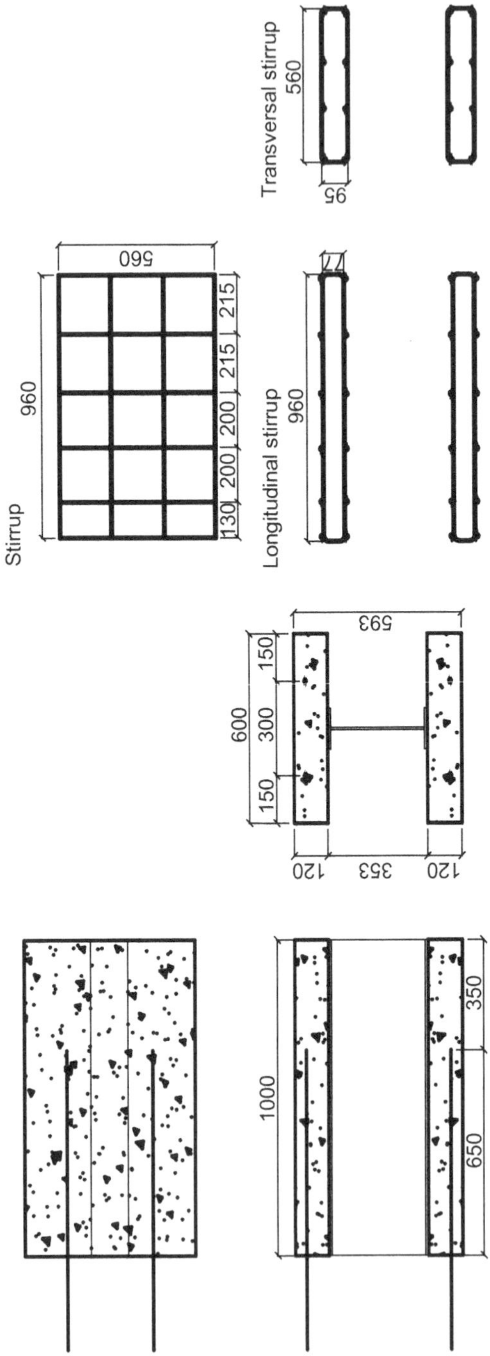

Fig. 2.165 Reinforcing bars.

reinforcing bars were made of longitudinal and transverse reinforcing bars. The 10-mm diameter stirrups adopted at the slab larger longitudinal dimension parallel to the studs were 960 mm long. The stirrups were proportionately distributed in each slab. The 10 mm diameter transverse reinforcing bars were 560 mm long, totalling six stirrups in each of the slabs.

A series of strain gauges, two at each point, measured strains without the concrete influence, whilst the other two pairs were used inside the concrete slab measuring the strains associated with the steel to concrete set under tension. The strain gauges' pair was positioned right next to the concrete slab, assuming this point to be under the bar maximum tension forces. The internal pairs were positioned in such a way as to obtain the behaviour in bars located in the region between shear connectors (Fig. 2.166).

A pressure transducer was necessary to enable the load cell centralisation, which in the initial tests slipped as the load application reached a considerable force magnitude. Fig. 2.167 shows the location of each group of LVDTs

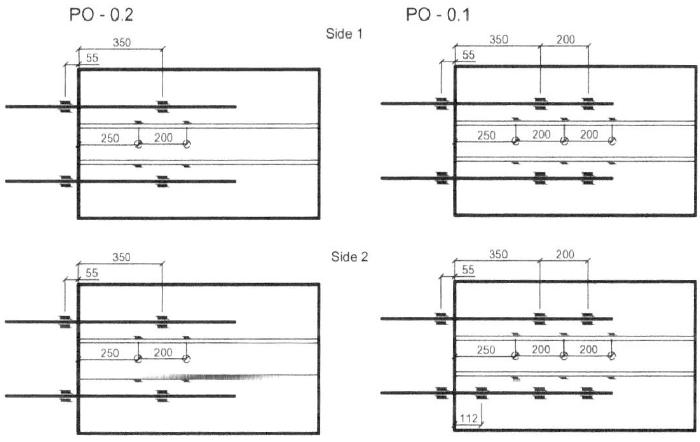

Fig. 2.166 Reinforcing bars strain gauges distribution.

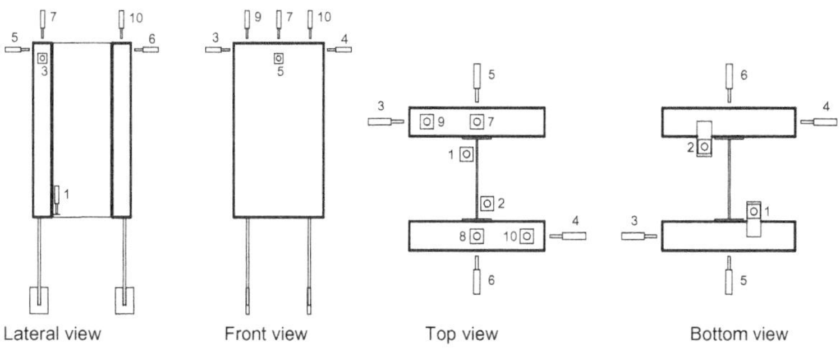

Fig. 2.167 LVDTs distribution and layout.

positioned at the test bottom part that measured the steel profile to concrete slabs relative displacements. Around all the horizontal perimeter of the pull-out tests LVDTs measured the lateral displacements to identify any structure imbalance in the horizontal plane. The upper LVDTs measured the pull-out vertical deflections, when the number 7 and 8 measured the steel profile to concrete slabs relative displacements. The LVDTs 1 and 2, at the other end of the pull-out, were adopted to verify if a differentiated behaviour between the two ends occurred. Finally, LVDTs 9 and 10 that acquired the test displacement in relation to the reaction floor level were placed at concrete slabs edges to also check for any structure imbalance.

The pull-out PO 0.1 collapse mode was the reinforcing bar tension rupture at 402.0 kN, being followed by their anchor rupture at 569.4 kN, not exceeding the jack maximum load capacity (i.e. 1000 kN). The PO 0.1 pull-out test started with a series of preliminary tests that mainly served to obtain initial results. These tests were also used to evaluate the efficiecy of the modified test setup now containing an efficient bracing system with the required strength to enable to execution of the planned experimental programme.

The first preliminary test involved the application of a preload on structure, seeking to reach around 20% of its collapse load, which occurred without any problem. On the next day, the definitive test of PO 0.1 was performed, following the recommendations of EC4 [64] for push out tests. First, the load and unload cycles are made, consisting of the application of a load of 20% of the collapse load 25 times. After this phase, the load was applied at a moderate speed until the pull-out test collapse. Unfortunately, the hydraulic jack moved as soon as the load reached around 30% of its planned load, interrupting the test. The jack displacement occurred due to its hinged head. Two other attempts were made, and finally, the removal of this hinge was concluded, Fig. 2.168, fixing the jack course for safety reasons. A main reinforcing bar rupture with an applied load of 513 kN was the ultimate limit state of this test.

The PO 0.2 test occurred as predicted, from the beginning, with the application of the loading and unloading cycles followed by the final test up to the

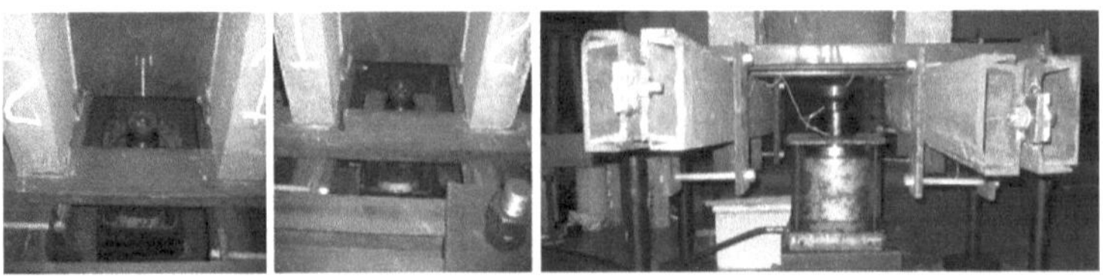

Fig. 2.168 Hydraulic jack without its hinged head.

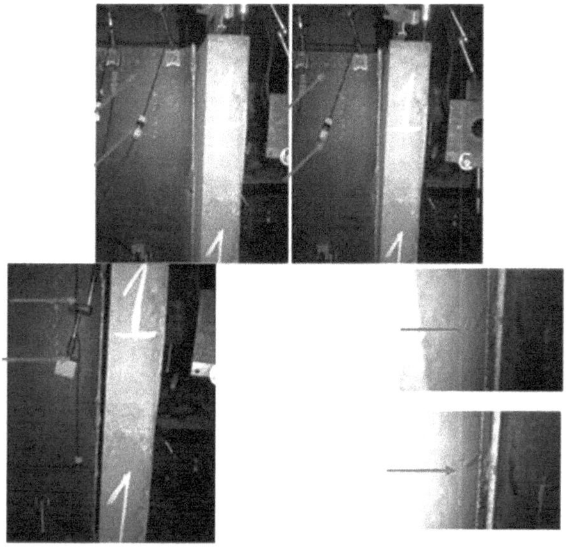

Fig. 2.169 PO 02 pull-out test rupture.

collapse. The test PO 0.2 was different from the PO 0.1 since only two shear connectors were used in each beam flange whilst the PO 0.1 had three shear connectors at each flange. The rupture prediction was also equal to the PO 0.1, a reinforcing bar tensile rupture at an applied force of 402 kN; however, rupture occurred in the shear connectors with a load of 473 kN (Fig. 2.169).

The definitive tests performed after the preliminary ones had identical dimensions of this latter test. The differences between these tests and the previous ones were in their forms, since they were made with plywood sheets, and the steel rods for measuring equivalent strains were not employed. The definitive tests also employed metallic compact steel profile, but kept the same height. In this series 4 different tests, PO1X, PO2X, PO7X and PO8X, were carried out, with a specific rupture mode where each series had three repetitions, totalling 12 tests.

In the definitive trials, the profiles between the concrete slabs have been modified, but remained with the same height by using a W410 × 60 along with the main reinforcing bars and bracing plates welded at their ends. The structure that composes the bracing system was kept, a new force distribution plate applied by jack was used. The concrete used was industrialised, due to the large amount of specimens to be concreted.

The required concrete f_{ck} was 30 MPa, which in their confirming tests that were carried out on the same days of the main pull-out tests, obtained average compression strength of 28 MPa. The reinforcing bars used in these tests were

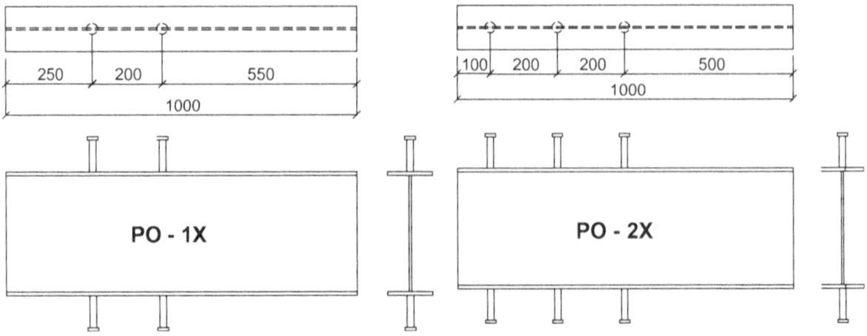

Fig. 2.170 Connectors distribution: PO-11, PO-12, PO-13, PO-21, PO-22, & PO-23.

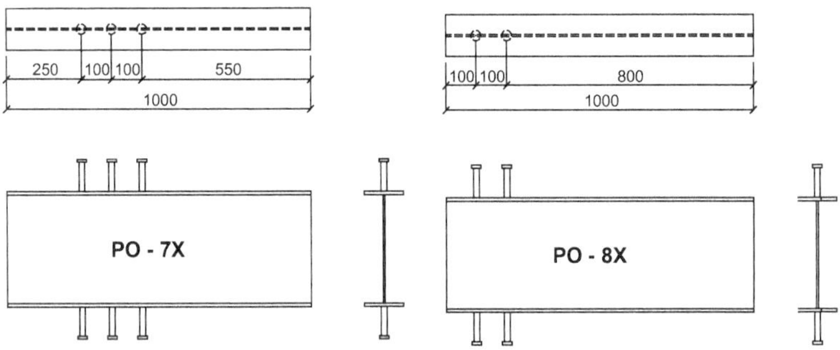

Fig. 2.171 Connectors distribution: PO-71, PO-72, PO-73, PO-81, PO-82, and PO-83.

the same of the preliminary test, only varying their anchor lengths, according to each model and collapse modes. Shear connectors were the same used in the preliminary tests, but their weld was performed with a specific equipment to weld shear connectors. The 12 tests used the same ASTM A-572 Gr. 50 W410 × 60 steel profile produced by AÇOMINAS. The used bracing system remained the same of preliminary tests. The connector's distribution varied according to each series, where four different distributions were used (Figs 2.170 and 2.171). The concrete slab kept the same dimensions; however, the composite geometry has changed due to the increased height of the steel girder (Fig. 2.172). In each series of tests the length of the main reinforcing bars varied (Fig. 2.172) to obtain an individual collapse mode.

The PO-1X test series had two 16-mm diameter main reinforcing bars with 1150 mm, in which 650 mm were embedded in concrete. The PO-2X test series had four 20-mm main reinforcing bars with 1000 mm, in which 5000 mm were embedded in concrete. The PO-7X test series had two 20-mm diameter main reinforcing bars with 950 mm, in which 450 mm were embedded in concrete. The PO-8X test series had two 16-mm diameter main reinforcing bars with 800 mm, in which 300 mm were embedded in concrete.

The instrumentation used for measuring strains and displacements was made of strain gauges and LVDTs similar to the preliminary tests, but with

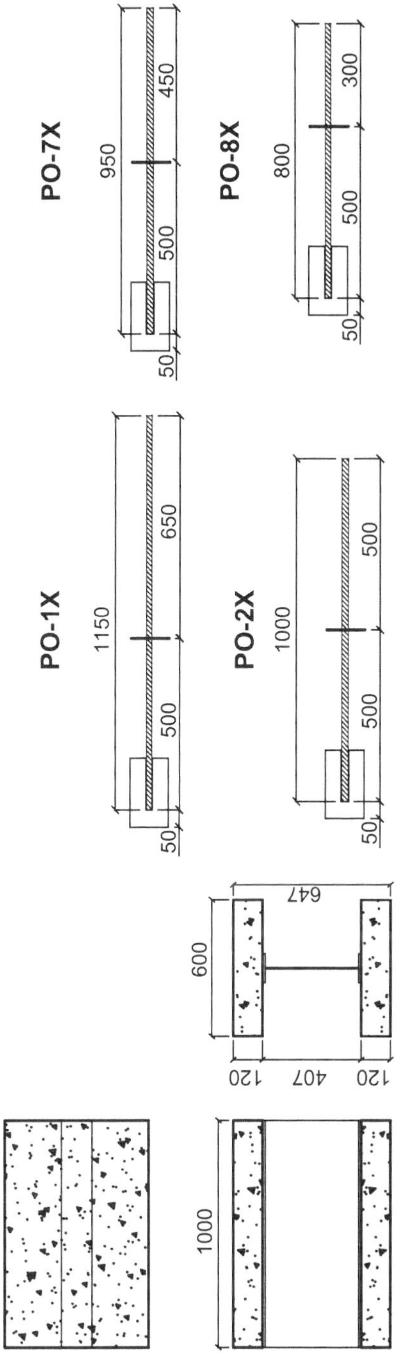

Fig. 2.172 Composite section & adopted reinforcing bars, 2nd stage tests.

some different locations between each of the test series. For each test series different locations for the strain gauges according to the distribution of the shear connectors were used.

In the test series PO-1X and PO-8X (Fig. 2.173), the use of four strain gauges per bar can be observed, with one pair at each point measuring per bar, and a couple was also positioned in the external and other at the internal parts, more precisely between the two lines of shear connectors. In the test series PO-2X and PO-7X (Fig. 2.174), the use of four strain gauges per bar is noted, with one pair at each point measuring per bar, and a couple was also positioned in the external and other at the internal parts in the central connector line. The positions of the LVDTs were distributed in a more simplified manner than in the preliminary test, in order to make better use of the results (Fig. 2.175).

The collapse load of PO1Xs was 402.12 kN, associated with the tensile rupture of the reinforcing bar. However, the collapse of tests PO-13 and PO-13 occurred differently from what was theoretically predicted. A modification was made in the structure by ensuring its capacity, more than a load

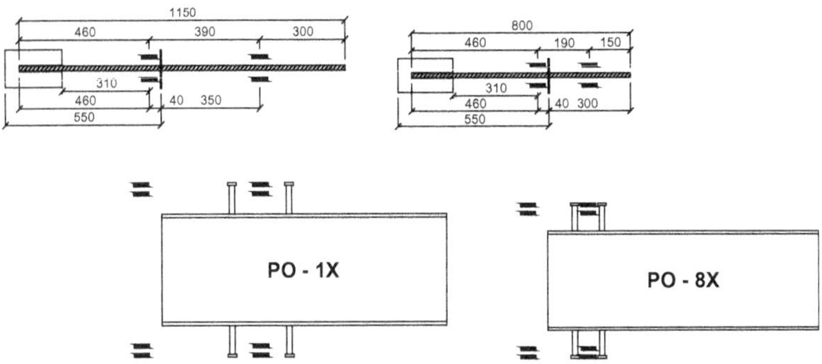

Fig. 2.173 Strain gauges layout: test series PO 1X and PO 8X.

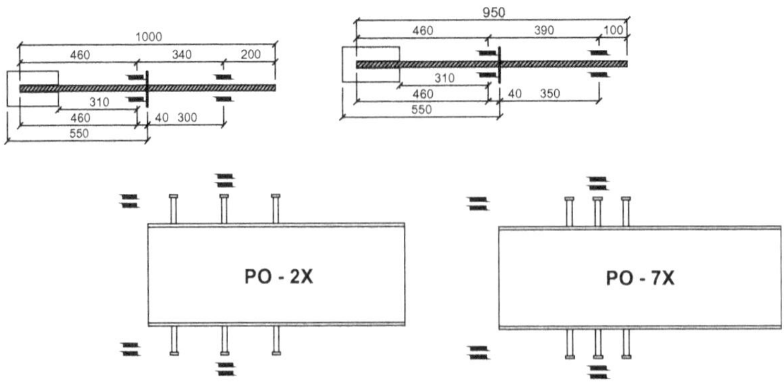

Fig. 2.174 Strain gauges layout: test series PO 2X and PO 7X.

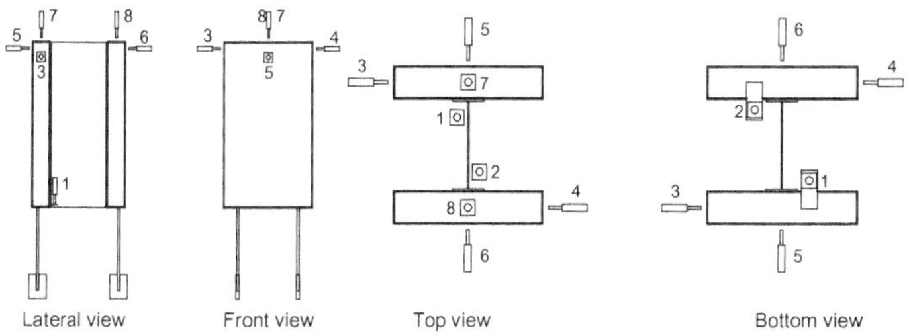

Fig. 2.175 LVDTs location's layout.

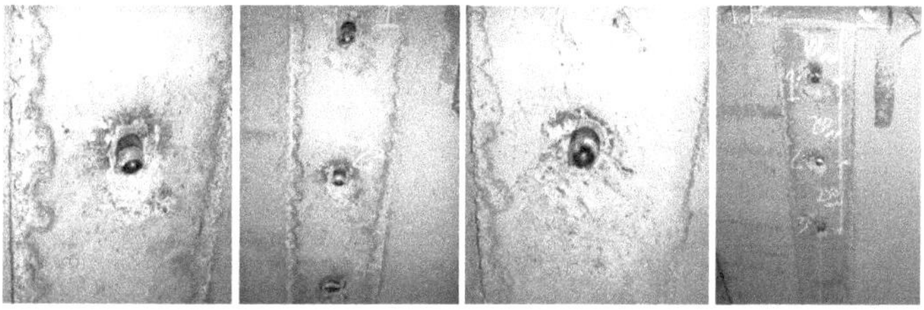

Fig. 2.176 Connectors' weld rupture details.

value higher than 516.21 kN, so that PO-13 test could reach its expected rupture. The problems were related to the tested elements manufacture, or more specifically, to the weld between the shear connector and the steel beam. This rupture mode was not foreseen in the tests, because it was considered that this weld would provide the required shear connectors strength for all the test imposed loads. Fig. 2.176 shows in detail the shear connector base, virtually intact, confirming the hypothesis of low amperage on the welding equipment, generating an insufficient fusion of the materials, and reducing the welding resistance to values smaller than the strength of the other test elements.

In Fig. 2.177, we can observe the modifications carried out to prevent the connectors weld rupture. Steel profiles were used between the steel profile and the reinforcing bars, braced to the concrete slabs and steel profile, completely diverting the load to the reinforcing bars, thus avoiding the connectors to be mobilised and their unexpected rupture.

The PO-13 test succeeded in determining the collapse mode (Fig. 2.178), due to the modification of its structure, presenting a rupture with 402.12 kN load. The collapse occurred in the reinforcing bar rupture with an ultimate load of 516.21 kN. In the series of tests PO-2X, tests PO-21 and PO-23 were again associated with a fragile rupture of the connectors weld. In PO-22 test, the local buckling of the steel profile web happened (Fig. 2.179) due to the change in the load distribution plate, which was located between the jack and the transition plate used to transfer the jack load to the specimens reaching a load of 522.02 kN. In the series of tests PO-7X, PO-71 and PO-73, the

Fig. 2.177 Test structure modification: test PO-13.

Fig. 2.178 Reinforcing bar rupture: test PO-13.

weld rupture occurred once again. In PO-72 test, reinforcing bars pull out was noted (Fig. 2.180) due to the rupture of the anchor between the bar and the concrete, collapse mode predicted, reaching an ultimate load of 448.56 kN, whilst the estimated collapse load was 480.95 kN.

In the PO-8X tests series, PO-81 had the collapse associated with the already mentioned connectors weld rupture. Tests PO-82 and PO-83 have been modified so that the rupture of the shear connector weld would not occur. These made the rupture be associate to a reinforcing bars pull, again due to rupture of the anchor between the bar and the concrete, reaching an ultimate load of 474.28 kN for the PO-82, 445.91 kN for the PO-83, whilst the estimated collapse load was 475.63 kN.

Fig. 2.179 Profile web collapse: test PO-22.

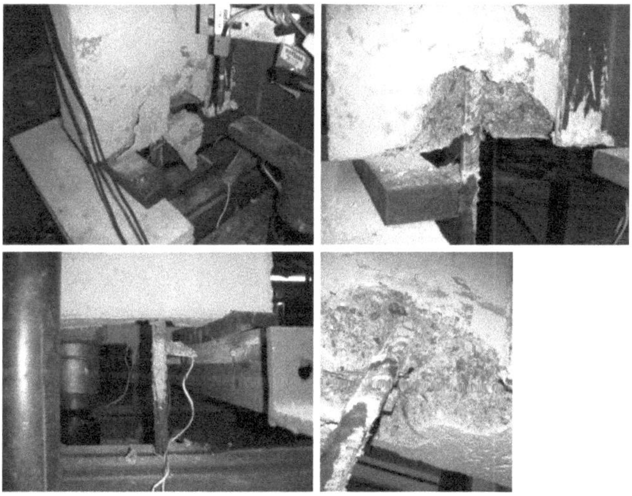

Fig. 2.180 Reinforcing bar anchorage rupture: test PO-72.

The test PO-11 used two main 16-mm reinforcing bars with a 650-mm anchor length on both sides of the slab, with two shear connectors welded in each one of the beam flanges. The collapse occurred with the rupture of the shear connectors weld at slab side I. The strain gauges of the group's g_3 and g_4 exhibit this rupture behaviour, at a load equal to 319 kN (Fig. 2.181). This figure shows their load versus strain curves for the vertical bar number 1, strain gauge group g_3, which is immersed in the concrete slab.

It is common to observe the development of applied load and its behavioural stages, where clear evidence of the steel yield, and the beginning the

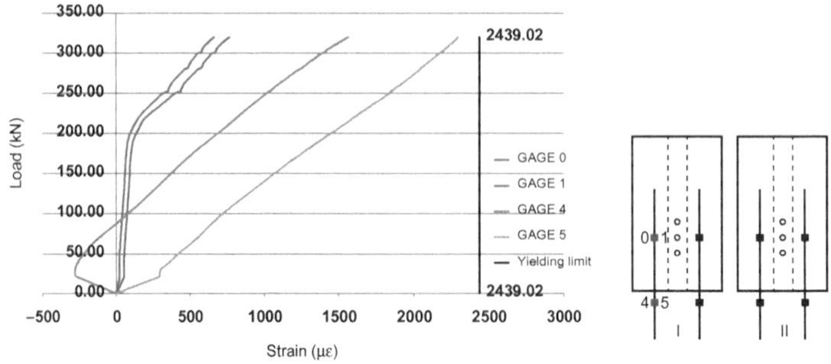

Fig. 2.181 Load versus strain curves: PO-11, group g_3.

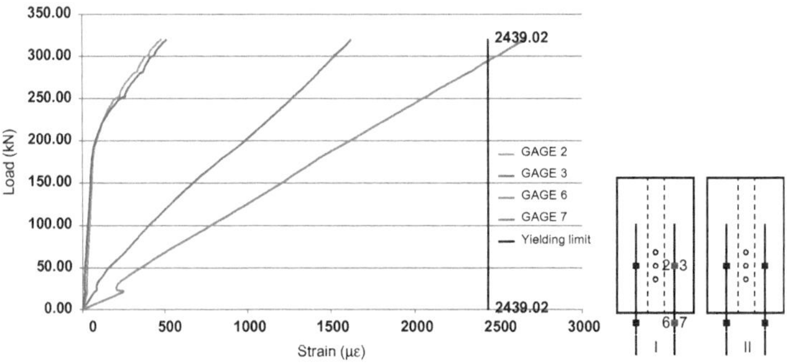

Fig. 2.182 Load versus strain curves: PO-11, group g_4.

rupture can be identified. In Fig. 2.181, it can be noticed that the strain gauges 0 and 1 indicated the start of the yield onset in the 200–250 kN load range, where an increase in strains occurs. The curves visibly show the characteristics in Fig. 2.182, where the tension responses of the steel bars inside and outside the concrete slab are compared. Fig. 2.182 shows the load versus strain curves for the strain gauge group g_4.

Fig. 2.183 shows the load versus displacement curves of LVDTs 1 and 2 located at the bottom that measured the relative displacement between the steel profile and the base of the concrete slab. LVDTs 7 and 8 present the pull-out test vertical displacements measured at the upper portion of the concrete slabs. The behaviour of the LVDT 8 clearly demonstrates the side where the collapse occurred, due to their displacement which is relatively larger than the equivalent LVDT 7. The behaviour of the LVDTs combines the results of the vertical displacements readings measured at the of concrete slab bottom by LVDTs 1 and 2, to the top by LVDTs 7 and 8.

The test PO-12 used two 16-mm main reinforcing bars with a 650-mm anchor length on both sides of the slab, with two shear connectors welded in each one of the beam flanges. The collapse occurred with the rupture of the

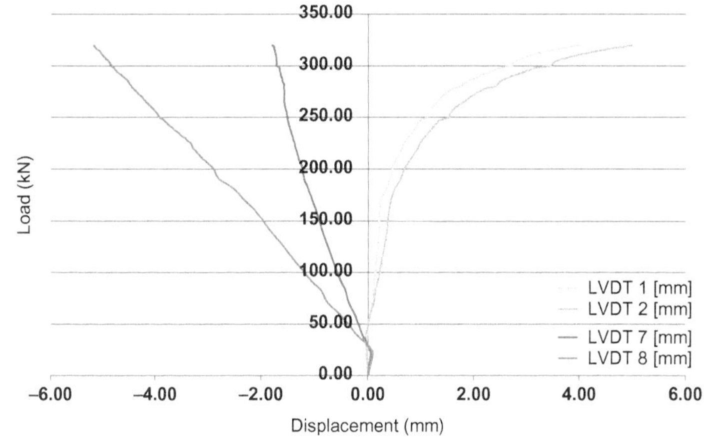

Fig. 2.183 Load versus displacement curves: PO-11, LVDTs 1, 2, 7, and 8.

shear connector's weld, more precisely on the slab side II with a load equal to 384 kN. The PO-13 test with two 16-mm main reinforcing bars with a 650-mm anchor length on both sides of the slab with two shear connectors welded in each one of the beams. Slab sides I and II contained the strain gauge groups g_3 and g_4, and g_5 and g_6,

In order to avoid the constant failure on the shear connector weld in this test, a structural alternative to achieve new results and determine other possible rupture modes was envisaged. These modifications have been discussed previously, and modified the load application point through the use of steel profiles supported on the concrete slabs, avoiding the rupture of the shear connectors' welds. The collapse occurred with the tensile rupture of the bar number 4, more precisely on the slab side II. The groups g1 and g_5 present the collapse behaviour of this bar at 516 kN load magnitude (Fig. 2.184).

Fig. 2.185 shows the load curves versus displacement of bottom LVDTs 1 and 2 indicating that a relative slip did not occurred between the concrete slab and the steel profile. This implies that only the reinforcing bar anchor

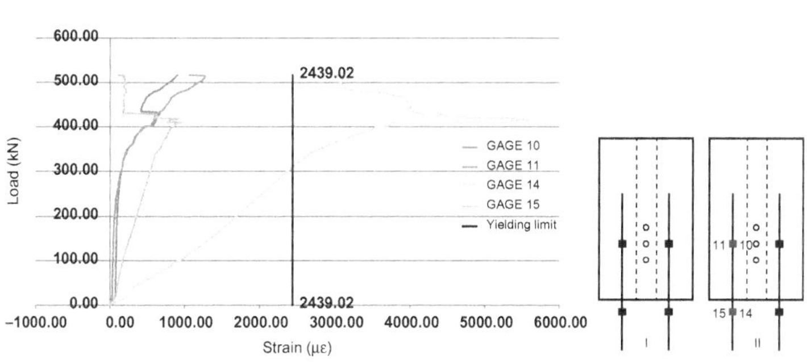

Fig. 2.184 Load versus strain curves: PO-13, group g_5.

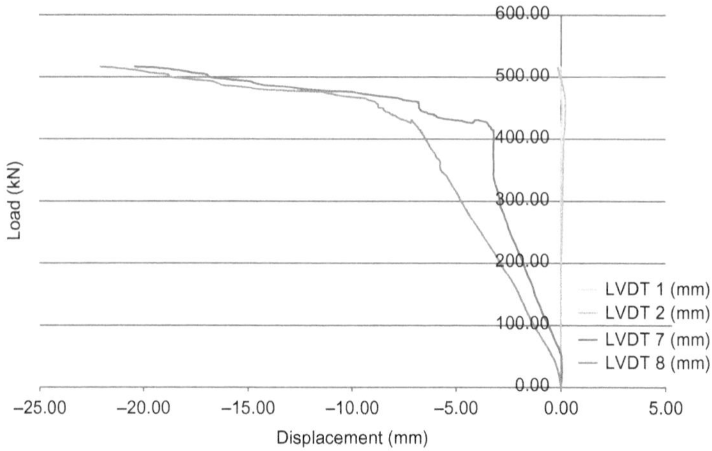

Fig. 2.185 Load versus displacement curves: PO-13, LVDTs 1, 2, 7, and 8.

and the tensile strengths supported the applied load. The collapse was forecasted, and occurred associated with reinforcing bars tension capacity. This figure also depicts LVDTs 7 and 8 load versus displacement curves associated with the pull-out vertical displacements measured at the top of the concrete slabs where it can be noted that after reaching 400 kN, the curves presented the yielding plateau.

The PO-21 test used four 20-mm main reinforcing bars with a 500-mm anchor length, and three shear connectors on each concrete slab. The collapse was again associated with the rupture of one of the shear connector's welds on slab side I, at a load magnitude equal to 437.90 kN. The PO-22 test adopted four 20-mm main reinforcing bars with an anchor length of 500 mm and three shear connectors on each concrete slab. Slab sides I and II were represented by strain gauge groups g_3, g_4, g_5, and g_6 and g_7, g_8, g_9, and g_{10}, respectively. The collapse was due to the steel profile local web buckling at a load magnitude equal to 522.02 kN. The PO-23 test utilised four 20-mm main reinforcing bars with a 500-mm anchor length and three shear connectors on each concrete slab. The collapse was due to the rupture of one of the shear connector's welds on slab side II, at a load magnitude equal to 413.00 kN.

The test PO-71 used two 20-mm main reinforcing bars with a 450-mm anchor length and three shear connectors on each concrete slab. The collapse occurred with the rupture of one of the shear connector's welds on slab side II, at a load magnitude equal to 415.00 kN. The test PO-72 utilised two 20-mm main reinforcing bars with a 450-mm anchor length and three shear connectors on each concrete slab. Slab sides I and II were represented by strain gauge groups g_3, g_4, g_5, and g_6 and g_7, g_8, g_9, and g_{10}, respectively.

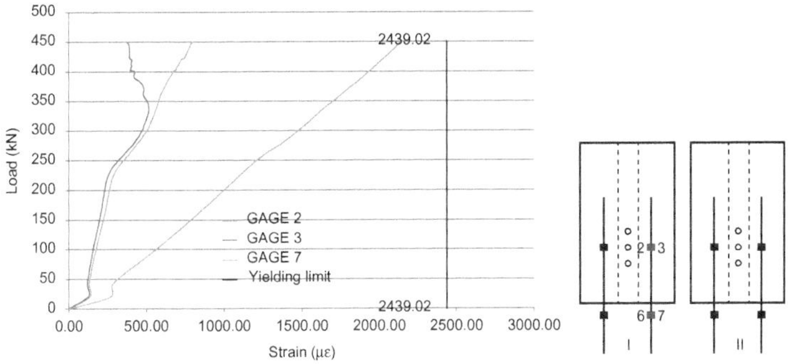

Fig. 2.186 Load versus strain curves: PO-72, group g_4.

The collapse occurred, as predicted, with the rupture of the anchor of the second reinforcing bar of slab side I. Strain gauge group g_4 presents the behaviour of this bar with an anchoring rupture at a load equal to 448.56 kN (Fig. 2.186). The PO-73 test used two 20-mm main reinforcing bars with a 450-mm anchor length and three shear connectors on each concrete slab. The collapse occurred with the rupture of the shear connector's weld on the slab side II with a load magnitude equal to 435 kN.

The test PO-81 adopted four 16-mm main reinforcing bars with a 300-mm anchor length and three shear connectors on each concrete slab. The collapse occurred with the rupture of one of the shear connectors on the slab side II with a load magnitude equal to 268.34 kN. The PO-82 test used four 16-mm main reinforcing bars with a 300-mm anchor length and three shear connectors on each concrete slab. A structural alternative was adopted to mobilise other possible rupture modes and obtain a better use of this test, avoiding another failure in the shear connector weld. The collapse occurred with the rupture of the anchor of one of the reinforcing bars on slab I instrumented by the strain gauge group, g_3, g_4, g_5, and g_6 with a load magnitude equal to 474.28 kN. Figs 2.187 and 2.188 show the load versus strain curves for strain gauge groups g_3, g_4, g_5, and g_6 located on bars immersed in the concrete slab I.

The test PO-83 adopted four 16-mm main reinforcing bars with a 300-mm anchor length and three shear connectors on each concrete slab. A structural alternative was again adopted to mobilise other possible rupture modes and obtain a better use of this test, avoiding another failure in the shear connector weld. The collapse occurred with the rupture of the anchor of one of the reinforcing bars on slab II with a load magnitude equal to 445.91 kN.

The test PO-0.1 used two 16-mm main reinforcing with a 650-mm anchor length and three shear connectors on each concrete slab. The collapse occurred, as predicted, with the rupture of the anchor of the first reinforcing

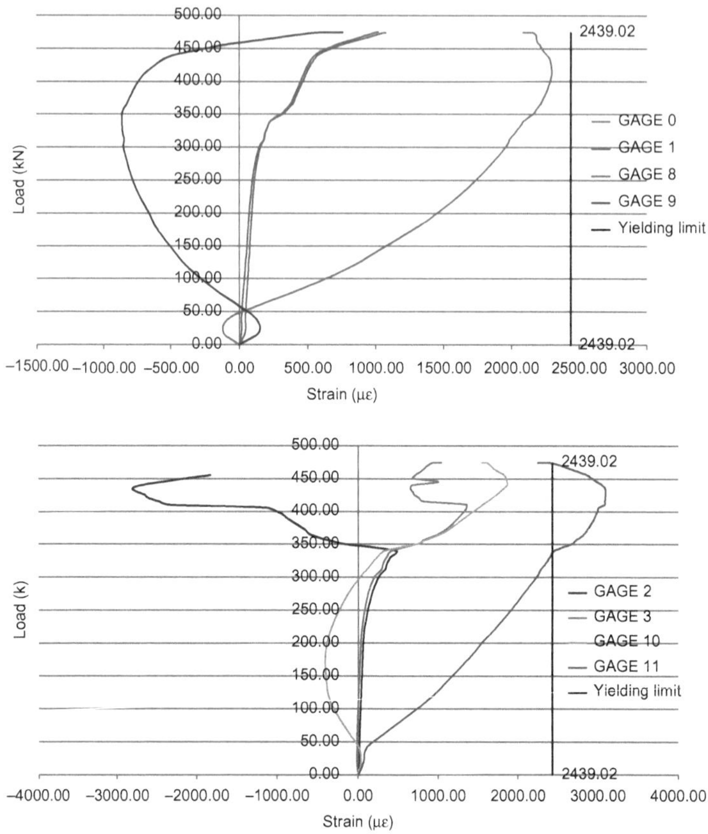

Fig. 2.187 Load versus strain curves: PO-82, group g_3 and g_4.

bar with a load magnitude equal to 530 kN. Fig. 2.189 shows the load versus strain curves associated with strain gauge group g_{2B}, that lies outside the concrete slab side II or b. It can be seen that the strain gauges 35 and 36 (Fig. 2.189) exceed the yield strength. Fig. 2.189 shows the load versus strain curves of strain gauge group g_3 where the response of the first bar immersed in slab side I, where the rupture occurred, is identified.

Fig. 2.190 shows the load versus displacement curves of bottom LVDTs 1 and 2 that measured the relative displacement between the steel profile and the concrete slab. The LVDT 1 depicted a horizontal plateau where the deflections increased for a constant load indicating the eminence of rupture. LVDTs 8, 9, and 10 present the vertical displacements of the pull-out tests, measured in the concrete slabs top.

The test PO-0.2 used two 16-mm main reinforcing bars with a 650-mm anchor length and three shear connectors on each concrete slab. The collapse

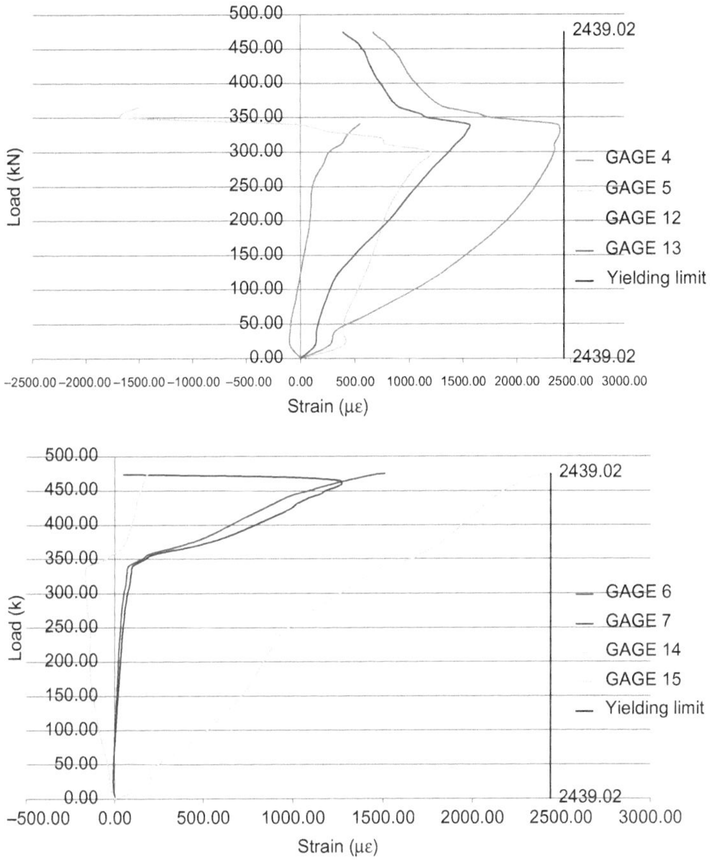

Fig. 2.188 Load versus strain curves: PO-82, group g_5 and g_6.

occurred, different from the predicted collapse mode, with a shear connector rupture on slab side II, with a load magnitude equal to 385 kN. In the curves shown in Fig. 2.191 it is interesting be observe the load versus strain behaviour of strain gauge group g_2, which is located outside the concrete slab.

Fig. 2.192 shows the bottom LVDTs 1 and 2 load versus displacement curves that measured the relative displacement between the steel profile and the concrete slab. Fig. 2.192 also shows the LVDTs 8, 9, and 10 load versus displacement curves corresponding to the lateral displacement of the pull-out tests, in the concrete slabs transverse direction, and the LVDTs 3, 6, and 7 curves associate to the vertical displacements of the pull-outs, measured at the concrete slabs top.

The development of a novel modified pull-out test was successfully carried out despite the unexpected failures of the weld between the shear connectors

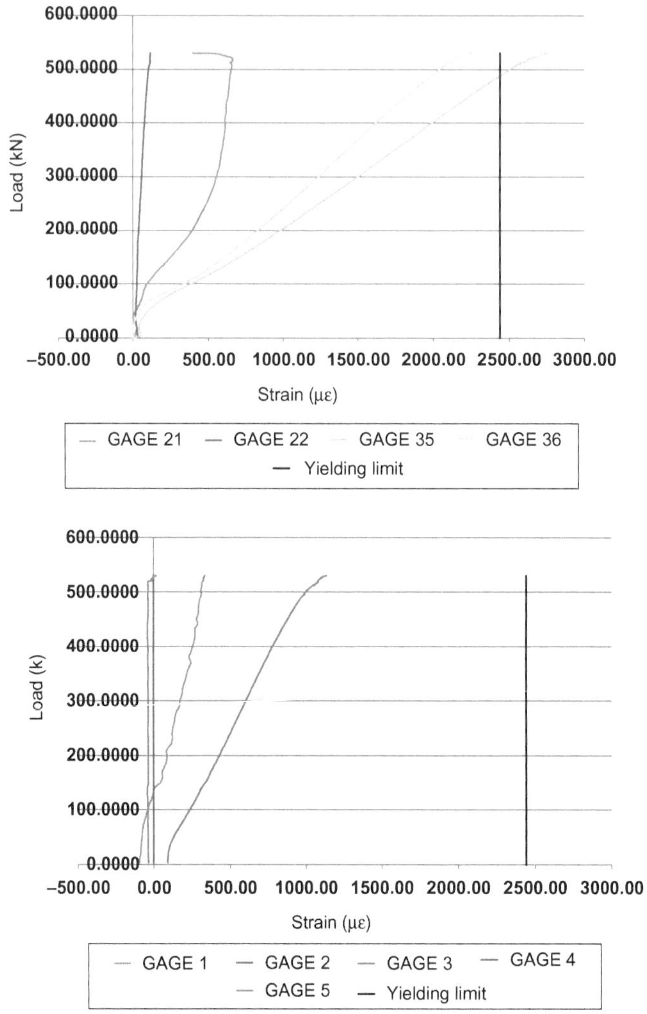

Fig. 2.189 Load versus strain curves: groups g_{2B} and g_3.

and the beam flanges. The conception of a bracing system for this pull-out test was an arduous task, due to various limiting factors that occurred throughout the development of the experimental programme. Despite all odds the result was an efficient bracing system that enabled, after numerous attempts and several tests, to achieve force magnitudes greater than 500 kN. The assembly and dismantling system became very simple, and reduce the time required to complete the tests. All these aspect contributed to the satisfactory result of the experimental programme where all tests were carried out successfully.

The preliminary test PO-0.1 succeeded in determining the collapse mode, i.e., tensile rupture of the reinforcing bars. However, the initially predicted

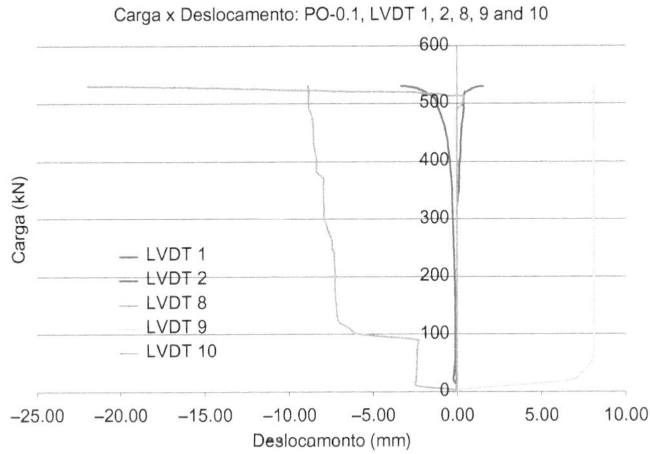

Fig. 2.190 Load versus displacement curves: LVDTs 1, 2, 7, 8, 9, and 10.

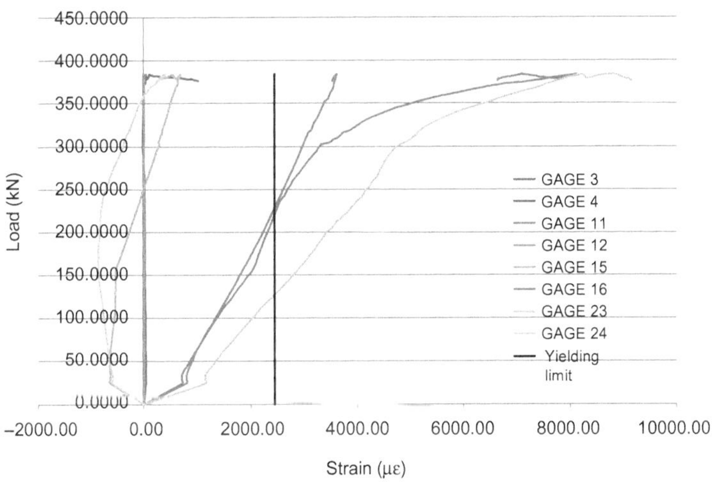

Fig. 2.191 Load versus strain curves: group g_2.

rupture load was lower than the experimental value, showing inefficiency in the resistance forecast formulation. The preliminary test PO-0.2 did not achieve the same success in determining the pull-out test failure mode, since it was forecasted to occur due to the tensile rupture of the reinforcing bars and, in fact, what happened in the experiments was the shear connectors rupture. This was expected because after the analysis of the test results concerning the pull-out test PO-0.1, there was a likely change in its rupture mode. This was due to an extra reinforcing bars strength capacity whilst the shear connectors, as the least resistant in this set, collapsed with a load almost equal to the

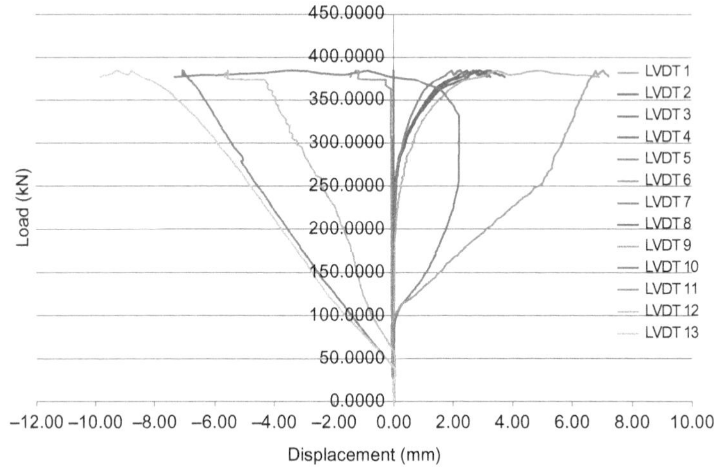

Fig. 2.192 Load versus displacement curves: LVDTs 1, 2, 3, 6, and 7.

theoretical prediction. It can be concluded that the shear connector resistance has not been changed in the hogging moment region where the concrete slab is under tension forces and is significantly cracked.

Table 2.11 presents a comparison between the responses of a strain gauge immersed and another outside the concrete slab for the PO-11 test. From

Table 2.11 Comparison Between Strain Gauge (G_2) Immersed in the Concrete and Strain Gauge (G_6) Located Outside the Concrete Slab, PO-11

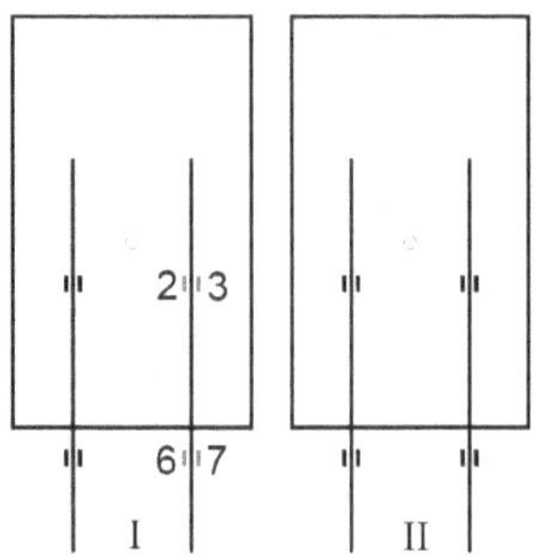

Load (kN)	Deformation ($\mu\varepsilon$)	
	G_2	G_6
50.59	28	359
100.81	44	790
151.0	58	1214
200.19	87	1618
249.96	201	2046
300.74	402	2501

this table it can be clearly noted the influence of the concrete that surrounds the reinforcing bars and its contribution to the bar stiffness and strength.

Shear connectors used in the joints would have performed efficiently when mobilised by a possible slip between the concrete slab and the steel profile if the shear connector's welds were properly made. The concrete slab, as expected, depicted a structural behaviour where cracks were formed in the perpendicular direction to the profile length, and main reinforcing bars with higher concentrations at the shear connectors' locations thus indicating that the concrete slab response was satisfactory. The reinforcing bars had a satisfactory behaviour when compared to the joint response since their anchor resistance was efficiently developed, increasing their stiffness and strength to sustain the high applied loads.

The natural continuation of this research programme consists of a full-scale test, which will generate results or behaviours similar to the 1% in ordinary structures found in the construction industry. This test can be described by observing Fig. 2.193 as follows: on one side of the beam, it is connected to a column, and on the other side, it is simply supported. This is an efficient way to determine the structure's forces, reactions, and displacements.

The practical/experimental model is also better understood by inspecting Fig. 2.194, where the presence of a simple support can be observed, it is a load cell, which monitors the updated reaction force due to the load applied at a certain point along the beam length. Using this configuration all the forces at the beam-to-column connection region can be easily determined.

Fig. 2.195 shows the final test configuration, where the slab effective width effects can also be investigated.

The designed test programme starts of simple models for understanding the forces trajectories present in end-plate semi rigid composite joints, in particular in the components area under tension. This region involves the beam top flange up to the main reinforcing bars passing by the shear connectors and the cracked concrete under tension. The implemented programme generated scouting possibilities of aspects such as the evaluation

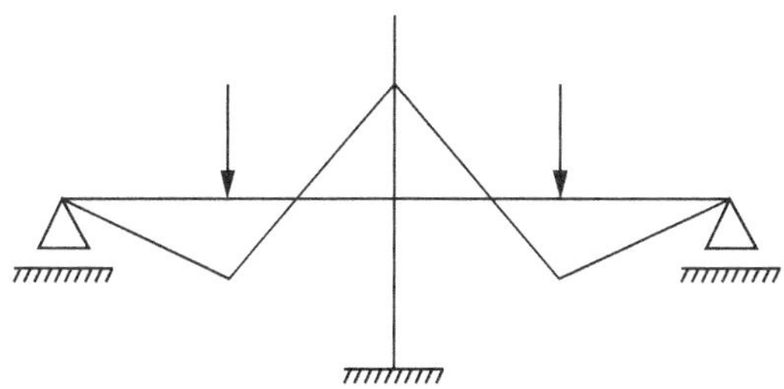

Fig. 2.193 Ideal structural model.

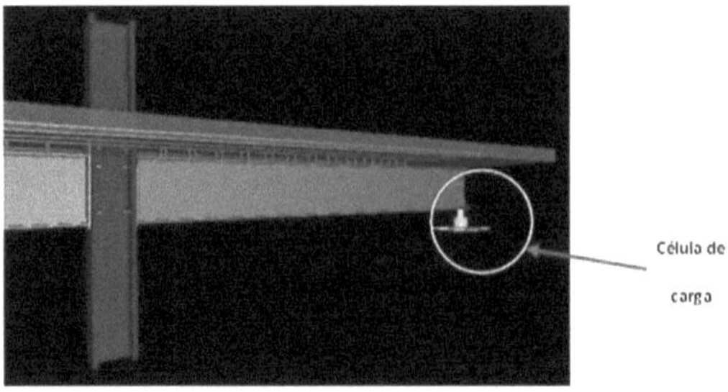

Fig. 2.194 Full-scale portal frame test with a load cell at the support.

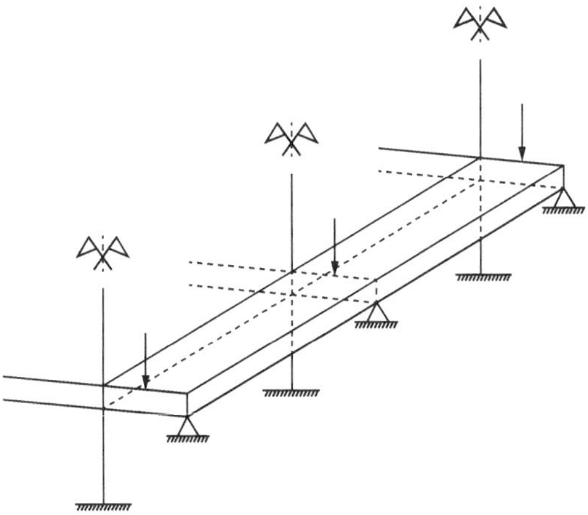

Fig. 2.195 3D full-scale portal frame test.

of the anchor resistance of the reinforcing bars, and concrete full, partial, and nonconfinement effects.

The natural evolution of this process goes through a symmetric system with one column and two beams under concentrated loads along their spans, simply supported one of their ends. In this system a correct specification of the distance between the load application and the column can provide a similar model to what occurs in a complete composite portal frame. However, a substantial advantage is generated with its adoption as a single load cell, used in each support, directly determines the moment in semi-rigid composite joint evaluated. Finally, if more than one of this set of elements is used (three columns with two beams each), not only the effects as the effective width in negative moment regions, but also a better understanding of the partial interaction in this region might have to be studied.

The adopted instrumentation in the tests has enabled the isolation of some component effects. Allied to it, the use of the Taguchi method to select the tests to be performed could also substantially reduce the total number of tests. The shear connector structural behaviour regarding the cracked concrete region still deserves a better evaluation, since in the present experimental programme, a greater emphasis was initially given to the reinforcing bar behaviour in terms of assessing effects like the different diameters and the number and anchor lengths. Another aspect that should be best investigated is directly related to the beneficial effect of the concrete compressed area confining the concrete area under tension forces.

2.6.11 Concrete-Filled Cold-Formed Beams' Resistance

The beams were tested with four hydraulic jacks, as shown in Fig. 2.196, to simulate a pure bending distribution. The experimental results will be presented in terms of deflections, deformations, and stress distributions for the two tested beams. An arrangement including a hinge, a load cell, and cylindrical rollers was utilised to ensure the correct boundary conditions between the jacks and the beam's top flange, as shown in Fig. 2.196.

Three preliminary tests were made in order to calibrate the hydraulic jacks, the supports, and the instrumentation gadgets used to measure loads, deformations, deflections, and slippage, as shown in Table 2.12.

The first preload was performed when the concrete had cured 53 days achieving a maximum load of 57.6 kN with a maximum deflection of 37.8 mm measured at centre span. A residual deflection of3.7 mm remained in the beam after the unloaded process. In the second preload the structure was loaded up to 100.8 kN with a maximum deflection obtained at the centre span of69.7 mm and a residual deflection of5.3 mm. In the third preload the structure was loaded up to 139.2 kN, with a maximum deflection of 99.5 mm and a total residual deflection of 20.3 mm (Fig. 2.197).

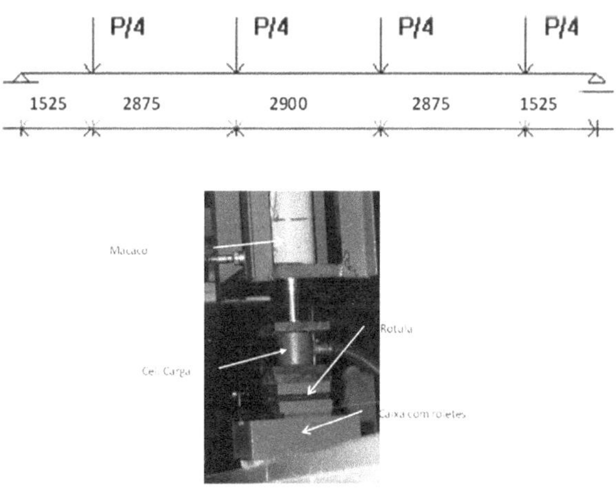

Fig. 2.196 Load configuration scheme and load application point detail.

Table 2.12 Test Results

Test		Maximum Load (kN)	Maximum Deflection (mm)	Residual Deflection (mm)
1	1st preload	57.6	37.8	3.7
1	2nd preload	100.8	69.7	5.3
1	3rd preload	139.2	99.5	20.3
1	Final test	240.0	225.1	–
2	1st preload	28.0	15.2	1.9
2	Final test	284.6	452.6	–

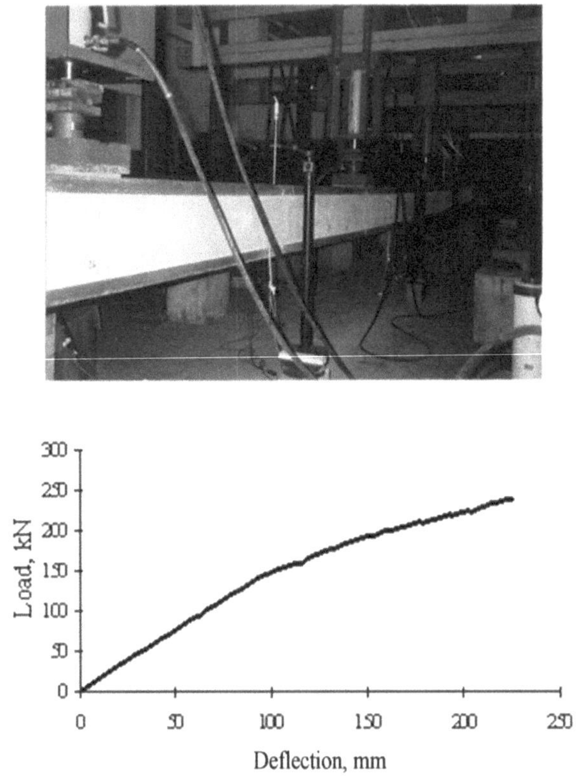

Fig. 2.197 Experiments general layout and 1st test load versus midspan vertical displacement curve.

The deflections were measured at the quarter and midspan of the beam by three LVDTs located on top flange. The midspan deflections are shown in Fig. 2.197 up to its collapse that occurred due to a weld rupture at midspan in the bottom flange. The load and the maximum displacement were 240.2 kN and 225.1 mm larger than the maximum allowed usually l/360 (i.e. 32.5 mm), as shown in Fig. 2.198.

Fig. 2.198 Bottom flange weld rupture, 1st test.

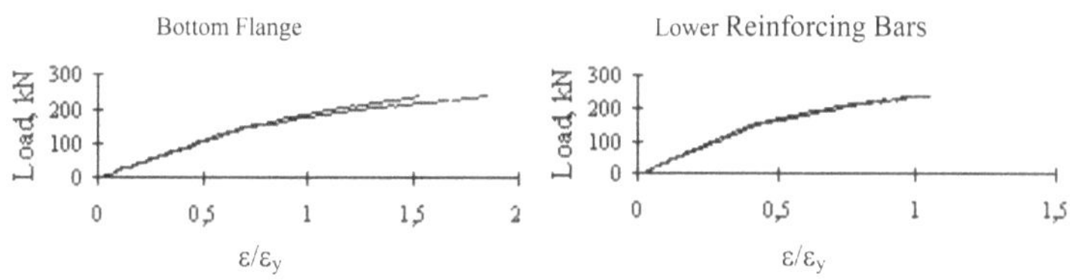

Fig. 2.199 Load versus normalised strain curve, 1st test.

Fig. 2.199 shows the beam's load versus strain curves, measured by two strain gauges located at the bottom flange, 200 mm far from the beam midspan. A nonlinear behaviour initiated for load values greater than 120 kN whilst the onset of steel yielding strains corresponded to load values of 176.8 and 187.5 kN, respectively, with a 6% difference. At the ultimate load, the strains reached 1.6 and 1.9 times the yield deformation, ε_y.

Load versus normalised strain curves for two reinforcement bars located near the tension part of the steel section are shown in Fig. 2.199. The strains were acquired by two strain gauges located at the bottom face of each bar, 200 mm far from the beam midspan. The curves are similar indicating a symmetric behaviour. The steel bar yield stress was reached at a load value of 253.0 kN.

When the concrete reached 44 days, a single preload was made in the second beam reaching a load of 28 kN and a maximum vertical central displacement of 15.2 mm and a residual deflection of 1.9 mm (Table 2.12). In the final test when the load was equal to 210 kN, the maximum central jack's course was reached leading to the need to anchor the structure to the laboratory reaction slab. When this task was completed the jacks were moved down and the test proceeded normally. Fig. 2.200 shows a deformed configuration of the second beam during the loading phase. When the applied load value was approximately 284.3 kN, a weld rupture at the single

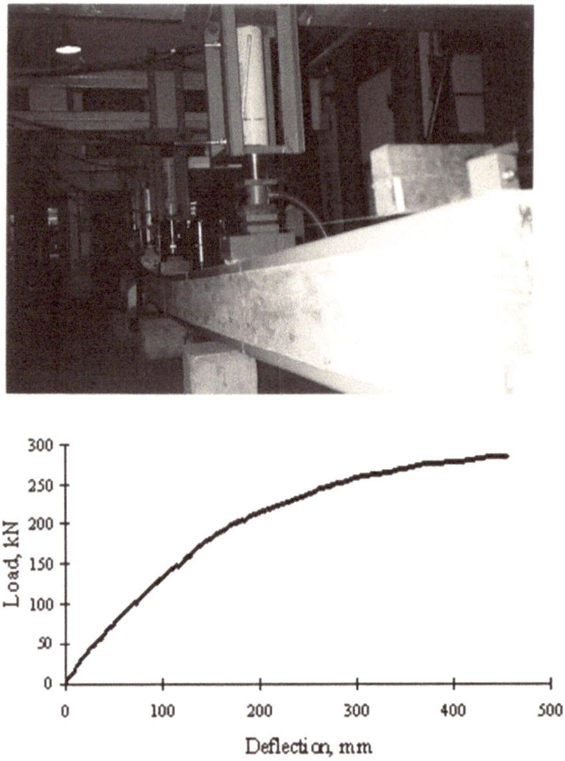

Fig. 2.200 Deformed configuration under loading and load versus midspan displacement, 2nd test.

cold-formed U of the bottom flange, at midspan led to the structure collapse. A maximum load of 284.6 kN was recorded shortly after the weld rupture.

Fig. 2.200 also shows the load versus vertical displacement curve measured by an LVDT positioned installed at midspan. A nonlinear behaviour was noticed for load values greater than 122.2 kN. From this load up, which corresponding to an 89.9 mm deflection, the displacements increased considerably. The maximum load was 284.6 kN, very close to the theoretical load of 284.3 kN. The maximum displacement was 452.6 mm, which is well over the serviceability limit of span over 360.

Fig. 2.201 shows two load versus strain curves for points located at the top flange of the cold-formed section, acquired by two strain gauges 150 mm from the beam midspan. The curves are almost identical with a linear behaviour for load values less than 109 kN. The yield strain was associated with load values of 196.4 and 193.5 kN. When the applied loading reached its maximum, the strains were five times the yield values. Fig. 2.201 shows the applied load versus strains curves for the two

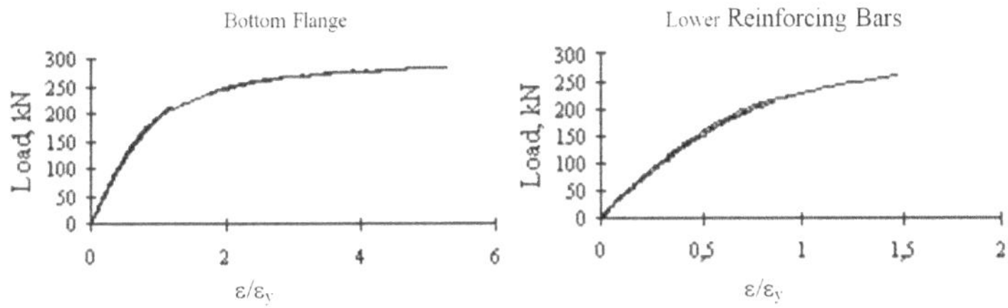

Fig. 2.201 Load versus normalised deformation curve, 2nd test.

reinforcement bars. Strains were acquired by two strain gauges located at their bottom side, 150 mm far from the beam midspan. One more time a similar behaviour was noted whilst yielding occurred at load of 220 kN.

The reinforced concrete presence increased beam's inertia making the structure more rigid and, consequently, leading to smaller displacements than a noncomposite solution. The theoretical and experimental values of displacements were above the maximum limits usually adopted in structural standards. With this restriction in mind, it is possible to recalculate the maximum allowed deflection for building structures where the tested beam will be suitable for spans up to 7.5 m.

The tests did not show sliding on steel versus concrete interface, confirming that shear connectors welded in the web of the steel section, usually adopted in rolled steel beams, are not required for tested cold-formed sections. This mainly occurred due to the additional resistance associated with the concrete confinement effect.

One of the most interesting aspects of this experiment consisted in the construction process of these elements and it was executed in three steps. The first one involved bending the steel sheets to form the desired shapes. This was followed by welding these isolated elements to form section I. Finally, there was the concrete casting. It was executed in two days, that is, one for each side of the profile. The concrete confinement proved to be extremely effective, eliminating the use of connectors at the beam web, which are common and necessary when rolled or welded steel profiles are used, instead of the adopted cold-formed profiles.

The adopted instrumentation used strain gauges, LVDTs and dial gauges, where the key point was the slip measurement at the steel to concrete interface. This was made with LVDTs positioned at steel profile by measuring the relative displacement of the concrete in this region and contributed to the understanding of the high concrete confinement within the steel profile.

Displacements and possible rigid body movements were also monitored in the proposed slab system with LVDTs and dial gauges. Strain gauges were used to determine the strain distribution in the steel profile to assess their real structural performance. Universal joints and rollers were used in the supports and load application points to ensure their proper performance within the structural model designed as simply support under the action of concentrated loads.

2.6.12 Composite Slabs With: Cold-Formed Profiles, Styrofoam Plates, and Self-Drilling Bolts

The use of precast and prestressed precast slabs in low-cost construction is becoming a routine procedure in Brazil. These systems are very efficient and economic, but they still present disadvantages related to transport and availability when they are utilised in remote locations. To surpass these difficulties and to improve their longer free-span-structural response, a composite slab made of a steel section and a styrofoam plate side by side filled with concrete were developed, as Fig. 2.202 shows.

The steel section is responsible for providing the necessary resistance to the slab system discarding the need for temporary shoring during construction. When the composite action is developed, the steel section sustains the main tensile component whilst the concrete cover slab works mainly in compression. To minimise the slab cracking a welded wire mesh was utilised at the cover slab mid height (Fig. 2.203).

It is well known that the shear connector geometry and type associated with the steel versus concrete interface bond directly influence the structural

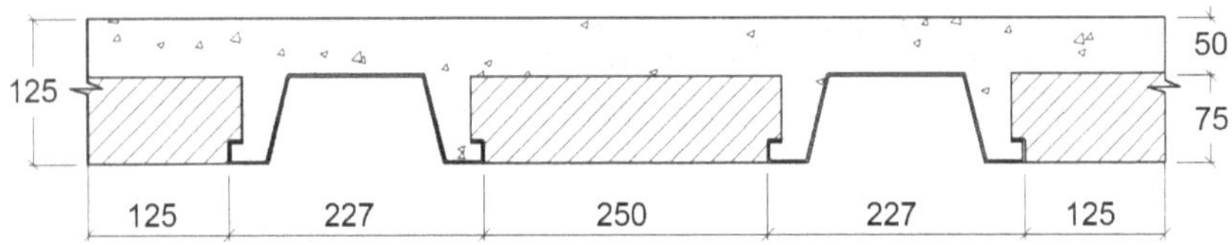

Fig. 2.202 Idealised composite slab system.

Fig. 2.203 Composite slab system before and after concrete casting.

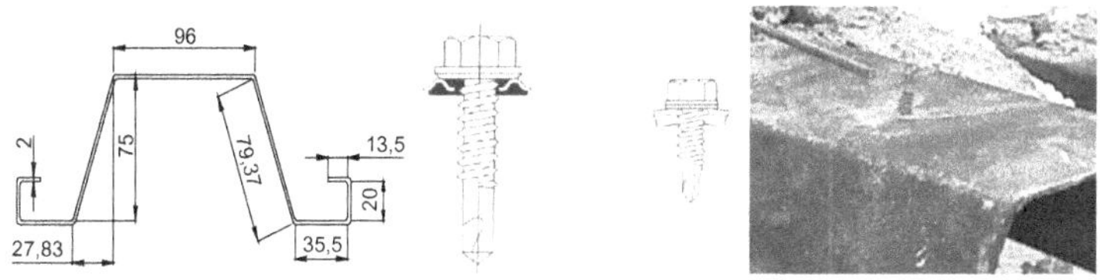

Fig. 2.204 Tested cold-formed steel profile and self-drilling bolts used as shear connectors.

system load-carrying capacity. In this test programme, self-drilling bolts, 1/4″ × 3/4″ (diameter × length), installed with a simple portable drill, shown in Fig. 2.204, were used as shear connectors. The experimental programme involved five tests composed of simple supported slabs with the shear connector configuration (Table 2.13).

The tested composite slab system had a 4.5 m length and it was 950 mm wide. The structural system was simply supported (Fig. 2.205). Two spans were tested. They were 3.0 and 4.3 m, respectively. The concrete used in the specimens had a compressive strength of 44 MPa, whilst the steel profile yield stress was 320 MPa.

The measurement of displacements in the vertical (R3, L2, L3, L4) and lateral (R1, R5) directions and the slippage between steel-deck and concrete at slab ends (R2, R4, L1, L5) was made with five LVDTs (L series) and five dial-gages (R series), as shown in Fig. 2.206. Strain measurements were

Table 2.13 Test Series Description

Test	**Span (m)**	**Shear Connector Configuration**
1	4.3	Without shear connectors
2	3	Without shear connectors
3	4.3	100-mm spaced shear connectors
4	3	100-mm spaced shear connectors
5	3	200-mm spaced shear connectors

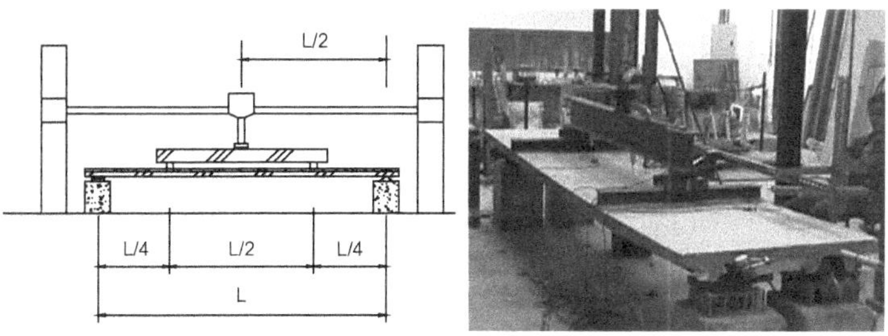

Fig. 2.205 Experimental layout.

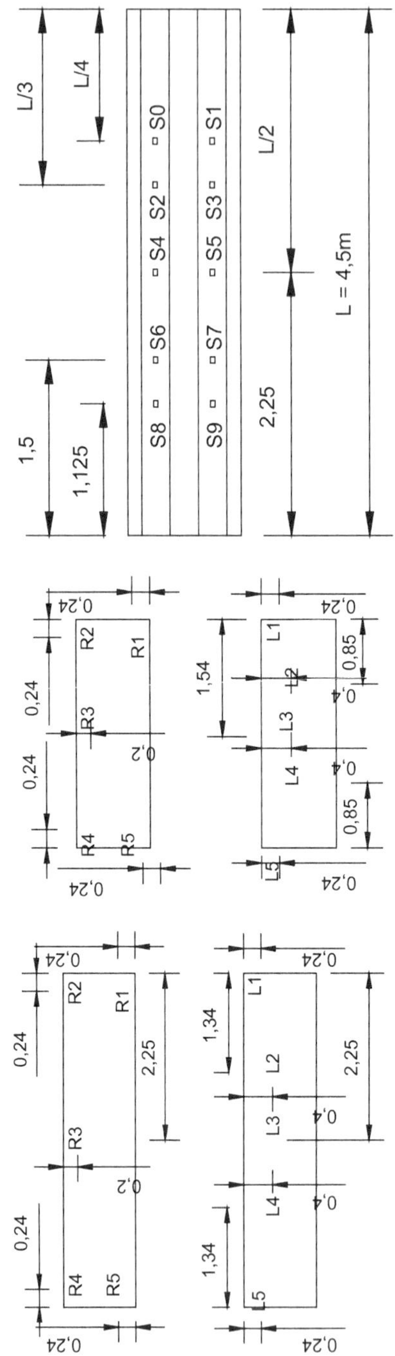

Fig. 2.206 LVDT & strain gauge distribution for the 4.3- and 3.0-m span slabs.

obtained with ten strain-gages located at positions (S0 to S9), as shown in Fig. 2.206.

The first and second tests used specimens with 4.3 and 3 m span without shear connectors. A load versus centre span deflection curve is shown in Fig. 2.207 where the obvious span influence is characterised by the larger stiffness related to the 3 m span test. Both tests had their collapse related to slippage in the steel/concrete interface. This was no surprise, since the chemical bond between the steel profile and the concrete cover is not enough to ensure the interface shear flow at high-load levels. The steel profile maximum recorded strain reached 60% of the steel yield strength. In the second test, without shear connectors, the failure was predominantly due to slippage between the steel deck and the concrete. Both first and second tests had the same shear bond failure mechanism shown in Fig. 2.207.

The third and fourth tests involved slabs with 4.3 and 3 m span with 100-mm spaced shear connectors. A load versus centre span deflection curve is shown in Fig. 25. Although the self-drilling bolts substantially improved the

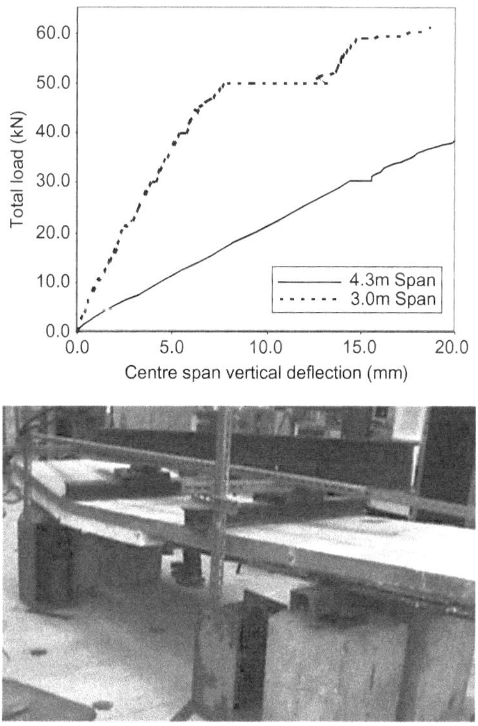

Fig. 2.207 Load versus deflection curves for slabs without shear connectors & 2nd test collapse.

shear resistance in the concrete steel interface, the failure was again related to the concrete to steel slippage. The slippage between the concrete and the deck (Fig. 2.208) represents the collapse reached in the third and fourth tests.

Fig. 2.209 shows a direct comparison between the 4.3-m span slabs without and with 100-mm spaced shear connectors. It should be emphasised that despite the fact that the slab initial stiffness was similar for both cases, the 100-mm spaced shear connector ultimate load was 86% higher than the test without shear connectors. It is also interesting to note that the strains attained in the 100-mm spaced shear connector test was significantly smaller than the strains from the test without shear connectors indicating that the full composite action was only achieved in the third test.

The final test was made of a 3-m span slab with 200-mm spaced shear connectors. The purpose of this particular specimen was to access the influence of shear connector spacing. This effect is clear in Fig. 2.210 where

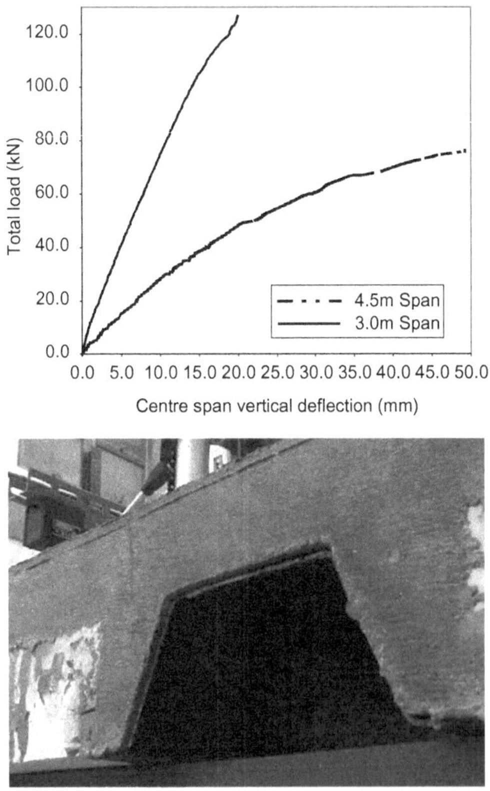

Fig. 2.208 Load versus deflection curves, 100-mm spaced shear connectors and slabs steel to concrete slippage limit state: 3rd and 4th tests.

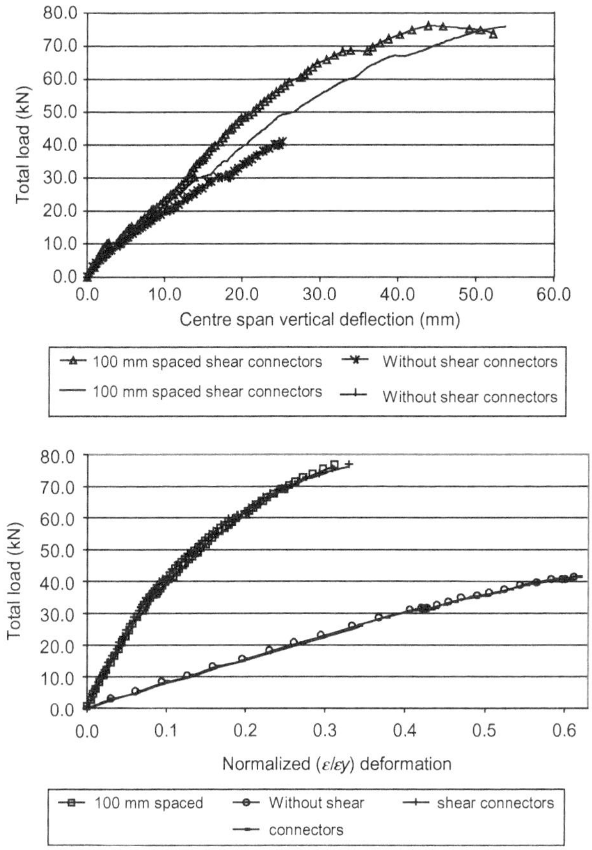

Fig. 2.209 Load versus deflection and load versus normalised deformation curves: 1st and 3rd tests.

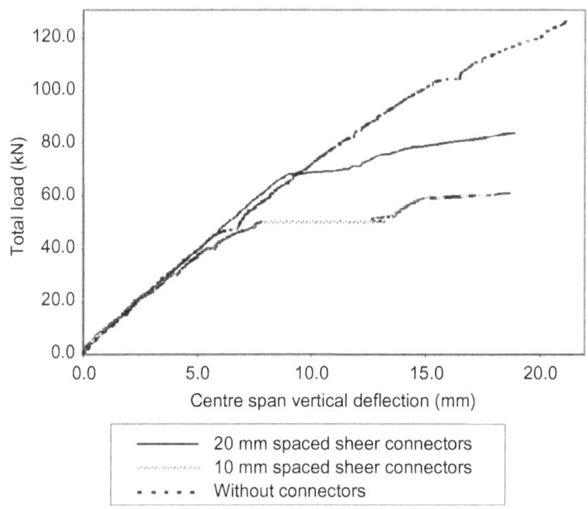

Fig. 2.210 Load versus deflection curves for specimens with 3 m spans.

load versus centre span deflection curves for 3-m span slabs possessing the shear connector arrangements are depicted. As expected the closer spaced shear connector slab was related to the higher collapse load. Another interesting point is that all the slabs had nearly the same initial stiffness.

The results confirmed the hypothesis that the composite slab ultimate load increases with a reduction of the shear connector centre-to-centre spacing. The use of self-drilling bolts was also studied and proved to be very efficient.

The developed slab system using of steel plates, styrofoam plates, and self-drilling bolts at the steel to concrete interface had a promising performance for popular housing structures. The tests design was based on established tests for composite slabs designed to evaluate their performance in terms of structural stiffness, strength, ductility, and collapse modes.

The adopted instrumentation involved strain gauges, LVDTs and dial gauges, where measurement of slip at the steel to concrete interface was performed with LVDTs positioned in steel profile measuring, in this region, the relative displacement of the concrete. Displacements and rigid body movements were also monitored in the proposed slab system with LVDTs and dial gauges. Strain gauges were used to determine the steel profile strain distributions in order to assess their real structural performance. Universal hinges and rollers were used in the supports and load application points to ensure their proper performance within the structural model designed as simply supported under the action of concentrated loads.

2.6.13 Structural Capacity of Composite Slab Systems With Cold-Formed Profiles

The objective of the current study was to demonstrate the viability of a composite construction slab system utilising cold-formed galvanised steel plates with indentations and mosses on the profiles web [63,70]. The use of cold-formed steel profile to substitute the slab reinforcement bars as well as acting as a shoring system at the construction stage will be studied. The study of this new steel profile was based on a good mechanical grip between steel and concrete and ductile behaviour characteristics. Variables such as weight, height, thickness, yield stress, load-carrying capacity, and maximum spans were considered. A typical slab system utilised precast concrete beams alongside styrofoam plates, or ceramic blocks, covered with a cast in place concrete with a minimum reinforcement mesh to avoid cracking. From this concept a novel composite slab system was developed (Fig. 2.211).

The current studies utilised the optimum profile geometry previously established and substituted the self-drilling bolts by plate with 50- to

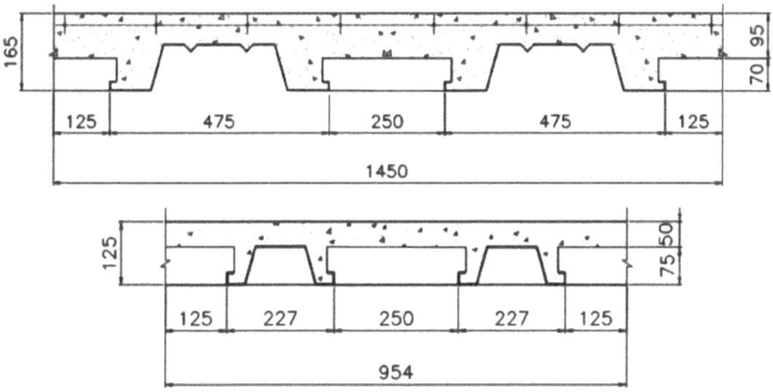

Fig. 2.211 Composite slab systems using: styrofoam plates and only steel plates.

100-mm spaced circular corrugations. Fig. 2.211 shows both composite slab systems and the newly conceived solution made of a folded 1.20 mm thick, 120 mm height steel profile and a concrete with a 25 MPa f_{ck} and previous experiments [65], which used 2 mm thick, 75 mm height steel profiles and a concrete with a 40 MPa f_{ck}.

A 6.3-mm diameter reinforcement bar mesh with 200 mm by 300 mm spacing, located 25 mm above the steel profile top flange was also used to reduce concrete cracking. All the full-scale composite slab tests were simply supported with two point loads simultaneously applied through a hydraulic actuator and a load-distributing beam at quarter span points. The concrete compressive resistance was designed to be limited to the range of 20–25 MPa. Twenty-eight days after casting nine concrete specimens were tested reaching a mean compressive strength of 25 MPa.

The instrumentation used for measuring vertical and lateral displacements, and the concrete to steel interface slip involved five dial gauges and five LVDTs. A dial gauge and an LVDT were located at the slabs ends with their bases fixed to the steel profile pointing to the concrete to measure the concrete to steel slip. One LVDT and one dial gauge were also utilised at the slab ends to record the occurrence of possible composite slab rigid body movements. The slab vertical displacements were monitored through one LVDT and one dial gauge at midspan and two LVDTs located 900 mm from the slab ends. The midspan steel strains were measured by six linear strain gauges located at the steel profile top flange, three in each profile. A strain gauge was installed at midspan whilst two others were used 775 mm from the slab ends. Fig. 2.212 shows the global configuration of dial gauges, LVDTs, and strain gauges used in the test and the load application points.

The test programme involved a 1.45 m × 3.0 m composite slab tested with a span of 2.91 m long. The slabs were made with two steel-folded plate

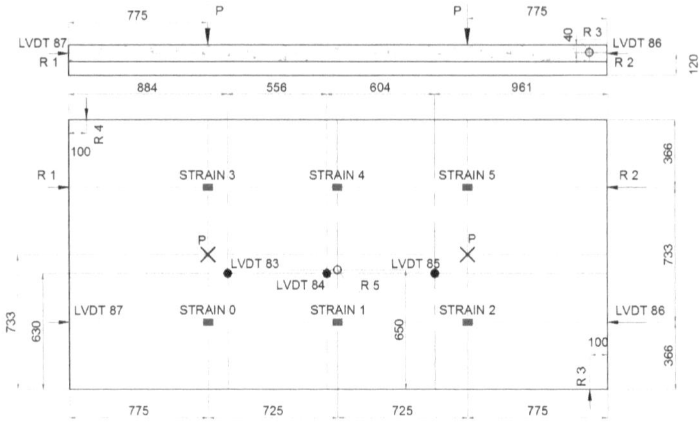

Fig. 2.212 Composite slab test instrumentation.

profiles with web indentations stamped every 100 mm and 310 mm (Fig. 2.213). The loads were equally applied in two points, by using a transition beam, at quarter span (Fig. 2.214). Three preliminary tests were conducted to check the instrumentation and gradually mobilise the structure. In all preliminary tests a 15-minute wait interval was respected before

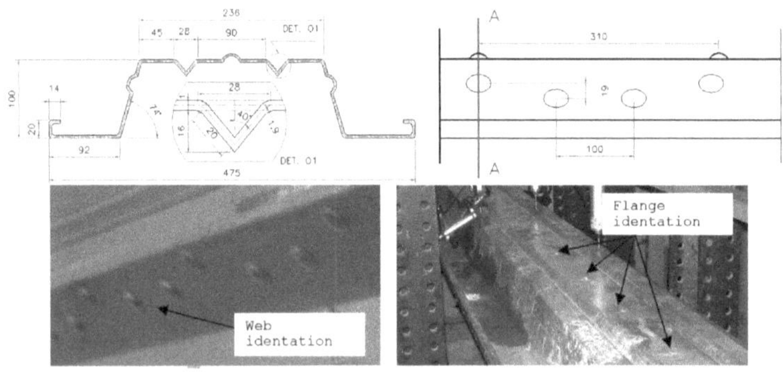

Fig. 2.213 Detail of indentations at the profile web.

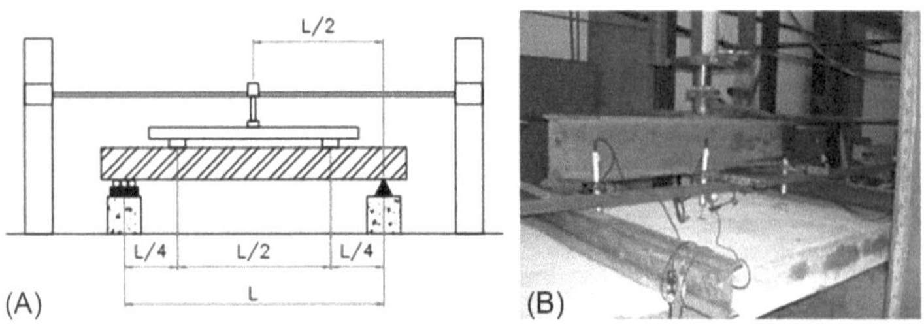

Fig. 2.214 Support & loads. (A) Slab midspan displacement. (B) Steel to concrete slip at the slab end.

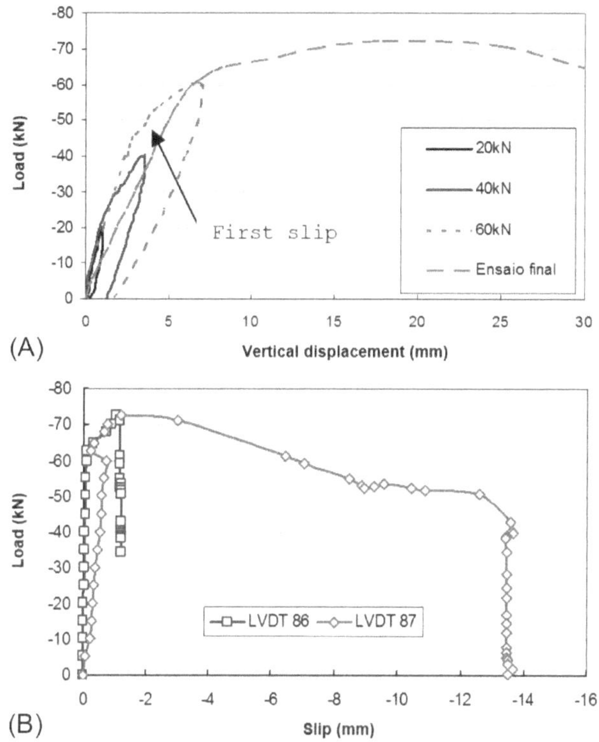

Fig. 2.215 Load versus displacement and load versus steel to concrete slip curves.

unloading the structure. Fig. 2.215A shows the load versus midspan vertical displacement curves for the preliminary tests (20, 40, and 60 kN) and final test. Fig. 2.215B shows the graph of load versus steel to concrete slip measured by LVDTs located at opposite slab ends during the final test.

The steel profile strains of the composite slab structure were obtained through strain gauges located on the underside of the steel profile top flange. Fig. 2.216 shows the load curves versus strain curves for the final test. It can be noted that yield was not reached during the test. An increase of the strain grow was observed in Fig. 2.216 due to the onset of slip at the steel to concrete interface.

Fig. 2.217A shows the initial slab configuration whilst Fig. 2.217B, the final slab presenting the concrete to steel slip that occurred at one of the slab ends. Fig. 2.217C shows the observed concrete cracks that occurred close to one of the load application points.

To analyse the results obtained from this test, a comparison between these results and the results of Takey [65] and Beltrão [69] was made. Fig. 2.218 shows the midspan load versus vertical deflection curves for all the above-mentioned tests with 3 m spans. Takey tested three composite slabs with

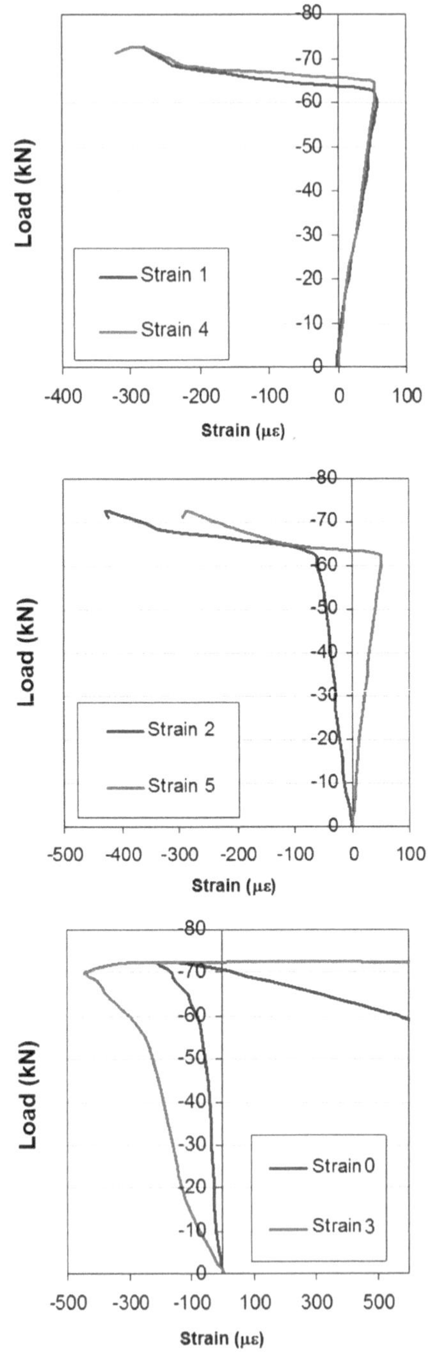

Fig. 2.216 Steel profile strains.

Fig. 2.217 Concrete to steel slip and concrete cracks.

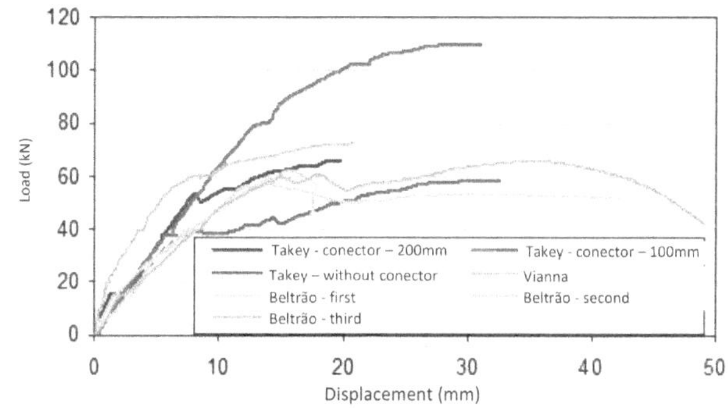

Fig. 2.218 Composite slabs comparisons.

identical cold-formed steel profiles. The first test had no shear connectors whilst the second and third utilised 200- and 100-mm spaced self-drilling bolts as shear connectors used by Takey [65]. Beltrão [69] tested three tests with 3-m span slabs and profiles adopting the same cross-section used by Takey [65], with web corrugations spaced every 50 mm profile in the first and second tests, and every 100 mm in the third test.

Takey results were significant stiffer that the present tests due to the use of two cold-formed steel profiles, resulting in a 1-m wide slab, and a higher concrete strength (40 MPa). In order to compare both test series, a first reduction factor of 0.5 was utilised in Takey results [65], to take into account that these tests presented two steel profiles, side by side. A second reduction factor was also used to consider the concrete compressive strengths difference (i.e. 25 and 40 MPa), according to the well-known concrete elastic modulus relationship.

The investigated composite slab system presented the highest stiffness amongst all the tests. This stiffness increase was due to the use of a slab with the highest thickens and a profile with the highest inertia. In addition, in the present test the steel profile presents web corrugations and an intermediate stiffener at the top flange (replacing the shear connectors). According to Eurocode 4 [64], the ductile behaviour of a composite slab is characterised when the ultimate load achieved by the system excels, in at least 10%, the load value recorded at the moment in which the steel to concrete slip reaches 0.5 mm. Otherwise, the system is considered fragile. As it can be seen in the curve shown in Fig. 2.218, a load of 40 kN with an associate 0.5 mm deflection corresponds to 55.5% of the ultimate load (72 kN) confirming that the studied system presented a ductile behaviour.

The main objective of this study was to present a contribution associated with a new alternative of a steel profile to be used in composite slabs. Variations of the same cross-section were made to theoretically and experimentally evaluate the structural behaviour of the steel profile, as well as the composite slab structural system.

Experiments have not only proven the suitable strength of the new structural system, but also the technical solution feasibility, which is supposed to be economically viable for the use in low-income constructions. The full-scale composite slab test presented a ductile behaviour and it was associated with the most common collapse mode, longitudinal shear, preventing the occurrence of the ultimate load associated with the structural system bending moment capacity.

This work was a first evolution of previously presented systems where the styrofoam plates and self-drilling bolts were replaced by additional steel plates containing indentations and mosses to increase the slip resistance at the steel to concrete interface. The tests were similar to the configuration previous adopted in this study and the tests have aimed to evaluate their performance in terms of structural stiffness, strength, ductility, and collapse modes. The adopted instrumentation also used strain gauges, LVDTs and dial gauges. The on-steel to concrete slip was measured with LVDTs positioned at the steel profile acquiring the concrete relative displacement in this region.

LVDTs and dial gauges monitored displacements and rigid body movements in the proposed slab system. Strain gauges were also used to determine the strain distributions in the steel profile to assess their actual structural performance. Finally universal hinges were used at the supports and load application points to ensure their adequate performance in the simply supported structural model under the concentrated load actions.

2.6.14 Structural Response of a Composite Slab System With a Cold-Formed

A typical slab system using precast concrete beams, styrofoam plates, or ceramic blocks, covered by a concrete layer with a minimal reinforcing bars mesh to prevent cracking was investigated. From this concept, a composite slab system was designed and developed, see Fig. 2.219 [65]. This system presents disadvantages associated with the use of styrofoam plates and

Fig. 2.219 Composite slab with styrofoam and steel plates side by side.

self-drilling bolts as shear connectors, increasing their operational costs, and encouraging the development of new structural solutions.

The connection at the concrete to steel interface in any composite solution is critical for a structural response. Consequently, it is expected that new solutions involve the materials according to their optimum structural responses. The lack of connection in the interface creates an undesirable slip that prevents the combined action, inducing the different materials to individual structural responses. Various devices and techniques can be used to ensure a good performance at the steel to concrete interface. One of the most common strategies is the use of dents and/or mosses in steel plate. This solution was used in the present investigation [69] to replace the self-drilling bolts previously adopted [65].

The steel sheet must be capable of transferring the shear stress at the different materials interface, enabling the development of the composite action. It is common knowledge that the natural chemical bond between the concrete and steel sheet is not sufficient to provide and properly transmit shear, therefore, requiring the use of mechanical devices. The shear transfer is obtained by two different ways: through a mechanical bond associated with concrete immersed in the steel sheet corrugations and an additional slip resistance due to the concrete confinement that occurs with a proper selection of steel sheet geometries. The corrugations and mosses induced in steel plate during its forming process are generally of a small depth; otherwise local fracture can be induced in the steel plate. The use of a geometry that favours the confinement in the adopted steel sheets also serves to prevent the

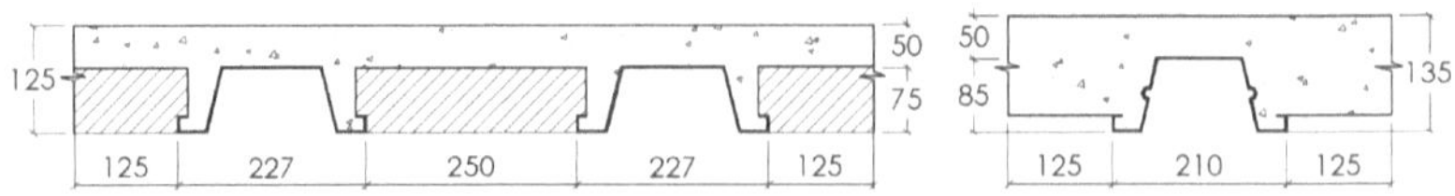

Fig. 2.220 Composite slab systems using: styrofoam and steel plates.

separation between the concrete and the steel sheet. The present study used a steel profile with dimensions given in Fig. 2.220 and replaced the self-drilling bolts by a plate with circular corrugations spaced every 50 or 100 mm. Fig. 2.220 shows these two composite slab systems.

Four slabs with 3 m spans were built from the new solution and subsequently tested. The mosses and corrugations were stamped in the steel plate with a tool made from a circular bar with a 50.8 mm diameter. The 10 mm original corrugations depth induced a fracture on the plate reducing its reduction to 5 mm. Simple steel plates, 1000 mm × 240 mm × 0.65 mm, replaced the previous system styrofoam plates and were tested with perpendicular loads to its plan. These lateral plates were 1 m in length due to the galvanised sheet coil standard width and reduce the waste (see Fig. 2.221). These plates are supported by cold-formed profiles through their stiffened edges that were designed to be compatible with the lateral plate's dimensions. The side lateral plates together with the steel profiles created a system that was robust enough to sustain the typical construction loads without requiring a temporary shoring.

The experimental programme consisted of three composite slabs tests measuring 0.46 m × 3.1 m. Two slabs adopted profiles with web corrugations spaced every 50 mm, whilst the third one used corrugations spaced every 100 mm. All full-scale tests composite slabs were simply supported under the action of two concentrated loads located in the beam quarter span and a hydraulic jack and a distribution beam simultaneously applied them. A 50-kN load cell located between the jack head and the distribution beam monitored the applied load (see Fig. 2.222).

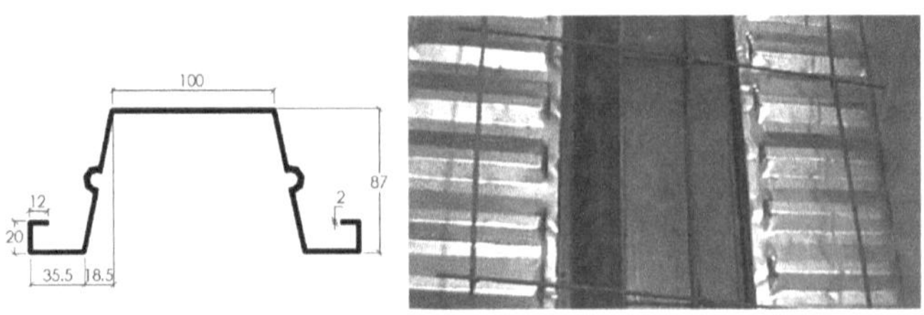

Fig. 2.221 Adopted cold-formed steel profile and corrugate plate detail.

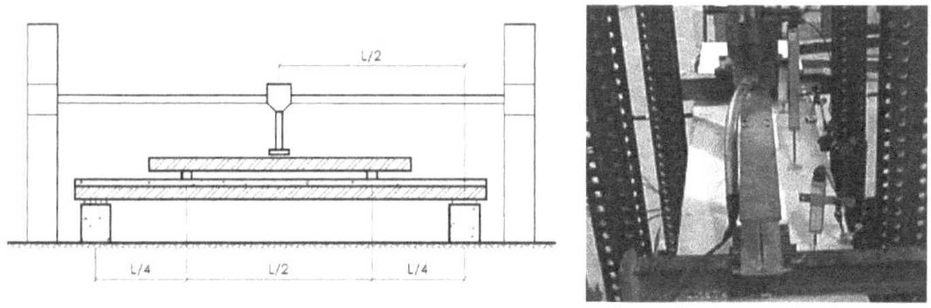

Fig. 2.222 Testing configuration.

Although the slab did not required any shoring, a concrete form was used (see Fig. 2.223). A reinforcing bar mesh (6.3 mm in diameter spaced every 200 in the span direction) was used. It was 25 mm above the steel profile top flange in order to reduce the concrete cracking.

Twenty-eight days after the concrete casting the specimens were tested reaching an average compression resistance to 25 MPa. The type of steel was A-36 with a 250 MPa yield stress. Fig. 2.224 shows the configuration of

Fig. 2.223 Composite slab minimal reinforcing bar mesh.

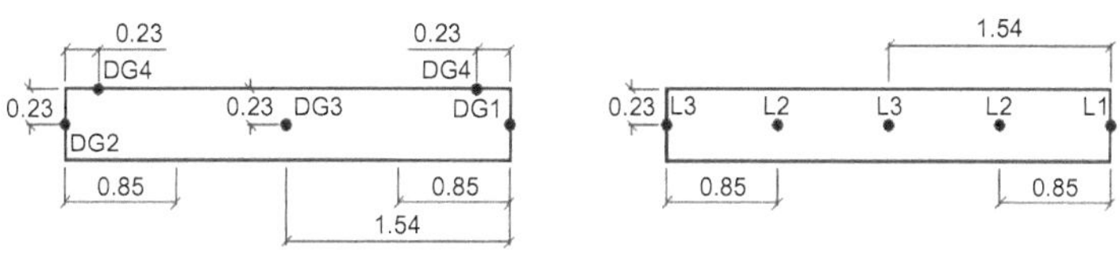

Fig. 2.224 Composite slab instrumentation.

adopted dial gauges and LVDTs. LVDTs (L1, L3) and dial gauges (DG1, DG2) were used in each of the slab ends to monitor the slip at the steel to concrete interface. The slab vertical displacements were monitored by an LVDT (L3) and a dial gauge (DG4) at the beam midspan and two LVDTs (L2) situated 700 mm from the slab ends. The stresses at the beam midspan were measured for two linear strain gauges located on the steel profile bottom flange.

Fig. 2.225 shows a comparison between the experimental programme and previous tests by means of vertical load versus displacement curves. Takey's results were significantly stiffer than the present tests due to the use of two cold-formed profiles steel, resulting in a 1-m wide slab with a higher concrete compressive strength (40 MPa). To compare both experimental series, a first correction factor equal to 0.5 was used in Takey results, to consider that these tests used two steel profiles, side by side. A second correction factor was also used to account for the used concrete compressive strength difference, that is, 40 and 25 MPa in accordance to the well-known elastic modules ratios:

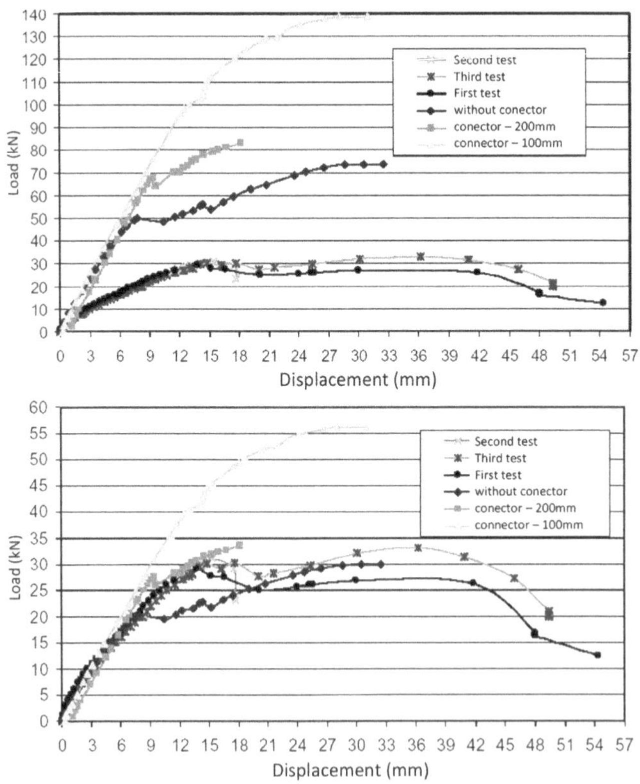

Fig. 2.225 Takey [65] comparisons: original and reduced layouts.

$$f_{\mathrm{corr}} = \frac{0.9 \times 600\sqrt{25 + 3.5}}{0.9 \times 600\sqrt{40 + 3.5}} = 0.81$$

The work continued with three modified push-out type tests measuring 210 mm by 650 mm (Fig. 2.226) [70]. These tests were conducted using jacks, located in the specimens centre simulate a confined plate using the same cold-formed profile adopted in previous composite slabs. The objective of these tests was to assess whether there has been an increase in shear strength on steel to concrete interface due to the adopted web profile. Two 50-kN jacks have been adopted since the predicted collapse load was equal to 60 kN. These two jacks were positioned in the slabs centre as shown in Fig. 2.226. A simple wood form was designed and built to separate the steel profiles and guide the actuators. Specimens were taken from the concrete and tested in the same days the push-out tests occurred reaching a compressive strength of 27 MPa.

A dial gauge and an LVDT were used in the centre of the top face whilst four LVDTs were utilised in each corner of the upper face to monitor the slip at steel to concrete interface. An LVDT was used at the end of the additional lateral surface of the specimen for measuring rigid body movements. The final tests had the same collapse mode, that is, concrete crushing, with an average maximum load of 45 kN. Unfortunately, the adopted instrumentation was not able to acquire the exact load value associated with the steel to concrete interface slip due to the phenomenon abrupt nature. Despite this fact maximum load and shear stress loads are presented in Table 2.14.

Fig. 2.226 Configuration of modified push-out tests.

Table 2.14 Push-Out Tests Ultimate Load and Shear Stress

Specimen	Ultimate Load (kN)	Ultimate Shear Stress (MPa)
1	51	0.24
2	44	0.21
2	41	0.19
Mean	45.33	0.21
Standard deviation	5.13	0.025

Fig. 2.227 Push-out tests typical collapse.

Fig. 2.227 shows a typical collapse configuration for specimens with web corrugations spaced every 50 mm where the concrete rupture and its complete separation from the steel plate are noticeable. It is also possible to observe that similarly to the full-scale composite slab tests, a sliding plan was formed close to the profile bottom web/flange close to the web corrugations.

The ultimate limit state associated with the investigated composite slab system was the on-concrete to steel interface slip. When these tests were compared it was possible to observe that the investigated composite slab was more efficient than slabs without shear connectors and equivalent to slabs with self-drilling bolts spaced every 200 mm. The load of the composite slab with self-drilling bolts spaced every 100 mm was higher than the system with web corrugations every 50 mm. The push-out tests have confirmed that the slip resistance on steel to concrete interface was fundamental to provide a suitable structural behaviour to the composite slab. Based on these results it was also possible to conclude that the profiles with web corrugations had a suitable steel to concrete interface resistance.

This work was a second evolution of the previously presented systems, where the steel profile contained indents and deformations to increase the resistance in the concrete steel interface. The initial tests were similar to the configuration used in the previous experimental programmes and aimed to evaluate their performance in terms of structural stiffness, strength, ductility, and collapse modes. The adopted instrumentation also used strain gauges, LVDTs and dial gauges, where the slip measurement in concrete steel interface was once again performed with LVDTs positioned in the steel profile measuring the relative displacement of concrete in this region.

LVDTs and dial gauges also monitored displacements and rigid body movements into the proposed slab system. Strain gauges were one more time used to determine the strain distributions in order to assess their real structural performance. Finally, universal hinges were used in the supports and load application points to ensure a suitable performance of the simply supported structural models under the concentrated loads actions.

A second test configuration to measure the slip resistance at the concrete to steel interface was also executed based on a modified push-out type tests. These tests were conducted using jack located at the specimens centre to simulate a confined plate and used the same cold-formed profile adopted in the previous composite slabs. The purpose of these tests was to assess whether there has been a shear strength increase at the steel to concrete interface due to the use of the web corrugations at the steel profile. Two jacks were positioned at the slabs centre and a simple wood form was designed and built to separate the steel profiles and guide the jacks.

In the second test configuration dial gauges and LVDTs were used to monitor the slip on the concrete steel interface. An extra LVDT was unlisted at the end of the specimen lateral surface for measuring rigid body movements. The final tests had the same collapse mode, concrete crushing. Unfortunately, the adopted instrumentation has not acquired the exact load value associated with the steel to concrete interface slip due to the phenomenon abrupt nature.

2.6.15 Structural Response of Perfobond and T-Perfobond Shear Connectors

The motivation for the development of new products for the shear transfer in composite structures is related to matters involving particular technology, economic or structural needs. In this context, some other alternative shear connectors are proposed for composite structures like the T-Perfobond (Fig. 2.228). This connector is derived from the Perfobond connector with the addition of a flange, working as a block. The motivation for developing this T-Perfobond

Fig. 2.228 T-Perfobond connectors. (A) IPN 340—University of Coimbra. (B) HP 200 × 53—PUC-Rio.

connector is to combine the high block type connector strength with some ductility and uplift resistance emerges from the Perfobond holes.

The T-Perfobond connector was designed in order to improve some characteristics of composite joints and to know how to increase the connector load and ductility capacities. The T-Perfobond is the combination of the Perfobond connector with the 'T' EUROCODE 4 connector [64] (Fig. 2.229). The Perfobond connector consists of a rectangular steel plate with holes welded to steel beam later immersed into the concrete slab. After the concrete curing, the shear longitudinal resistance and the prevention of the vertical split between the steel beam and the concrete slab are achieved. This is due to the contribution of concrete dowels that are formed in the Perfobond holes and may be even more significant with the use of additional reinforcing bars passing through the connector holes.

The first part of the instigation involved Perfobond connectors push-out tests carried out at the University of Coimbra. The studied parameters were the presence and number of holes and the number of reinforcing bars reinforcement at the connector holes. A total of eight types of configurations with repetition were performed, as shown in Fig. 2.230. The Perfobond connectors were made without holes, with two or four holes, respectively, in one or two rows in the direction of load transfer, for slabs with 120 and 200 mm thickness. In all cases of this test series the concrete had a constant compressive strength, from the same mixture and tested at an age after stabilisation of its strength, corresponding to a nominal C50/60. Slab thickness of 120 and 200 mm was used. The adopted geometrical characteristics for the connectors are shown in Fig. 2.230.

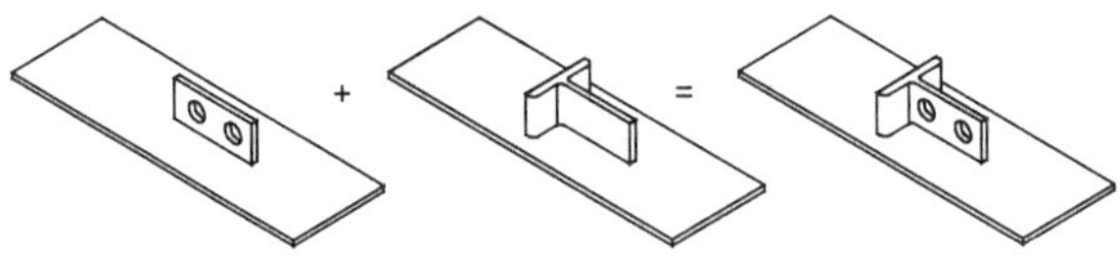

Fig. 2.229 T-Perfobond connector origin.

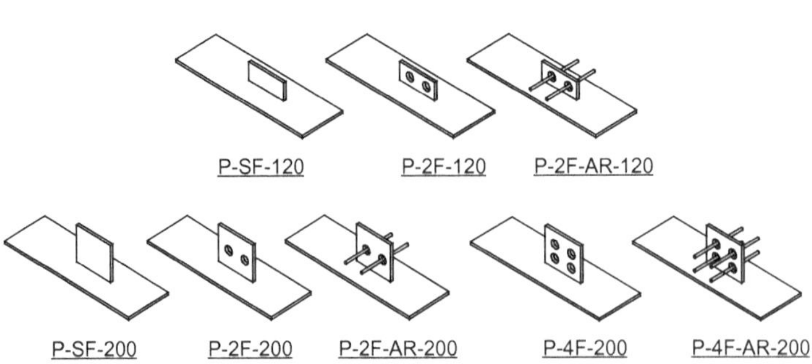

Fig. 2.230 Perfobond-rib connector tested configurations.

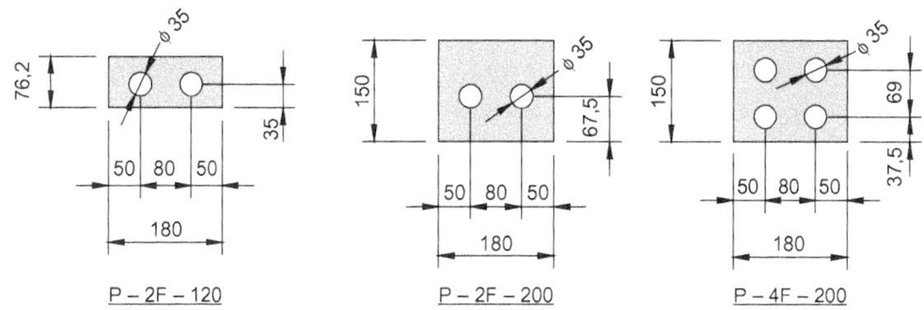

Fig. 2.231 Layout of tested connector.

The second series consisted of 16 push-out tests (eight different tests with a repetition: specimens A and B). Fig. 2.231 shows an overview of the specimen's characteristics. The dimensions of the Perfobond type connectors were established as a result of the required slab thickness and the holes spacing, respecting the minimum distance of 2.25D. The 76.2 mm height connectors were utilised for 120 mm thick slabs and the 150 mm height connectors for the slabs of 200 mm thickness. The Perfobond connectors were made from grade S355 (nominal yield stress of 355 MPa) plates 12.5 mm thick.

Fig. 2.232 shows the specimens' instrumentation. LVDTs were used to measure the relative displacement (slip) between steel and concrete and the uplift. An LVDT was used at the specimen upper part, close to the hydraulic jack for control purposes. In the 2 F-120-A, P-2 F-200-A, and P-4 F-AR-200-A specimens, strain gauges were used to acquire the stress state at the connectors and at the steel section flange. In the P-2 F-120-A and P-2 F-200-A specimens, four strain gauges were adopted. Two linear strain gauges

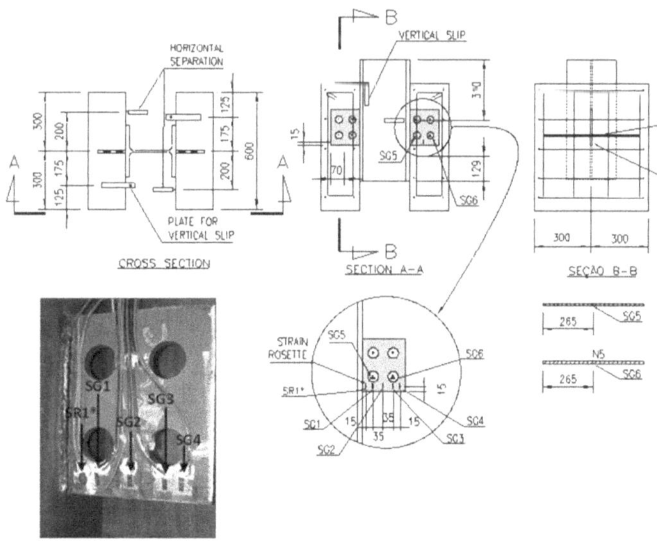

Fig. 2.232 Test instrumentation layout: P-4 F-AR-200-A strain gauge arrangement and LVDTs.

and one rosette were used on the connector and one linear strain gauge was utilised on the reinforcement bar located at the connector hole. A total of seven strain gauges were adopted: four linear strain gauges and one rosette on the connector, and two linear gauges at each of the reinforcements bars located at the bottom connector hole.

The reinforcement bars utilised in the concrete slab and bars passing through the connector's holes were made from 10 mm S500 corrugated bars (nominal yield stress of 500 MPa). For the P-2 F-AR-200-A specimen, the strain results are shown in Fig. 2.233, where the results from linear strain gauges indicate that the gauge located near the hole (at central position) stopped working before the connector capacity was reached (549 kN). Alternatively, the remaining gauges located at the connector and the reinforcement

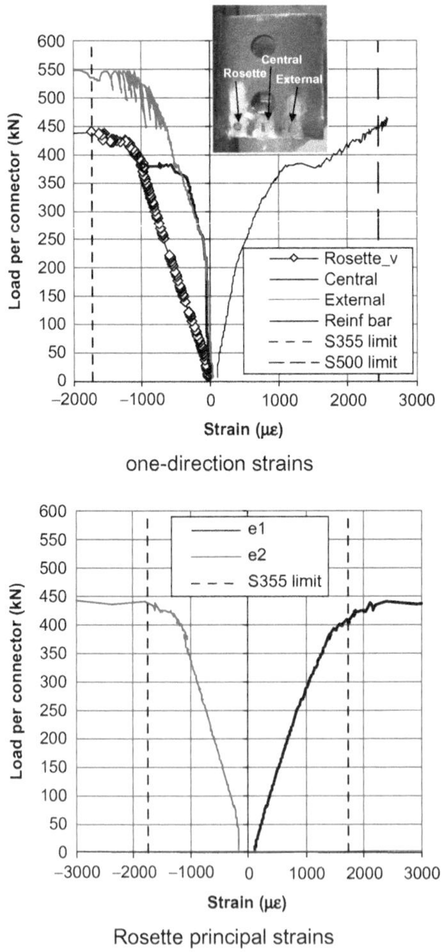

Fig. 2.233 T-Perfobond strains.

bar kept working until the maximum test loading, showing yielding in both, the connector plate and the reinforcing steel bars. Rosette results portray a roughly symmetrical response for directions 1 and 2, but with direction 1 indicating a slightly larger deformation. The connector plate yielding was reached well before the maximum test load contributing for the connector ductility.

For the P-4 F-AR-200-A specimen, the strain results are shown in Figs 2.234 and 2.235. Six linear strain gauges (four at the plate and one at each of the two reinforcement bars passing through the connector holes) and one rosette were utilised, Fig. 2.234 indicates that the strain distribution along the connectors height follows a growing pattern from the top to the bottom, where the maximum value is next to the weld. This fact was confirmed from a larger strain of the reinforcement bar located in the bottom connector hole, compared to the bar located at the top hole. Fig. 2.235 shows that the former has a greater strain than in rebar 6, and both yielded

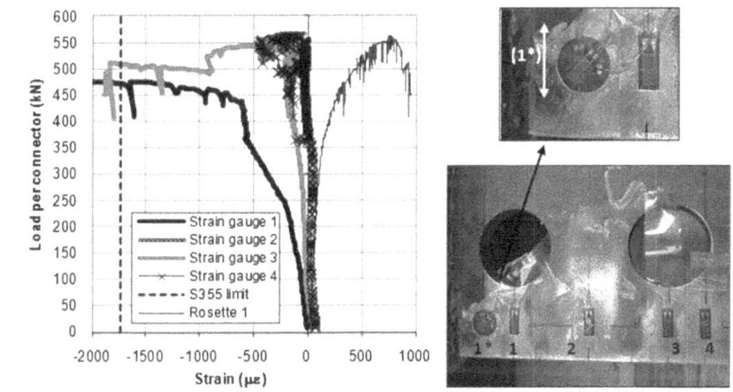

Fig. 2.234 Strains at the connector P-4 F-AR-200-A.

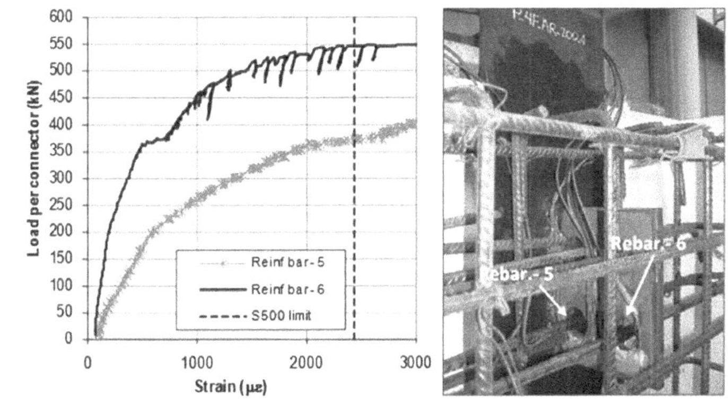

Fig. 2.235 Strains at the connector P-4 F-AR-200-A and reinforcing bars' details.

during the test, before maximum load was attained for 562 kN. Finally, the fact that gauge 4 indicated compression revealed that this was the area where the system anchorage was effective.

Fig. 2.236 shows the force–slip curves of the Perfobond connectors for 120 and 200 mm thick slabs. The 120 mm slab used connectors with a height of 76.2 mm without holes, and with two holes and reinforcing bars passing through them. Fig. 2.236 shows the equivalent results for the 200 mm slab, and connectors with a height of 150 mm. It can be concluded that adding the holes leads to a connection resistance increase of 8% for the 120-mm thick slab. Providing 12 mm reinforcement bars inside the holes leads to a resistance increase of 28% for the 120-mm thick slab and of 9% for the 200-mm thick slab. Summarising, the resistance gain due to the presence of holes or reinforcement proved to be larger for the smaller connectors embedded in 120 mm slabs, than for the larger connectors, embedded in 200 mm slabs.

As far as slip capacity is concerned, an increase of 17% was noted when reinforcement bars were added to the two holes connector in a 120-mm thick slab (i.e. from test P-2 F-120 to test P-2 F-AR-120). The connectors without holes depicted a high ductility and a large vertical uplift, since the holes were the only mechanism that opposed to this displacement. Finally, it should be observed that all tested specimens depicted a ductility capacity of at least 6 mm, thus enabling to classify them as ductile according to Eurocode 4. Similarly to the first test results, the collapse has involved, in all tests, the formation of a longitudinal crack along the slab that progressed and opened with further loading. This was followed by concrete crushing at the connector's front face. Again, the connector yielding was, as already mentioned, present in advanced load stages.

Additional experimental tests were carried out to assess the performance of the T-Perfobond connector. The first part was held at the Civil Engineering Department of the University of Coimbra whilst the second was performed at the Civil Engineering Department of the Pontifical Catholic

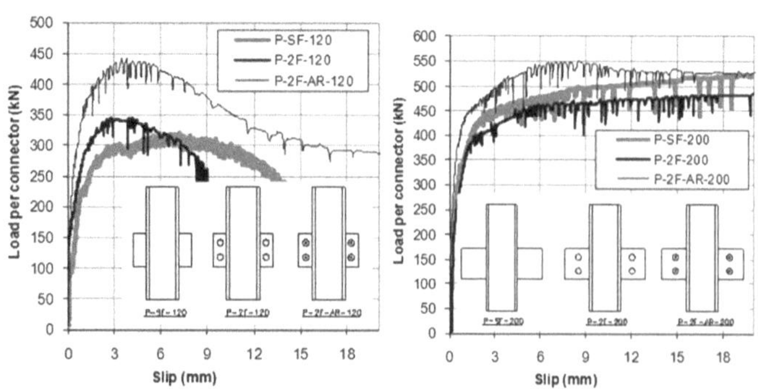

Fig. 2.236 Perfobond connectors' results for 120 and 200 mm thick slabs: 2nd series.

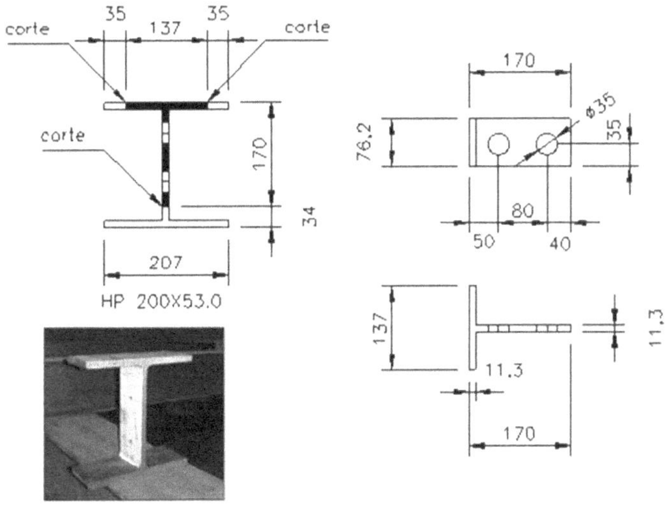

Fig. 2.237 T-Perfobond connector geometry.

University of Rio de Janeiro, PUC-Rio. In the first part, the T-Perfobond connector was made from S275 rolled IPN 340 profiles (Fig. 2.237). On the second part, the T-Perfobond connector presented a new geometry, made from an ASTM A572 Grade 50 steel, equivalent to S355, HP 200 × 53 profiles (Fig. 2.237). This new connector geometry was based on the idea of a connector using the same web and flange thicknesses (Fig. 2.237). T-Perfobond connectors were manufactured for 120-mm thick slabs with compressive strength corresponding to a concrete class C30/37.

The experimental programme, divided into two stages, consisted of two tests in the first stage, followed by three more tests on the second. All connectors had two holes, 76.2 mm, height, 170 mm length, as shown in Fig. 2.237. In order not to have results affected by the chemical adherence between the steel profile and the concrete slab, the steel surface except on the connectors was previously treated with oil. Vertical displacement transducers were used in the two horizontal slabs to measure the relative concrete to steel slip, as well as the vertical uplift (Fig. 2.238).

In the second stage, the test system was adapted to a reaction frame with two 1000-kN hydraulic using a transition beam to simulate a single centred load (Fig. 2.239). A hinge was also used between the transition beam and the steel profile to ensure a concentrated load, used a rod 'I' push-out. From the curves present in Fig. 2.240 it possible to conclude that the T-Perfobond connectors made from HP 200 × 53 profiles, with flange and web equal thickness, presented a better ductility. The connector that presented the best load-carrying capacity was the connector manufactured from the HP

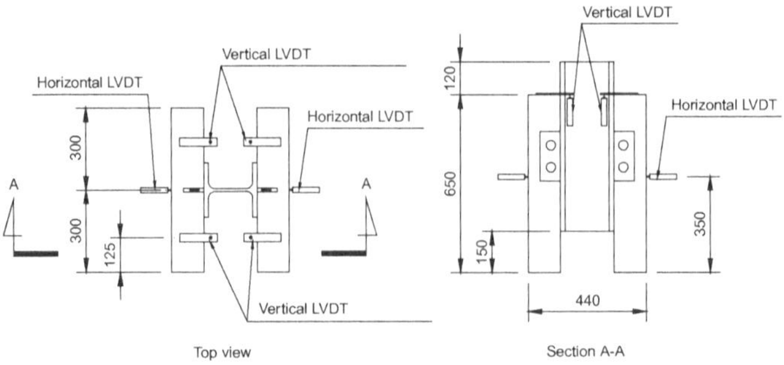

Fig. 2.238 Models instrumentation.

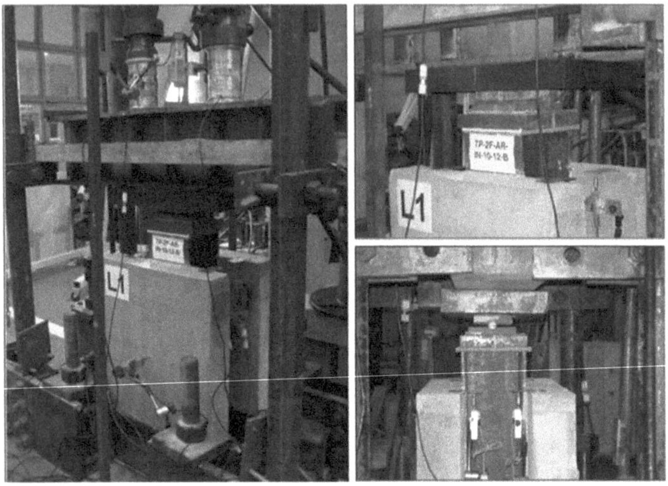

Fig. 2.239 Push-out test reaction frame, instrumentation and hinge.

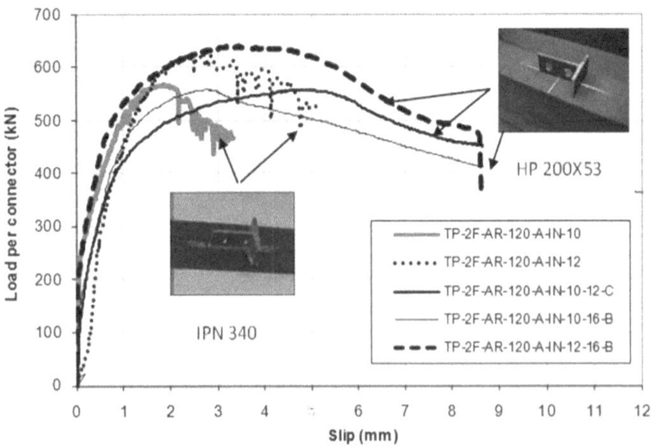

Fig. 2.240 T-Perfobond comparison IPN 340 versus HP 200 × 53.

200 × 53 profiles that used 12 mm stirrups and 16 mm reinforcing bars in the holes, TP-2 F-AR-120-A-IN-12-16. Compared to the TP-2 F-AR-120-A-IN-10 connector, from the first stage, it presented a 13% resistance and 137% ductility gains. The connectors manufactured from the IPN 340 presented a higher load-carrying capacity when compared to the other two second stage tests, but presented very little ductility. Fig. 2.240 shows the first and second stage of tests evaluating the contribution of the more flexible connector of the second stage with equal flange and web thicknesses.

In the first stage tests, the collapse started with the emergence of a longitudinal crack in the slab where is located the connector front and is gradually extended along the slab, followed by the concrete crushing. In the second stage tests no longitudinal cracks were detected in the slab. As the load increased a slip between the steel and concrete gradually occurred. This was followed by cracks in the internal part of the slab, gradually spreading on the concrete element and later by the concrete crushing. Fig. 2.241 shows the TP-2 F-AR-IN10-12-C prototype configuration after the concrete removal.

From an analysis of the push-out test results it was possible to select a test with optimum ductility and load-carrying responses to be adopted in the full-scale test. The selected test was the TP-2 F-AIR-IN-12-16-B, which used armour of 16-mm reinforcing bar passing through the connectors holes and 12 mm stirrups. The push-out tests simulated the behaviour of the steel and concrete in the composite beam. However, the load distribution along the beam is not the same that occurs in push-out tests (i.e. direct shear).

In a composite beam, shear flow on the steel to concrete interface varies along the beam and is dependent on the load distribution. In addition, in a composite beam, the connector's deformations alter the shear force flow, inducing a decrease of the maximum shear force and its redistribution along the beam span. Therefore, it is important to verify if the push-out test results are representative of the full-scale composite beam to be used in future analysis.

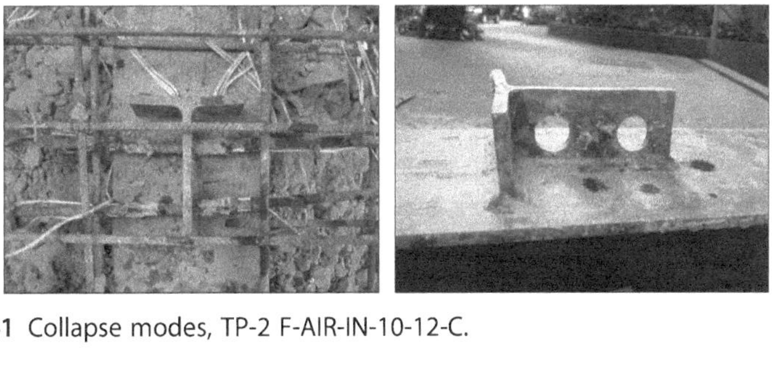

Fig. 2.241 Collapse modes, TP-2 F-AIR-IN-10-12-C.

A full-scale test was performed to investigate the structural behaviour of the T-Perfobond connector in an actual structure simulation, investigating its deformation and load-carrying capacities. The test, in which dimensions were a 9 m span, a slab with a 2.3 m width and a 0.12 m thickness, was conducted in the PUC-Rio Laboratory under a uniformly distributed load. A W410 × 60 profile, equivalent to a W16 × 40 section, was used to compose the steel beam, whilst an HP 200 × 53 profile was used to manufacture the connector, the same previously used in the push-out tests. All these profiles were made of an ASTM A572 grade 50 steel from AÇOMINAS. The adopted connectors spacing was 1600 mm (Fig. 2.242). Between the T-Perfobond connectors hook splicers were welded to the 6.3 mm reinforcing bars. The purpose of the hook splicers was to avoid the concrete slab buckling.

The connectors were welded with 18 mm E70XX fillet weld. The thickness of the slab was the same used in the push-out test. The assembly of the supports was made before positioning of the beam in its final location. The beam was simply supported with a fixed and a mobile support. To ensure the composite beam stability in the supports, they were assembled in pairs as shown in Fig. 2.243. An 'I' beam was positioned on these support serving as the support for the main composite beam. The resulting span

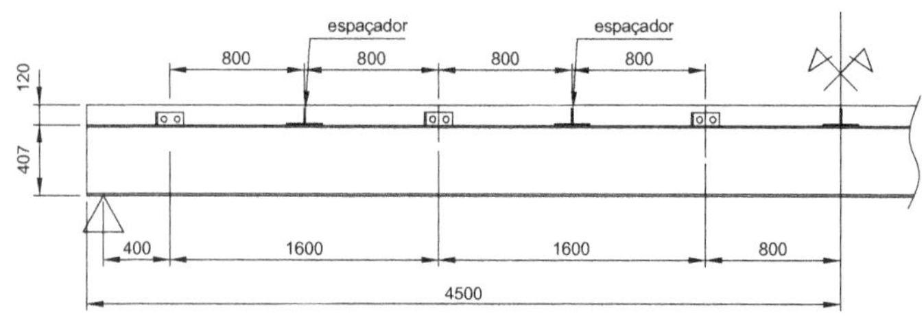

Fig. 2.242 Spacing between connectors.

Fig. 2.243 Support system: fixed & mobile—8.8 m span.

between the supports was equal to 8.8 m. The same reinforcing bars used in the push-out test were utilised around the connectors. A cone-shaped distribution was adopted with an approximately 60° spreading angle depending on the compression zones and possible cracks that could occur during the test. The slab reinforcing bars were distributed every 320 and 200 mm in the longitudinal and transversal directions. A 6.3-mm diameter reinforcing bars mesh was used to minimise the concrete surface cracking.

Fig. 2.244 shows the global configuration of the full-scale test [71]. A distributed load application system simulated by four concentrated load points along the beam was used. Two 50-kN hydraulic jacks were used as well as two-500 kN load cells were utilised under each actuator. This load cell was installed between a universal hinge and the support.

Two LVDTs at each end of the slab were used to measure the slip at the concrete to steel interface. Two LVDTs were also installed on the slab transverse sides to control the slab horizontal translation. Four LVDTs were used to measure the beam vertical displacements. The connectors and steel profile strains were monitored with linear strain gauges. Four strain gauges were positioned in the connectors located at the slab. The intermediate and central connectors used two strain gauges in the web and flange (Fig. 2.245).

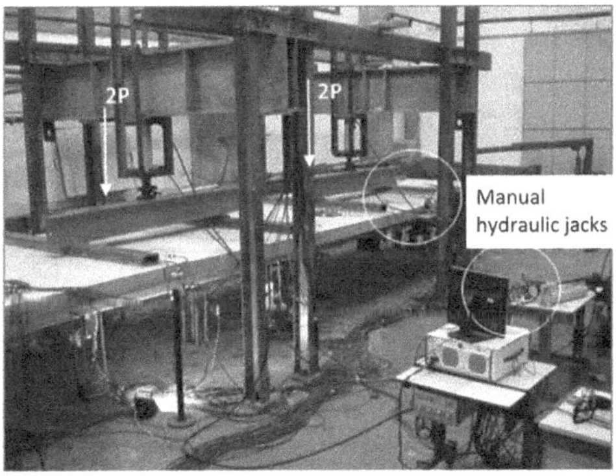

Fig. 2.244 Instrumentation and loads.

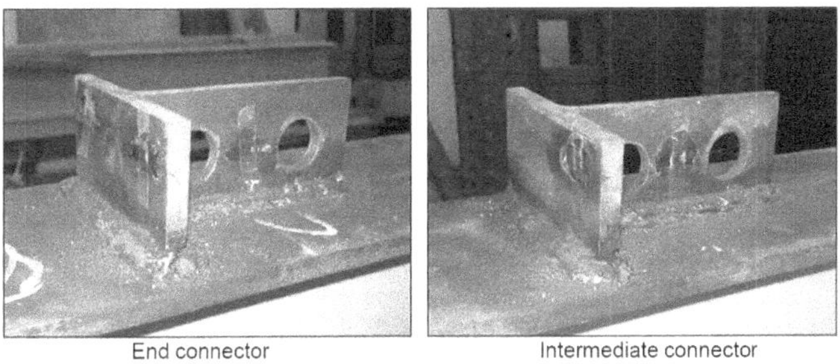

Fig. 2.245 Strain gauges at the T-Perfobond connector.

A previous evaluation of the capacity of the studied composite beam was made indicating that it was able to sustain a 21.25 kPa stress. The tested slab area was 20.7 m^2. This load is equivalent to two equally spaced 220 kN concentrated loads applied to the beam. Three preliminary tests were conducted for calibrating the instrumentation and gradually mobilise the structure. These preliminary tests were associated with the following maximum load values of 100, 155, and 200 kN. In the first, second, and third tests the loads were maintained for 15, 30, and 30 minutes before the loads were removed. The final test was performed after the preliminary test described earlier. The maximum load values listed earlier correspond to the loads per jack, referred to as load 2P (Fig. 2.246). The composite beam was designed, considering a partial interaction, for a maximum moment of 664.82 kN m.

Fig. 2.246 shows the total moment vs vertical displacement curves for the three preliminary and the final tests. The total moment is the sum of the moments due to the self-weight and the moment produced by the applied load. The calculated moment due to self-weight was 83.2 kN m. The composite beam self-weight considering the slab weight, the profile weight, reinforcing bars, connector, concrete, and load distribution beams was equal to 8.59 kN m.

The composite beam test presented an almost linear behaviour up to a load level of $2P=100$ kN (imposed in each actuator, where the total beam load was 200 kN) remaining in the elastic range. From this point onwards, a stiffness loss associated with a nonlinear response was detected. In the second preliminary test ($2P=155$ kN), the beam entered the plastic range and a residual vertical displacement of 12 mm was recorded. In the third preliminary test ($2P=200$ kN) a 35-mm residual vertical displacement was measured. Up to this stage, no cracks were detected in the composite beam.

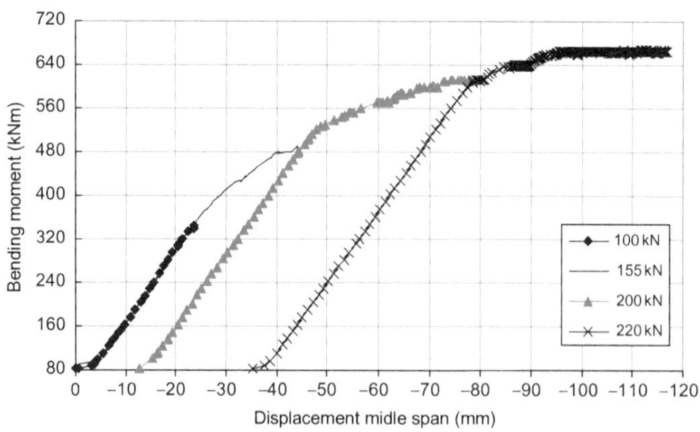

Fig. 2.246 Maximum vertical displacement versus midspan vertical deflection.

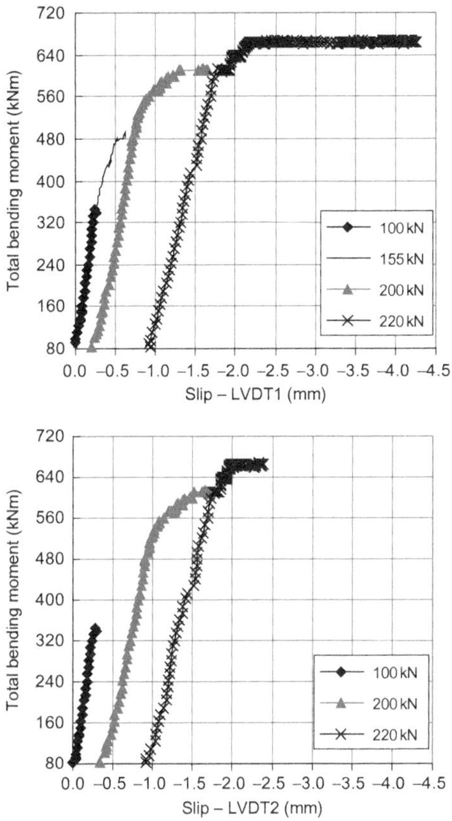

Fig. 2.247 Maximum moment versus midspan slip.

Fig. 2.247 shows the slip at the concrete to steel interface monitored with the LVDTs installed at the beam ends. In this figure, all the slips from the preliminaries and final tests are presented. One of the LVDTs depicted higher slip values implying that one of the composite beam ends slipped more than the other. The final test was conducted to a maximum load of 220 kN because the slip has reached more than 4 mm, value enough for a comparison with the push-out tests.

Fig. 2.248 shows connector flanges strain gauges values based on their position in the beam to evaluate the beam deformation capacity. It can be noticed that the connectors at the ends are the most solicited.

In the final full-scale test, a shear failure between concrete slab and steel beam occurred. This connector failure and was associated with the loss of the load-carrying capacity of the structure. A significant slip between the steel and the concrete was observed during the test, finally provoking the composite

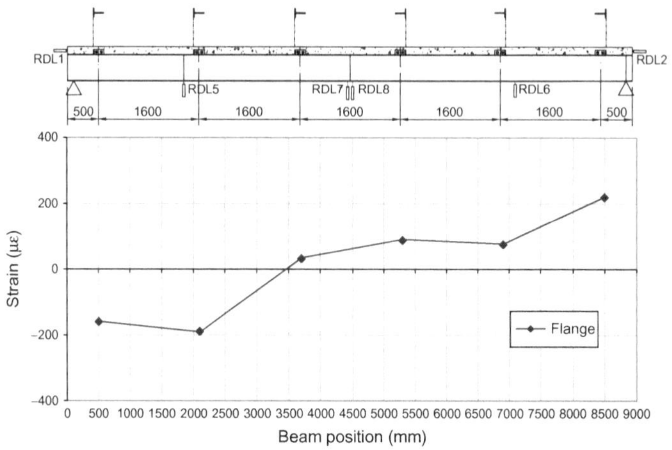

Fig. 2.248 Connectors flange deformation in relation to their beam location.

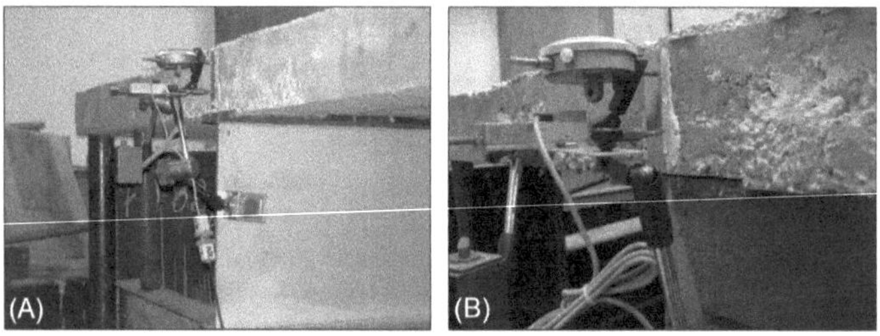

Fig. 2.249 Collapse by shear. (A) RDL 1—Slip. (B) Slip and vertical split.

beam failure. This collapse essentially occurred in one of the beam sides whilst the vertical split between the steel and concrete section was visible near the support (Fig. 2.249). The concrete cracks occurred vertically at the beam midspan.

The results of the push-out tests experimental campaign were used to evaluate the structural behaviour differences between Perfobond and T-Perfobond connectors in terms of their load and deformation capacities, as well as their collapse modes. The T-Perfobond connectors presented a higher load-carrying capacity and stiffness than the Perfobond connectors. The advantage of using this type of connector is related to its fabrication made from rolled profiles leftovers, not requiring the production of a new specific connector. The T-Perfobond connectors produced in Portugal, from the IPN 340 profile, unlike the remaining connectors, did not present a ductile behaviour, therefore requiring a less economical elastic design. Alternatively, the T-Perfobond, connectors made in Brazil, from HP 200 × 53 profiles, adopting reinforcing bars in the holes, showed a ductile behaviour, enabling the use of a plastic design.

The full-scale test enabled the evaluation of the T-Perfobond connector behaviour under a bending test. The composite beam was designed with partial interaction. In the test, as predicted, the failure was conditioned to the slip in the concrete steel interface in one of the beam sides, and depicted a high vertical deformation capacity. The interface slip showed a great influence on the beam stiffness, since as the slips increase, a structural system stiffness loss occurs.

The test indicated that highest stress concentrations occurred at the beam ends, since the connectors installed on these points were the most deformed. It was also observed that the experimental moment strength was practically equal to the one forecasted by a plastic design analysis. With this, it was possible to conclude that the overall stress block model was satisfactory for the beam design with T-Perfobond shear connectors.

The present experimental programme initially contemplated a series of push-out tests performed on Perfobond and T-Perfobond connectors. In a first phase developed at the University of Coimbra, the control was done through deflections, enabling the monitoring of ultimate loads and deformations in the loading and unloading ranges to determine their ductility according to Eurocode 4 criterion [64]. The second series of push-out tests performed at PUC-Rio adopted load control strategy by hindering the monitoring of loads and deformations in the unloading range.

With these tests it was possible to determine the influence of the number and arrangement of holes in one and two lines, and the assessment of the individual effects of the resistant portions present on these connectors design formulas. It was also possible to improve the performance of T-Perfobond connectors in terms of ductility from the reduction of their flange thickness and their use with the connectors flanges always positioned towards the closest beam support. This development has led to its use in composite beams with uniform spacing, according to the plastic design allowed for shear connectors with ductile behaviour. The adopted instrumentation in these tests has been commented in previous cases and contemplated LVDTs and strain gauges used in slab, profile, and connector to determine displacements, rigid body movements, slip, and stress distributions in the shear connectors.

In a final phase, a full-scale test using T-Perfobond connectors was also executed using partial interaction. These tests made possible to determine the actual distribution of the forces acting on the connectors, as well as the effect of large spacing between connectors in a composite beam. The full-scale test presented a behaviour compatible to the expected, showing a considerable ductility and having its performance conditioned by the serviceability limit state associated to the excessive displacements. This test confirmed the T-Perfobond connector's ability to redistribute forces, serving also as an indicator of the expected stress concentrations in regions close to the connectors, due to the fact that their small number have not reduced the load carrying ability of the investigated structural system.

2.7 FINAL CONSIDERATIONS

The purpose of this chapter was to present the necessary steps for the correct design, implementation, and results interpretation of steel and composite structures experimental models. Every day, new systems for acquisition and real-time monitoring with fibre optics, laser measurements or with piezoelectric materials are created, fact which facilitate part of this process.

On the other hand, nothing can replace the experience of technicians and researchers who have been dealing with these types of modelling for a long time and, even with limitations of space, time, equipment, and financial resources, have produced excellent experimental models that have contributed to a better understanding of complex structural engineering problems.

The next chapter contains a similar description to what was done in this chapter for numerical simulation models. Consequently, main steps for the design, fault detection, course corrections, surpassing convergence problems, and the correct treatment and interpretation results obtained will be presented and discussed in detail.

CHAPTER THREE

Finite Element Modelling

3.1 INITIAL CONSIDERATIONS

The objective of this chapter is to present the necessary steps for the design and results interpretation of numerical models for steel and composite structures simulation. It should be once more emphasised that the present book was conceived from an idea where the most important aspects of numerical modelling are introduced, as computer simulations of practical cases of numerical experiments are presented. Initially, some aspects and general questions will be addressed to elucidate key points of numerical modelling planning to ensure the validity and reliability of the performed computer simulations.

A very simple idea, but extremely effective, is usually presented as the master key for the numerical model design of steel and composite structures simulations. This idea is based on the concept that the model must be continually developed, starting from a simple initial, yet representative model of the studied problem. Based on the complete understanding of this initial model, refinements and more complex modelling strategies can be gradually introduced, and calibrated to ensure its reliability. This continuous development allows a more reasoned decision making and enables a consistent adoption of various modelling strategies to be made until the accuracy required for the studied model is reached. Although this strategy at a first glance looks like a more time-consuming process, in medium and long terms, this procedure is fundamental to overcome difficulties in complex models, where the main focus can be easily lost or where the next steps in the process are unknown, due to: inconsistencies, errors, lack of consistency, or even an improper model calibration. Very complex initial models are usually too difficult to be evaluated and calibrated, due to the presence of several variables and factors interacting with one another, making it difficult to identify what steps should be taken to correct directions or to improve the efficiency of the investigated structural model.

Modelling Steel and Composite Structures
http://dx.doi.org/10.1016/B978-0-12-813526-6.00003-9

The numeric modelling described in this chapter will be based on the finite element method, traditionally adopted by structural engineers to simulate complex problems that usually cannot be solved with the suitable accuracy with mechanical, or simple analytical models. The first step towards an efficient numerical modelling is to establish the support conditions, imperfections, and type of material that are compatible with the assessed problem. This is generally followed by a mesh refinement study necessary to ensure the correct structural model performance, usually evaluated in terms of ultimate and serviceability limit states.

Another fundamental issue is associated to the numerical models calibration. When these simulations are performed, users should be aware that results are always reached at the end of the process. Situations where: improper support conditions, poorly established symmetry conditions, badly defined material, among many other things, can compromise the model that often took weeks or months of working time with countless computer processing hours.

The result's reliability can only be guaranteed with the calibration of the model with the experiments, and when these are not available, with the results of nonquestionable classical theoretical formulations. In the literature, many models are mistakenly calibrated only with other numerical simulations, generating a vicious cycle, where it cannot be assured, with suitable accuracy and clarity that the models represent the reality of the steel and composite structures modelled. This strategy also inhibits the determination of the validity and accuracy limits of the designed numerical models.

3.2 TYPES OF STRUCTURAL ANALYSIS

3.2.1 Introduction

In a finite element method modelling, several types of analyses can be performed, for example, static or dynamic analysis, where the latter could be modal or involve forced vibrations. In this chapter, for scope and space limitation reasons, the focus will be only on static analysis. Within this perspective, a static linear elastic analysis is the simplest and default case, but with limited use and application, due to its restricted scope.

This requires the use of alternatives to consider the possibility of structural instability, better known as global buckling, that may be initially studied based on differential equation solution processes that solve this problem through an eigenvalues and eigenvectors analysis. This type of analysis is also largely used for determining the model critical imperfections, adopted to represent, with greater fidelity, the actual structural response and surpassing, for example, possible bifurcation points at the examined structures equilibrium paths. However, when local and/or lateral buckling, residual stresses or even phenomena involving yielding arise in the evaluated structures, other types of more refined analysis become necessary. The next section presents in detail the main types of static analyses including its nonlinear effects to summarise the reader on their scope and proper application to steel and composite structures modelling.

3.2.2 Nonlinear Analysis

Structural analysis with different levels and degrees of refinement can be used, due to the fact that the structural behaviour is usually nonlinear. These analytical techniques require different degrees of complexity and can only be simulated with computer programs, due to the large work involved. The methods are summarised in Fig. 3.1 on the basis of a structure load versus displacement curves.

Linear analysis is still the most used method in practice. In this analysis, it is assumed that the displacement (of a reduced magnitude) is proportional to the applied force leading to negligible second-order effects (associated to structure geometry changes). It is also assumed that the structural member's stiffness is constant and independent of the presence of axial forces, which reflects not considering P–δ and P–Δ effects, as well as the material's nonlinearity. In such cases, an alternative to take into account the effects of some nonlinearities, uses an amplification factor to magnify the bending moments acting on the structure.

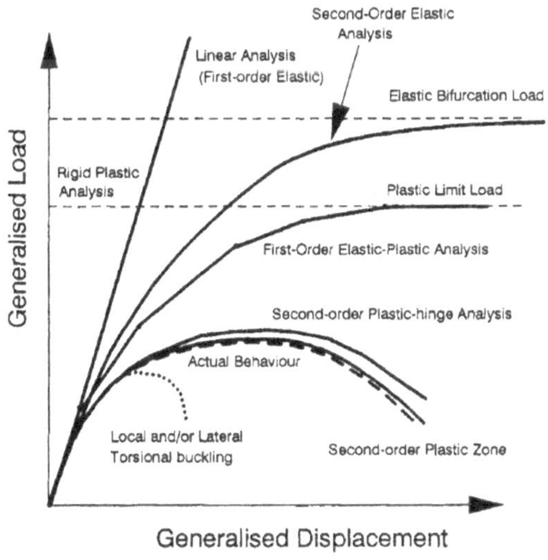

Fig. 3.1 Portal frame analyses, Paulino et al. [79].

A more realistic and accurate assessment of the structure must be performed with analyses that consider the various nonlinear effects that can usually occur. Two classes of nonlinearities are usually identified. The first class is the material nonlinearity, which is originated from the changes in the material physical response to the applied stresses, and that is considered with trajectory-dependent variable constitutive laws. The second class is the geometric nonlinearity that is produced by finite displacements accompanied by structure stiffness changes under an acting load.

When the material nonlinearity is considered in the analysis, this is usually done in the material constitutive matrix. In the cases studied in this chapter, i.e. structural steel and reinforced concrete, numerous formulations can be used to model their behaviour. In General, the steel is modelled with the von Mises criterion and an isotropic hardening flow rule, where the stress versus strain curve is linear elastic up to yielding and perfectly plastic between the elastic limit and the start of the strain hardening phase. This procedure is based on verifying the behaviour of each point of the structure, so that it is contained within the isotropic hardening yield surface. When this condition is violated, a radial surface return procedure is usually adopted to ensure that, during the analysis, the material behaviour initially established continues to be valid. More details about the material nonlinearity can be found in Crisfield [89].

On the other hand, a suitable concrete modelling uses a multilinear isotropic flow rule described by linear segments of the stress versus total strain curve, starting at the origin, with positive values of stress and strain, considering the concrete compressive strength (f_c) corresponding to a 0.2% compressive strain. The stress versus strain curve also assumes a total increase of 0.05 N/mm^2 of the concrete compressive strength up to a 0.35% compressive strain to avoid numerical problems due to an unrestricted flow rule. Normally, the concrete compressive strength is assumed to be equal to the values obtained in compressive strength cylindrical tests, while the concrete tensile strength is assumed as 1/10 of its compression resistance. The examples of this chapter will contain, if applicable, a detailed description of how the material nonlinearities were incorporated into the developed numerical models.

In a linear analysis of a structure, the force is assumed to be proportional to the displacement and the superposition effect principle can be applied to obtain the structure's force and moment distributions. For geometrically nonlinear problems, two types of nonlinearities are considered: the $P\text{–}\Delta$ and P-δ effects. The $P\text{–}\Delta$ effect occurs due to a change of geometry structure, while the $P\text{–}\delta$ effect results from the element stiffness change in the presence of an axial force and the bar deformed shape along its length. In these cases, the Newton–Raphson method provides an exact structural response to the structure ultimate or service load. On the other hand, the exact modelling of a structure behaviour including all sources of nonlinearity is not a simple process. Even when the buckling effects are considered, the load versus displacement trajectories associated to the various buckling modes such as the global Euler buckling, local buckling, and lateral torsional buckling, are not easy to be obtained.

Moreover, the second-order elastic analysis method automatically includes second-order effects due to the geometry changes and the element stiffness variation in the presence of axial force. Additionally, the use of empirical methods for amplifying moments in the presence of axial force used in various standards, and the need for an evaluation of the elements effective buckling length in sway frames become unnecessary.

In the context of a geometric nonlinear analysis, some assumptions are usually made to consider the buckling effect. A perfectly straight column under the action of a concentric axial force will not deform laterally until the bifurcation load is reached, point where the column deflects sideways infinitely without any further load increase. At this point, the structure stiffness reaches zero, due to the cancellation of the linear stiffness matrix caused by the geometric stiffness matrix. These approaches for the analysis, including bifurcation points, are often found in practice, but they represent a simplification and/or idealisation of the buckling phenomenon.

In a second simplified approach, changes in geometry are allowed, i.e. the structural geometry is continuously changing, enabling updated structure geometry to be performed in terms of their load versus displacement trajectories. This method better reflects the structural behaviour, being known as a load versus displacement analysis and it is usually adopted for assessment of numerous problems with geometric and material nonlinearities. The bifurcation analysis is based on a mathematical idealisation where the structure only deforms in load direction until it reaches the bifurcation load. This way, the element stiffness matrix is only a function of structural member forces, material, and geometric properties before the load is applied. The tangent stiffness matrix, which relates to the displacements of the load increments will be independent of the displacement. Additionally, the structure is assumed to be submitted to deformations only in the load direction, consequently leading to insignificant geometry changes effect. Under these assumptions, the structure will be assumed as perfect, once it is under a perfectly centred load without any load component or eccentricities in the other directions.

The validity of the solution for the structure buckling load is based on the hypothesis that the structure is perfect, i.e. the column is perfectly straight and the axial force is ideally centred. Strictly, this strategy is a mathematical idealisation that rarely occurs in practice, because all structures have initial imperfections, while the loads have eccentricities. On the other hand, the structure deflections, although small, can occur in directions out of the loading plan. Generally, the bifurcation analysis results should not be directly used in a practical design and should only provide an upper limit to the structure buckling load. The advantages of this approach are associated to its simple formulation and small computational time.

Load versus displacement curve analyses are necessary to generate a more accurate structural response. In this simplified approach, the structure's equilibrium path is determined while the strains and deflections are continuously updated and monitored. The largest difference between the load versus displacement curves analysis and bifurcation analysis are the considerations of the large displacements effect in the stiffness equation. The influence of residual stresses and the structure deformed geometry can also be incorporated into the stiffness equation.

In general, the difficulty associated with a nonlinear analysis results from interdependencies of the stiffness, load, and displacement. In contrast to the linear analysis, the stiffness matrix in a nonlinear analysis depends on the deflection, as well as the forces. The applied force will produce deflections that will change the geometry and thus the structural strength against external loads. Using the force and an initial stiffness matrix, the deflections are obtained and, along with the applied forces, change the stiffness matrix and invalidate the previous analysis results. In order to obtain a solution that satisfies the previous equation, a linear incremental approximation, i.e. a trial and error process or an incremental-iterative method is necessary. Generally speaking, a direct iterative method, an incremental method, or procedure predictor–corrector process can be used for the previous equation solution. The following paragraphs will contain a brief discussion of some of these methods like: direct iterative method, pure incremental method, incremental-iterative method, and Newton–Raphson method.

In the direct iterative method, the adopted stiffness matrix is based on the applied forces, while the deflections are calculated and used to reevaluate the new stiffness and the new deflection. This process continues until the input data are sufficiently close to the output results, case where the convergence is reached. The graphical representation of this process is shown in Fig. 3.2. The process is repeated until the next and the last deflection are equal or sufficiently close to each other, characterising the case where the adopted stiffness, deflections, and forces have been correctly evaluated, Fig. 3.2. This process requires only the establishment of successive secant stiffness relations associating the total forces and the total deflections for the solution, but its convergence is usually slow. However, the method has an advantage of

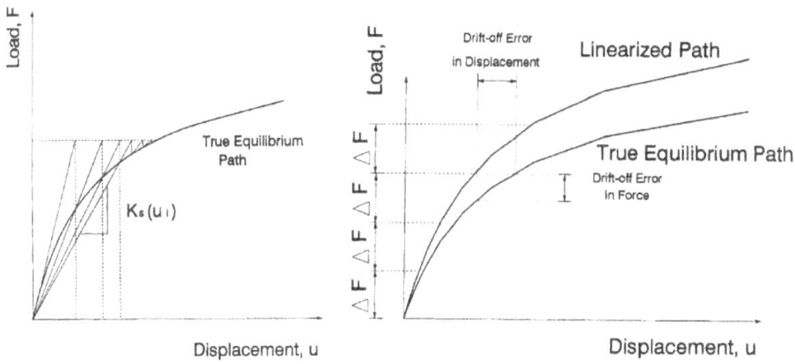

Fig. 3.2 Iterative methods Secant and pure direct incremental—constant increments [79].

simply only using the secant stiffness that makes the complex formulation of tangent stiffness matrix unnecessary.

In the pure incremental method, a curve is usually approximated by a series of line segments. As these segments become smaller, the better the approximation turns out to be, a greater computational demand would also be necessary. The pure incremental method is based on a similar concept in which the load is divided into a series of small load increments, as shown in Fig. 3.2. In contrast to the direct iterative method, this procedure only requires the tangent stiffness, while the secant stiffness becomes unnecessary. The tangent stiffness should be able to accurately correlate the incremental displacement to the load increment, while the method precision depends on the tangent stiffness accuracy and on the step size. This procedure, in principle, can determine the equilibrium path pre- and post-buckling ranges. The load increment can be monitored by the sign of the tangent stiffness matrix determinant. This method requires load steps small enough for the accurate solution of highly nonlinear problems, where the numerical error is unknown.

In the Newton–Raphson method, the tangent stiffness is used to estimate the displacement increment due to a small, but finite, force. The secant stiffness is then used to guarantee the balance. If an unacceptable error is generated, iteration is performed. In the full Newton–Raphson method, the tangent stiffness is corrected at every iteration (Fig. 3.3) while in its modified method version the tangent stiffness is only corrected at a specified number of iterations, usually between 3 and 5 (Fig. 3.3).

The Newton–Raphson load control method is used to determine forces and moments on elements, as well as the structure displacements up to its ultimate and/or service loads. If the structural behaviour is linear, or very close to it, the first prognosis is enough, and it can be then used to evaluate forces, displacements, moments, and stresses. For problems with small

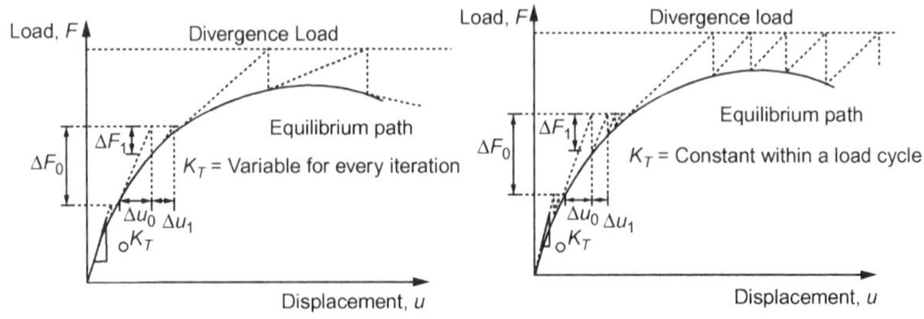

Fig. 3.3 Newton–Raphson method. (A) full; (B) modified, Paulino [79].

displacements, the forces and moments in the structure, after transformation, will balance the applied forces, so that the conditions of equilibrium can be satisfied. However, when the structure behaves nonlinearly, the forces and moments calculated from the displacements will not be able to balance the applied loads. This is the reason for the need to reevaluate the structure's displacements. The element nodal displacement increment is then calculated by extracting the corresponding displacement in its corresponding degree of freedom, while transforming the local element axis. The forces and moments on this element will then be determined by the element stiffness equation. The total or accumulated forces for the element can be then updated. Transforming the element forces and moments from the local axis to the global axis, the resistance of the entire structure can be assembled and evaluated. Comparing the resistance against the applied load, the equilibrium equation for the unbalanced forces is generated. After this step, the evaluation and update of the total displacements is performed.

Graphically, the error is illustrated in Fig. 3.3 as $[\Delta F]_i$. In order to eliminate the error, it is necessary to revisit the structure displacements and check the equilibrium conditions one more time. In the Newton–Raphson method, the error in the force $[\Delta F]$ is calculated and then substituted back in the analysis for determining the displacement increment. The deflections imbalance, $[\Delta u]_{i+1}$, at this instance now, will then be added to the final total displacement to update their values $[u]_{i+1}$. As shown in Fig. 3.3, this method assumes a small load increment parallel to the applied load while the corresponding displacement increase is then determined and accumulated in the total displacement updating the structure geometry. Based on this updated and on the new geometry and forces, the structure strength is then determined and compared to the applied load. When this unbalanced force or equilibrium error is large, this value will be substituted back in the incremental equilibrium for the evaluation of another displacement increment, and the process is repeated until the equilibrium error is small enough. After the equilibrium was reached, a new load increment is imposed to the structure and a new configuration is calculated by repeating the previous processes.

The basic idea of displacement control method (DCM) is to use a fixed displacement component as the control parameter to determine the equilibrium path. Usually, the control parameter is intuitively or empirically selected remaining fixed throughout the process to determine the equilibrium path. For this reason, the DCM is very suitable in regions near the limit load points (snap-through behaviour), but may fail near the displacement limit points (snap-back behaviour). Fig. 3.4 shows the DCM behaviour of near limit points with snap-through and snap-back.

The work control method (WCM) is a potentially powerful method to solve nonlinear problems. It was independently proposed by Bathe and Dvorkin [87] and Yang [88]. Instead of using load increments (as in the Newton–Raphson method) or deflection increments (as in DCM), this method uses work increments, i.e. simultaneous increments of load and displacements. The good performance of this method for limit points and snap-through behaviour is widely documented by Yang [87]. However, for some difficulties, the WCM with bifurcation points and snap-back behaviour are still frequently reported in the literature.

The arc length method (ALM) that considers simultaneous iterations of the load and displacements was originally proposed by Wempner [91] and Ricks [92] for geometrically nonlinear structures. Subsequent modifications performed by Crisfield [90] and Ramm [93] transformed the method, enabling a simple implementation in finite element programs. The ALM basic idea consists in the introduction of a constraint equation such that the increment, measured in the iteration *j* of the increment *i* (evaluated, for example, by the Euclidean norm) that could keep the norm (or length). There are several methods to define the constraint equations and countless possibilities to establish the iterated values sequence that originate the various ALM versions. Fig. 3.5 illustrates the incremental-iterative process adopted in the ALM.

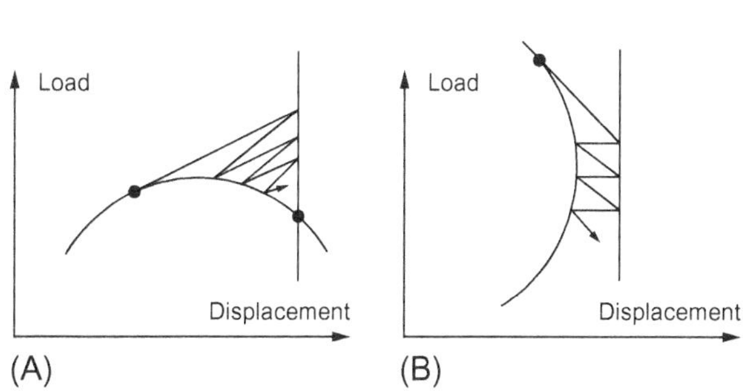

Fig. 3.4 Displacement control method (DCM): (A) snap-through and (B) snap-back [79].

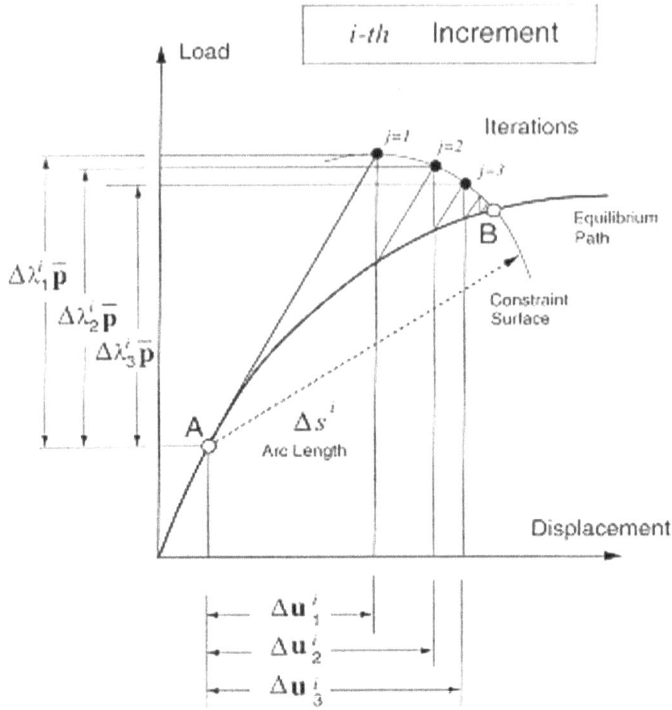

Fig. 3.5 Arc length method [79].

Another important variation of ALM, due to Wempner [91] and Ricks [92], is the ALM linear version o. The main characteristic of this version is that the iterations are processed in a plan. When the iterative vectors are orthogonal to the incremental initial vector, the method is known as Fixed Normal Plan. In addition, when the vector is orthogonal to the immediate preceding incremental iterative vector, the method is known as Updated Normal Plan. Although these linearised versions are easy to implement, they are not as robust as the ALM original version.

The generalised DCM (GDCM) was proposed by Yang and Shieh [94] and is based on a parameter known as GSP (General Stiffness Parameter), expressed by:

$$\mathrm{GSP} = \frac{\delta u_{I1}^{1} \cdot \delta u_{I1}^{1}}{\delta u_{I1}^{i-1} \cdot \delta u_{I1}^{i}} \tag{3.1}$$

where δu_{I1}^{1}, δu_{I1}^{i-1}, and δu_{I1}^{i} are the first components of the displacement vector, evaluated in the first iteration of steps 1, $i-1$, and i, respectively. The GSP parameter is the fundamental idea behind the GDCM, because its use allows

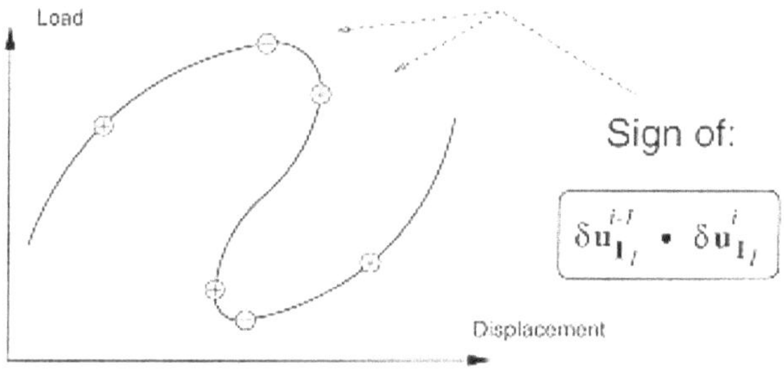

Fig. 3.6 GSP parameter, Paulino et al. [79].

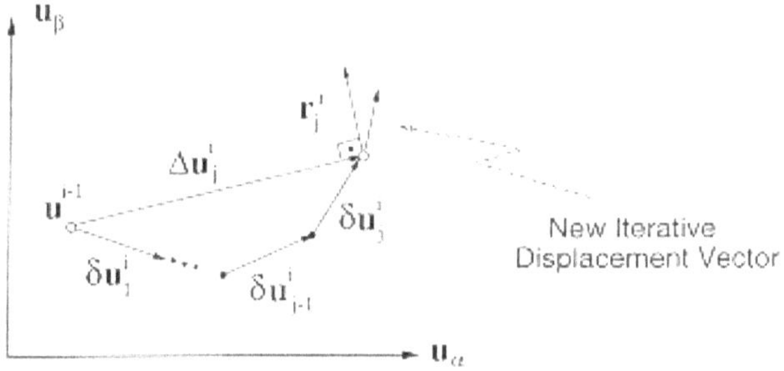

Fig. 3.7 Orthogonality condition [79].

and predicts the load direction sign changes in limit and snap-back points, as shown in Fig. 3.6, making this method highly computationally efficient.

A new scheme for the nonlinear systems solution has been proposed by Krenk [80] usually known as orthogonal residual procedure (ORP). The fundamental idea of this method is based on the condition of orthogonality among the conjugate variables, i.e. the residual vector (unbalanced loads) and the incremental displacement vector, as shown in Fig. 3.7. This is the mathematical interpretation of the physical condition in which the residual orthogonal load will not reduce or increase the magnitude of the current incremental displacement vector.

3.3 MODEL ANALYSIS

As has been previously mentioned, the numerical models calibration is a critical step to ensure the reliability of the performed analyses. It is expected that the calibration be made throughout comparisons to experimental results, or in the absence of these, to analytical models. These comparisons, in turn, cannot ignore the possible numeric and experimental errors that may be present on both sides of the problem causing the need for always keeping a critical view to early detect possible disagreements on the various analysis types comparison.

As a simple example, the case where very thin plate thicknesses are used in beam webs under the patch loads can be cited. Beam experiments with webs less than 1 mm can generate doubts about imperfections and the weld quality, leading to nontrustworthy results. When these tests are part of a larger database containing numerous simulations with several experiments associated to the most diverse characteristics, the structural engineer's responsibility is to evaluate and discard the tests with doubtful character form the database to ensure greater result reliability.

A great part of the examples that will be later presented had several convergence issues during the solution process. Overcoming these problems is not associated to a unique solution, varying in a case-by-case basis, and is directly related to the experience and skill of the structural engineer that is responsible for their modelling. These nonlinear problems require several processes to surpass these difficulties, going through numerous analyses restarts, decreasing loading step sizes, and altering the type and magnitude of the convergence criteria during analysis, among others. More details on the effective strategies to overcome these convergence problems will be subsequently introduced in the application examples presented in the next sections.

3.4 PARAMETRIC ANALYSIS AND RESULTS PRESENTATION

One of the main goals of the numerical modelling of steel structures is, once properly calibrated, to expand the database for a specific structural phenomenon. This expansion is made using a parametric analysis, which enables a substantial reduction of the solution time and cost when compared to conventional laboratory tests.

Despite this fact, some care must be taken in specifying the scope of these analyses, because if they were calibrated to a specific range of variation of their main controlling variables, parametric analyses performed for variables outside these limits, and/or extrapolating their validity, will always be questioned and often invalidate future developments of design equations based on these findings.

A consistent results analysis using established statistical techniques is recommended for assessing the developed parametric analysis reliability. These parametric analyses in turn, are a valuable tool to use in order to assess the relevance of the main variables that control the investigated structural response. This step is extremely important for the correct and reliable generation of procedures, models, and design equations that, in the future, will be incorporated in structural design standards.

The next section of this chapter will present fourteen examples of steel and composite structures numerical models, as well as details regarding their design, analysis, modelling, and strategies for surpassing convergence problems, and also some ideas for the correct interpretation of their main outcomes and findings.

3.5 APPLICATION EXAMPLES

3.5.1 Local Web Buckling in Composite Tapered Beam

An investigation was made using a nonlinear elasto-plastic finite element program and covered practical ranges of the important variables, including the tension flange area and the position of the plastic neutral axis. Using the FINAS program, a simulation of a panel of a tapered beam was developed [90,12], and validated with both series of tests and some additional runs [10–12], using different parameters to isolate the buckling effect.

The main goal of using the finite element method was the production of a comprehensive set of results, covering the full practical ranges of a different number of variables, a so-called 'parametric study', which could be used for design purposes or to validate any proposed simplified design rules. Preliminary finite element simulations were performed prior to the experiments to develop ideas on the design and outcome of the experiments. The obvious way to carry out this validation was by comparisons with the experiments.

Full and modified Newton–Raphson incremental methods were used to cope with the nonlinear characteristics of the solutions, and convergence strategies including load, displacement, and arc-length control were also utilised. The finite element used in the present modelling is an eight-node doubly curved shell with six degrees of freedom per node.

The displacement relationship was modelled by an updated Lagrangian technique while the material nonlinearity was treated by a von Mises analysis through two layers in the shell thickness. A two by two Gauss numerical integration is used and proved effective in avoiding shear-locking problems. The numerical model exploited symmetry and was composed of 112 shell elements distributed over one half of the beam as shown in Fig. 3.8. The length of the specimen panel was equal to 1.3 m and the total length was 1.95 m.

The beam was subjected to uniform bending using three concentrated forces spread across the top of the upper flange and was resisted by three vertical supports across the bottom flange, Fig. 3.8. The force applied directly

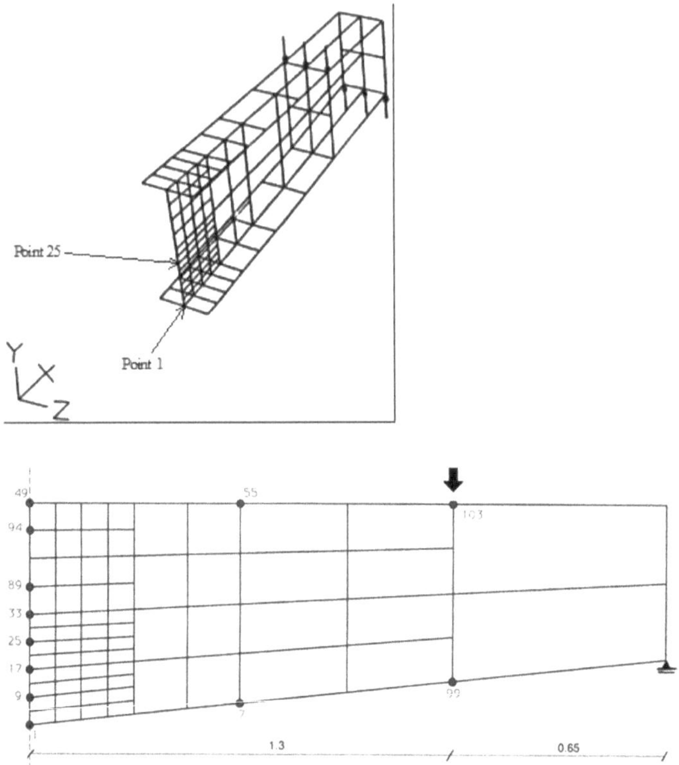

Fig. 3.8 Finite element model mesh.

over the web was 50% of the total load. As in the experiments, the composite (steel/concrete) top flange of the section was modelled by an appropriately increased steel thickness. Lateral buckling of the compression flange was prevented by restraining the movement in the out-of-plane z direction of all the nodes on the line that joined the web and the top flange. Web stiffeners with thickness equal to 50 mm, were utilised at the loaded points and the supports. They prevented local failure of the web at those points and contributed to the loading scheme in a very efficient way. The flanges and the web between the stiffeners were also designed not to buckle or yield, using very high yield stresses and substantial thicknesses to avoid any adverse influence on the load carrying capacity of the specimen panels.

In order to standardise the tests, an imperfection shape in the form of a half-sine wave in the longitudinal and vertical directions was used with a maximum amplitude of approximately 2.5 mm, i.e. $h/250$. A yield stress of 265 MPa was assumed for the flanges and the web following the results obtained for a grade 43 steel. At a later stage, some other simulations were made including the experimental measured imperfections and true yield stresses for calibration and comparisons with the assumed values.

The number of load steps to achieve and pass the peak load varied from case to case. Due to the great nonlinearity of the problem, the full Newton–Raphson method had to be utilised in some cases from very early stages of loading, and the number of increments reached up to 400, which made the process very time consuming. Consideration of the vertical web-buckling phenomenon identified the slope change and the bottom flange area as being other significant parameters.

All the beam models span and the moment lever arm were respectively equal to 4.8 m and 800 mm. The top and bottom flanges widths were 250 mm and their thicknesses were 25 and 12 mm respectively, with the extra thickness of the top flange modelling the effective section of the composite beam as observed earlier. The maximum depth at the beam midspan was 800 mm. The web, top, and bottom flanges yield stresses were assumed as: 420, 391, and 398 MPa, respectively. These relatively high values were adopted to induce the local web buckling most critical condition. The four adopted web thicknesses were 3, 5, 7, and 10 mm. The taper angles varied from 0° to 14.84° ($\alpha = 0$ to 1/4). Zero angle cases, i.e. a parallel flange beam, were also investigated because they characterise an upper bound to the problem.

In the first series, all of the beams with 10 and 7 mm webs had their failure associated with a plastic hinge mechanism. The exact plastic hinge location varied from beam to beam, moving towards the side vertical stiffener when the taper angle increased. The main conclusion for the 7 mm web is that only the steepest angle, 14.84° ($\alpha = 1/4$), depicts, at the slope change, an incipient web-buckling tendency.

In the 5 mm beam with parallel flanges, the failure occurred due to the formation of a plastic hinge near the end stiffener. The beam with an angle of 1.15° ($\alpha = 1/50$) collapsed by plastic vertical buckling of the web. Its deformed shape at the maximum load point is illustrated in Fig. 3.9A where insignificant web deformation is observed. However, at subsequent stages on the unloading path, the web became very distorted, as can be observed in Fig. 3.9B. The same collapse type happened for the girders with angles of 1.43° and 2.86° ($\alpha = 1/40$, 1/30). Three larger angles produced elasto-plastic collapses, and by 14.84° ($\alpha = 1/4$) the web stresses and the buckle turned to be more concentrated near the bottom flange.

The 3 mm web parallel flanges beam buckled in the longitudinal direction, due to the high longitudinal compressive bending stresses in the web. The failure at the ultimate loading stage is portrayed in Fig. 3.10. The girder with an angle of 1.43° ($\alpha = 1/40$) collapsed by a combined longitudinal and

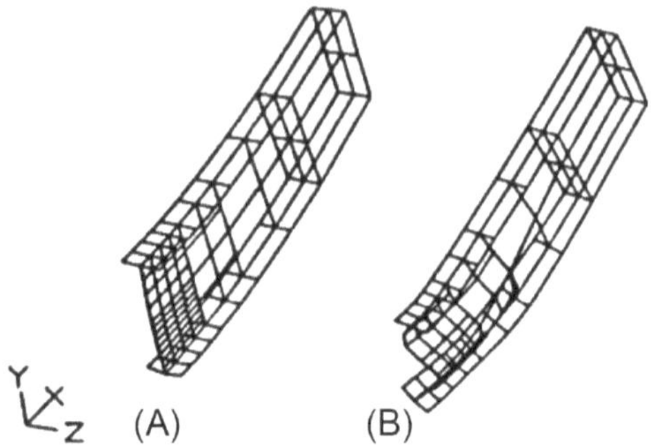

Fig. 3.9 Deformed shapes for beam with $\alpha = 1.15°(1/50)$ at ultimate and ultimate loading stages.

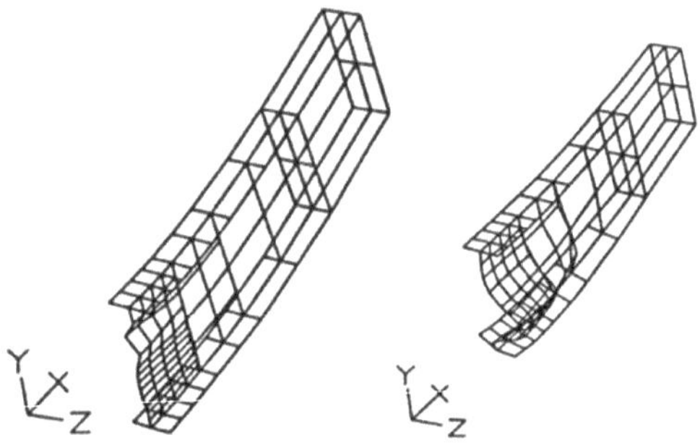

Fig. 3.10 Deformed shape of the beam with $\alpha = 0°$ and $2.86°$ (1/20).

vertical buckling of the web. A vertical buckling of the web in the inelastic region was the main cause of failure for the beam with an angle of 2.86° ($\alpha = 1/20$). The other angles were associated with an elastic vertical web buckling failure. It became evident that if a 3 mm web is to be utilised, vertical stiffeners are needed at the web centre for any taper angle.

In the second series of simulations, the models were very similar to previous series having the same loading, support, initial imperfections, and materials properties. The adopted web thicknesses were 3, 5, 7, and 10 mm, but the taper angle was fixed at 5.71° ($\alpha = 1/10$). The only other change in relation to the previous series beams was the top and bottom flanges thickness that increased to 50 and 25 mm, respectively. The bottom flange thickness was changed to increase the bottom flange area from 3000 to 6250 mm^2 while the top flange thickness had to be also increased to inhibit the local flange buckling ultimate limit state.

The 10 mm web beam failed due to the development of a plastic hinge mechanism close to the edge vertical stiffener. The 7 mm web girder collapsed by a plastic vertical web buckling. Its deformed shape at the point of maximum load is illustrated in Fig. 3.11A, where only small lateral web deflections are

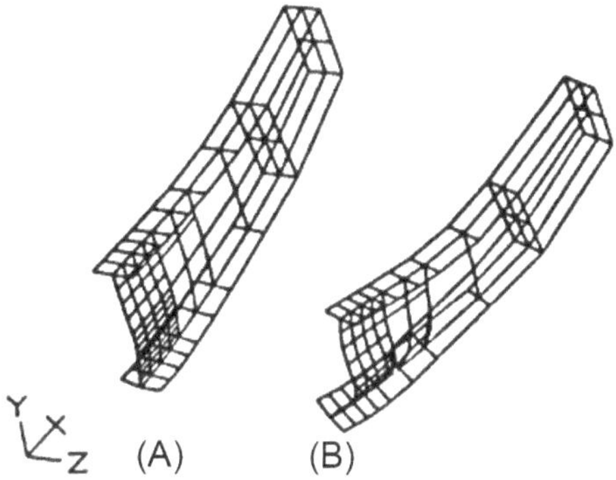

Fig. 3.11 Deformed shapes of the beam with $t_w = 7$ mm at penultimate and ultimate load stages.

observed. However, at subsequent points down the unloading path, the web became very distorted, as depicted in Fig. 3.11B. In this case, the utilisation of a vertical stiffener at this particular point would substantially enhance the beam's performance. In both cases, the 3 mm beam collapsed by an elastic vertical web buckling. The 5 mm beams collapse was associated to a plastic vertical web buckling while the 10 mm beams failed by the development of the plastic hinge at a point close to the panel edge. The main difference of the two different flange sizes occurred in the 7 mm web beams. The beam with a 3000 mm^2 tension flange collapsed by a plastic mechanism, while the other case failed by plastic vertical web buckling.

From these results it is possible to conclude that vertical stiffeners at the centre of the web are required for the web thicknesses less than or equal to 5 mm in the 3000 mm^2 cases, and 7 mm in the 6250 mm^2 specimens. The taper angle and the web thickness influences can be better observed with Fig. 3.12 where it is evident that the bending moment resistance for beams with different web thicknesses, with the same slope increase as the slope becomes steeper. These results confirmed that the taper angle is a very significant parameter in the girder's load carrying capacity. Vertical web buckling only happened in one case for the standard flanges for a h/t_w ratio less than 155. This occurred when the 7 mm beam was associated with the very large angle of 14.84° ($\alpha = 1/4$) bur practical cases with these high h/t_w ratios are rare. A small increase in the web thickness would be the most effective and least costly solution to avoid this kind of collapse.

The tension flange area effect on the moment capacity of the girders is also depicted in Fig. 3.12. When the tension flange area doubled, the decrease in moment resistance as a proportion of the beam fully plastic moment at midspan was up to 45%. These results also showed the strong influence of this variable over the buckling phenomena. Vertical web buckling only happened

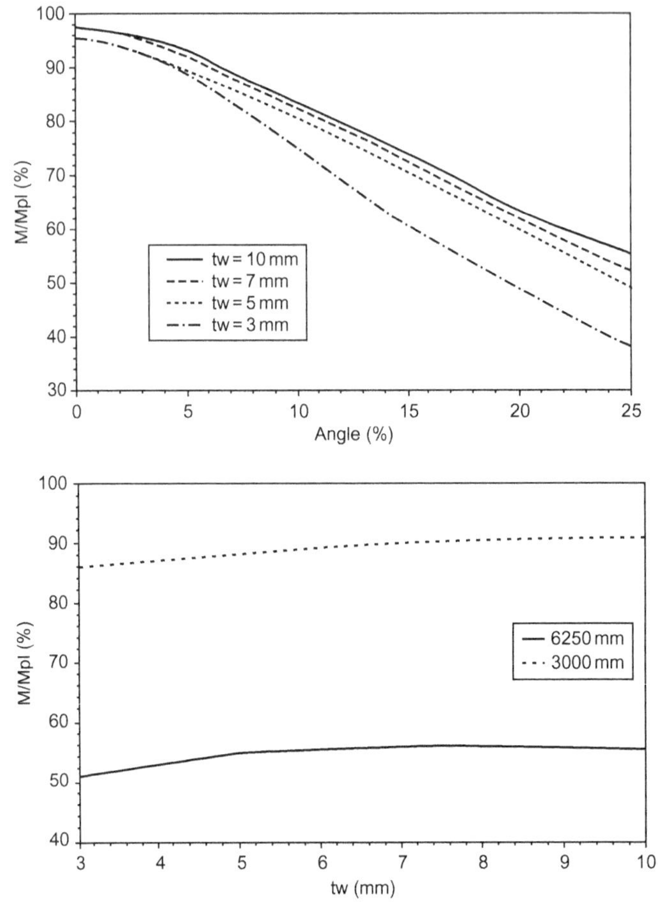

Fig. 3.12 Taper angle and web thickness and bottom flange area influences over the collapse loads.

in one case for the standard flange for a h/t_w ratio less than 155. The finite element simulations have reached a good agreement to the experiments by predicting the final deformed shape and the web stress patterns at the girder ultimate loading stage. The results enabled the development of a simple design model, Fig. 3.13, based on the elastic theory of plates, which was duly compared with numerical and experimental results [9,11,12,90].

Given the high cost of fitted stiffeners, a slightly web thickness increase to avoid vertical buckling seems to be the most satisfactory design expedient in limiting cases. If the web thickness could not be increased, then a vertical stiffener should be utilised on both sides of the web over approximately one third of the web's depth where these stiffeners should be designed as normal load bearing stiffeners.

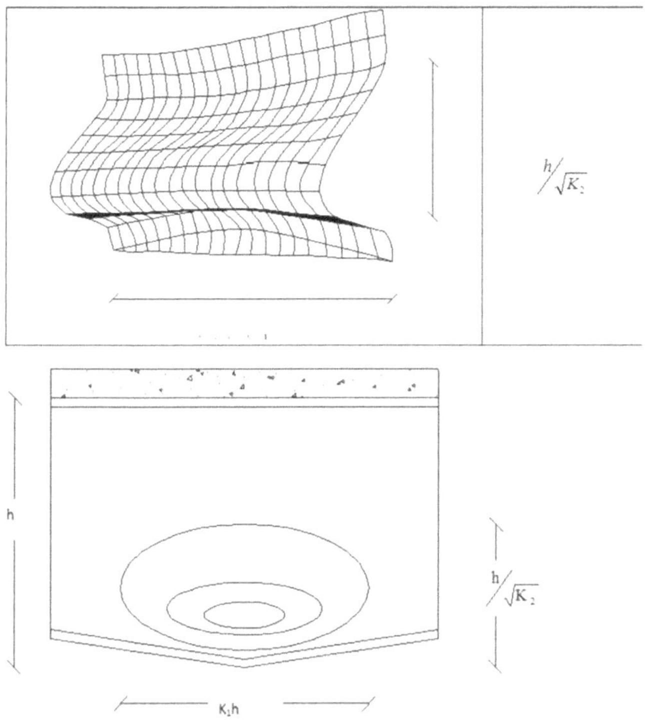

Fig. 3.13 Deformed web shape model and experiments.

The numerical modelling involved ultimate limits states related to local buckling, crippling, and yield, as well as all possible interactions between these phenomena. The collapse was many times associated to the already mentioned interactions, generating the need for a precise modelling of the model support conditions and applied load. The supports were designed to prevent rigid body movements, simulating pure bending with simple supports while the beam top flange was laterally braced. This bracing condition combined with a thicker top flange simulated the concrete slab that, in addition to the creation of an extra resistance component related to the concrete strength, also inhibits the occurrence of lateral torsional buckling.

Similar to the experimental model already mentioned in the previous chapter, the load application was made in a region with an enhanced strength to avoid undesirable collapses. Transversal stiffeners were utilised at the support and load application points to prevent local buckling and better distribute the loads between the web and flanges. The symmetry was used to reduce the model DOF (Degrees of Freedom) and facilitate its convergence. The maximum allowed imperfections present in the fabrication standards and the measured experimental imperfections have been incorporated in the investigated numerical models.

Convergence and the ultimate load determination, as well as the model response on the unloading path, varied from case to case. Various models demanded a high computational force requiring more than 200 load steps, with several restarts and changes in convergence strategies. The adopted convergence criteria were based on varying the square of the displacement norm magnitude under a fine-tuning control to overcome the ultimate load and reach the unloading stage. Another extremely relevant fact is associated to the central issue of the web local buckling at elastoplastic transition. Numerous models at the ultimate load point only showed an almost imperceptible vertical local buckling of the web. In these models, this buckling phenomenon was only visible in subsequent steps of the model unloading path highlighting the need to reach this load stage.

At this point, it is also imperative to stress that many numerical investigations found in the literature do not reach the unloading stage by early stopping, due to the lack of convergence and incorrectly stating that this point is the ultimate load. Of course, this is not the structure ultimate load where various strategies are needed to overcome these convergence problems to surpass these problems.

This lack of convergence can be related to very large load steps, initial, or during the process,, improper use of convergence criteria that are too tight or too loose, among many other possibilities that only an experienced researcher in structural modelling of the elastoplastic buckling phenomena is able to identify and to take the necessary measures for surpassing them.

3.5.2 Numerical Modelling of Semirigid Portal Frames

Despite the substantial boost in structural design knowledge, the semirigid connection design is still facing resistance from structural engineers. This is partly due to the lack of information of the semirigid design, philosophy fact that motivated the present study. This study was centred on a parametric analysis made on a standard four-storey residential building considering all the necessary steps for a semirigid design [91,92].

The study variables were: beam-to-column joint stiffness and strength, structural system (steel or composite), and lateral frame stability (sway and nonsway frames). The parametric analysis is based on the Eurocode 3 [19–21], procedures for semirigid connection design. Based on this methodology a simple design model was proposed and tested on the ANSYS program [32]. The model incorporated geometrical and material nonlinearities and considered the connections semirigid response.

The studied steel building was based on low-cost residential buildings made by Usiminas, a Brazilian Steel Mill [93]. The building had four pavements with a total of sixteen residential units. Each unit had a 46.69 m^2 area with a 2.8 m clear height. The building consisted of three typical pavements, a ground floor, and a roof pavement containing a water reservoir. The architecture consisted of a two block modulated structures with columns distributed in an orthogonal 6.70 m x 12.80 m grid. The 6.30 m span beams made the block interconnection over the staircase shaft. The structure had twelve columns, six per block, and is portrayed in Fig. 3.14.

The beams with spans between 6.30 and 6.70 m, were connected to columns or primary beams. The beam-to-column connections considered three possibilities: rigid, pinned, and semirigid behaviour according to the used structural system. The beam-to-beam joints were considered pinned. The wind loading was resisted by rigid, semi-continuous, or braced frames. The column section disposition was chosen to optimise the sway portal frame lateral strength. With this philosophy the adopted portal frame structures had two, three, or four columns, Fig. 3.15.

Whenever the use of bracing was required, two different systems were utilised. An '*X*' bracing system was used in wall without openings. Alternatively, a modified '*K*' bracing system was used in all other cases, Fig. 3.15. Composite and noncomposite designs were utilised to assess the beams flexural strength. Beams supported a one-way 120 mm thick precast slab. The steel sections consisted of a 250 MPa yield stress steel grade (similar to A-36). The used concrete compressive strength was 20 MPa for the slabs and 15 MPa (25 kN/m^3) for the foundations. The used bolts were ASTM A 325 while the circular steel bars utilised an ASTM A36.

The slab self-weight, plus finishing layers were expected to be 2.5 kN/m^2. The masonry walls utilised in the building walls imposed a distributed load of 2 kN/m^2 in the slab so-called L1, Fig. 3.14 and 1.5 kN/m^2 in the slab L2 of the standard pavement. The wall load was equal to 4.5 and 2.52 kN/m for the standard and roof pavements, respectively. The structure self-weight was 0.25 kN/m^2. The live load was equal to: 3 kN/m^2 to corridors free for public access, 2kN/m^2 for service areas, laundry, and general store rooms, 0.5 kN/m^2 for the roof floor, and 1.5 kN/m^2 for all other building areas. The water reservoir imposed a load of 13 kN/m^2 to the structure.

The structures were studied with plane frame models made of linear elements. Based on suitable load combinations, all the required strength and stability design checks according to the Eurocode 3 [20], were performed. The structural analysis was performed on the ANSYS [32], finite element program. The adopted finite element model encompassed geometric and

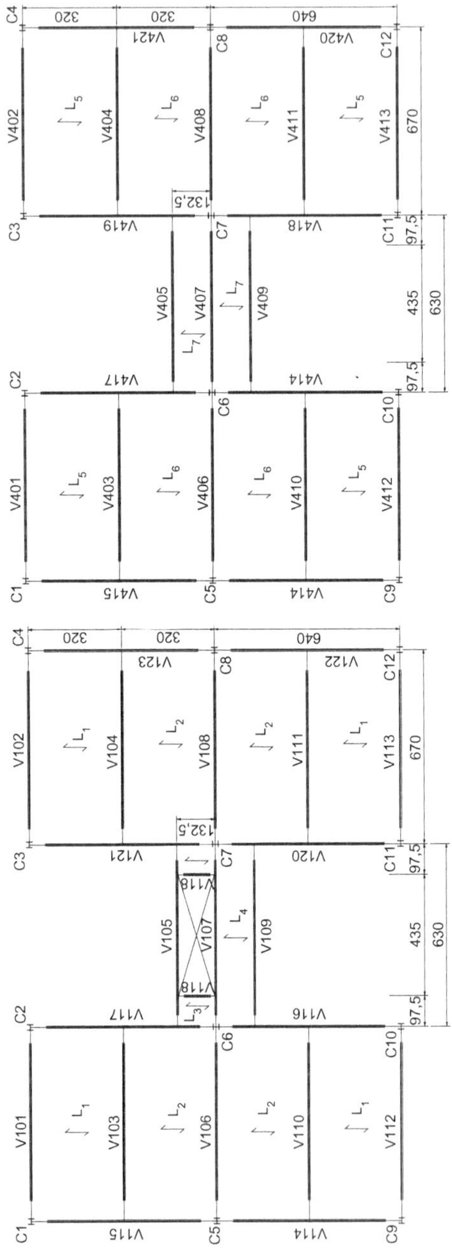

Fig. 3.14 Typical framing plan.

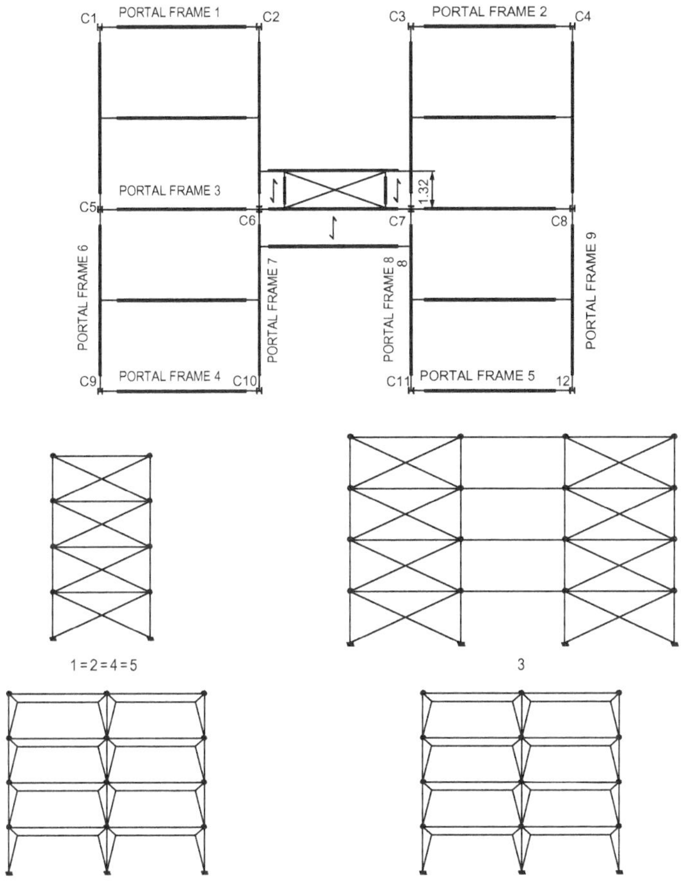

Fig. 3.15 Portal frame idealisation and bracing schemes.

material nonlinearities and a semirigid joint response. The adopted steel constitutive response is depicted in Fig. 3.16, while the updated Lagrangean formulation was utilised to model the geometric nonlinearity. The semirigid joint used in this study, Fig. 3.16, was tested by Carvalho [33].

The Eurocode 3 model was used to define the connection initial stiffness to be utilised in the parametric analysis. A joint initial stiffness value of $S_{j,ini}$ = 50 kNm/rad was utilised to assess the connection secant stiffness. This value was used in the elasto-plastic finite element modelling leading to the values illustrated in Table 3.1.

Since a simple beam finite element with the necessary rotation stiffness to simulate a semirigid joint was not found in the ANSYS finite element library

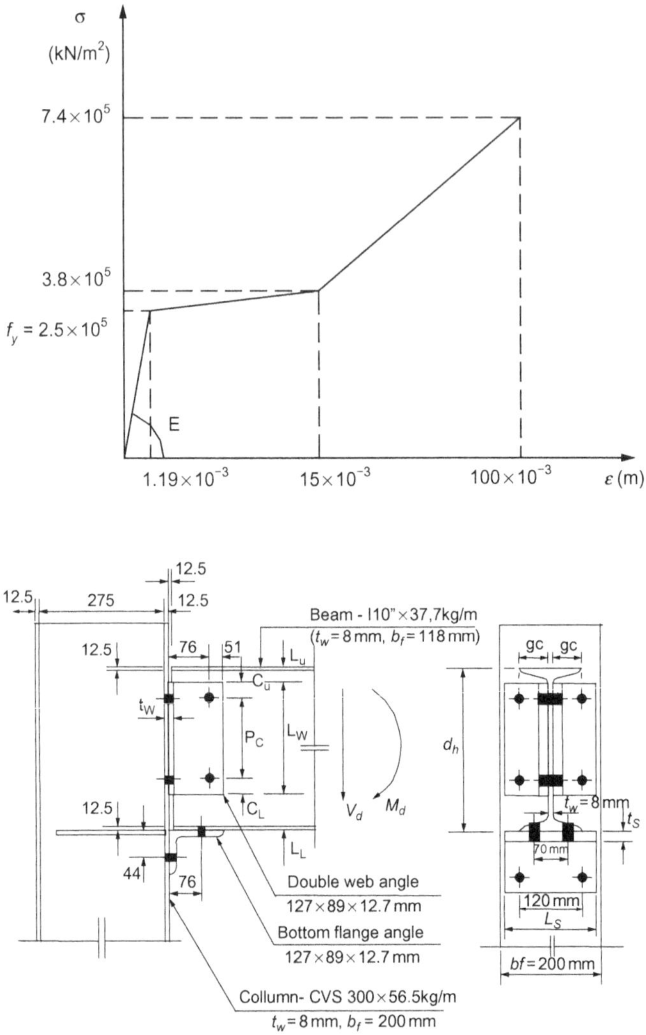

Fig. 3.16 Steel stress versus strain curve and adopted semirigid connection [16].

Table 3.1 Joint Initial Stiffness Parameterisation $S_{j,ini}$

Type	Secant Rotational Stiffness (kNm/rad)	M_u (kNm)	ϕ_0 (rad/1000)
1	Rigid $S_j = \infty$	–	0
2	2. $S_j = 50.0$	60.0	1.20
3	$S_j = 25.0$	60.0	2.40
4	$S_{j/2} = 12.5$	60.0	4.80
5	$S_{j/4} = 6.25$	60.0	9.60
6	Hinged $S_j = 0$	0	-

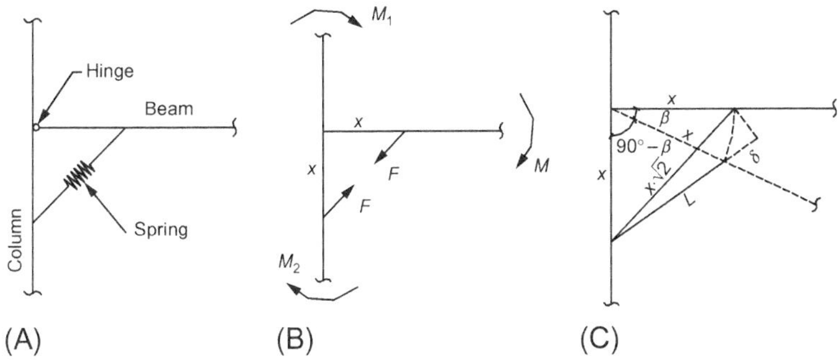

Fig. 3.17 Semirigid finite element model description.

a simple mechanical spring model, illustrated in Fig. 3.17, was conceived, implemented, and calibrated. Eq. (3.1) is reached from the imposition of force equilibrium in Fig. 3.17:

$$F \cdot \cos 45^\circ x = M \Rightarrow F = \frac{M}{x \cdot \cos 45^\circ} = \frac{M \cdot \sqrt{2}}{x} \qquad (3.2)$$

where M is the joint applied moment and F is the spring force. Using a distance from the joint, x, equal to 0.30 m, Eq. (3.1) becomes:

$$F = \frac{M \cdot \sqrt{2}}{0.30} \qquad (3.3)$$

Using simple geometry concepts:

$$L = 2 \cdot x \cdot sen\left(\frac{90^\circ - \beta}{2}\right) x\sqrt{2} = L + \delta \qquad (3.4)$$

where β and δ are an angle and a distance defined in Fig. 3.17. Eq. (3.5) can be determined replacing Eq. (3.4) in (3.3):

$$\delta = x \cdot \sqrt{2} - 2x \cdot \cos\left(45^\circ + \frac{\beta}{2}\right) \qquad (3.5)$$

A simple transformation can change the force versus displacement results obtained from Eqs (3.2)–(3.5) into the more common moment versus rotation variables.

The model calibration was performed based on comparison to other semirigid model and experiments found in literature. The semirigid modelling was performed by Steenhuis [94], who studied the joint stiffness influence over the forces, moments, and displacements of sway frames with semirigid joints, Fig. 3.18. The frame span and height were equal to 6 and 4 m, respectively. The beams and columns consisted of A-36 IPE

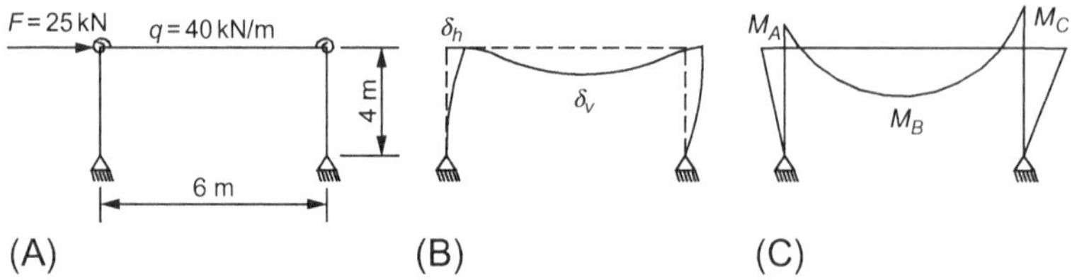

Fig. 3.18 Portal frame 2 (PF2) outline. (A) Geometry. (B) Deformed shape. (C) Bending moments.

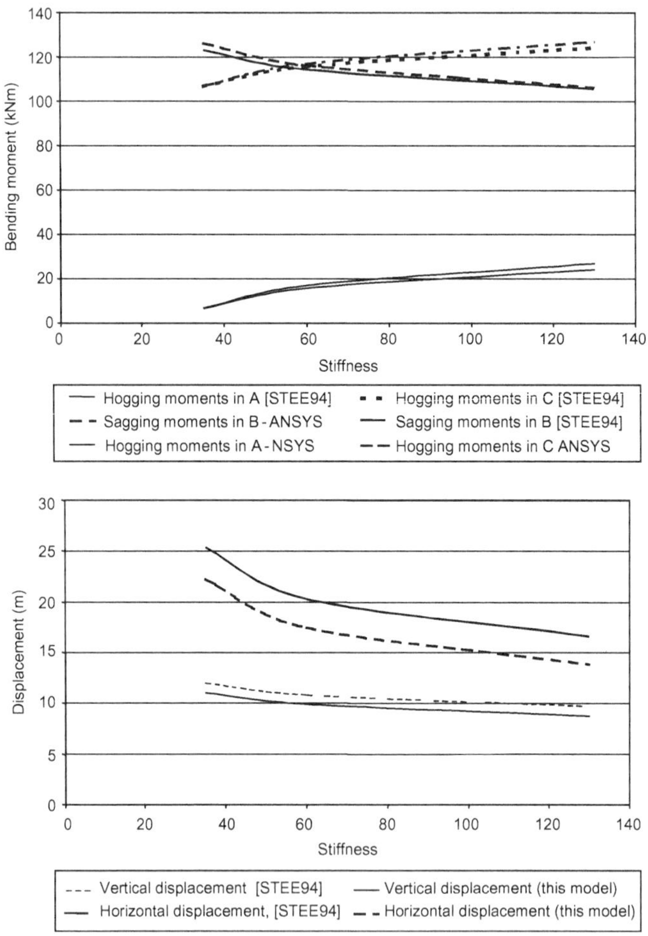

Fig. 3.19 Portal frame 2 (PF2) model calibration.

360 steel sections. The acting loads were: a 25 kN concentrated horizontal load and a 40kN/m uniformly distributed load, Fig. 3.18. This structure was also modelled on the ANSYS program, utilising the suggested connection rotation stiffness, Fig. 3.19. The very small difference found in both analyses

indicated that the semirigid proposed model can be adequately utilised to model portal frames with semirigid joints.

The parametric study nominal loads due to permanent, live, wind, and initial geometric imperfection loads that were evaluated for the studied residential building and are illustrated in Fig. 3.20. The structural design of the various beams and columns was performed according to the Eurocode 3. The beam steel sections are depicted in Fig. 3.21.

Only five beam and four column sections were utilised to standardise the structure fabrication and erection process. A pinned column base was initially utilised. This hypothesis led to horizontal displacements greater than the maximum design recommendation for portal frames with semirigid joint stiffness equal to S_j, $S_{j/2}$, and $S_{j/4}$ leading to the use of a fixed column base. The sagging and hogging moment variation plotted against the connection stiffness S_j, for the controlling load combination of the sway and nonsway frames, so called 1, 2, 4, and 5 in Fig. 3.20 are depicted in Fig 3.22 (standard pavement girder, V101).

A simple inspection in Fig. 3.22 sway frame results indicated that for V101 girder joint stiffness equal to $2S_j$, S_j, and $S_{j/2}$ has not produced any significant change in structural response. When the nonsway frame solution was studied the V101 girder, Fig. 3.12, it had its moments substantially

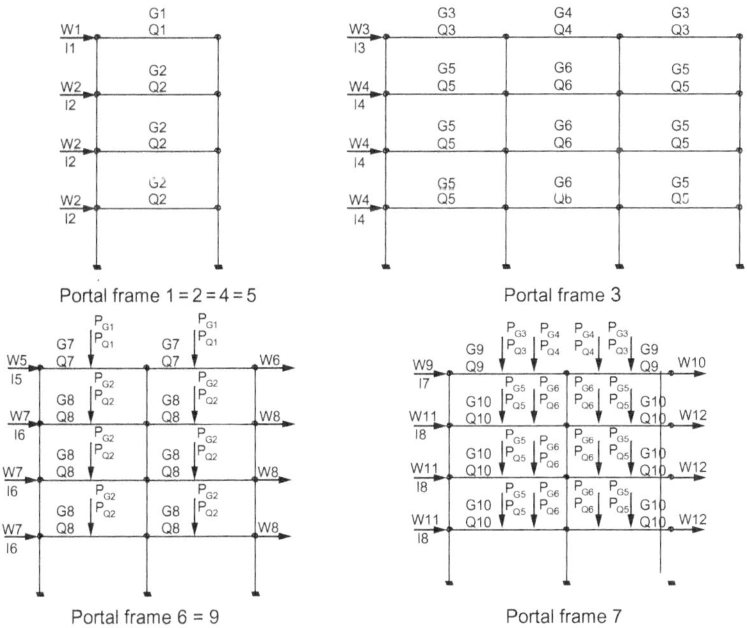

Fig. 3.20 Typical portal frame loading.

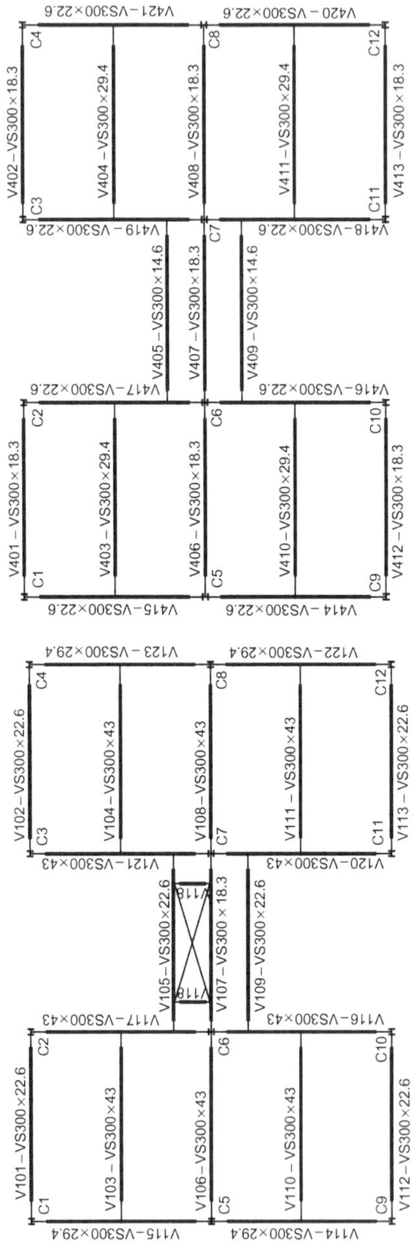

Fig. 3.21 Typical and roof floor beam sections.

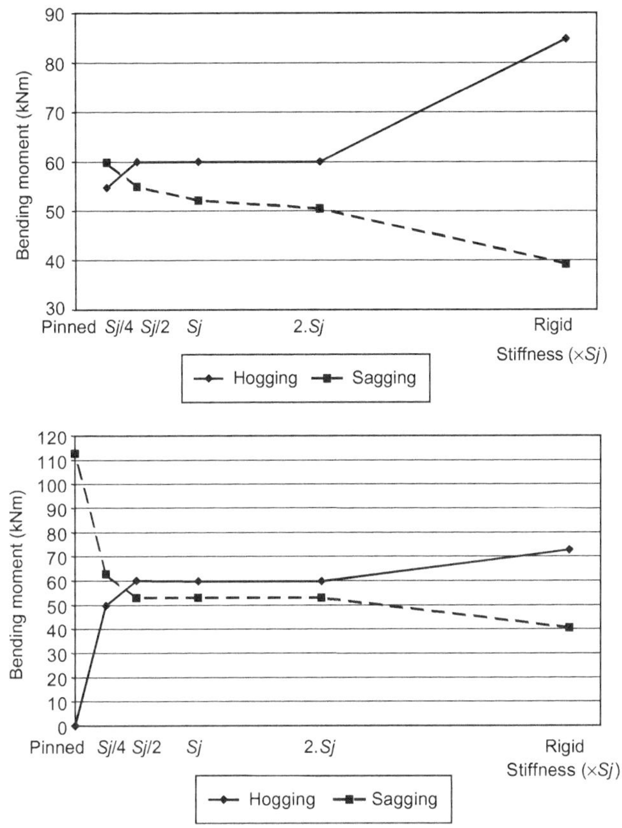

Fig. 3.22 V101 moments, sway and nonsway frames, 2nd and 1st load combinations.

changed from the rigid response up to a joint stiffness value of $S_{j/2}$. Lower stiffness values lead to less severe modifications on the structural moments. No significant change in the structure weight was observed for the V101 and V401 girders in a sway frame when the stiffness joint was modified.

When a nonsway solution was investigated this optimum point was shifted to the stiffness interval from $S_{j/4}$ to $S_{j/2}$, pointing out for a possible weight reduction for frames with this semirigid stiffness. When composite beams are utilised in sway frames, the optimum weight is found in an interval between the pinned up to the $S_{j/4}$ joint stiffness. This point has shifted from the non-composite solution due to the fact that the composite solution optimum is related to beams with higher sagging moments. The same optimum weight point range was encountered in composite nonsway structural frames. Other similar analyses were performed and can be found in Brito Junior [92].

Considering that as the joint stiffness changes, the hogging moments can also change the entire column used in the analysed portal frames adopted a standard steel section. On the other hand, as the controlling force for the column design, the axial component, was only slightly influenced by the joint variation no change in the column sections was needed. The assessment of the portal frame structure weight variation according to the used joint stiffness was also made. Table 3.2 summarises this assessment based on the hypothesis that all frame joint initially assumed a single standard stiffness.

The curves presented in Fig. 3.23 delineate the structure weight distribution for the portal frames studied when different beam-to column joint stiffness was assumed. Two additional structures developed by Usiminas [93], were also included in this graph. The first Usiminas solution utilised semiindustrialised structures where the 212 beams consisted of 175 mm height double reinforced U cold-formed sections. The 34 used columns, almost three times the columns used in this study, consisted of 140x70mm back-to-back hat cold-formed sections. The structure utilised

Table 3.2 Structural Solutions Global Weight Comparison (kN)

Solution	Sway + Steel	Sway + Composite	Nonsway + Steel	Nonsway + Composite
Hinged			302.48	229.64
$S_{j/4}$	262.47	209.71	265.02	211.86
$S_{j/2}$	256.66	215.51	264.21	216.57
S_j	261.15	215.51	264.21	216.57
$2S_j$	263.40	217.76	265.15	217.51
Fixed	257.82	231.64	257.85	229.30

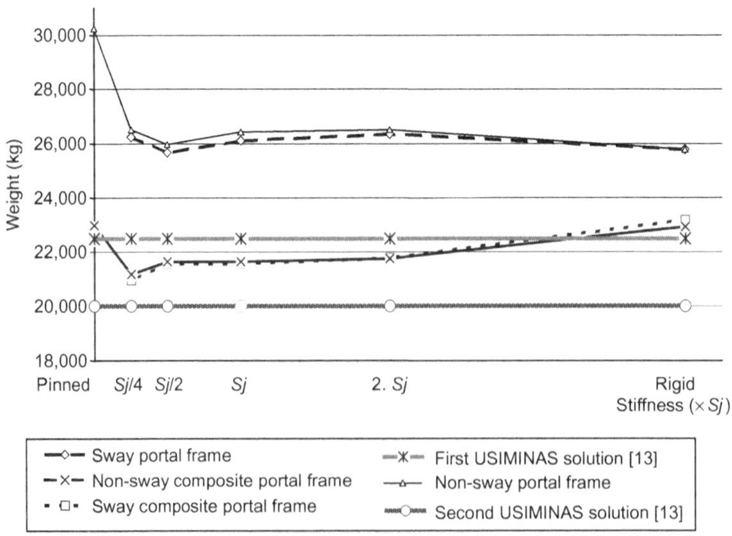

Fig. 3.23 Structural solution assessment.

a 80 mm cast in place slab and ceramic bricks masonry walls. Braced frames in one direction and rigid frames in the other direction totalling a weight of 225 kN of steel were utilised to guarantee the lateral stability.

The second Usiminas solution used an industrialised structure. The beams consisted of 84 rolled '*I*' 350 × 130 mm, 300 × 130 mm, 250 × 150 mm, and 250 × 130 mm sections. The 12 utilised columns used rolled 'H' 220 × 180 mm sections. The slabs utilised steel decks and the external walls consisted of precast concrete panels and drywall internal partitions. The structure lateral stability was guaranteed by the use of rigid frames totalising 200 kN of steel. The curves portrayed in Fig. 3.23 indicate that no significant change in weight occurred for sway and nonsway structures. This was mainly due to the small building height, four floors where the bracings had little influence over the structural response. An interesting conclusion is that the structures with pinned connections represent the heavier solutions among the noncomposite alternatives.

Savings up to 15% in terms of steel weight was achieved for the studied structures, even when compared to the traditional most economical solution. As these structures are utilised in large-scale construction, the achieved 15% weight reduction can generate an even greater economy. The studied four storey structures indicated that the nonsway portal frame solutions are slightly lighter (around 2%) than the equivalent sway portal frame structures. The use of composite construction and semirigid connections significantly reduced, 23% of the structure weight.

One of the most interesting aspects of this numerical modelling consisted in the use of a linear extension spring positioned diagonally to the beam and column to simulate the semirigid joints. The solid adopted model behaved in a satisfactory manner and, by increasing their stiffness, was able to simulate the effect of concrete in composite beams. The results evaluation was based on comparisons in terms of deflections, rotations, and forces aiming to determine the most efficient and economical structure. Complete and simplified processes of nonlinear geometric second-order analyses were also compared defining the validity range of the simplified procedures.

3.5.3 Structural Evaluation of Minor Axis Semirigid Joints With Angles

The use of semirigid joints has increased significantly in recent years. In an attempt to represent the joints actual behaviour, many models have been proposed, mainly for the major axis inertia. When the minor axis inertia is considered, the knowledge is still very limited. A finite element nonlinear analysis of minor axis inertia was performed using the ANSYS program.

The finite element simulations focused on the column web thickness and the influence of the use of column web stiffeners over the joints structural response. The present investigation [95,96], adopted a shell finite element for modelling the web and the flanges of the studied column. This strategy leads too much lighter models instead of solid elements while coping with the relevant deformations: bending, and membrane. The column flanges were considered coupled with the common nodes of the column web. The adopted finite element program, ANSYS [32], enables the material and geometric nonlinearity to be considered. The finite element mesh utilised to model the first test column is depicted in Fig. 3.24.

The boundary conditions were: all base nodes fixed in z direction, one base node full fixed and another, in opposite side, fixed in y and z directions. The load was applied through prescribed displacements in the location of the four bolts in the column web. The material nonlinearity adopted a von Mises yield criterion associated with a three-linear stress–strain relationship to incorporate strain hardening.

The geometrical nonlinearity assumed an updated Lagrangean formulation. A validation of the finite element model simulation was made by a comparison with experiments, Fig. 3.26 [37]. In this graph, load versus displacement curves for three points (A, B, and C) located in the column web, Fig. 3.24, showed a good agreement between experimental and finite

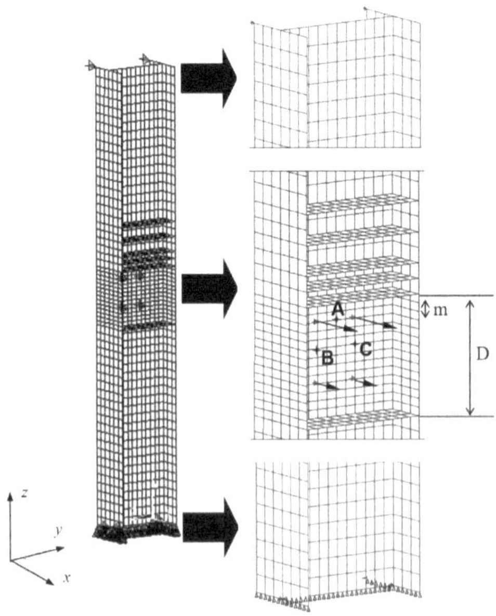

Fig. 3.24 Finite element mesh, load application points, and boundary conditions.

element simulations. The experimental curve was obtained with the aid of displacement transducers as it can be seen in Fig. 3.25.

The principal stresses at the ultimate load step, evaluated through a von Mises yield criteria, is presented in Fig. 3.26 where it is possible to note that a large part of the web panel has yielded.

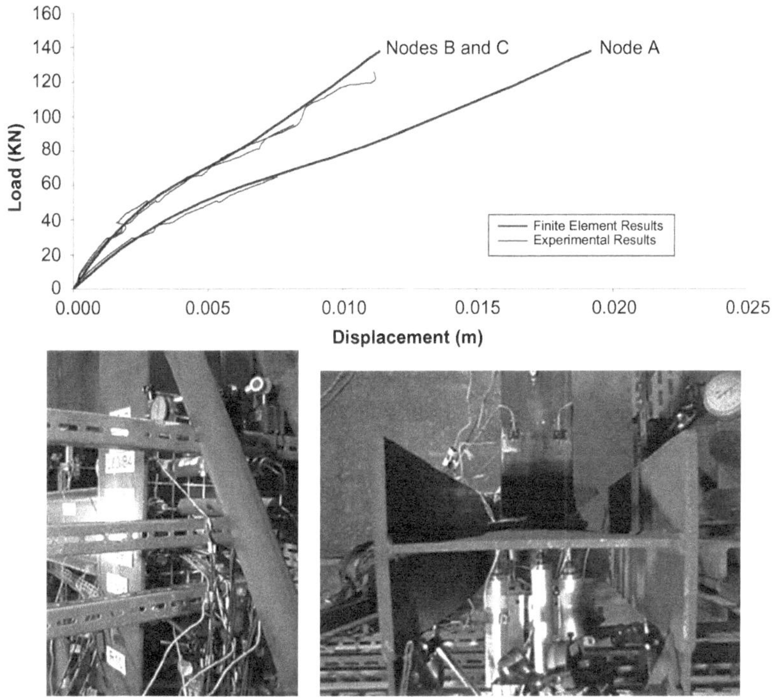

Fig. 3.25 Experimental results calibration and displacements transducers layout.

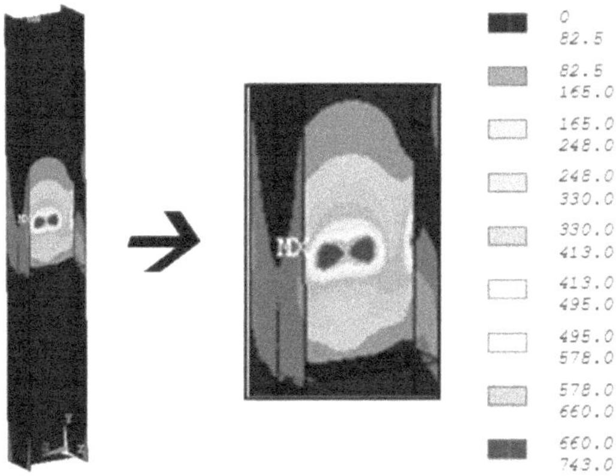

Fig. 3.26 von Mises stress distribution (in MPa).

The experimental column web thickness, 8 mm, was changed to 6.3 and 10 mm keeping all the other profile test dimensions (corresponding to 43.65, 37.37, and 27.5, column web slenderness) to study the influence of the column web thickness. The bilinear model was applied to the finite element results generating k_e and k_{pl} stiffness values for the three studied column web thickness, Table 3.3.

The values of k_e and k_{pl} were determined as the slope of the straight lines that best fit the curve points, for the elastic and plastic parts, respectively. The column web in bending component stiffness, k_{21}, was evaluated dividing the k_e stiffness, determined in Fig. 3.27 curves, by the Young's Modulus.

The simulations proceeded focusing on the distance between the stiffeners. Six combinations were made with the ANSYS program keeping all the other dimensions, Fig. 3.28. The load versus displacement curves obtained are depicted in Fig. 3.28. With the aid of these curves, it is possible to determine the value of k_{21}, i.e. the elastic stiffness of the column web loaded chord in bending.

Experimental work involving photoelasticity already commented in this book [96], was also modelled with finite elements. A comparison between the finite element model and the experiments in terms of the principal strain differences $(\varepsilon_{\mathrm{I}} - \varepsilon_{\mathrm{II}})$ in the top bolt region measured along the dashed line

Table 3.3 Influence of the Column Web Thickness Over the Joint Stiffness

t_w (mm)	h/t_w	k^e (kN/m)	k^p (kN/m)	k_{21} (m)
6.3	43.7	3715.0	2312.1	0.018
8.0	34.4	6529.4	2087.5	0.032
10.0	27.5	11933.3	1786.4	0.058

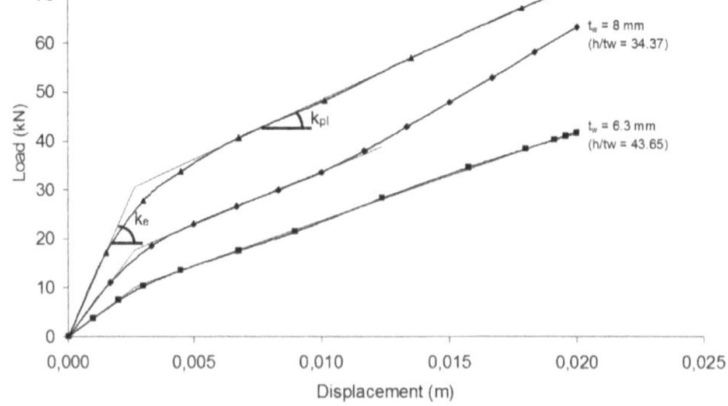

Fig. 3.27 Influence of the column web thickness over the connection structural response.

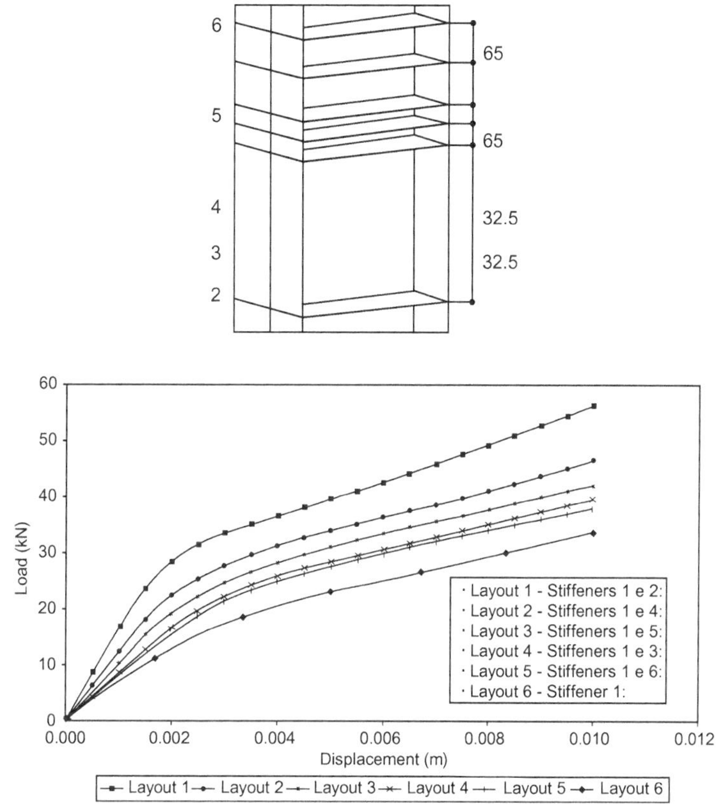

Fig. 3.28 Column web stiffeners location layout and Influence of the column web stiffeners over the connection structural response.

and the effective width of the more stressed joint zone was also performed. A detailed analysis of the finite element results indicated that the model was able to identify the points where the principal strain difference is zero, represented by a dark blue region, Fig. 3.29. These regions are also identified in the photoelastic model by dashed circles. The evolution of the strain pattern with the increase of the applied load can be noted, mainly in Fig. 3.29. An inspection of Fig. 3.29 shows a difference between the finite element and the experiments. These differences were due to the finite element simplifications adopted in the modelling. Using the point where the principal stress distribution is zero, it was possible to determine the effective width depicted in Fig. 3.30. Observing the numerical and experimental fringe patterns, Fig. 3.30, these distances are approximately equal to 39.1 and 41.2 mm, respectively.

The results of the present study regarding the assessment of the effective width of out of plane column web also demonstrated a good agreement

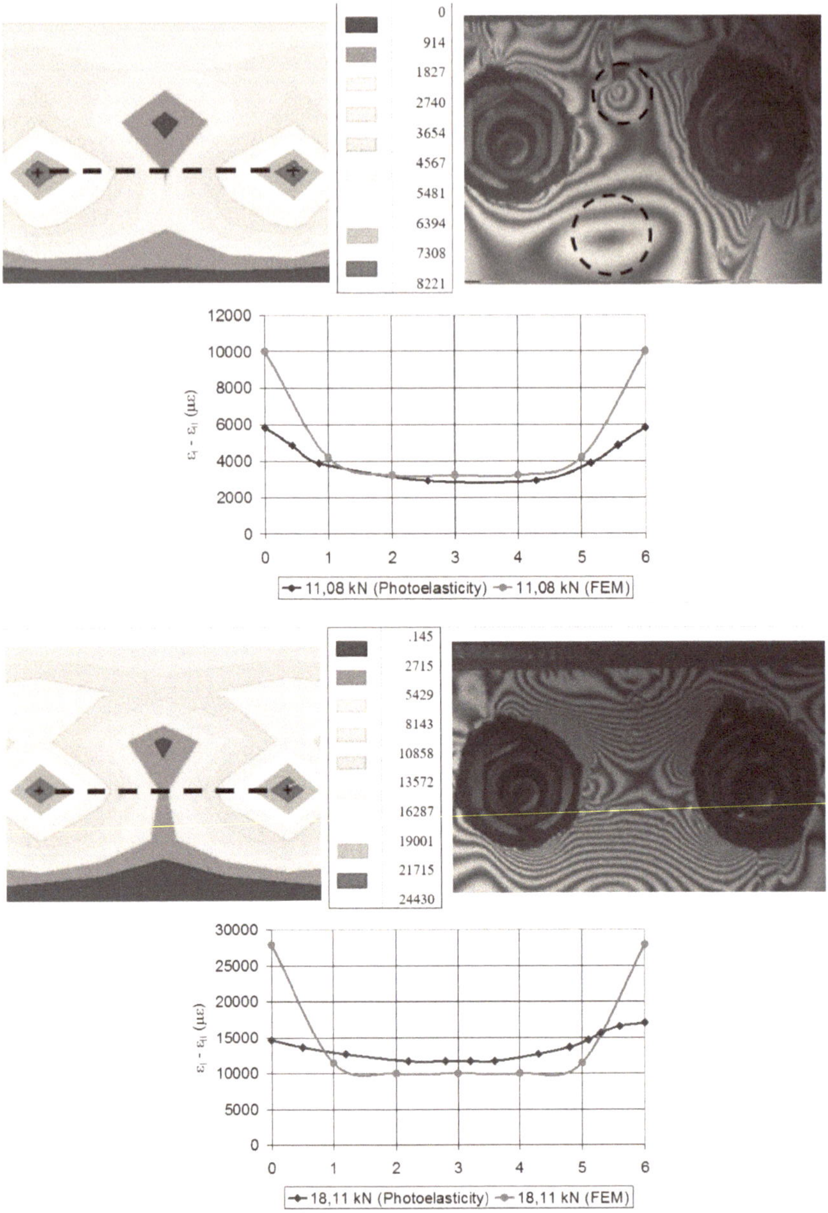

Fig. 3.29 $\varepsilon_{\mathrm{I}}\ \varepsilon_{\mathrm{II}}$ FEM and test distribution (applied loads: 11.08 and 18.11 kN).

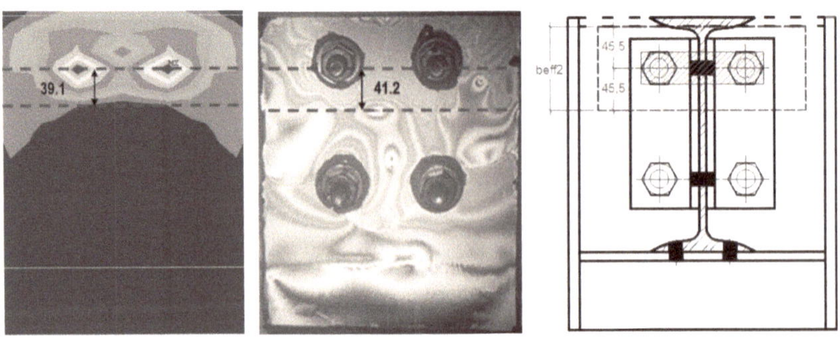

Fig. 3.30 Effective width evaluation—numerical, experimental and analytical values (in mm).

between the experimental, numerical, and analytical analysis presenting differences less than to 10%, being in accordance with the acceptable margin of errors of structural engineering problems.

The investigation focused on the column web in bending component present in beam to column semirigid minor axis connections. The numerical simulations were able to detect the column web plastic hinges development, allowing a comparison to equivalent stress distributions determined in experimental analysis already presented in the previous chapter. The high nonlinearity of this phenomenon required a rigid step size control, limits and various strategies of convergence were adopted to enable the determination of its real behaviour around and after the ultimate load.

Once again, full and modified Newton–Raphson methods were used coupled with a displacement control. The loads have been applied to the column web simulating the action of the bolts through equivalent deflections to facilitate the convergence control in the elastoplastic region. The developed model proved to be in agreement with the experiments and enabled a parametric analysis to be performed, as well as a better understanding of the joints structural behaviour.

3.5.4 Structural Evaluation of Major Axis Semirigid Endplate Joints—Bending Moment to Axial Force Interaction

This work objective is the study and computational implementation of geometrically nonlinear formulations of semirigid portal frames through a numerical approach [97–101]. Seeking a refined analysis of bi-dimensional frame, the consideration of the joints behaviour cannot be neglected since they play a fundamental part in the structure global response. The adoption of an accurate and representative model is important because in the structural analysis the joints stiffness loss is a nonlinearity factor that can have a great influence on the final result. These ideas motivated the implementation of a technique that updates the joints stiffness every load step, irrespective of the adopted model.

A semirigid connection can be modelled as a spring element inserted at the point of intersection between the beam and column, as shown in Fig. 3.31. For the vast majority of steel structures, the effects of axial and shear forces in deformation of the connection are small when compared to those caused by the bending moment. For this reason, only the rotational deformation of the spring element is considered in practice. For simplicity the joint spring element has, by hypothesis, a negligible size, as presented in Fig. 3.31.

In the procedures presented by Chan and Chui [102], the joint spring and beam-column elements are combined to form a hybrid element. Fig. 3.31 depicts the idealised hybrid element to model portal frames with finite

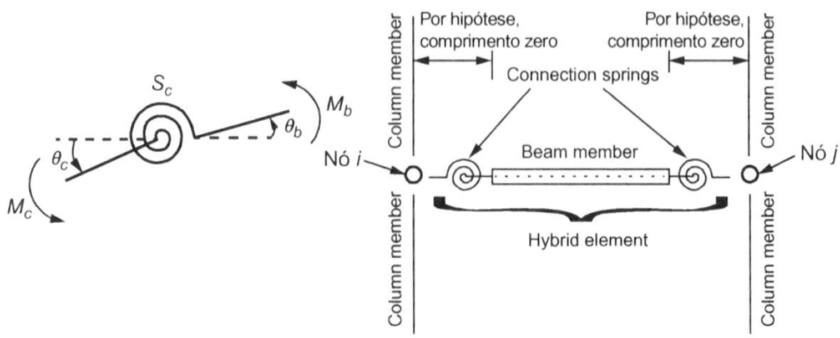

Fig. 3.31 Joint spring element and idealised semirigid frame element model (Chan and Chui [102]).

elements. One side of the spring element is connected to the beam-column element while the other side is linked to the global node. The deformed configuration of the hybrid element with springs at its ends is illustrated in Fig. 3.32. The element is initially straight in its nondeformed configuration and later deform to a curvilinear configuration by translation and rotation movements.

The details of the internal forces and deformations in the connection spring are portrayed in Fig. 3.32. The connection rotation is defined as the difference between the rotation angles of the sides connected to the global node and the side linked to the beam-column element. In the presence of joint springs added to the beam-column ends, the element conventional stiffness matrix should be modified to consider, in the formulation, the semirigid joints effects. The resulting stiffness matrix can be then used in subsequent analyses.

A spring element was proposed to model semirigid connection. This element should be inserted at the beams to columns intersection points where the semirigid joints are located. The main motivation for the implementation of this spring element was due to the fact that the hybrid element proposed by Chan and Chui [102], described earlier, presents a bending moment inappropriate behaviour, when the element is under distributed loads.

The need to have an independent element to couple the stiffness effects (axial, rotational and translational), was also another motivating factor, as these effects were not considered in Chan and Chui [102] hybrid element, where the only variable is the rotation stiffness and this variable is coupled to the element.

Many investigations have been conducted regarding the effects of axial [47], and shear forces in joint deformations, in addition to those caused by the bending moment. This spring element, even in this first implementation, considers all the stiffness effects decoupled, leaving space for further work where the interaction between the stiffness could be considered, just by

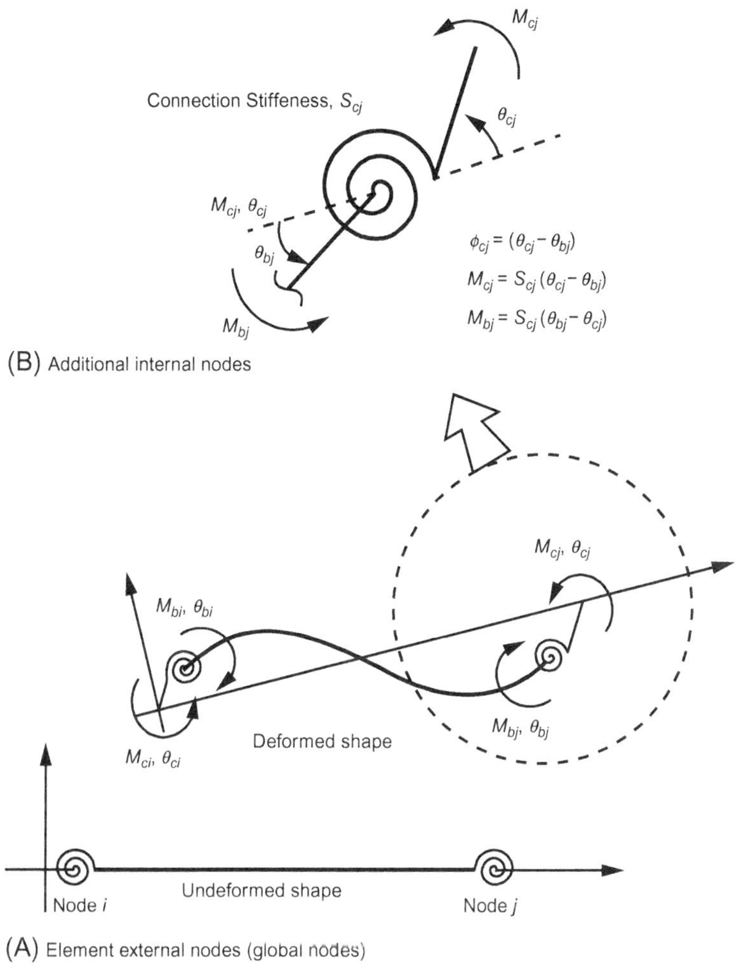

Fig. 3.32 Beam to column joint spring element at the element ends (Chan and Chui [102]).

modifying the element stiffness matrix. This new spring element behaves properly for any load type, making it possible to model the structure behaviour more precisely without the need for any type of load discretisation.

With this spring element, it is possible to simulate elastoplastic analysis of joints, from their moment versus rotation curve that describes the joint behaviour. It is also possible to perform elastic plastic analysis of structures by inserting spring elements at the points where plastic hinges are expected. Regarding the spring element stiffness matrix formulation, this is basically resumed to a stiffness matrix incorporating the axial, translational, and rotational stiffnesses. By definition, the length of this element is equal to zero. Thus, the spring element stiffness is given by:

$$\begin{Bmatrix} P_i \\ Q_i \\ M_i \\ P_j \\ Q_j \\ M_j \end{Bmatrix} = \begin{bmatrix} S_A & 0 & 0 & -S_A & 0 & 0 \\ 0 & S_T & 0 & 0 & -S_T & 0 \\ 0 & 0 & S_R & 0 & 0 & -S_R \\ -S_A & 0 & 0 & S_A & 0 & 0 \\ 0 & -S_T & 0 & 0 & S_T & 0 \\ 0 & 0 & -S_R & 0 & 0 & S_R \end{bmatrix} \begin{Bmatrix} u_i \\ v_i \\ \theta_i \\ u_j \\ v_j \\ \theta_j \end{Bmatrix} \quad (3.6)$$

where S_A, S_T, and S_R are the components of axial, rotational, and translational tangent stiffness, respectively, of the spring, u_i, u_j, v_i, v_j, θ_i, and θ_j are the translational and rotational axial deformations, respectively, of the two spring element ends and P_i, P_j, Q_i, Q_j, M_i, and M_j are normal and shear forces and the bending moment, respectively. Fig. 3.33 depicts the spring element model designed to simulate a semirigid connection. Fig. 3.34 illustrates the axial deformation, the rotation, and the translation of each spring that composes the idealised model, as well as the forces in each spring.

Aiming to test the best hybrid element linear implementation and spring element, a portal frame proposed by Steenhuis et al. [94] was selected. The results were also compared with Brito [92], that proposed a simplified model for semirigid beam-column connections using the ANSYS program [32]. The semirigid joints, for this example, were considered with constant stiffness. The

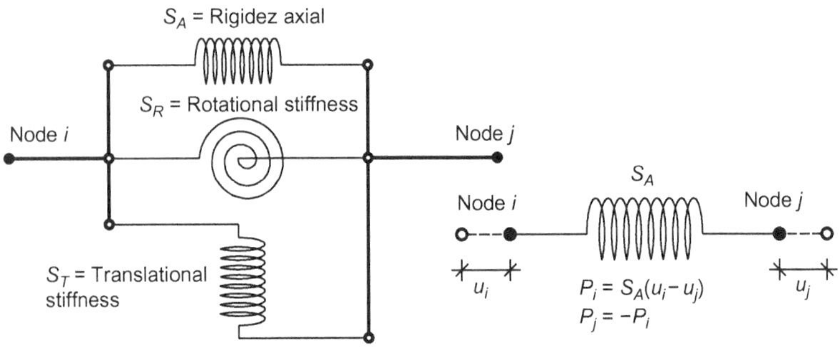

Fig. 3.33 Idealised spring element model and axial stiffness spring.

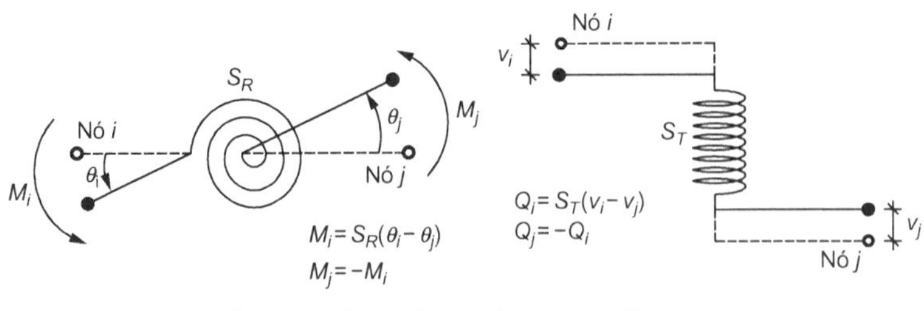

Fig. 3.34 Rotational stiffness and translational spring stiffnesses.

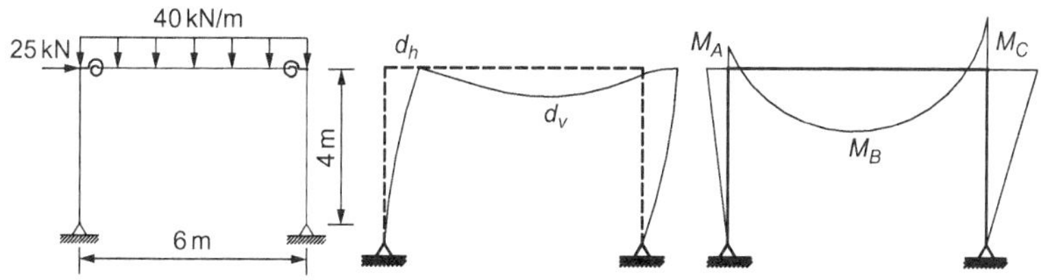

Fig. 3.35 One storey portal frame, deformed shapes and bending moments [62].

example in Fig. 3.35 was proposed by Steenhuis et al. [94] to illustrate the effect of semirigid joints on the deformation and distribution of forces in a nonsway structure. A first-order elastic analysis was used. The portal frame span was 6 m, and the height, 4 m. Na IPE360 profile was used for the beam and columns. The load consisted of a horizontal load of 25 kN and a uniformly distributed vertical load of 40 kN/m. The deformed configuration and the bending moment distribution of the studied portal frame are presented in Fig. 3.35. The portal frame was analysed, adopting the secant stiffness, which is half of the tangent stiffness. The analysed tangent stiffness values were 35, 60, and 130 kNm/mrad. The rigid joint hypothesis was also investigated by adopting a large value for the spring stiffness. In order to better study the hybrid element behaviour under distributed loads, two portal frame discretisations were made. The first used a finite element for the beam, while the second adopted two finite elements. For both cases, the column only used a single finite element. Comparisons between the results are presented below.

For best viewing of the obtained results and calculated error magnitude, three bending moment versus stiffness diagrams are presented, Fig. 3.36, for the upper left corner (MA), for the midspan (MB), and for upper right corner (MC), respectively. These graphs indicate the good approximation obtained with the use of the proposed spring elements.

Fig. 3.37 presents a comparison between the obtained displacements and the values of Steenhuis and Brito, for the investigated cases. In general, for all cases examined, the displacements obtained were close to the values presented by Steenhuis and better than those obtained by Brito. For better viewing the obtained results the calculated error magnitude are shown in two graphs: horizontal displacement versus stiffness and vertical displacement versus stiffness, Fig. 3.37 were made for the upper left corner (dh) and midspan (dv), respectively. These graphics also indicated that a close approximation was obtained with the use of the hybrid spring elements.

The work continued with the validation of the formulations for the spring element, and implemented in this work, for a nonlinear semirigid portal frame analysis using curves full moment versus rotation curves.

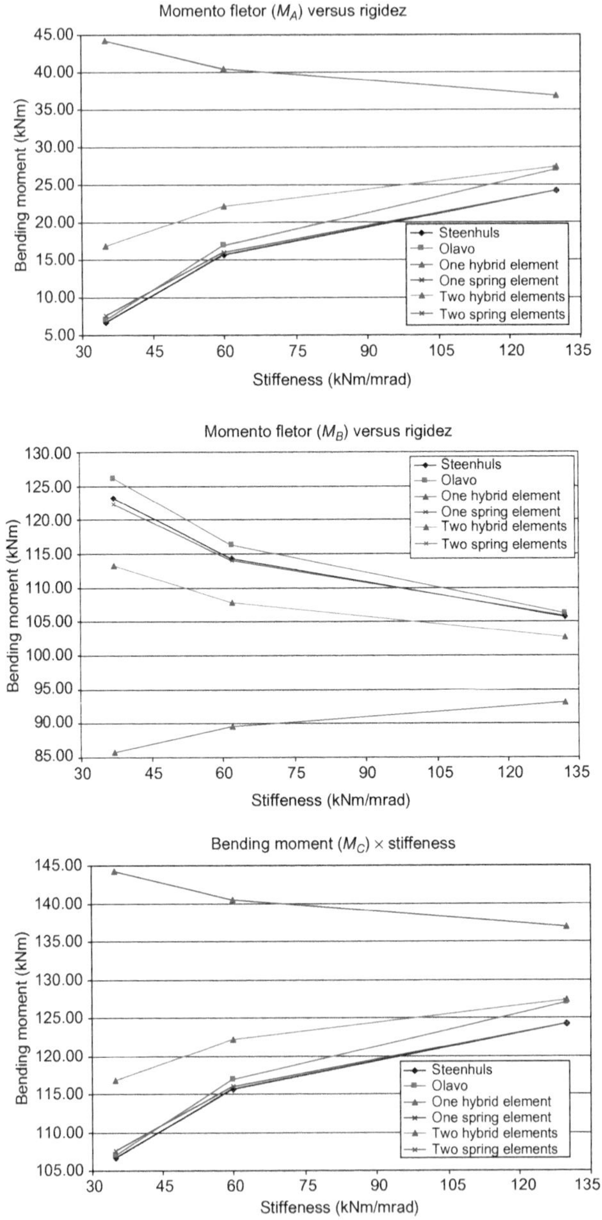

Fig. 3.36 Comparison of moments (MA, MB, and MC), proposed and simulated models.

Comparisons were made with the analysis performed by Nethercot [103], using full moment versus rotation curves and a bilinear approximation. This last hypothesis is a simplification of the joint modelling and is known as the half initial secant stiffness method. A 7.2 m long portal frame, 3.6 m high, as

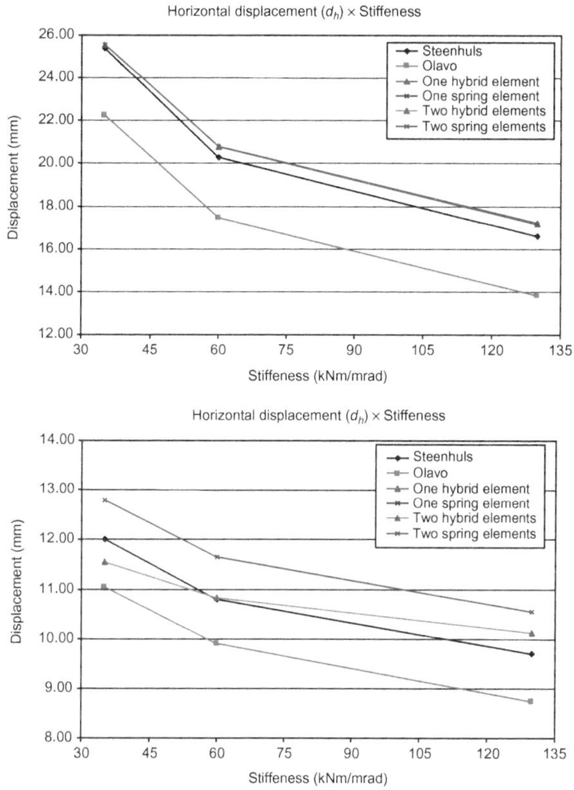

Fig. 3.37 Horizontal and vertical deflections comparison: proposed and simulated models.

shown in Fig. 3.38 was studied [103]. The beam used a IPE360 while a HEA260 profile was utilised on the column. The characteristic curves used in the analysis are presented in Fig. 3.38. The column and beam are connected with the bolted flush endplate connection shown in Fig. 3.38. The horizontal load αF usually represents the load imposed by the wind, but also considers the imperfections that generate second-order effects. The α factor was taken as: 0.1, 0.15, 0.2, 0.3, and 0.5. The vertical load w was equal to 1/6 of the vertical load *F*.

The diagrams present in Figs 3.39–3.41 illustrate the three test results. The first two were performed by Nethercot and the third was carried out using spring elements, proposed in this work. The analyses carried out by Nethercot i.e. Reference Analysis were performed using ANSYS, version 5.5 [32]. A second order elasto-plastic analysis was used to obtain a reference solution. The plastic hinges in beam were modelled using rotational spring elements at the points where the plastic hinges occurred. These spring

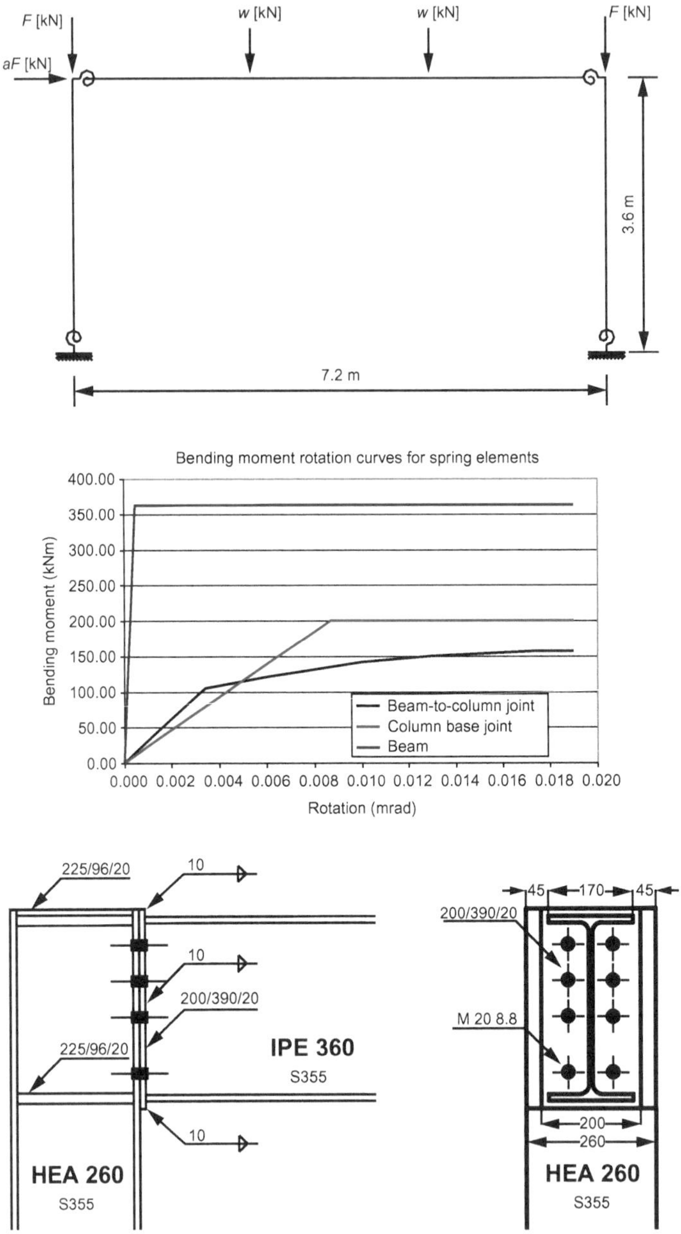

Fig. 3.38 One storey portal frame, characteristic curves and beam to column joint [60].

elements had rigid plastic characteristics that neglect the influence of the normal and shear forces on the bending plastic resistance. The Half Initial Secant Stiffness Approach was accomplished using a bilinear moment versus rotation curve considering half of the secant initial stiffness.

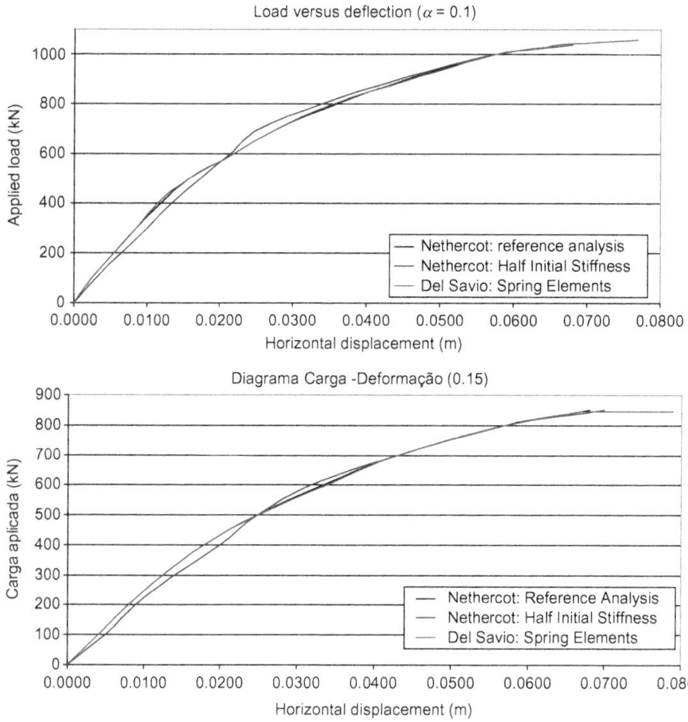

Fig. 3.39 Load versus deflection curves ($\alpha = 0.1$ and 0.15).

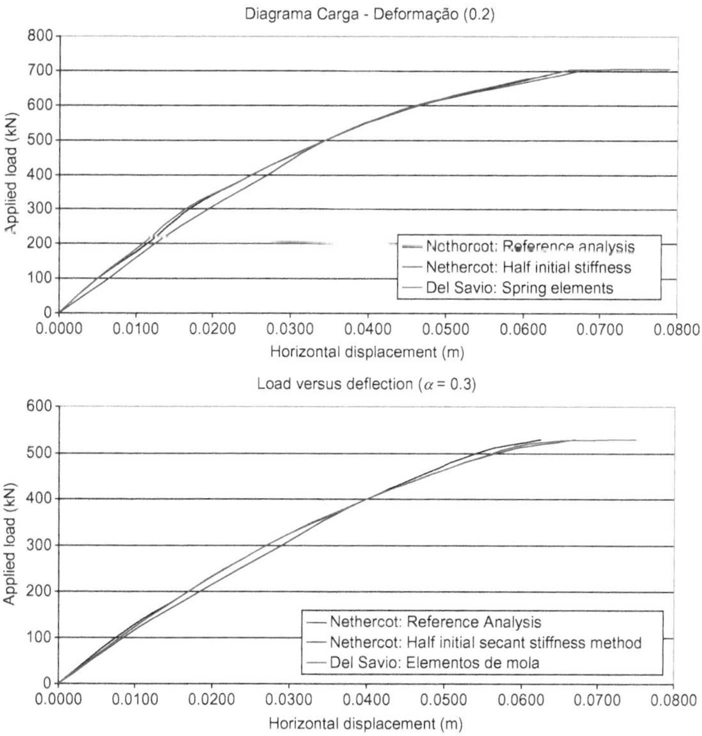

Fig. 3.40 Load versus deflection curves ($\alpha = 0.2$ and 0.3).

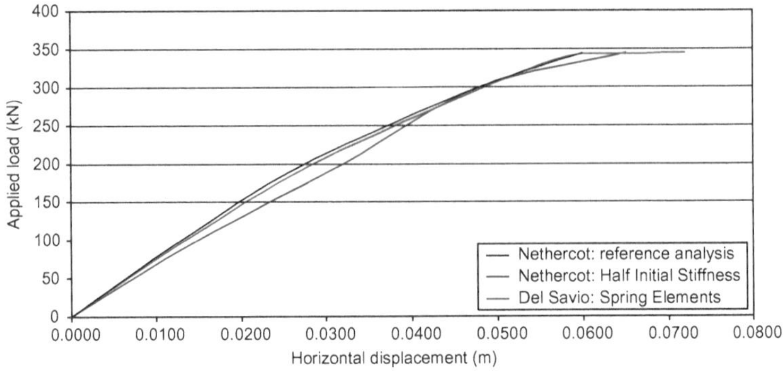

Fig. 3.41 Load versus deflection curve ($\alpha = 0.5$).

For the values of the α factor (0.1, 0.2, 0.3, and 0.15; 0.5) analysed, it could be observed that the values of displacements were practically equivalent to the Nethercot reference analysis., thus demonstrating the efficiency of formulations (spring elements) for the nonlinear analysis, as well as the efficiency of computational implementations.

Complementing the analysis, a 3-storey portal frame 7.2 m long and 10.8 m high, as shown in Fig. 3.42 was evaluated. For the beams adopted IPE 360 profile and the columns utilised HEA260 profiles. The characteristic curves used in the analysis are presented in Fig. 3.42. The columns and the beams are connected with a bolted connection shown in Fig. 3.42. The diagrams in Figs 3.43 and 3.44 illustrate the results of three tests. The first two were performed Nethercot and the third was carried out using the proposed spring elements.

Regarding the analyses performed by Nethercot: the Reference Analysis was performed using the finite element program ANSYS. A second order elasto-plastic analysis was adopted to obtain the reference solution. The plastic hinge in beam was modelled using rotational spring elements at the points where it was expected that the plastic hinges would occur. Fig. 3.42 illustrates the springs properties used at beam-column joints, in the column-base connections and in the beam spring. For such analysis a bilinear moment versus rotation curve was considered, Fig. 3.42, using half of the initial secant stiffness. One more time, the horizontal load αF usually represents the loading imposed by the wind. The α factor was taken as 0.15. The vertical load w was equal to one sixth of the vertical load f.

For the analysed values, it can be observed that the displacements are very similar to those of the reference analysis performed by Nethercot, until the emergence of the first plastic hinge, Fig. 3.44. After the formation of the first plastic hinge, the values differ a little, but still feature a good

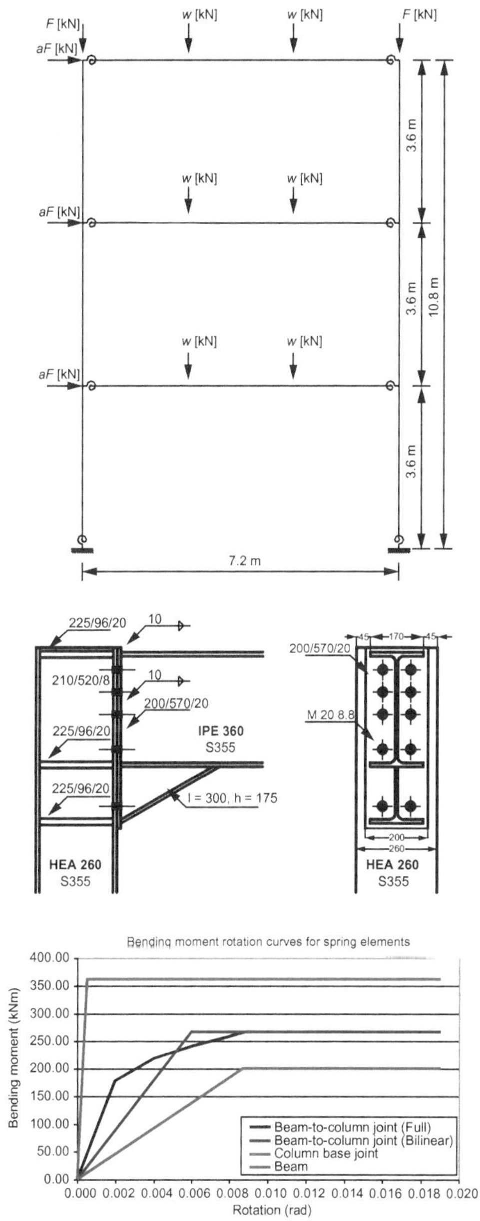

Fig. 3.42 Three storey portal frame, characteristic curves and beam to column joint.

approximation. The divergence after the start of the yield is mainly associated to the slope of the last segment of the moment versus rotation curve. So, with a small increase in that inclination, which a priori is null, causes the results of the approach shifts more and more to the results presented by Nethercot.

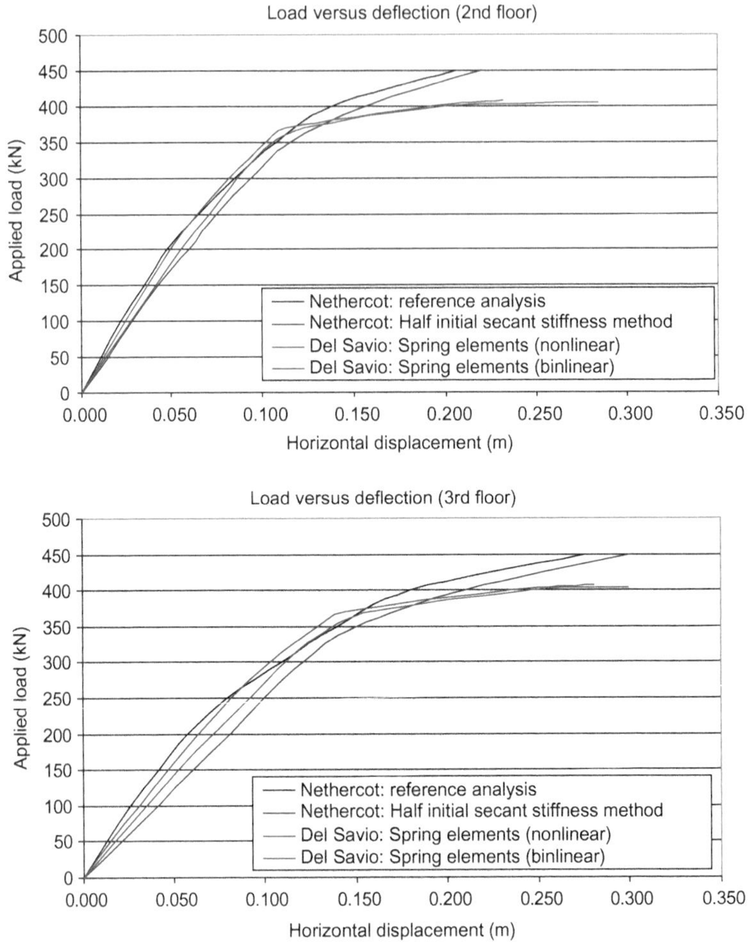

Fig. 3.43 Load versus deflection curves, 2nd and 3rd storeys.

This demonstrated the efficiency of formulations (spring elements) for the nonlinear analysis, as well as the efficiency of the computational implementations for structures involving a higher degree of complexity. It is worth noting that with the implemented spring element, a plastic analysis can be made by inserting springs at the points where the plastic bearings are expected to occur. For distributed load cases, where the location of the plastic hinges in unknown, the identification and automatic insertion of spring elements in the possible plastic hinge points, eliminating the need for the user to identify them in advance.

Simulations were also made in Vierendeel type girders that had the great advantage that its elements in portal frames are associated to bending moments

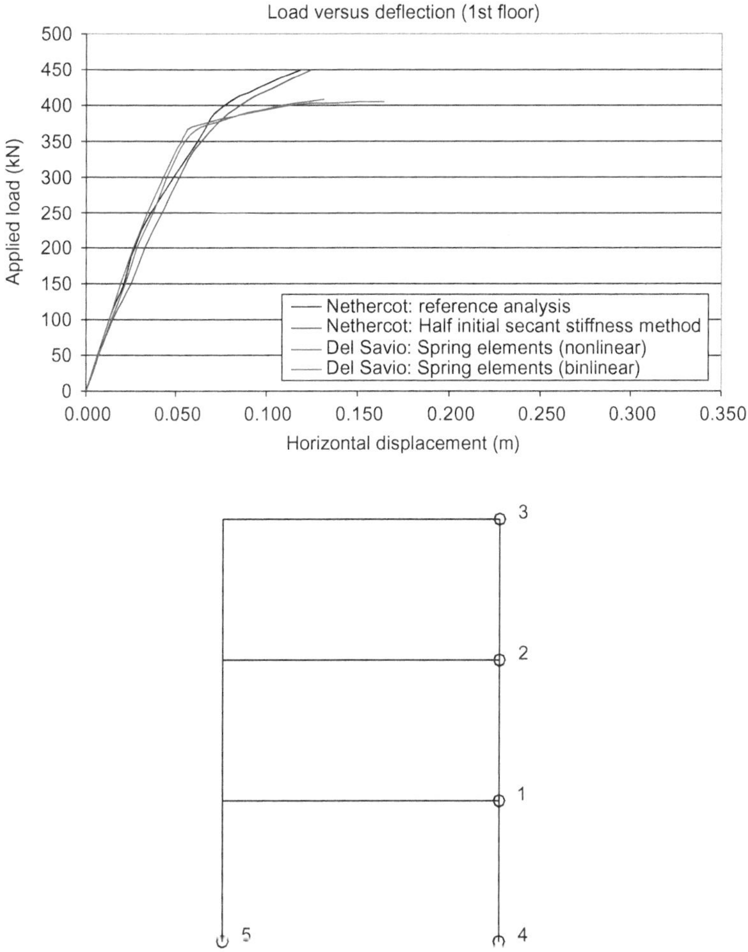

Fig. 3.44 Load versus deflection curve, 1st storey and plastic hinge evolution.

and axial forces behaviour by eliminating the presence of diagonals. This way, this structural solution enables a greater flexibility in installations, enabling the passage of pipes, people, etc. In addition, they may, for example, be used in buildings with alternated floor trusses at every floor ('staggered-truss' systems) as shown in Fig. 3.45.

This joint finite element model was formulated and implemented to model the semirigid connections. This model was initially based on the independence of the forces acting on the semirigid connections (bending moment, axial, and shear). This motivated the development of a nonlinear semirigid joint finite element formulation including the interaction between axial forces and bending moment. The consideration of this interaction did

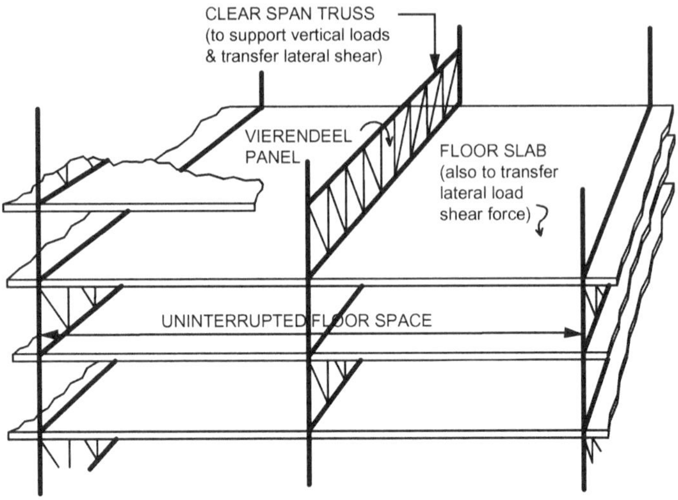

Fig. 3.45 Typical truss stagger pattern and transverse load-distribution mechanism.

not change the basic formulation of the nonlinear joint finite element. This was only possible because the consideration of the interaction between axial force and bending moment was performed in the joint moment versus rotation curve. The idea to consider the interaction between axial force and bending moment, without modifying the nonlinear joint finite element formulation is not complex. It can be summarised; basically, in finding a 'correction factor' that applied to the joint bending moment versus rotation curve considers the referred interaction effect.

The main basis for the formulation of the above mentioned 'correction factor' starts in limiting the area of the moment versus axial force diagram where the connection response is valid. This area was constructed by defining a convex dominium, formed by straight line segments, from the points generated by Lima et al. [47], numerical model, Fig. 3.46. Considering the symmetry present in the numerical model only the positive bending moment region considered. This strategy leads to the region definition by three lines expressed by:

$$f(N) = \begin{cases} 0.0963N + 104.380 & (-1084.54 \leq N \leq -257.00) \\ -0.0274N + 72.477 & (-257.00 < N \leq -105.20) \\ -0.0962N + 65.328 & (-105.20 < N \leq 678.00) \end{cases} \quad (3.7)$$

where the load range valid for each line segment is presented inside the parenthesis and $f(N)$ is a function that calculates the maximum bending moment that could be utilised for a certain applied axial force N. With these limits in hand

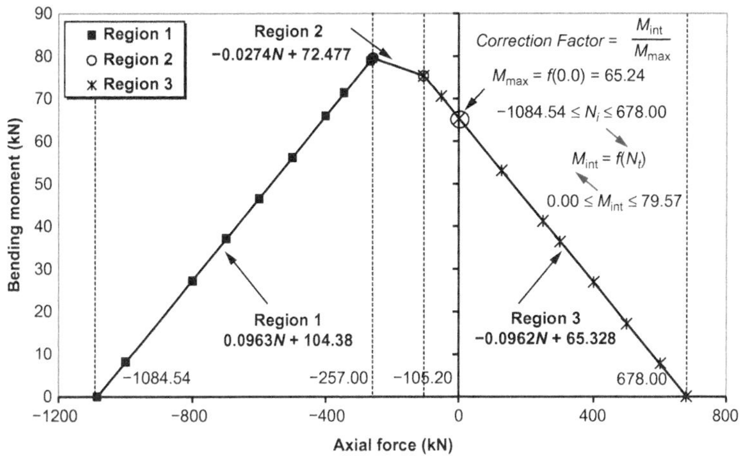

Fig. 3.46 Moment versus axial force diagram and required data for determining the 'correction factor'.

the 'correction factor' can be calculated as the ratio of the bending moment considering and discarding the moment versus axial load interaction, i.e.:

$$\text{Correction Factor} = \frac{M_{\text{int}}}{M_{\text{max}}}; \quad \begin{array}{l} M_{\text{int}} = f(N_i) \\ M_{\text{max}} = f(0.0) \end{array} \tag{3.8}$$

where N_i is the axial force load magnitude acting in the current i iteration. Fig. 3.46 depicts all the required information necessary for the assessment of the 'correction factor'. This figure also portrays the bending moment versus axial force iteration diagram, defined by the line segments expressed in Eq. (3.1), as well as their respective validity range (highlighted by vertical dotted lines).

The proposed 'correction factor', that considers the moment versus axial force interaction, can be incorporated into the connection structural response by multiplying its value to the original moment values present in the moment versus rotation curves. This strategy shifts this curve up or down, depending on the acting axial force level. An idealised moment versus rotation curve, constructed from experiments, can be represented by a 6th grade polynomial up to the connection maximum acceptable bending moment capacity evaluated by the Eurocode 3, bi-linear approximation, as portrayed in Fig. 3.47. From this point onwards the curve follows the horizontal yield path suggested by the Eurocode 3.

The Vierendeel girder semirigid structural system involved a 12-m span divided in four, 3 m long and one meter high, segments. The beams were under four concentrated loads of 35, 30, 10, and 20 kN, respectively

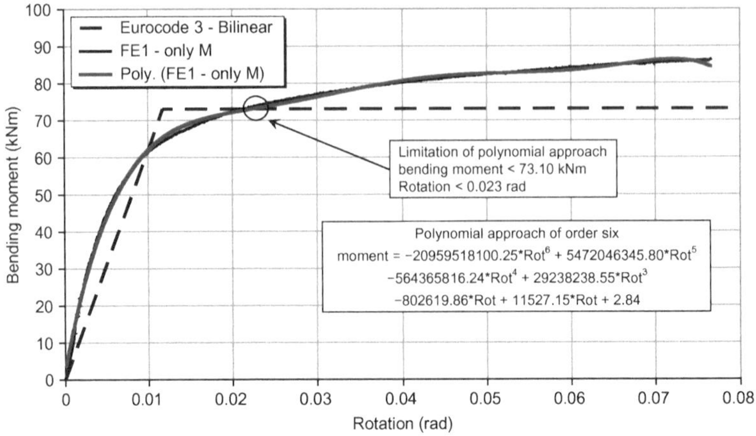

Fig. 3.47 Idealised moment versus rotation curve adopted in the nonlinear finite element analysis.

applied to the nodes 3 (P3), 4 (P4), 8 (P1), and 9 (P2). The purpose of this nonsymmetric loading was to create nonlinear geometric disturbances in the structure and to 'overload' element 12, which will be the main target for the comparisons among the results obtained in the variations of the joints' stiffness values. A similar analysis can be made with the use of lateral loads, which are often found in these structures due to wind forces.

Fig. 3.48 depicts the structural model conceived for the Vierendeel beams. This figure portrays the applied load, the numbers of the nodes and elements (inscribed in a rectangle), the dimensions, and identifications of the connections represented by S_i (i varying from 1 to 16). The beams elements (horizontal members), utilised an IPE 240 steel profile, while the columns used a HEB 240 steel section. The steel grade utilised in all elements of the beam has a Young's modulus of 205,000 MPa and a specific weight of 78.5 kN/m^3.

All the analysis involved geometrical nonlinear finite element simulations up to the structural system ultimate load capacity. The simulations were performed based on a generalised displacement control strategy, with a maximum of 1000 load steps, step factor of 0.02, and maximum number of iterations equal to 100.

Four different cases were considered using the connection finite element previously mentioned, i.e.: 1st used a constant stiffness equal to 6000 kNm/rad (initial bi-linear curve stiffness) for all structural connections; the 2nd one utilised an updated stiffness bi-linear curve, for all structural joints. The 3rd one used an updated stiffness according to the 6th grade polynomial curve, for all structural connections. The 4th one utilised an updated stiffness

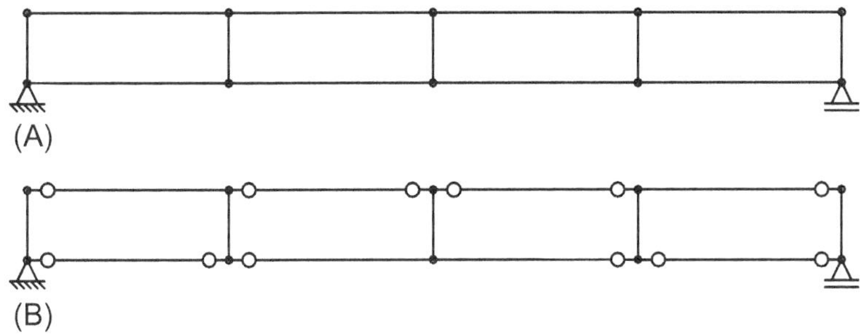

Fig. 3.48 Idealised structural model: Vierendeel girder system: (A) rigid; (B) pinned.

according to the 6th grade polynomial curve considering the moment versus axial force interaction effects for all structural connections; The results for each analysed case are depicted in Fig. 3.49, respectively, in terms of forces and moments of element in number twelve, and limit loads and equilibrium paths of node number three.

Only the first case, with constant connection stiffness reached the maximum load step multiplier equal to 20 times the applied load, with a linear load path and did not present any maximum load limiting value. This strategy has also produced an unreliable set of force and moment distributions when compared to the other simulations.

For better visualising the load path results, for all the simulated cases, the maximum vertical displacement of node number three was utilised and was limited to 0.2 m, Fig. 3.49. This strategy was utilised since cases 2, 3, and 4

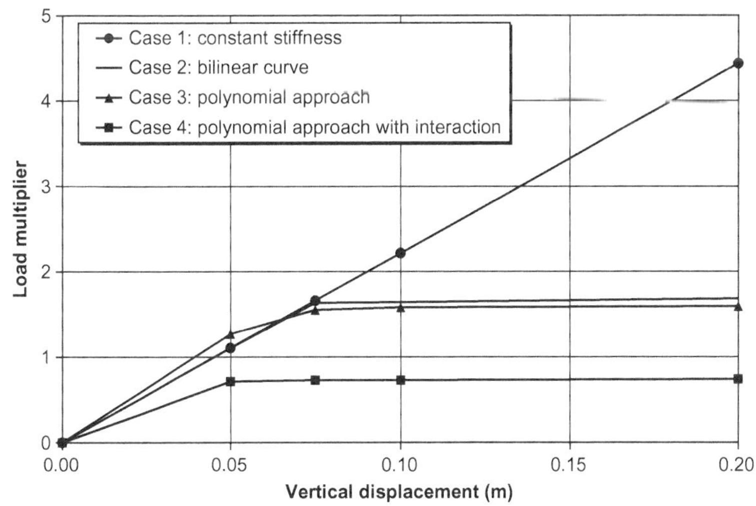

Fig. 3.49 Equilibrium path, node number 3, vertical displacements up to the ultimate load.

not even reached a load multiplier equal to 2 and the first case presented a linear elastic response. As expected when the semirigid joint response was considered, by means of its associated moment versus rotation curve, a limiting value for the applied load was reached, Fig. 3.49.

The second and third cases (without considering the moment versus axial force interaction) presented limiting load multipliers equal to 1.68 and 1.59 times the maximum applied loads. This similar response can be easily explained by the use of the same maximum connection bending moment capacity calculated by the Eurocode 3 bi-linear approximation. The fourth case that takes into account the moment versus axial force interaction depicted a substantial reduction of the system ultimate load (0.74) when compared to the second and third cases (1.68 and 1.59). This fact only confirmed the idea that neglecting the moment versus axial force interaction could mislead structural designers to perhaps, even unsafe solutions.

As it could be clearly observed in the fourth case, the consideration of the connection moment versus axial force interaction effects significantly reduced the Vierendeel system ultimate load bearing capacity up to 74% of the original applied loads. This fact indicated that these interaction effects cannot be neglected, especially in structural system where high/significant levels of axial force could occur. If this interaction was not considered, as for example in the third case, the system ultimate load would be estimated to be 159% of the original applied loads.

Thus, the efficiency of the proposed spring element for linear and nonlinear analysis, as well as the efficiency of computational implementations was demonstrated. The hybrid element implemented only for linear analysis, presented a limitation in relation to the loading, i.e. cannot adequately represent the bending moments in case of distributed loads. For this reason the hybrid element was not implemented for nonlinear analysis. For all the analysed examples, it could be concluded that for small displacements, considerations of linear and nonlinear stiffness were almost equivalent. As for nonlinear solution algorithms, the one which proved to be more appropriate was the GDCM. This method was used in all nonlinear analyses carried out in this work.

One of the main contributions in terms of modelling of the present work was the implementation of a discrete finite element spring contemplating simply bending interaction versus axial stress by applying interaction loads of between these forces. The simple uniaxial modelling of the problem allowed a quick and simple assessment of the effects of this interaction highlighting the cases where care must be taken in design.

A model of springs and rigid bars was also developed in the Ansys program based on the components method present in Eurocode 3. With the calibration of this model with experimental results it was possible to perform a parametric analysis aiming at a better understanding of the role of each component in the overall performance of the connection in terms of initial stiffness and bending transmission capacity.

The convergence of these analyses was not simple giving rise to a series of measures such as reducing the step size, restarts with appropriate steps, various types and maximum values to ensure convergence. Often inadequate step sizes in the analysis steps generate a deviation trajectory of equilibrium that requires the system to be restarted in its early stages. This whole process is not of a simple understanding and implementation, requiring experience and time to overcome these difficulties.

3.5.5 Beams Under Patch Loads

Concentrated loads applied to the main beam flange, so-called patch loads, can lead to an instability phenomenon. One of the solutions to surpass this problem is the use of transversal stiffeners welded to the beam's web. The state of the art studies in the patch load phenomenon still did not produce an efficient formula for the forecast of this local buckling. Many theoretical and experimental studies were performed, but all the proposed equations include a 20% error margin when compared to experiments.

The modelling of this phenomenon was executed with the aid of the finite element method taking into account the material and geometric nonlinearities associated to the patch load problem due to the plastic local buckling involved [105–107]. Good results in terms of critical loads, deformed shapes and web's stress distribution were obtained with the aid of a nonlinear finite element program, Saloof [108]. The geometry of the finite element model was made of a simply supported beam under a load uniformly distributed along a short length c. This model was made of 260 elements and 290 nodes. The flanges were modelled with 40 elements (4 vs. 10). The web was composed of 120 elements (10 vs. 12). Transversal stiffeners were modelled with 24 elements (12 vs. 2). The model geometry was completed with flange stiffeners to simulate the load distribution length, Fig. 3.50.

The adopted Semiloof finite element was developed for thin shell applications. A nonlinear geometric and material analysis was made using von Mises yielding criterion and full Newton Raphson iterative scheme. The boundary conditions were established to represent the actual experimental constraint conditions performed by Bergfelt [109]. In order to calibrate the finite element model, five beams were simulated and compared to the experiments [109]. The material and geometric characteristics of these beams are depicted in Table 3.4.

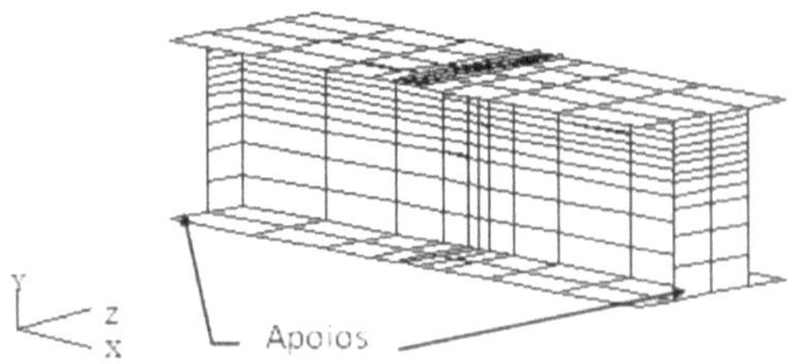

Fig. 3.50 Adopted finite element mesh.

Table 3.4 Beam's Geometric and Material Characteristics

Beam	b (mm)	h (mm)	t_w (mm)	b_f (mm)	t_f (mm)	c (mm)	σ_w (MPa)	σ_f (MPa)
b14	2400	400	2	100	8	180	294	294
R01	800	800	2	300	15	40	266	295
R03	800	800	2	120	5	40	266	285
b08	800	800	2	120	5	40	285	290
b83	800	800	3	250	12	40	328	298

Unfortunately, no data regarding the beam imperfections was found [109]. An investigation of the influence of the initial imperfections over the beams critical collapse loads was made. Beam b14 was modelled utilising different amplitude and positions of initial imperfections. In this study, maximum imperfection amplitudes of: $h/1600$, $h/800$, $h/650$, and $h/560$, applied at node point 140, 0.2 h away from the top flange were adopted. The influence of the initial imperfection's amplitude over the ultimate load is depicted in Fig. 3.51. It is noted that this influence is negligible (minimum and maximum values for the ultimate load ranged from 91.5% to 92.7%. The maximum amplitude position of the initial imperfection was also studied. The imperfection's application point was located at: 0.25, 0.20, 0.15, and 0.10 h (nodes: 139, 140, 141, 142, respectively) away from the top flange, Fig. 3.51. The collapse load varied 90–92.5% proving that for this beam the influence of the location of the maximum imperfection was not relevant.

A standard imperfection of $h/500$ applied at a node $d=0.2$ h far from the beam's top flange was utilised. Four more beams were modelled to calibrate the model. They were beams: R01, R03, b08, and b83. The behaviour of beams R01, R03, b08, and b83 in terms of load versus lateral displacement of web nodes: 140, 141, 142, and 143 (0.20, 0.15, 0.10, and 0.05 h away from the superior flange) is portrayed in Figs 3.53 and 3.52.

The buckling configuration related to the principal web stress distribution for beam R01 and b83 is illustrated in Figs 3.54 and 3.55. In Fig. 3.54, the loading stages correspond to 83.81%, 83.85%, 105.3%, and 103.3% (unloading path) of the experimental ultimate load. In Fig. 3.55, the loading

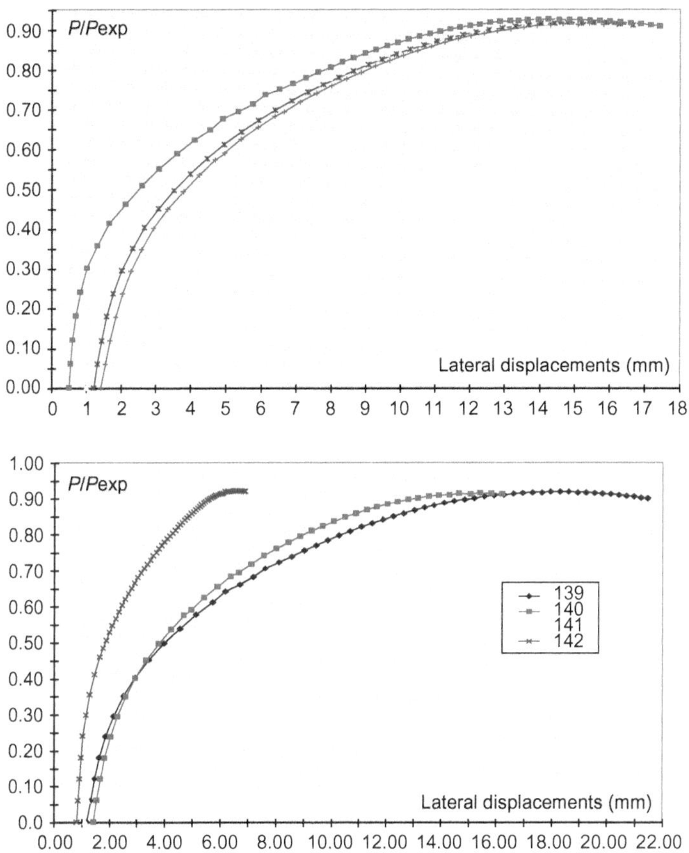

Fig. 3.51 Load versus lateral displacements, b14 and Imperfection amplitude and position effects.

stages correspond to 40.9%, 93.6%, 93.8%, and 91.9% (unloading path) of the experimental ultimate load. An inspection of the stress evolution related to the deformed shape reveals that the onset of web buckling has begun at a load levels corresponding to 83% and 93% (beam r01 and b83) of the ultimate load. All the other modelled beams had similar deformed shapes and web stress distribution.

Interesting results in terms of collapse loads, deformed shapes, and web stress distribution were obtained with the performed numerical analyses. The amplitude and position of the initial imperfections over the collapse load was insignificant. These results were also compared to theoretical and experimental investigations, validating the structural behaviour of the model and guaranteed the consistency of the proposed solution.

The modelling of beams under the action of concentrated loads proved to be an extremely complex process due to the interaction between the

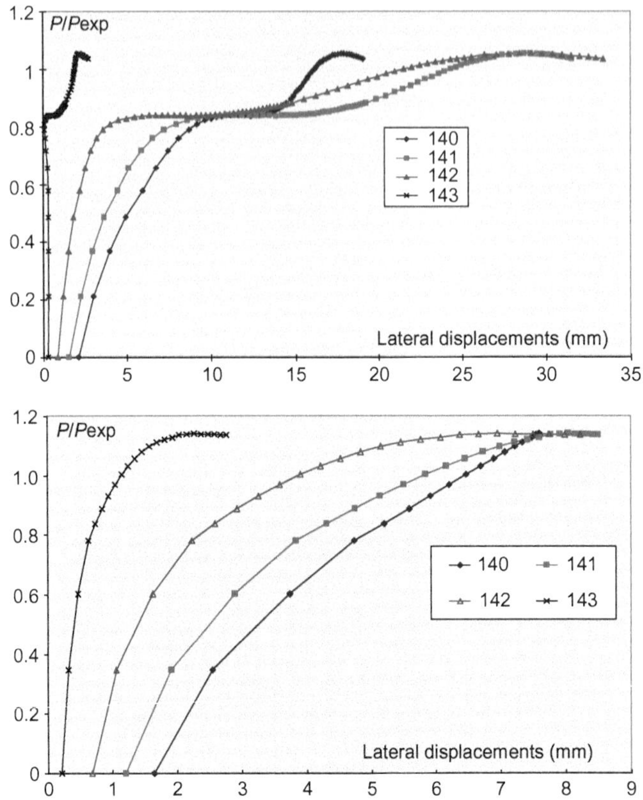

Fig. 3.52 Load versus lateral displacement, R01 and R03.

ultimate limits states that control their structural behaviour: yielding, crippling, and local web buckling. The model adopted the usual conditions of symmetry and support to remove rigid body movements and represent the real conditions found in practice.

The load control was made by applying an equivalent displacement to enable its better control in the regions close to the structural collapse. The convergence has proved to be difficult in some cases, generating the use, from the beginning, of the full Newton–Raphson method and a reduced load step size. The convergence criteria also had to be adapted in terms of their maximum tolerance and of which variable should control the analysis (displacement, force, rotation, energy etc.).

3.5.6 Prestressed Stayed Steel Columns Resistance

The present investigation assessed the effect of initial imperfection, stay axial stiffness, and prestress force magnitudes on stayed steel columns [110,111]. One of its main focuses is related to the search for a prestress force magnitude

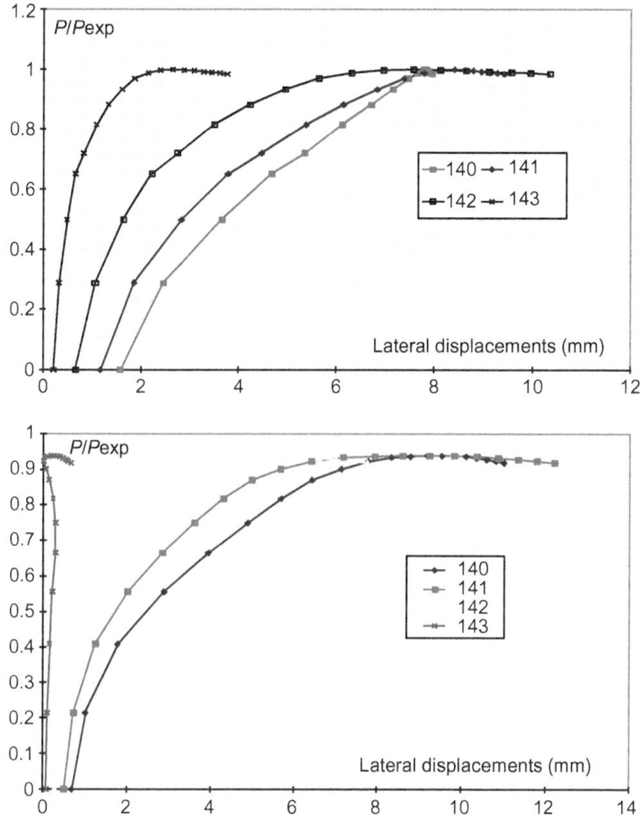

Fig. 3.53 Load versus lateral displacement, b08 and b83.

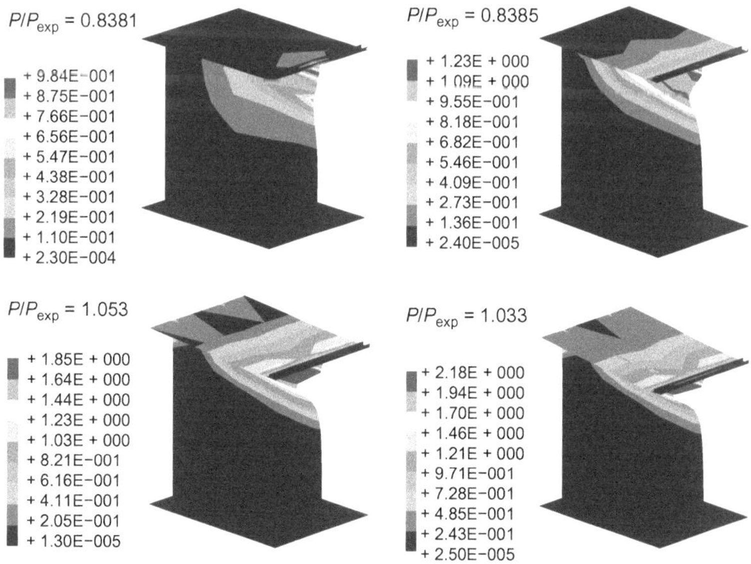

Fig. 3.54 Principal stress distribution evolution, R01.

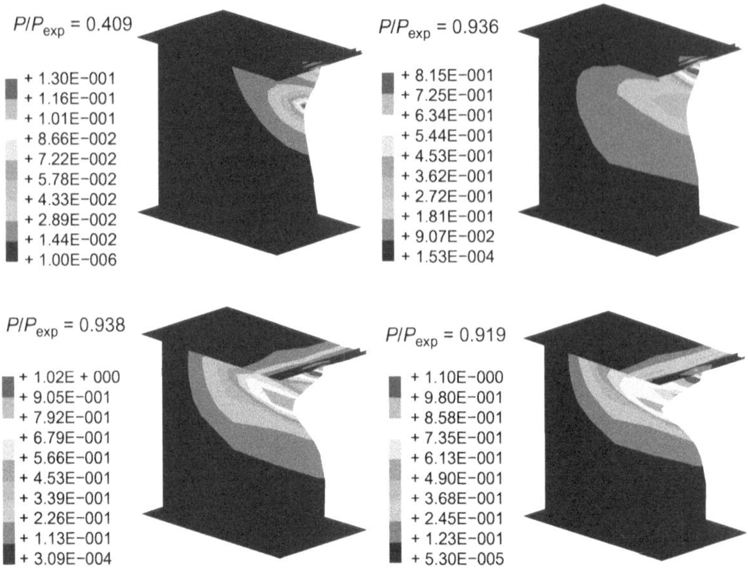

Fig. 3.55 Principal stress distribution evolution, b83.

that will lead to the system optimum load carrying capacity. The investigation encompassed an assessment of the influence over the type of initial imperfection, as well as the stay axial stiffness. The initial imperfections measured in the experimental model motivated a strategy where three levels of initial column imperfection were studied.

An assumption adopted in the computational model was that the maximum bow imperfection occurred at the stayed column centre. The three performed numerical analyses adopted maximum bow imperfection amplitudes of: L/1500, L/500, and L/300 (i.e.: 8, 24, and 40 mm). The L/1500 imperfection had a similar amplitude of the experimental imperfection, while L/500 is the maximum imperfection permitted in fabrication tolerance standards. At this point, it is important to remind the readers that when the experimental column was supported only at its ends, a 115 mm vertical displacement was recorded, due to the self-weight being a substantial initial imperfection.

Three levels of stay axial stiffness were adopted, when two were modelled with cable, while the other used a steel bar (a typical reinforcing bar). The cables had cross-section areas of 16.2 and 31.7 mm^2, while the rebar area was 126.7 mm^2. The 31.7 mm^2 cable was designed according to the procedure proposed by Belenya [112] while the 16.2 mm^2 cable was equivalent to the stay used in the experiments.

The numerical analyses were performed assuming: simply supported ends; a rotation restriction around the column axis and a perfectly centred load. As mentioned in the previous chapter, in order to solve the problem of excessive deformation caused by the Colum self-weight, a cable was used at the column midspan to support the structural system. A numerical model presenting the same characteristic of the tested specimen was developed to determine the cable influence over the structural system performance.

The adopted Ansys library elements were the PIPE20 element for the main column and cross bars and the LINK10 element for the stays. The PIPE 20 element is a uniaxial element capable of dealing with tension-compression, torsion, and bending with six degrees of freedom per each node: translations in x, y, and z nodal directions, and rotations about the x, y, and z axes. The element has as characteristics: plasticity, deformation, and expansion. The LINK10 is a 3D cable element with a bilinear stiffness matrix capable of representing axial forces (tension or compression). With the 'tension-only' option, the element stiffness is removed if the element is under axial compression (simulating a loose cable or a loose chain).

Two types of materials were used in the stayed column modelling. One for the main column and cross arms, and a second for the stays. The main column was divided in 12 1000 mm long elements, while the cross arms were built with 300 mm long two elements. The stays were modelled by a single element. The column was initially built as perfect. Subsequently, a force was perpendicularly applied to the column at its centre. The force magnitude corresponded to a value capable of imposing to the column the required imperfection magnitude. Fig. 3.56 illustrates the modelled stayed column with its initial imperfection.

In the computational model, restrictions were applied at the column ends to simulate a simply supported system. The bottom end displacements have been restricted in the global axis x, y, and z, to eliminate rigid body motions, while the rotation around the global y axis i.e. the column axial direction,

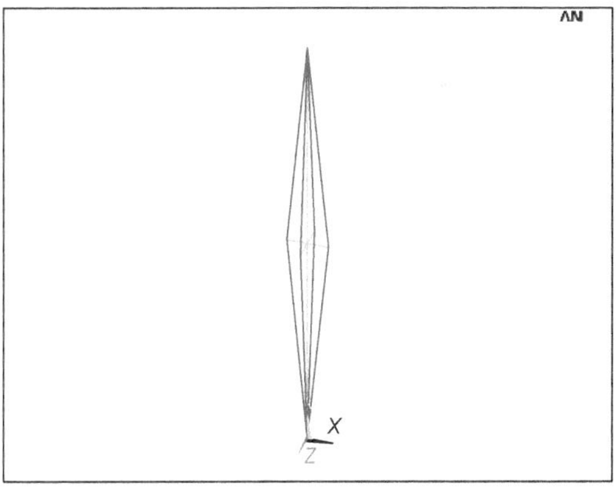

Fig. 3.56 Cable-stayed column numerical model.

was also prevented. The top end displacements have been restricted only in the global axis x and z. The LINK10 element selection has been very useful, since one of its data input characteristics is associated to its initial deformation. Using this initial deformation, various prestressing force magnitudes have been applied to the stayed steel column system.

The next step was the load application in the cable-stayed column. In order to facilitate the load control in the nonlinear phase near the critical load, displacement control strategy was adopted. The load was applied via a nodal displacement at the column top. Two main load phases were used: first the prestress forces were applied to the system by means of an initial deformation of the LINK10 element; this was followed by the application of a nodal displacement in the global y axis direction to simulate the axial loading. Fig. 3.57 presents the relationship between the stays prestress force load and the stayed column critical load without prestress forces. For a better understanding of Fig. 3.57, 'B' corresponds to a reinforced bar material, 'C' corresponds to a cable material while the numbers represent the column imperfection magnitude. The captions Imp 8, Imp 24, and Imp 40 correspond to columns without stays and their respective imperfections magnitudes.

An inspection of Fig. 3.57 shows that when stays are used in a slender column, there is a significant increase of their critical buckling load. For an 8 mm imperfection column without stays, a 160% difference is observed in relation to a stayed column with cables and a 485% difference is noted for a stayed column with bars. It is concluded that, for both, the bar and the cable systems, the higher the column imperfection magnitude, the lower the prestress force effect over the column critical load. The optimum critical load gain by placing reinforcing bar by cables is equal to 83% for an 8 mm imperfection, 85% for a 24 mm imperfection, and 88% for a 40 mm imperfection. The prestress force to critical load ratio for structural system with cables is of nonlinear response, while in systems with bars this ratio is almost linear. Another interesting comparison can be seen by observing the ratio of the overall structural system critical load by the ultimate strength of the adopted cables and bars, as seen in Fig. 3.57.

The nominal cable strength was 23.8 kN while the 12.7 mm reinforcing bar CA-50 was 69.7 kN. From Fig. 3.58, it is possible to observe that the curves are quite similar, but a difference can be noted in the adopted percentage value (horizontal axis), which is more evident in the curve associated to the cable with a 24 mm imperfection. The following results show the maximum and minimum jack load versus midspan lateral displacements curves and versus axial displacement at the column end, where the applied load was controlled (Fig. 3.58).

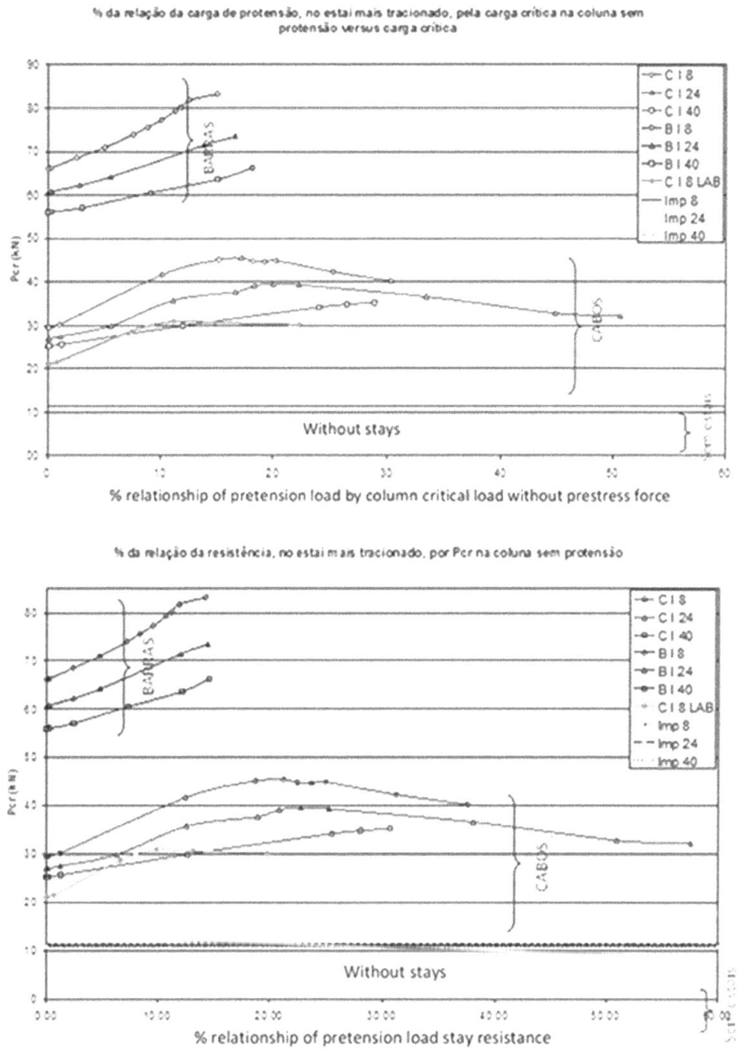

Fig. 3.57 Critical column loads (with and without stays, with and without prestress forces) and prestress force to ultimate load ratio versus column critical load without prestress force.

Fig. 3.59 depicts the jack load versus column midspan lateral displacement for the investigated imperfection magnitudes, where it is possible to note that the higher the stays axial stiffness, the higher their associated critical loads and corresponding deflections. It could also be noted that the larger the initial imperfection magnitude, the higher the ultimate load corresponding displacement, and less sudden the system collapse.

Fig. 3.60 shows the jacks load versus column top end axial displacement for the studied imperfection magnitudes. The critical load gain can be clearly

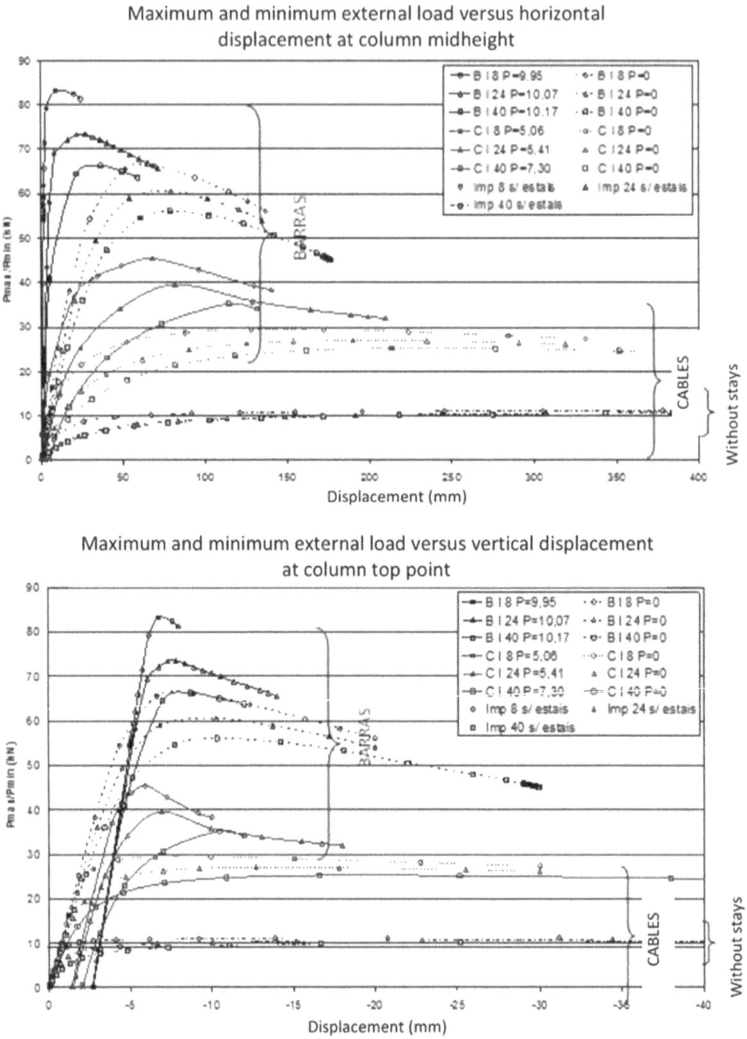

Fig. 3.58 Minimum and maximum applied load versus column horizontal and vertical midspan displacements (prestress force values in kN).

noted for stayed columns when compared to columns without stays, as well as the subsequent resistance increase when prestress forces are used in the stayed column systems. It is worth mentioning that this load gain occurs up to a certain prestress magnitude, after which there is a reduction of the overall system ultimate load.

Fig. 3.61 contains the applied load versus load on the most tensioned bar and cable load curves (which is in the imperfection direction) for each studied imperfection magnitude. A very distinct behaviour can be observed between cases with reinforcing bars or cables. In cases with reinforcing bars,

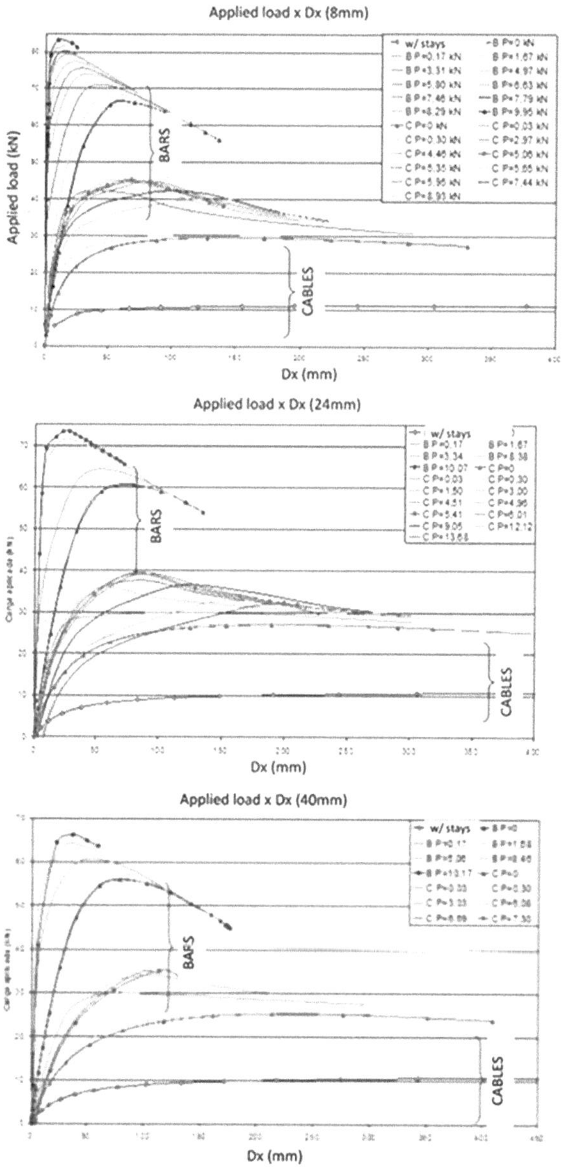

Fig. 3.59 Applied load versus column horizontal midspan deflection (imperfections: 8, 24, and 40 mm).

the stay loads remain constant up to a value where it begins to grow until the structure collapse. In the reinforcing bars system the load initially decrease and later increase. These observations are more evident in columns where smaller imperfection magnitudes are attenuated. This implies that the cables system increase their load earlier and that the h load reduction of the bars system in the initial load stages is smaller.

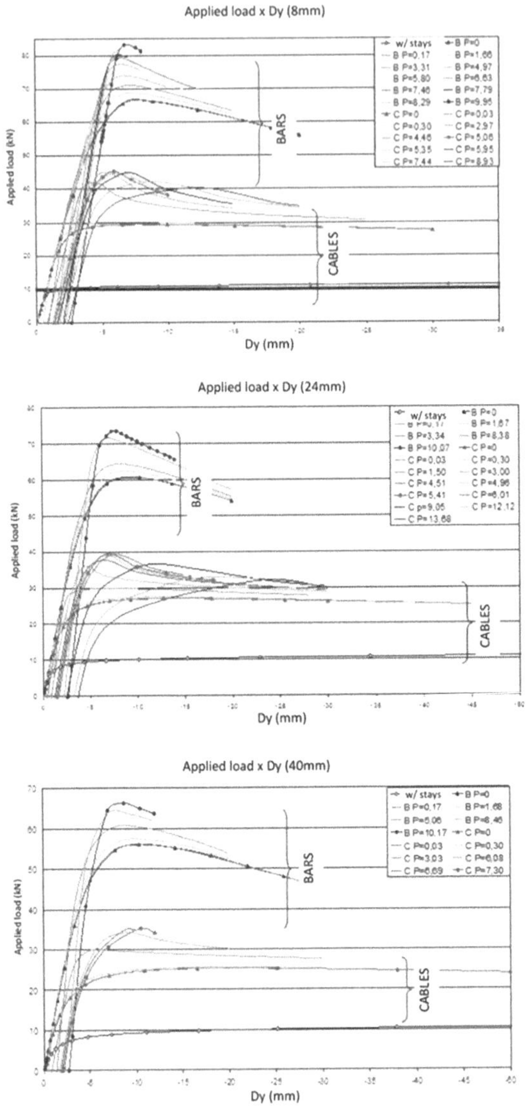

Fig. 3.60 Applied load versus column vertical midspan deflection (imperfections: 8, 24, and 40 mm).

It was also possible to observe that, although the analyses considered the material nonlinearity (related a possible yielding of the adopted cables, bars and tubes), all investigated cases remained in the elastic regime. Naturally, this can be explained with the great slenderness of the adopted stayed column (approximately 400 for a 12 m length or 200 for half of this size). Fig. 3.62 shows the typical deformation configuration generated by the ANSYS program. In the model analysed with cables, the deformation

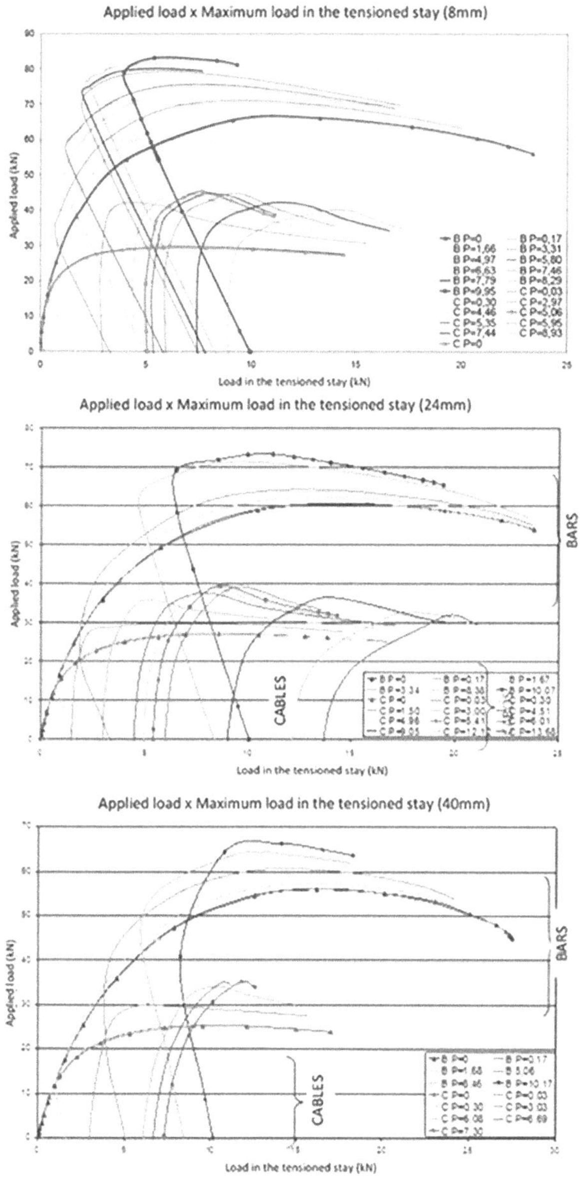

Fig. 3.61 Applied load versus most stressed stay load (imperfections: 8, 24, and 40 mm).

was associated to the first buckling mode, while in the model with bars (greater stiffness) it occurred in the second buckling mode, represented by two waves on cable column.

After an extensive numerical study of the behaviour of stayed columns with the ANSYS program, it could be concluded that by using stays, a significant increase of the critical load is noticed. It could also be observed that the stiffer

Fig. 3.62 First and second buckling modes.

the stays, the greater its influence on the critical load increase. For the investigated axial stiffness levels, it could be concluded that the higher the imperfection magnitude, the lower the prestress force effect over the critical load. The stay axial stiffness also influences the stayed column structural behaviour.

Having a low stiffness, column behaves in a nonlinear manner, while increasing this stiffness, the behaviour becomes linear. It is interesting to note that for the evaluated columns, the simple use of cables without prestress increases the system ultimate load in 100%, and when correct level of prestress is used this gain can rise to 200%. It was also possible to observe that optimum prestress force was higher than 1% of the column design value per stay, as suggested by Belenya [112].

Two alternatives were created to assess the structural system optimal load. The first is made by using the ratio of the stay prestress force by the stayed column critical load without prestress. This alternative depends on the adopted structural system being particular to a type of structural configuration. The second is based on the ratio of the stay prestress force, when the stay strength is independent of the structural system and easier to be used in construction.

Modelling the stays was fundamental to assess their influence on the structural behaviour of the studied prestressed system. The prestress force application was made by attempts using stay length reductions till the stay force achieved the required magnitude. A preliminary study indicated that the use of the cable to reduce the initial system displacement due to the self-weight effect in the horizontal configuration have not influenced the model structural response. This could be evidenced by a very similar structural behaviour between the numerical models with and without this cable.

The adopted nonlinear modelling was not trivial, demanding numerous restarts and changes in the steps size, strategies, and convergence criteria. The supports were adopted to simulate the conditions present in the experiments by restricting rigid body movements and simulating a simply supported system. The load was applied by means of an equivalent axial displacement at the column top for a better convergence control of the phases close to the structural collapse.

Imperfections present in experimental models were incorporated in the numerical simulations in terms of shape and amplitude, always respecting the manufacturing tolerances usually specified in their associated standards. Some maximum imperfection amplitudes were investigated demonstrating their influence over the stayed columns structural behaviour. The numerical models also prove to be useful for the simulation of partial rotation restriction at one of the column ends present in the preliminary tests. Boundary cases considering fixed and hinged supports were evaluated and served as limits for the already mentioned partial rotation restriction.

Numerical simulations were also made in the developed stay loads cells to investigate possible problems caused by unavoidable load eccentricities that could be related to the fact that the holes diameter was larger than the bars used to apply the stay prestress forces. These simulations have shown that these eccentricities had a negligible effect on the load cells strain distribution of the points where the stresses are acquired (for measuring the forces acting on the stays).

The effect of the prestress over the structural behaviour enabled the distinction of two different behaviours between systems with cables and reinforcing bars. While cable systems had a optimum prestress force magnitude where higher prestress forces lead to a reduction of the stayed column ultimate load, this optimum point did not occur on reinforcing bars systems, where a continuous increase of the system load occurred till the point where only the application of the prestress force directly led to the system structural collapse.

3.5.7 Wind Tower Modelling

The wind energy is an abundant source of renewable energy, clean and widely available. The use of this energy source to generate electricity on a commercial scale started just 30 years ago, based on the application of aviation industry knowledge. The equipment for wind power generation has quickly evolved in terms of ideas and preliminary concepts for high-tech products. This work has developed a computational model that satisfactorily represents the structural behaviour of the Repower MM92 wind Tower. Some simplifications of the structural model were considered, in order to allow a satisfactory assessment of the static and dynamic response of the investigated wind tower. The material and geometric nonlinearities effects, the self-weight of the tower, and the equipment needed for its functionality, as well as the wind action over the tower blades, were considered in the methodology analysis developed in this study [113–115].

One of the most important loads to be considered in the wind steel towers analysis was originated from the wind. The NBR 6123 prescribes that the wind action can be interpreted as a dynamic effect due to atmospheric turbulence. Fig. 3.63 represents, in simplified form, the wind action adopted in this study. The wind action on the propeller blades can be seen as a distributed load along its acting area, Fig. 3.63. In order to simplify, but also maintain the results consistency, a resulting force equivalent to distributed load is adopted. In Fig. 3.63 the self-weight is also considered.

The steel wind tower studied in this work refers to the MM92 Repower, present in several countries such as Spain, Portugal, and Germany, having a capacity to generate 2 MW of electricity. The electrical energy production is directly linked to the wind speed in the region where the towers are installed. The MM92 begins to produce energy from a speed of 3 m/s and stops its production when it reaches a speed of 24 m/s.

The MM92 has a shape of a hollow tapered stem divided into three segments in order to facilitate transport and assembly. The first segment has a 21.77 m height, base diameter of 4.30 m, and top diameter of 3.917 m. The second segment has a 26.62 m height, base diameter of 3.917 m, and

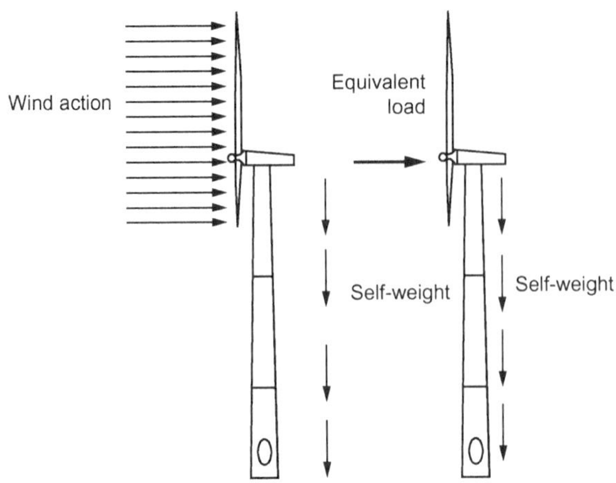

Fig. 3.63 Wind action on a wind Tower.

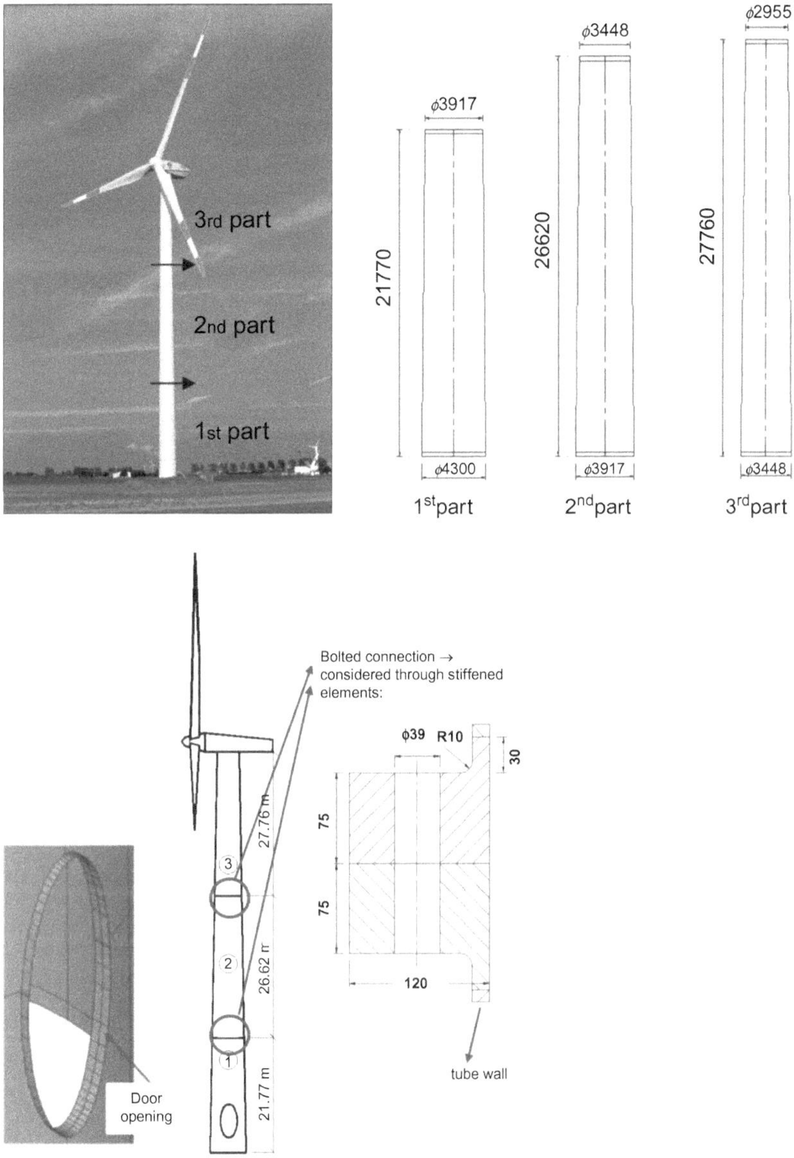

Fig. 3.64 MM92 Repower, Tower, joint and stiffener layouts.

top diameter of 3.45 m. Finally, the third segment has a 2.78 m height, with base diameter of 3.45 m, and top diameter of 2.96 m totalling 76.20 m high. Fig. 3.64 illustrates the divisions of Tower and a model of the Tower MM92 with a maximum external diameter of 4300 mm located at the base and 2955 mm at the top. The tower wall thickness varies along its height between 30 mm at its base and 12 mm at the top. Fig. 3.64 depicts an equality of the tower outer surface diameters to generate continuity, i.e. a flat surface.

The bolted joint system is used to connect the tower segments. The investigated tower employed a total of 464 bolts with: 45 mm diameter for connecting the first segment to the foundation, 39 mm diameter for connecting the second segment to the third segment, and finally, 30 mm diameter for connecting the third segment to the top. Stiffeners are used on connecting the parts of the Tower, since the joint flanges' thicknesses are greater than the tower wall thickness. Fig. 3.64 illustrates one of the joint flanges used in the tower and the stiffener used at the joint, the tower parts are connected by the flanges and, as they have larger thicknesses than the tower walls, stiffeners are necessary. Fig. 3.64 schematically represents a stiffener used in the tower. There are two openings, one for internal access (larger) and another for ventilation (smaller). Both have the shape of an ellipse with a stiffener perpendicular to the opening, Fig. 3.64.

The tower top is made of three propeller blades measuring 45.20 m each. They are manufactured with a plastic resin reinforced with fibre glass and weigh 8000 N each. The paddles are transported separately and fixed to the bolt rotor creating, when in operation, a varied surface of 6,720 m^2. The tower rotor is responsible for rotating the turbine and therefore producing electricity. The nacelle is the set of all mechanical and electrical equipment located at the top of the tower, weighing 6900 kg.

The numerical models were developed based on the finite element method using SHELL181; shell elements are present in the Ansys library [32]. This element is suitable for the analysis of shell structures with thin and medium thickness and has four nodes of six degrees of freedom, being a suitable element for modelling structural problems involving linear analysis and physical and geometric nonlinearities.

Geometric and physical nonlinearities were incorporated into finite element models, in order to fully mobilise the tower capacity to axial and shear forces and bending moment caused by the global forces actions. Additionally, the use of geometric nonlinearity enables the prediction of large deformations, considering the load redistribution in the tower after the onset of initial yielding.

The boundary conditions of the wind steel tower were simulated in the numerical models by restricting the appropriate degrees of freedom, i.e. the rotation and translation of the x, y, and z axes at the tower fixed base. The finite element model meshes were defined through validation tests, i.e. modal analyses of the models were made with various mesh densities. When the results converged to the experimental results, i.e. they did not show significant variations, and the meshes were considered as acceptable. The final model was composed of 17,094 elements and 17,124 nodes, as shown in Fig. 3.65. In modelling the tower openings, all their geometrical

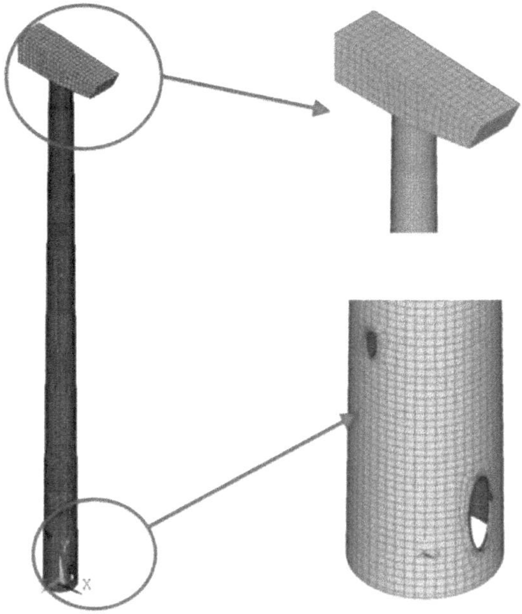

Fig. 3.65 Structural detail and finite element model.

characteristics and stiffeners were considered. The propellers blades, rotor, and nacelle were represented by shell elements, as shown in Fig. 3.65, with a density equivalent to their weights.

The effects of warping and distortion of the sections were not considered in this modelling; it was considered that the imposed stresses do not cause yield in the elements cross sections. However, second order effects were taken into account in the analysis. The tower connections were not under shear deformations, while the material of the nacelle, rotor, and propeller were considered as a linear-elastic and isotropic materials. A bilinear elastoplastic behaviour with a 5% strain hardening was associated to the tower walls, flanges, and stiffeners used in the openings.

The purpose of static analysis was to determine the maximum wind Tower displacement caused by a concentrated load applied at the rotor centre, representing the resulting force of the wind acting on the propeller blades. Fig. 3.66 portrays the tower von Mises stress distribution due to the action of the wind with an average speed of 36 m/s causing a load of 308.45 kN applied at the rotor centre.

In this analysis, it was noted that the higher von Mises stress (97.2 MPa) was reached at the tower opening, but it did not exceed the 355 MPa steel yield stress. The largest displacement occurred at the load application point, i.e. at the rotor centre, i.e. 510 mm. Fig. 3.66 also illustrates the stiffeners von

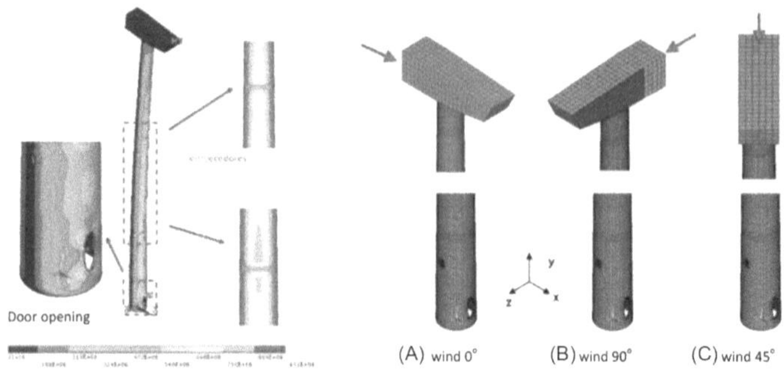

Fig. 3.66 von Mises stress distribution, 308.45 kN applied load, and load application points.

Mises stress distribution tensions where the smaller von Mises stresses occurred. The higher von Mises stress was located in the openings, whereas the smallest at the tower top. This fact is justified by the difference in the material stiffness adopted for the nacelle with respect to the tower wall.

Smaller values were also observed in the stiffeners, fact which validated their need. The serviceability limit state of the tower according to Eurocode 3, part 3-2 is checked by comparing the maximum displacement at the tower top with the allowable value given by $h/50$ where h represents the tower height. Using the data from the wind tower Repower MM92, a value of 1.53 m higher than the 0.51 m caused by wind is reached.

A nonlinear analysis was made to assess the actual wind tower global behaviour. The full nonlinear analysis was performed on the structural model using the material and geometric nonlinearities. The load was applied in terms of equivalent displacement at the tower nacelle centre. The nonlinear analysis principle is to cause a shift in the tower rotor centre. In the static nonlinear numerical analysis performed, the material nonlinearity was based on von Mises yield criteria using a bilinear constitutive stress versus strain law, in order to represent an elasto-plastic behaviour with a 5% strain hardening.

A 205 GPa elasticity modulus and 355 MPa yield stress were assumed for the steel. The geometric nonlinearity was introduced in the model through the updated Lagrange formulation. The Ansys program uses the Newton–Raphson method for solving the nonlinear equations system. This method is based on the division of the loads in a series of increments, applied in several load steps. In each step an equilibrium configuration is generated and a new step is applied and a new equilibrium configuration is reached until the total number of increments is completed. The nonlinear analysis was performed through the application of displacement at the Tower rotor centre

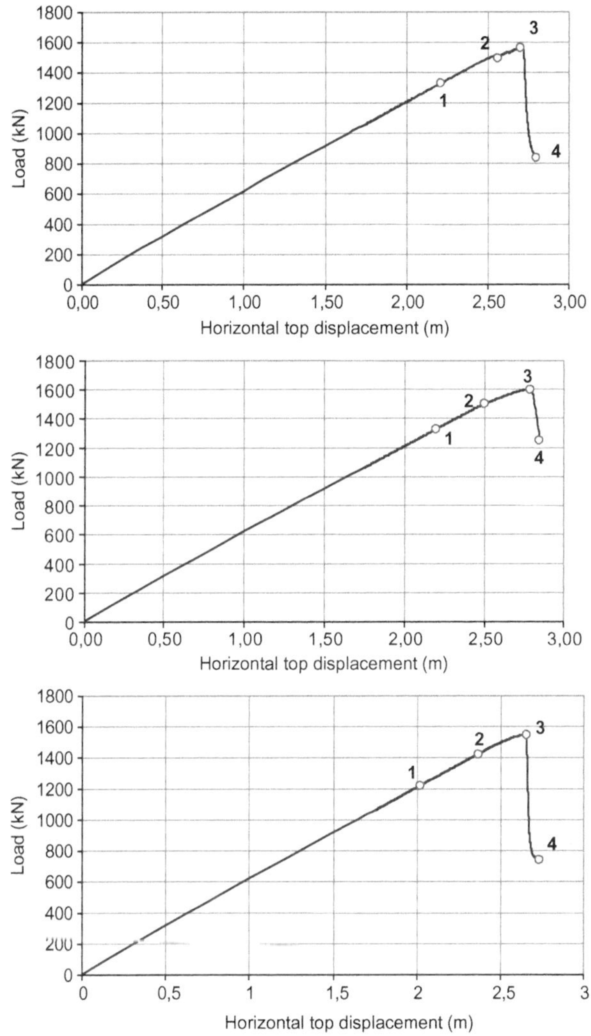

Fig. 3.67 Wind load versus displacement curves: 0°, 90°, and 45° wind load combinations.

in the x-axis direction (0° wind), in the direction of the z-axis (90° wind) and in a direction with 45° degrees angle between the x and z axes (45° wind). Fig. 3.67 illustrates the three different positions of the nacelle adopted for the load application. This is justified by the fact that the tower nacelle has the same direction of the wind. The analysis considered, for the three studied locations, compression stresses in the opening areas. Fig. 3.67 depicts the curves associated to the loads: 0°, 90,and 45° acting in the tower rotor centre against the load application point displacement resulting from the wind action at the propeller blades.

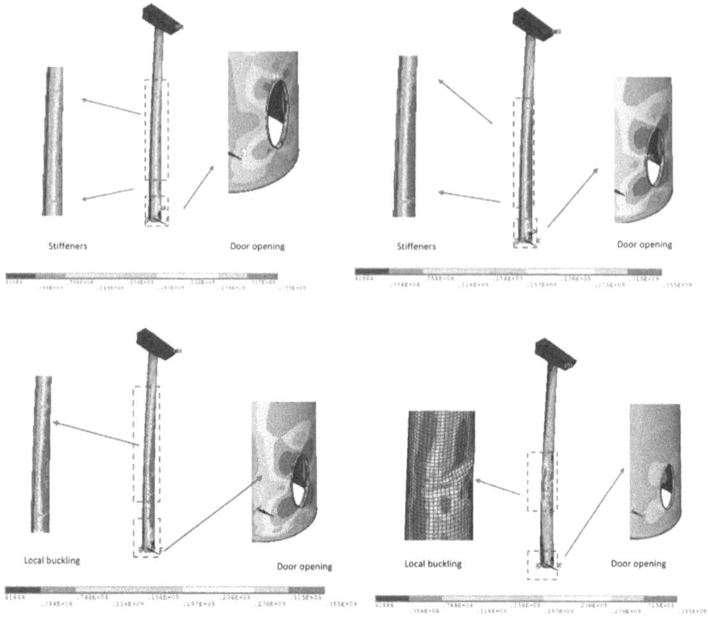

Fig. 3.68 von Mises stresses distribution, 0° wind load combination, points. 1 2, 3, 4.

Fig. 3.68 illustrates in the 0° wind case of von Mises stress distribution corresponding to points 1, 2, 3, and 4 present in Fig. 3.67, where the emergence of stress values close to the yield stress at a region near the stiffeners and the openings is observed, corresponding to a 1327.65 kN applied load. Point 2 corresponds to a 1491.00 kN applied load and 2.56 m deflection where it can be observed that an increase of regions with von Mises stresses equal to the material yield stress, indicating the start of the yield on the tower critical section, but still not compromising the structure. The buckling occurrence can be observed for a 2.703 m applied deflection corresponding to a 1559.92 kN load, point 3. The yielding around the openings continues to increase in point 4 with a 2.80 m the maximum deflection and a 832.38 kN load. The region where the local buckling deformation occurred presented high strains, inducing the development of a new von Mises stress distribution that caused a relief in regions where high stress concentrations had occurred before. It is also observed that in the openings detail, there was a region in compression, due to the wind direction, i.e. 0°, coinciding with the direction parallel to the x axis.

Fig. 3.69 presents the 90° wind case for an applied load magnitude of 1497.06 kN with a 2.50 m corresponding displacement, Fig. 3.67, point 2, where it was observed that the Tower, in the same regions described earlier, presented an increase in areas with occurrence of yield. Later, when the applied

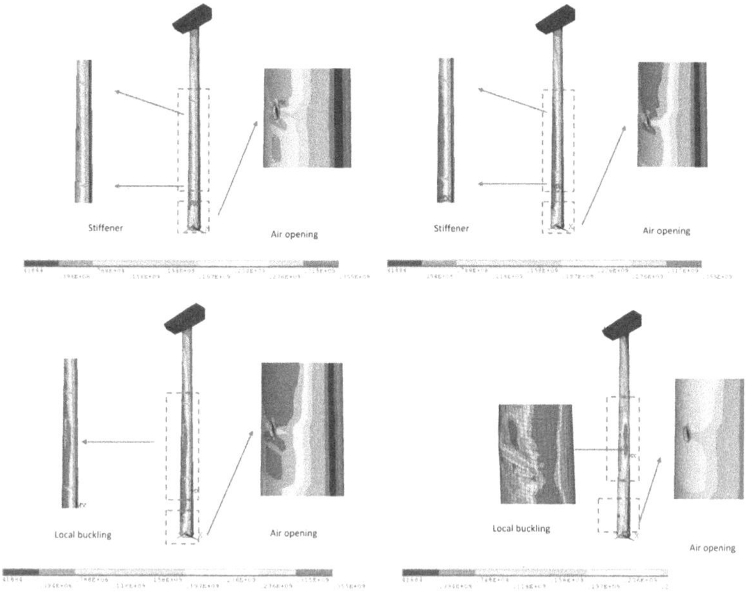

Fig. 3.69 von Mises stress distribution, 90° wind load combination, points 1 2, 3, 4.

deflection reached a value of 2.79 m, corresponding to the point 3 of Fig. 3.67, a local buckling in the region between the stiffeners was observed. When a 2.85 m maximum displacement corresponding to a 1245.33 kN load was reached, the region where the local buckling started presented significant von Mises stresses, creating new stress distribution in other areas, causing a relief in other tower parts, for example, the ventilation opening.

Fig. 3.70 depicts the 45° wind case von Mises stress distributions corresponding to the highlighted points presented in Fig. 3.67. Point 1 is associated to an applied load magnitude of 1217.30 kN and a 2.02 m displacement where high von Mises stresses appeared on the tower stiffeners. Point 2 is associated to an applied load magnitude of 1419.93 kN load and a 2.37 displacement, where an increase in the von Mises stress distribution between the stiffener and the tower can be noticeable. In point 3 there is the appearance of a local buckling associated to a maximum load of 1542.73 kN and a corresponding 2.66 m displacement. In point 4 a local buckling and stress redistribution was observed in the tower for a load of 738.28 kN with a corresponding 2.74 m displacement. In the opening areas the same high stress concentration depicted in the previous examples was not observed in this wind load case.

From Fig. 3.67 it can be noted that the 90° wind load case action was associated to the highest ultimate load, followed by the 0° finally by the

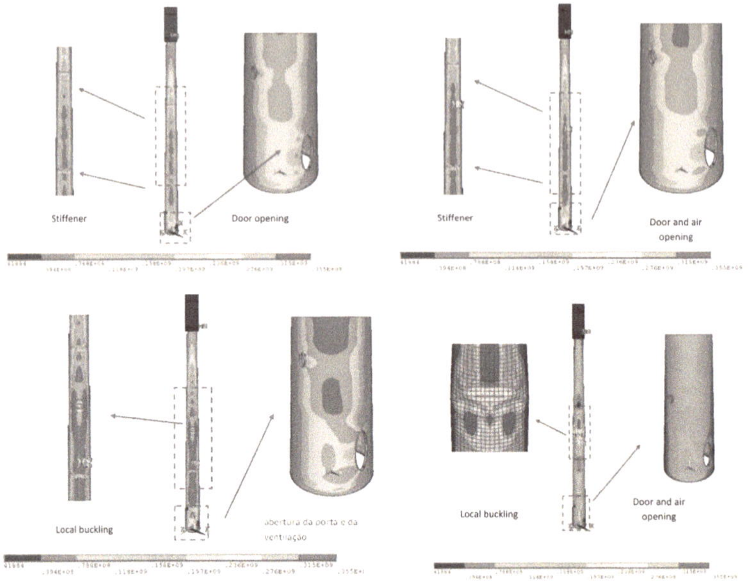

Fig. 3.70 Von Mises stresses distribution, 45° wind load combination, points 1 2, 3, 4.

45° wind load cases. The differences between the maximum resistances were of 3.52%. A suggestion to address the emergence of local buckling on the Repower MM92 wind tower could be the use of stiffeners or an increase in the tower wall thicknesses in these regions. The tower openings were located in the regions where the tower yielding started and to solve this problem, a new configuration of stiffeners or an increase of the tower wall thicknesses around the openings must be performed.

This work aimed at studying the static behaviour of a Repower MM92 tower wind. This evaluation was performed through the use of usual discretisation techniques via the finite element method. Nonlinear static responses were analysed through the application of loads that simulate the action of the wind on the tower blades. The finite element numerical method proved to be very useful and accurate for assessing the wind tower structural behaviour.

Its use proved to be effective for predicting linear and nonlinear static analyses, when compared with experiments. It was verified that the investigated load cases did not generate stresses higher than the yield stress, fact that, in principle, ensures their structural safety. However, this statement must be taken into consideration with caution, given the approximations made concerning the wind load actions. It has been shown that the wind tower, in a nonlinear analysis, presented its collapse associated to local buckling. The maximum stress values did not vary significantly when the applied load direction changed, simulating the possible wind directions acting on the tower.

First, the tower modelling involved the determination of imperfections to be incorporated in the model. For this purpose, the deformed configuration of first buckling load obtained in an eigenvalue analysis was used. In this particular case, despite knowing that this is not always the imperfection that would lead to the critical structural performance, since this is not always true in the plastic range, there was no other evident way that could motivate its adoption. The maximum adopted imperfection amplitude was established to comply with the maximum tolerance limits prescribed by steel structures manufacturing standards. A mesh refinement sensitivity analysis was also performed, while the adopted support types used the same principles described in the previous examples.

As the wind direction is unknown, load cases associated to 0° and 90° directions were adopted. Once again the load was simulated by an equivalent displacement to facilitate the convergence control. One of the most interesting aspects was the evaluation of the opening influence over potential stress concentrations in their surroundings. One more time, the convergence near the collapse was complex, leading to the use of strategies solutions already discussed.

Once again, the load application control was made by applying an equivalent displacement to enable a fine tuned control in regions close to their collapse. Convergence also proved to be difficult in some cases, leading to the use of a reduced load step size. Some examples also required the use of numerous convergence criteria adopted in terms of their maximum tolerance, which variable should be controlled (displacement, force, moment, rotation, energy, etc.).

3.5.8 Tension Capacity of Carbon and Stainless Steel Plates With Staggered Holes

The present investigation studied the tension capacity of cover plate joints with the aid of the Ansys program [32]. The numerical model used solid elements (SOLID45) characterised by eight nodes with three degrees of freedom per node: translations in the nodal x, y, and z directions. The finite element mesh was selected so that the elements had a proportion and size to avoid numerical problems.

Contact elements (CONTA174 and TARGE170) contained in the Ansys Elements Library were used between the plates and between the holes and the bolt shanks. The load was applied by means of axial displacements in the load plate such as illustrated in Fig. 3.71, where it is possible to note that the bolt head and nuts were simulated by UZ displacements restraints at the hole adjacent area.

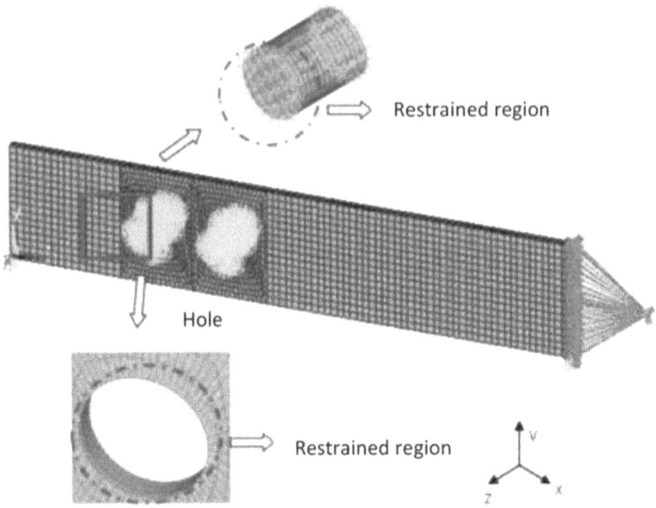

Fig. 3.71 Adopted finite element model and contact elements [24].

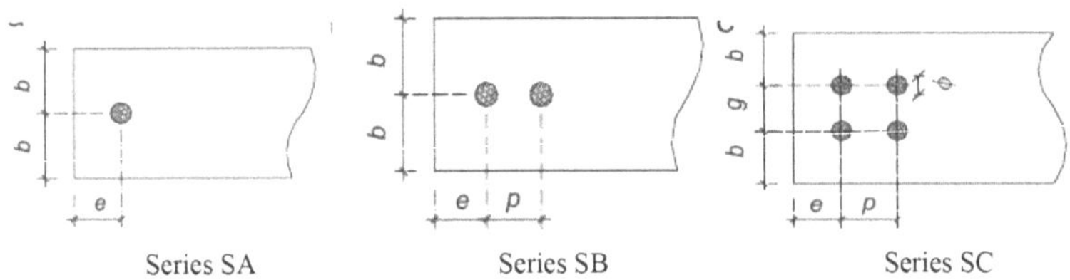

Fig. 3.72 Geometry of the joints [116].

Kim and Kuwamura [116], modelled a stainless steel bolted joint with contact elements. The results obtained in this study were calibrated with experiments obtained by Kim and Kuwamura [116]. The experimental work performed by Kim and Kuwamura [116], considered three different bolt arrangements, as shown in Fig. 3.72, characterised by only one shear plane; The test were made with SUS304 austenitic stainless steel plates, 1.5 mm and 3 mm thick, and 12 mm bolts (A2-50; SUS bolts or 10 T-SUS, high-strength bolts). The two plate ends were fixed at an Amsler universal testing machine through which the load was gradually applied to the tested plates.

The stress versus nominal strain curves of the tests used to calibrate the present SUS304 stainless steel model is illustrated in Fig. 3.73, along with their associate true stress versus true strain curves.

SA2-2 test, associated to an experimental ultimate load of 48.05 kN, and the best numerical ultimate load match by Kim and Kuwamura [116] equal to 44.85 kN Fig. 3.74 for the joint configurations containing a single bolt. The results indicated that there is no influence of the plate length when

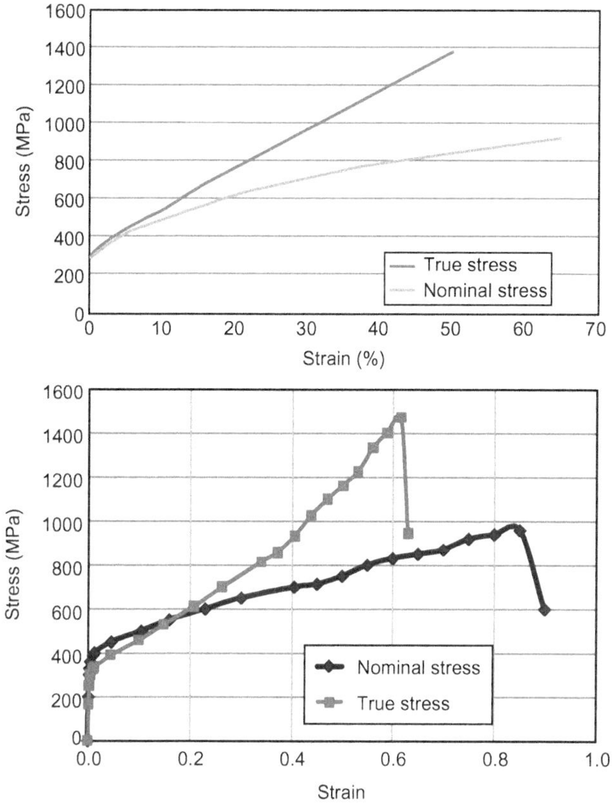

Fig. 3.73 *Stress* versus strain curves, A304 stainless steel.

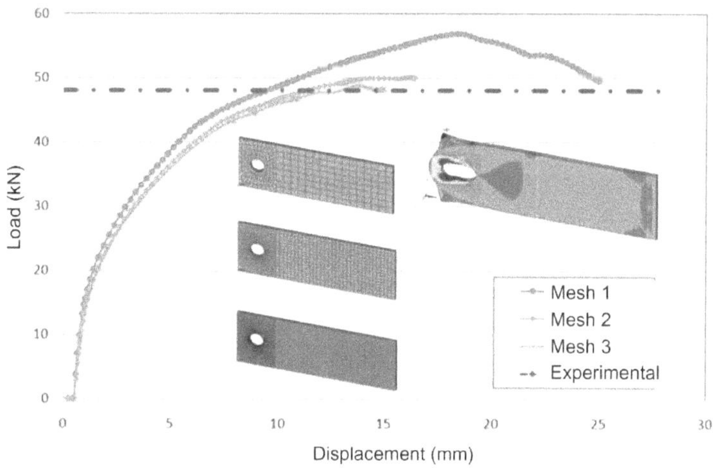

Fig. 3.74 Load versus slip curve, SA2-2.

values equal or above 150 mm are used in the analysis, since the ultimate load and collapse mode of the joint remain unchanged when this length was varied. Thus, for the investigated model, a 150 mm length was used, while the load was applied by means of a displacement control strategy with convergence tolerance of 0.01.

Fig. 3.74 presents the results of second and third models where the mesh size varied indicating that they had similar structural responses. On the other hand, the first model showed a similar behaviour to the other two up to slip of approximately 5 mm. From this point on, this model presented a different behaviour, resulting in an ultimate load much higher than the experiments. The third model presented an ultimate load closer to the experimental load with an identical failure mode, Kim and Kuwamura [116].

SB2-4 test associated to an experimental ultimate load of 85.62 kN, and the best numerical ultimate load match by Kim and Kuwamura [116] equal to 86.81 kN in Fig. 3.75, was used for the joint configurations containing two bolts. Again, Kim and Kuwamura [116] show that there is no plate length influence if values equal or above 150 mm are used, thus, the adopted investigated model length was 230 mm.

Fig. 3.75 depicts the results of all studied models where the mesh size varied indicating that they had similar structural responses up to a 30 mm slip. But the first model presented the larger slip associated to the ultimate load, while the third model presented the smallest slip. These results may have been influenced by the first model finite element mesh that was made of larger elements than the others, resulting in a less accurate representation of high stress concentration areas.

SC2-4 test, associated to an experimental ultimate load of 162.32 kN, and the best numerical ultimate load match by Kim and Kuwamura [116] equal to 152.79 kN Fig. 3.75, were used for the joint configurations containing four bolts. As in the previous models, a length was 240 mm higher than the 150 mm limit was adopted in the investigated model.

It can be observed in Fig. 3.75, that the three models presented a similar response up a 10 mm slip, but the mesh used in the second model led to the best results. An inspection of Fig. 3.75 enables the detection of the curling effect, which is a large one localised out of the plan deformation at the plate edge. This effect occurs when the distance between the hole and the plate edge is too large. When the ultimate loads are assessed it can be concluded that they were not influenced by this effect because it only occurred in the model unloading phases, after the ultimate load. It can also be noted that the proposed model has managed to reproduce the failure mode of Kim and Kuwamura [116] with an ultimate load close to the experimental value.

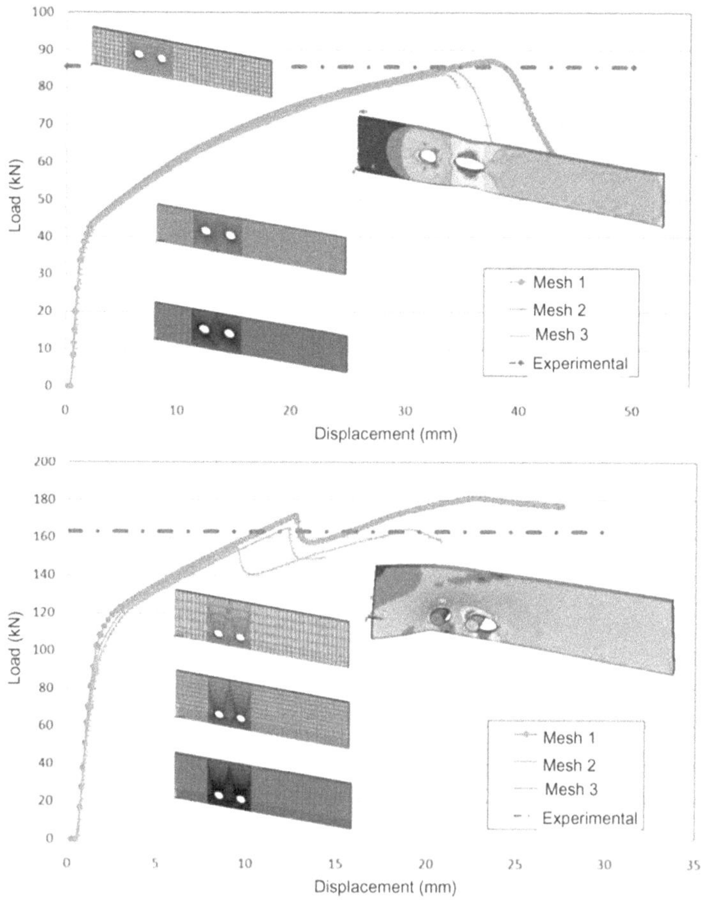

Fig. 3.75 Load versus slip curves, SB2-4 and SC2-4.

An innovative experimental program was utilised to assess the tension capacity of carbon and stainless steel plates with staggered bolts [22]. The experiments comprised bolted cover plate connection made of stainless steel A304 and carbon steel USI300 denominated EX_INOX_Y and EX_CARBON_Y [23], respectively. The bolted connections consisted of two 3 mm thick stainless and carbon steel plates and two 15 mm thick carbon steel plates utilised to transfer load to the 3 mm plate with a 5 mm gap. The horizontal bolt pitch, *s*, was changed in each test while the vertical bolt pitch, *p*, was of 55 mm.

The curves obtained in tensile coupons tests depicted a nonlinear expected response, mostly for the stainless steel. The stainless steel yield stress was obtained by using a straight line parallel to the initial stiffness at a 0.2% strain, i.e. 350.6 MPa while the ultimate tension stress was 710.7 MPa. For

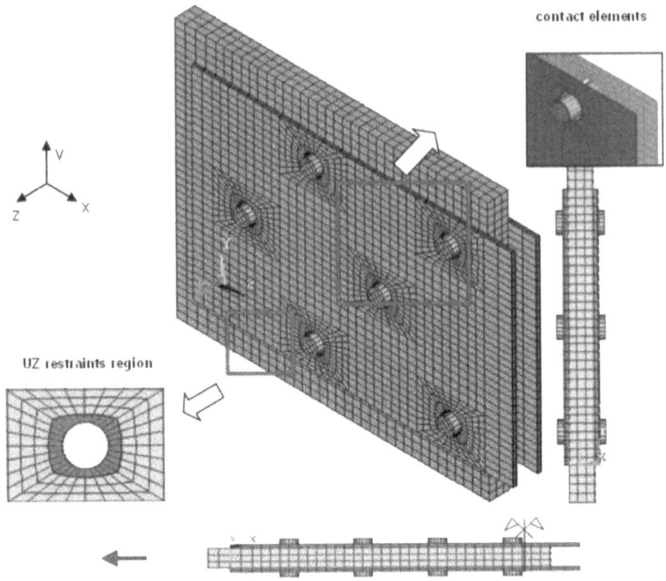

Fig. 3.76 Finite element model and contact elements.

the carbon steel, these values were equal to 386.8 and 478.7 MPa for the yield and ultimate stress, respectively. This curve was utilised in the finite element models, due to the large strain and stresses related to the problem where σ_t, ε_t, f_y, and ε_n depict the true stress, true strain, yield stress, and original measured strain, respectively.

$$\sigma_t = f_y(1 + \varepsilon_n)\, \varepsilon_t = \ln(1 + \varepsilon_n)$$

Fig. 3.76 portrays a typical mesh configuration of the complete model. It should be stressed that only half of the model was considered by using the symmetry conditions being sufficient to represent the connections ultimate limit states. The adopted material properties were: Young's modulus of 210 GPa and a Poisson's coefficient of 0.3. As mentioned earlier, stainless steel true stress versus true strain curves with a nonlinear behaviour were adopted using tensile coupons tests data [22].

A full nonlinear analysis was performed in which the material nonlinearity was considered utilising a von Mises yield criterion associated to a multilinear stress–strain relationship and isotropic hardening response. The geometrical nonlinearity was introduced in the model by using an Updated Lagrangean formulation. This procedure represents the full structural assessment of the studied bolted connections, and may be summarised in several outputs, namely the stress distribution (that detects, among other data, first yield), or the force–displacement curve for any node within the joint.

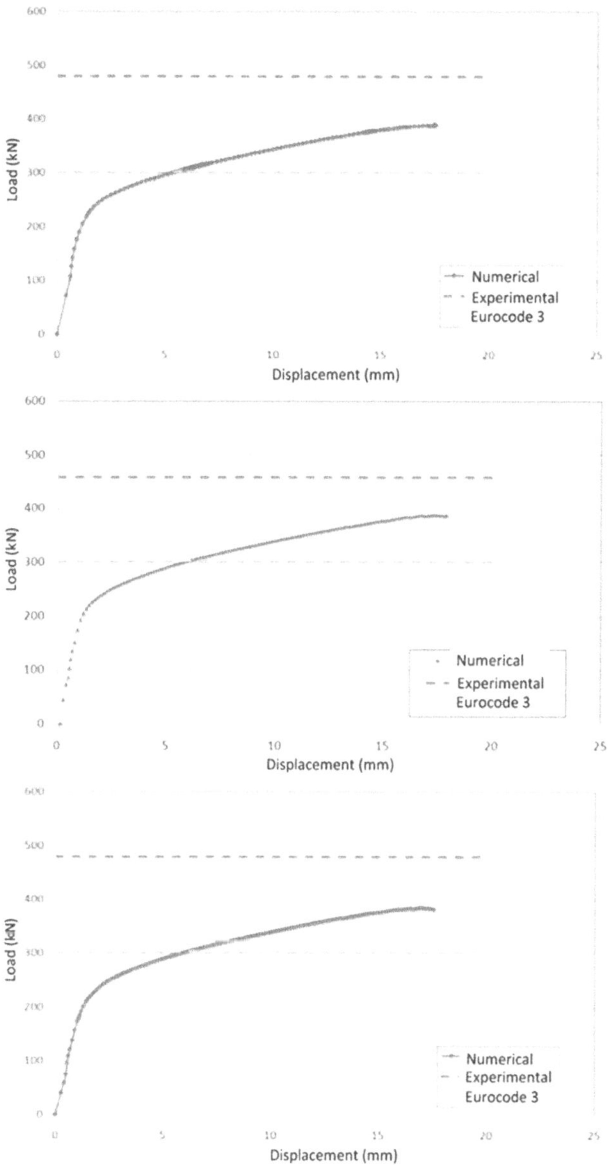

Fig. 3.77 Experimental load versus displacement curve.

Fig. 3.77 presents individual load versus deflection curves for each test, where the ultimate loads for E5_INOX_S50, E7_INOX_S30, and E9_INOX_S23 tests were 389, 389, and 385 kN, respectively. It can also be observed that in all numerical models, the numerical models ultimate load is situated among the Eurocode 3 [21], predictions and the experimental loads.

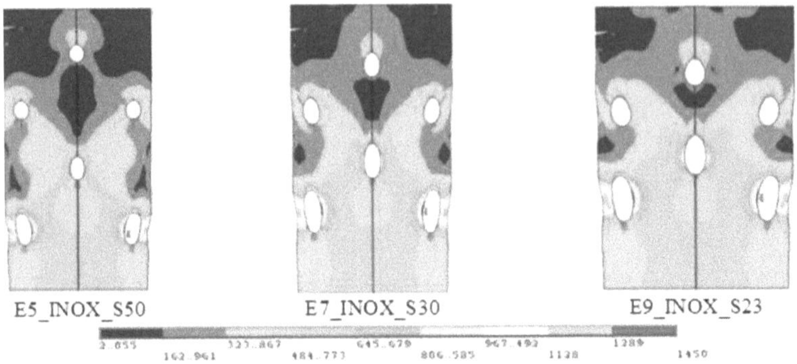

Fig. 3.78 Ultimate load von Mises stress distributions (MPa).

Fig. 3.78 presents the von Mises stress distribution for the three numerical models, in which high concentrated stress are detected in all models within an area between the hole and the plate edge. The associated stress distribution of the E5_INOX_S50 numerical model signalise a net rupture failure in an area passing through two holes, being in line with the test experimental collapse mode. The same trend did not occur in the E7_INOX_S30 numerical model where the stress distribution did not clearly identify if the net section failure passed through two or three holes. In the case of the E9_INOX_S23 model, the stress distribution identified the net section failure passing through three holes, fact also confirmed in the experimental findings.

In order to analyse the collapse mode associated to the cross section net rupture, a series of load versus strain curves of a point located in a section with two holes in the horizontal symmetry axis were constructed, as Fig. 3.79

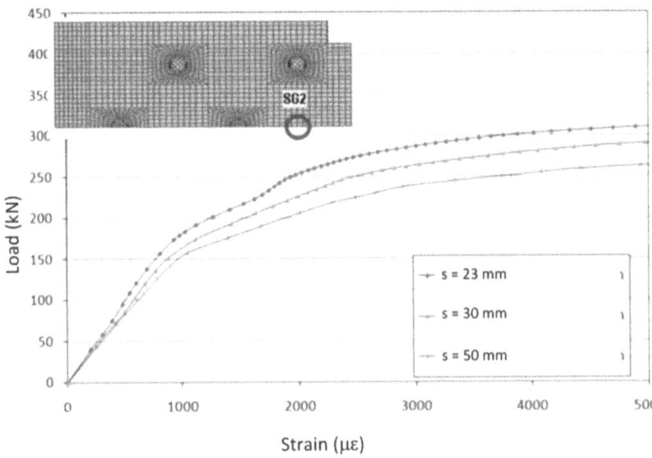

Fig. 3.79 Numerical models' load versus strain curves.

shows. From these graph it is possible to observe that the higher the horizontal distance between holes, parameter s, the lower the adjacent section bolt forces. For example, for a 250 kN applied load magnitude, the curve at a section further to the left presents the lower strains to the other indicating the net section failure passing through three holes. Alternatively, at this load level, the numerical model associated to E5_INOX_S50 test, shows that collapse involves a net section failure passing through two holes.

The present work had as objective the evaluation of the tensile strength of structural stainless steel elements with staggered bolts. The methodology consisted in the development of a numerical model based on the finite element method. The ultimate load values for all numerical models were lower than those obtained in the experiments. This can be attributed to the fact that the numerical models were conceived, in an idealised form, to represent the joints without incorporating any kind of imperfections.

The stainless steel modelling required that the material is simulated with the aid of the true stress versus true strain curves previously presented. This was due to the fact that the austenitic steel reaches large rupture load strains, around 60%. The boundary conditions had to be carefully implemented, so as not to generate additional stresses in the models under tension that could change their actual behaviour. The contact between the elements has been implemented assuming a friction coefficient of approximately 0.25 to simulate the transmission of forces between the plates and between the plates and the bolts. At the interface, the degrees of freedom were restricted to the axial load direction so as not to generate asymmetries in the plate. It is also interesting to note that the confinement effect at plate regions under the bolt heads and under the nuts was simulated by out of the plane restrictions of the plates. This strategy was used to avoid the curling effect on the plates at these regions since this had not occurred in experiments due to bolt and nut confinement.

The load application was made by using a single external node displacement attached to all nodes of the loaded surface plate. This strategy allowed a better control at stages close to the structural collapse. The models convergence was not simple in advanced loading stages, generating the need for numerous restarts, load step reductions and changes in the adopted convergence criteria (displacements, rotations, forces, moments and energy). The adopted nonlinear solution strategy varied between the full and modified Newton Raphson methods, depending on the loading stage.

The various collapse types associated to the effective net area rupture under tension in sections passing through two or three holes was identified throughout the stresses distribution at these regions. In the studied models, there was no collapse associated to the plate bearing, the gross section yield or bolt shear. However, the adopted modelling proved to be able to identify when these ultimate limits states would control the structural design.

3.5.9 Structural Response of Beams With Web Openings

In this study, finite element models were developed to simulate the structural response of steel girder with web openings [120,121]. These models were calibrated according to numerical studies performed by Chung et al. [122–124] and to experiments performed by Redwood and Mccutcheon [125]. Physical and geometrical nonlinearities were incorporated in the model to simulate the failure as accurately as possible. These nonlinearities enable the development of large strains at the steel beam web due to the stress redistribution that occurs along the web opening after first yield. This strategy enables a detailed assessment of the Vierendeel failure mechanism represented by the successive formation of the plastic hinges on the 'tees'.

The models utilised shell finite elements (SHELL181) present in the ANSYS library [32] featuring four nodes with six DOF per node. The loads and supports of the experiments were considered by imposing restrictions in some of the numerical model nodes. A finite element mesh sensitivity analysis was first executed to guarantee that it was properly refined to represent the studied phenomenon. This investigation enables the determination of a mesh with the minimum possible element number to represent the steel beam with web opening experiments. The model material nonlinearity was considered through a bilinear elastoplastic response with a 5% strain hardening and a 205,000 MPa Young's modulus. The rupture and yield stresses, as well as the simulated steel beams geometric properties with a single opening, utilised Redwood and Mccutcheon experimental values [125]. All the beams investigated adopted a simple support configuration. During the numerical study it was necessary to ensure that the model failure was not associated to lateral torsional buckling, nor with a local buckling at the load application point or supports. This was made by restricting the lateral displacements of the top flange along the beam span, and using transverse stiffeners at the load application point and supports.

The present numerical results were developed with the material and geometry utilised by Chung et al. [122–124]. A direct comparison with the results of Chung et al. [122–124], confirms the accuracy of the proposed finite element model that was later utilised in a subsequent parametric analysis of steel beams with web openings. The models adopted in the present study are depicted in Fig. 3.80.

The finite element models are similar to the models proposed by Chung et al. [122–124]. The mesh was refined near the web opening, enabling the

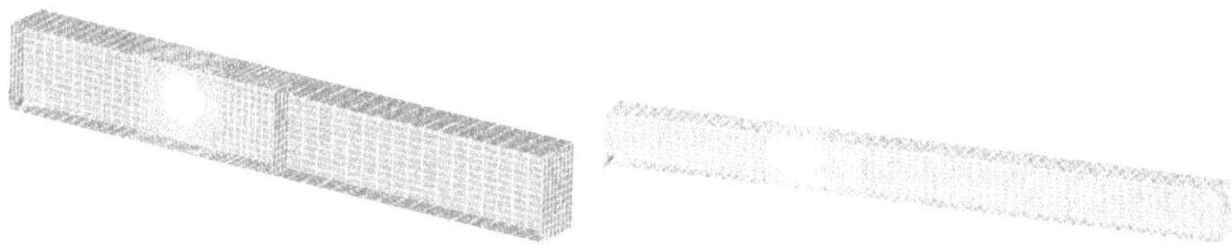

Fig. 3.80 Beam 2A and 3A finite element models.

model to accurately consider the effects related to the stress concentration, the onset of yield in the web and flanges and the hinge formation that characterises the Vierendeel collapse mechanism. The beams supports were created to reproduce simple support conditions. These figures also present the lateral bracing supports utilised along the beam to the flange to prevent the lateral buckling. A comparison of the present results with those of Chung et al. [122–124] in terms of the yield distribution patterns on the beam web near the openings with an increase of the applied load can be observed in Fig. 3.81. Chung et al. [122–124], models for beam 2A are very similar to the present study results. Both distributions are also in agreement with Redwood and McCutcheon [125] experiments, validating the numerical modelling.

The developed models calibration was also performed through comparative graphs containing the experimental results produced by Redwood and McCutcheon [125]. Fig. 3.82 portray the bending moment at the centre of the web opening versus the central span vertical deflection curves for beams 2A and 3A. The curves indicated that a reasonable agreement with the experiments was reached, especially in the elastic range. On the plastic range, a slight

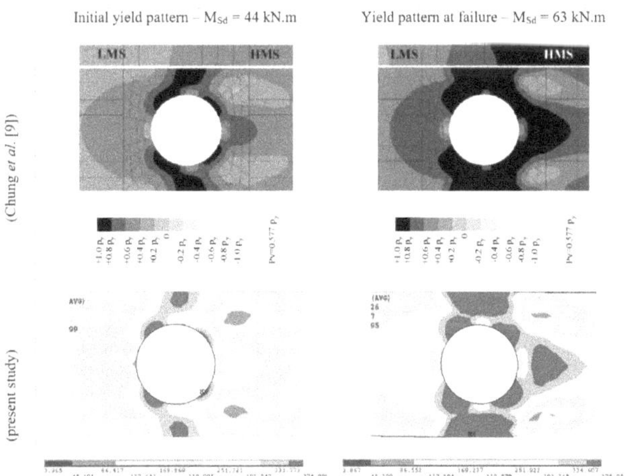

Fig. 3.81 Beam 2A von Mises stress distribution at the perforated section (MPa).

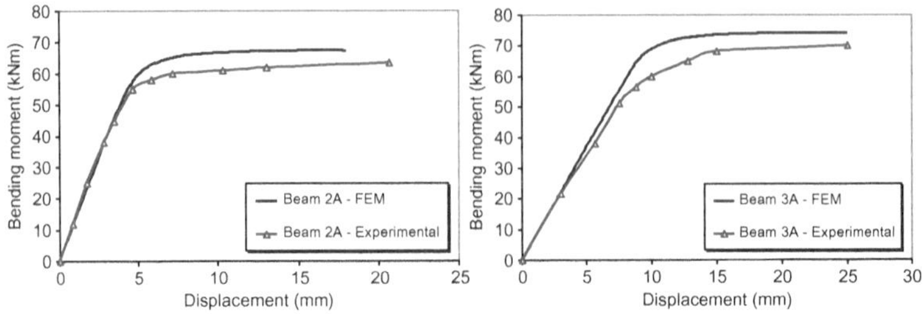

Fig. 3.82 Beam 2A and 3A finite element and experimental model results comparison.

difference between the numerical and experimental results was noted. This difference can be attributed to the 5% strain hardening Young modulus adopted in the numerical models. Another reason could be associated to the residual stresses induced by the welding/rolling processes, or by geometries that were not incorporated into the numerical models. Geometric imperfections present in these beams could also be the reason for this difference.

The accurate results achieved in the simulations enable a comprehensive parametric analysis of steel beams with web openings to be made. An overview of the parameters studied in the parametric analysis for the IPE500 and IPE750x137 where the studied parameters were: (a) the steel profile size, (b) the web opening location in relation to the beam span, (c) the web opening diameter relation to the profile height, (d) the beam span, (e) the web opening geometry, and (f) the concentrated load position in relation to the beam span. It should be noted that beams with rectangular openings were simulated by setting the opening length to be twice the size of its height.

Fig. 3.83 depicts the variation of the bending moment at the web opening centre versus the load application point vertical displacement for specimens 1, 3, and 5. From this figure, it is possible to conclude that the higher ultimate moment is related to the beam with a square opening closely followed by the beam with a circular opening. In addition, a 35% reduction of the ultimate moment capacity was noted in the beam with a rectangular opening. The beams with square and circular openings showed a similar structural response and showed collapses related to an interaction of the Vierendeel mechanism and the load application point bending collapses. Alternatively, the beam with a rectangular opening presented a typical Vierendeel collapse mechanism with an abrupt failure without the presence of a yield plateau.

Figs 3.84 to 3.86 present an evolution of the von Mises stress distribution for specimens 1, 3, and 5. The load steps compared are presented in Fig. 3.83 by the letters 'A' to 'I', indicating the load step magnitudes related to 9-mm vertical displacements ('A', 'B', and 'C'), 18-mm vertical displacements

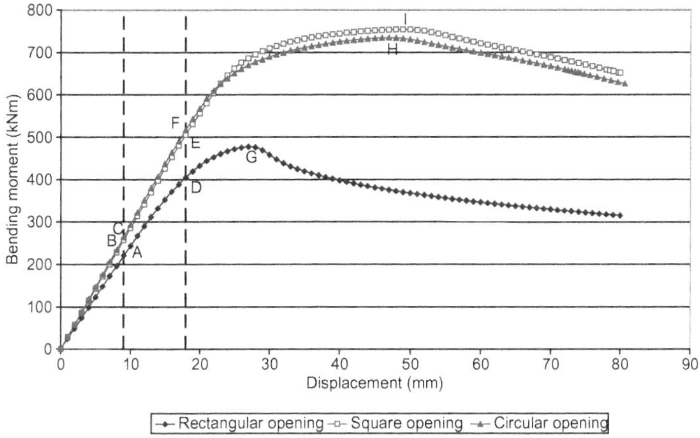

Fig. 3.83 Bending moment versus vertical deflection curves.

('D', 'E', and 'F'), and the ultimate bending moment ('G', 'H', and 'I') for beams with square, circular, and rectangular openings, respectively.

Figs 3.84–3.86 showed that despite the adopted corner concordance radius being equal to twice the web thickness, stress concentration points still formed at the corner of the openings of beams with square and rectangular geometries. The stress concentration increases and is redistributed throughout the web with an increase of the applied load until one or more plastic hinges are formed. The beams with rectangular opening geometries depicted collapse associated with simple Vierendeel mechanisms. The collapse of beams with square openings was related to an interaction of the Vierendeel mechanism and the load application point bending collapses.

The beam with a circular opening tended to have a collapse associated with a load application point bending, but could also have a collapse related to an interaction of the Vierendeel mechanism and the load application point bending, if the load application point was near the web opening region. Fig. 3.87 depicts the evolution of the von Mises stress distribution at the ultimate bending moment for specimens 13, 15, and 17. From these results, it is possible to conclude that in all the beams, the collapse was related to the formation of four plastic hinges at the opening extremities, clearly characterising the Vierendeel collapse mechanism.

Fig. 3.88 presents a variation of the bending moment at the web opening centre versus the load application point vertical displacement for specimens 13, 15, and 17. From this figure, it is possible to conclude that all the beams depicted a considerable yielding before collapse due to a crescent stress redistribution that happened in the beam web opening edge region up to the development of the plastic hinges. It is also possible to note that the beam

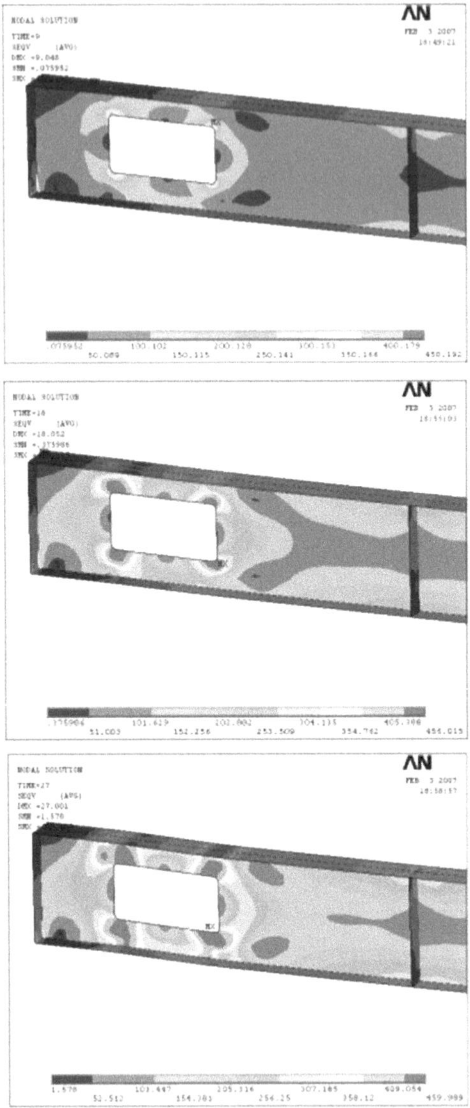

Fig. 3.84 Von Mises stress distribution associated to 9, 18 mm vertical deflection and collapse due to a concentrated load (rectangular opening, points A, D, and G) (MPa).

with a circular opening had an ultimate load five and two and a half times the ultimate load related to beams with rectangular and square openings.

These results can be related to the better stress redistribution achieved with the circular opening configuration that led reduction on the stress concentration at the beam web opening edge region. It was also possible to observe that beams with rectangular and square openings were related to

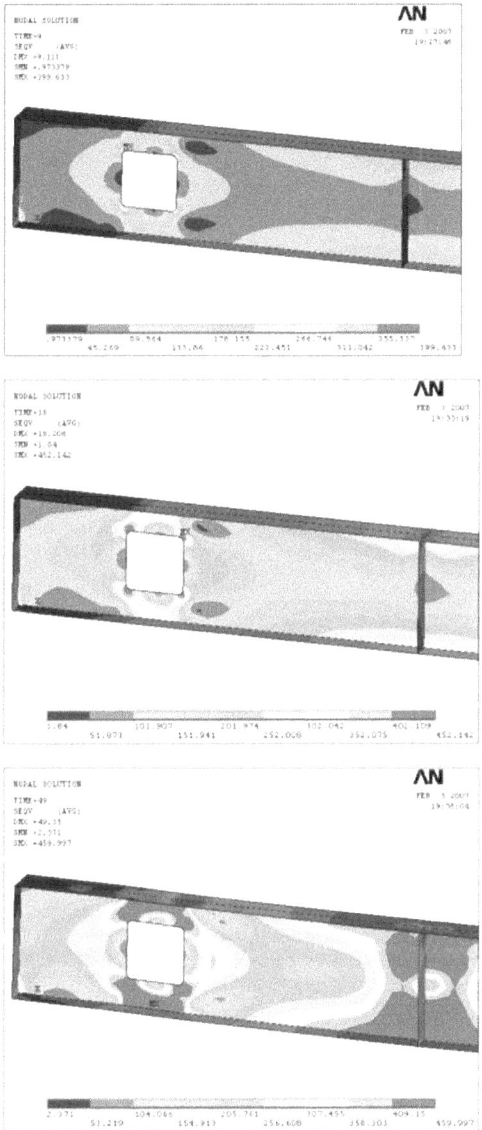

Fig. 3.85 Von Mises stress distribution associated to 9, 18 mm vertical deflection and collapse due to a concentrated load (square opening, points B, E, and H) (MPa).

high stress concentration at the beam web opening edge region, confirming the idea that a concordance radius equal to two times the web thickness, as utilised in those regions, was appropriate.

Beams with square and rectangular opening heights equal to 0.75 H had their structural capacities severely reduced. Despite this fact, there are

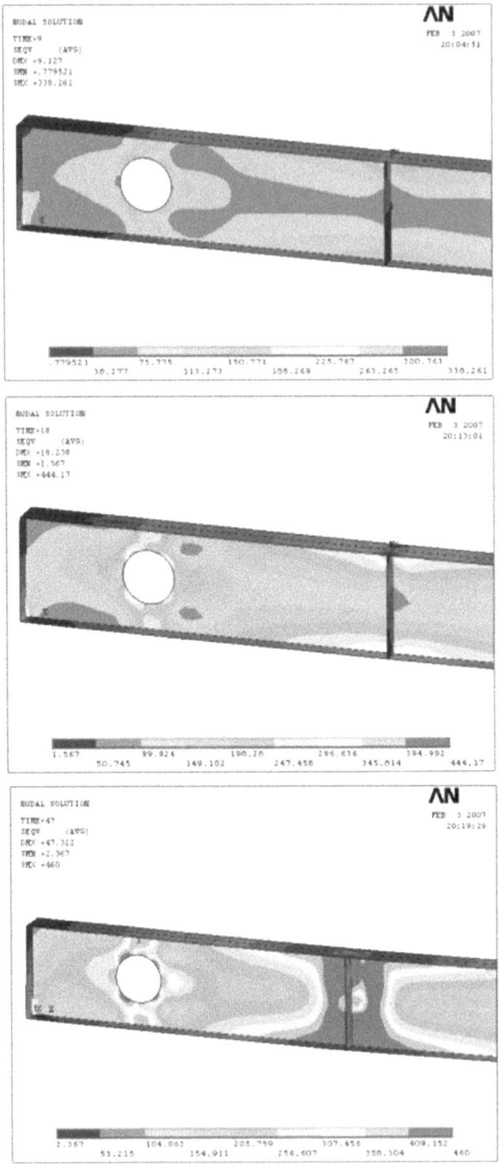

Fig. 3.86 Von Mises stress distribution associated to 9, 18 mm vertical deflection and collapse due to a concentrated load (circular opening, points C, F, and I) (MPa).

situations where the use of these openings cannot be avoided. For such cases, the use of longitudinal stiffeners welded at the horizontal edges of the web openings are strongly recommended by the structural best practices and design standards. Three beams were simulated with and without stiffeners to verify the structural efficiency of these stiffeners.

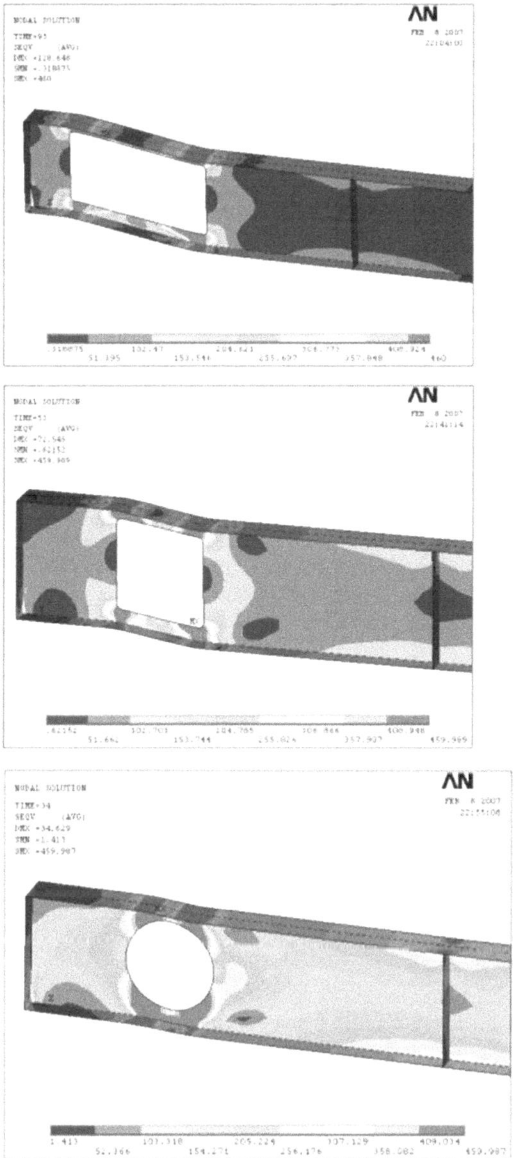

Fig. 3.87 Von Mises stress distributions associated to a collapse due to a concentrated load (rectangular, square and circular openings without stiffeners) (MPa).

They presented the same characteristics of some of the simulations performed in the parametric analysis (i.e. models 13, 15, and 17). The longitudinal stiffeners were simulated with the geometrical characteristics presented in Fig. 3.88. Fig. 3.89 indicates that the presence of the longitudinal stiffeners provides a better stress redistribution along the web openings (compared to

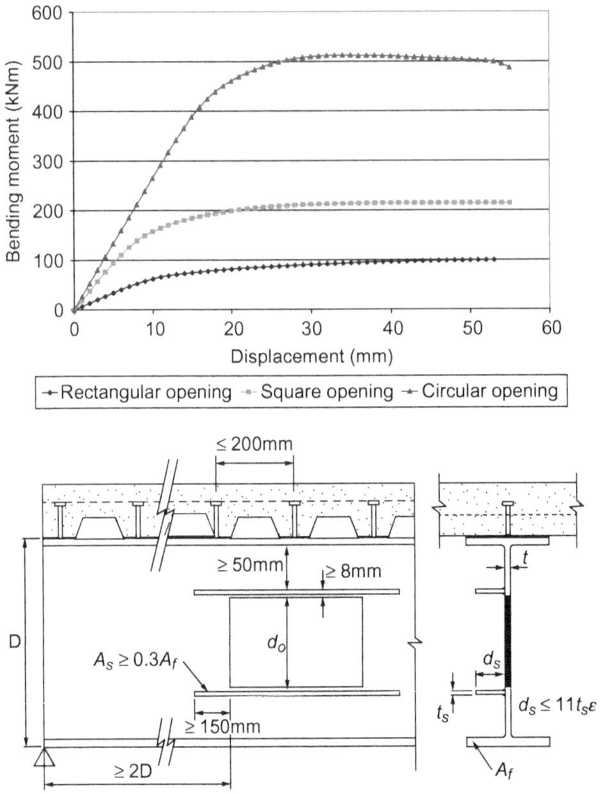

Fig. 3.88 Moment versus deflection curve and longitudinal stiffener requirements.

the results presented in Fig. 3.87) and increases of the beams load carrying capacity. Despite this beneficial effect, the Vierendeel mechanism was still the controlling failure mode for these beams.

It could be noted that the use of longitudinal stiffeners along the opening region presented a considerable advantage when compared to beams with nonstiffened web openings. The stiffeners increased the beam ultimate bending capacity around three and two times the unstiffened case for beams with rectangular and square openings, respectively. When a comparison of the stiffened and unstiffened solutions was performed, even the beams with circular openings were related to a 20% increase of the beam ultimate bending capacity. The present study confirmed that their use can double or even triple the ultimate load of beams with rectangular and square opening heights equal to 0.75 H.

The numerical study initially contemplated a sensitivity study of the mesh and the imperfections to be adopted in the model. The model was able to accurately represent the plastic mechanisms development that usually

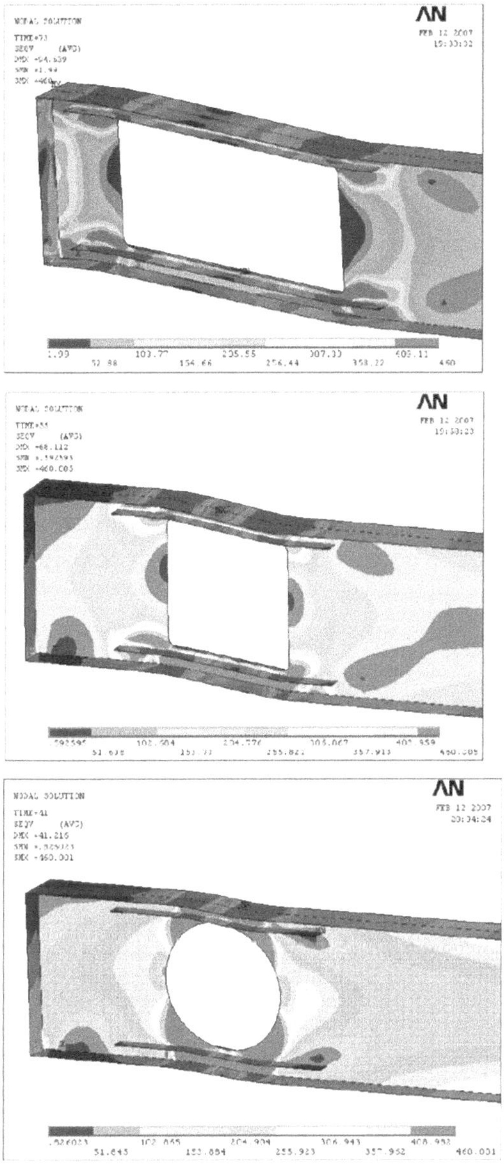

Fig. 3.89 Von Mises stress distributions associated to a collapse due to a concentrated load (beams with rectangular, square, and circular openings with longitudinal stiffeners) (MPa).

controls its design in qualitative and quantitative terms. The modelling also included the assessment of the dimensions and shape of the openings as well as their relative positions to the point of load application. An interesting study has also been made on the influence of using stiffeners near these openings and what was the strength gains obtained with their use.

3.5.10 Structural Evaluation of Major Axis Semirigid Endplate Joints—Bending Moment to Axial Force Interaction

The experiments that were used to calibrate the numerical model were made by Lima [47–51], and were the basis for the comparison with the numerical results [126–128]. Three load types were used: first only bending moments were applied to the joint; second, tests were made with bending moment and compression axial forces; the third, test was made with bending moment and tension axial forces. The experiments were made with a beam section not very large since the applied axial load were functions of a percentage of the beam plastic capacity and was also limited by the laboratory existing equipment capacity. After some preliminary studies, a IPE240 profile was adopted for the beams, made from a S275 steel grade, i.e. with yield and tension rupture stresses equal to 275 and 430 MPa.

For the column, a profile with a sufficient bending moment capacity caused by the imposed axial force was selected. Another limitation was related to the need for the column components not to reach the yield limit before the remaining joint components. Considering these restrictions, the S275 HEB240 profile was adopted on the column. The tests also utilised a 15 mm thick S275 endplate. The beam to endplate connection was made with 8 mm thick butt welds. M20 10.9 full thread bolts were also adopted at the endplate to column flange connection. The test layout adopted a reaction frame made of a HEB300 beam that supported the hydraulic jack attached to the lab reaction wall by an extended 30 mm thick endplate and eight DYWIDAG threaded bars. Fig. 3.90 depicts the tests load application frame.

Table 3.5 presents a summary of the tests performed along with their applied axial loads where these values represent a percentage of the beam axial plastic capacity (1084 kN).

The simulations adopted an updated Lagrangian formulation with large displacements and rotations to take into account the system geometric nonlinearities. In this formulation, the involved stresses, deformations, outside forces, etc. are defined in relation to the deformed structure configuration. The material nonlinearity was considered through a constitutive law defined by the stress versus strain curve using a logarithmic strain ($\sigma_n - \varepsilon_n$), instead of the conventional constitutive law commonly used ($\sigma - \varepsilon$). This was due to the fact that the actual material curve was obtained in a uniaxial test, always considering the initial area, while neglecting the strain hardening effects.

Fig. 3.91 presents the materials constitutive laws curves and the corresponding nominal curves applied to the nonlinear physical analysis that was considered in this study. Through this analysis, it was possible to obtain the joint global response by making a coherent comparison between the

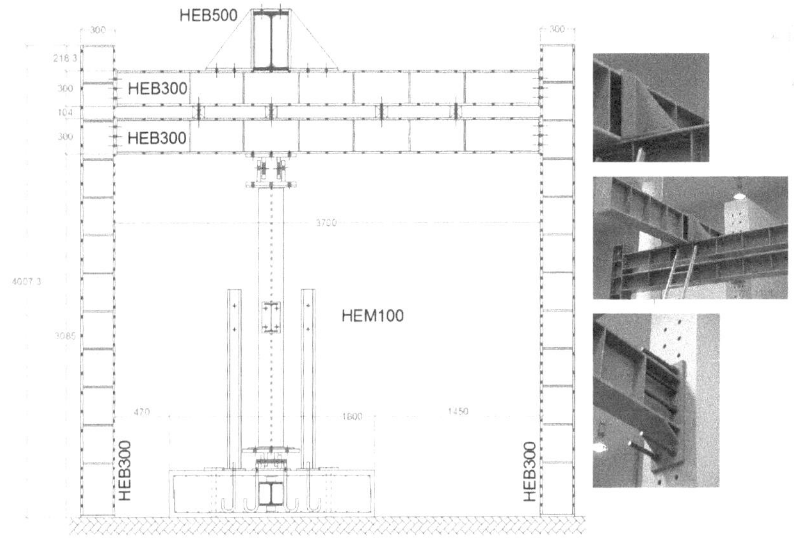

Fig. 3.90 Load application frame.

Table 3.5 Tests Applied Loads Description

	Axial Load	
ID	**%$N_{beam\ plastic}$**	**(kN)**
FE1	–	–
FE3	−4%	−52.7
FE4	−8%	−105.6
FE5	−20%	−265.0
FE6	−27%	−345.0
FE7	−20%	−265.0
FE8	+10%	+130.6
FE9	+20%	+264.9

Eurocode 3 [19], results and the experiments and numerical models in terms of the joint limits states.

The numerical model was developed with the aid of the Ansys 11 [32] finite element program. The adopted elements were: SOLID185 for the beam, column, endplate, and bolts i.e. an element with eight nodes and three degrees of freedom per node. This element was selected to maintain the compatibility with the utilised contact elements to be described later. The stiffener adopted the shell element SHELL181, which has four characteristic and six degrees of freedom at each node. The material nonlinearity was characterised through the von Mises yield criterion. Fig. 3.91 presents the

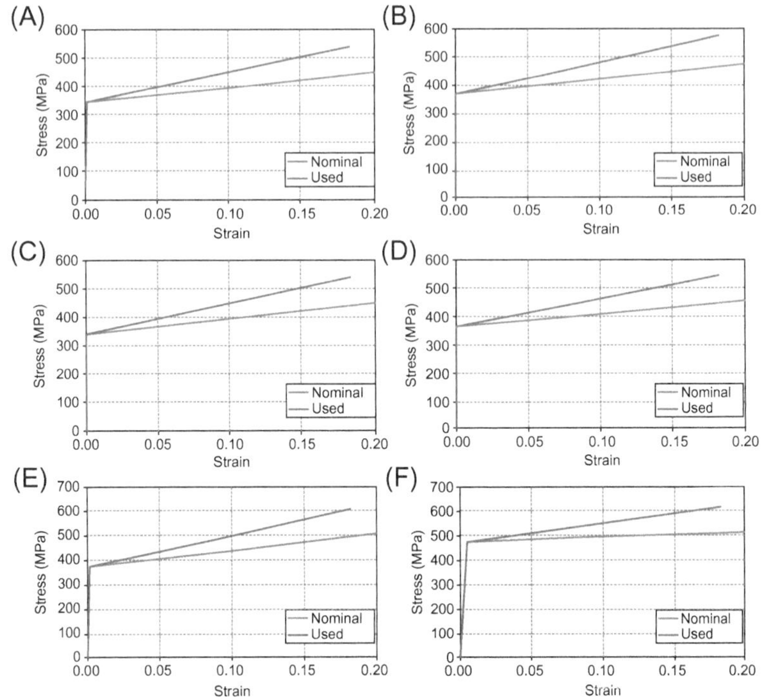

Fig. 3.91 Adopted materials stress versus strain curves. (A) Column flange. (B) Column web. (C) Beam flange. (D) Beam web. (E) Endplate. (F) Bolts.

adopted model materials properties for each element, there is a specific value for the modulus of elasticity. A 0.3 Poisson's coefficient was assumed for all the adopted model materials.

M20 10.9 bolts were simulated with a body defined through the threaded cross section area with its length equal to the sum of the thicknesses of the connecting elements: endplate and column top flange, Figs 3.92 and 3.94. Loading, for all cases studied, was applied by displacements that varied according to the required force magnitude. The bolt prestress was incorporated into the model, through the application of a deflection in the bolt head and nut (Δ_1 and Δ_2). The boundary conditions assumed restrictions on the column web and flanges in the x and y axes. The vertical displacement (y axis) was also restricted in the endplate, while the girder top flange was restrained laterally (x axis), Fig. 3.92.

Different models were evaluated to characterise the different applications of normal forces, as described earlier. First, a preliminary mesh study was made so that the results' reliability could be ensured. The selected mesh had elements with sizes to avoid numerical problems. Figs 3.92–3.94 portray details of the developed model components.

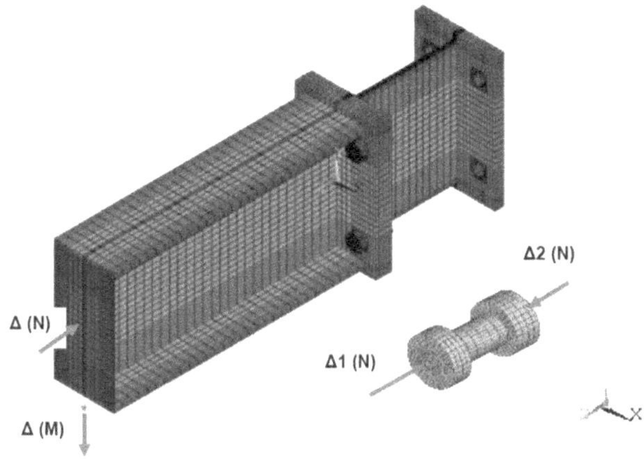

Fig. 3.92 Numerical model-flush endplate joint.

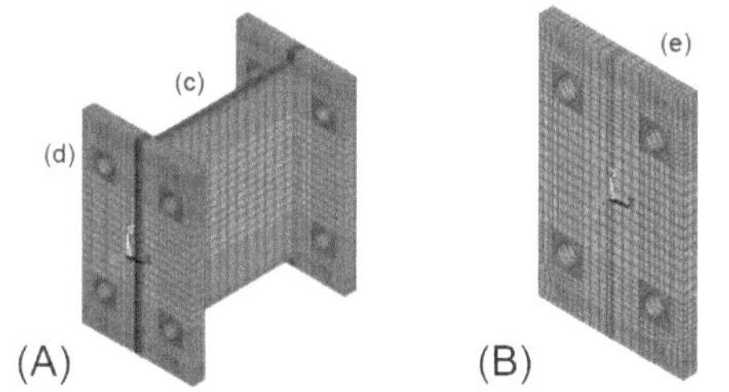

Fig. 3.93 Numerical model- column and endplate. (A) Column detail. (B) Endplate detail.

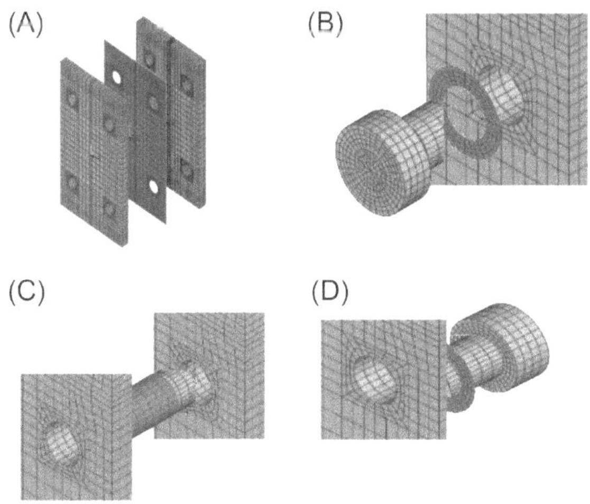

Fig. 3.94 Effective contact Area. (A) Endplate to column flange. (B) Bolt head to endplate. (C) Bolt shank to endplate and column flange. (D) Nut to column flange.

The contact element of the numerical simulation was used to represent the interface phenomenon of elements in a finite element analysis. Contact elements were used in these regions with the objective of considering the contact areas through the interaction between the surfaces. Fig. 3.94 presents the regions where contact elements were adopted. Contact elements were considered in the following regions: endplate and the column flange; bolt head and endplate, Fig. 3.94B; bolt shank, endplate and column flange, Fig. 3.94B; nut and column flange.

The contact element used a coefficient of friction (μ) between the contact areas defined as base area, and the contact area, which will cause friction between the surfaces. For the characterisation of the contact element in the numerical model, 0.25 friction coefficient was utilised. The contact elements used were CONTA173 and TARGE170 for the base areas and contact areas, respectively.

Test FE01 with only bending moments was used to calibrate the model by a comparison of the experiments and the obtained numerical results. Since the calibration has been carried out satisfactorily, the study proceeded with other analyses considering the calibrated model as comparative base. Fig. 3.95 presents the results of the model calibration in terms of moment versus rotation curves of the experiments and numerical models.

From these curves it is possible to observe that the two curves are similar. The highlighted points in the experiments are related to the components yield sequence from the strain gauge readings: endplate in bending, beam flange in compression, and bolts in tension, respectively. The highlighted points in the numerical curve, Fig. 3.96, show that the yield sequence was the same that occurred in the experiments. For point 2 ($M=65.9$ kNm), Fig. 3.97, it is possible to observe the start of the yield of the beam flange in compression component.

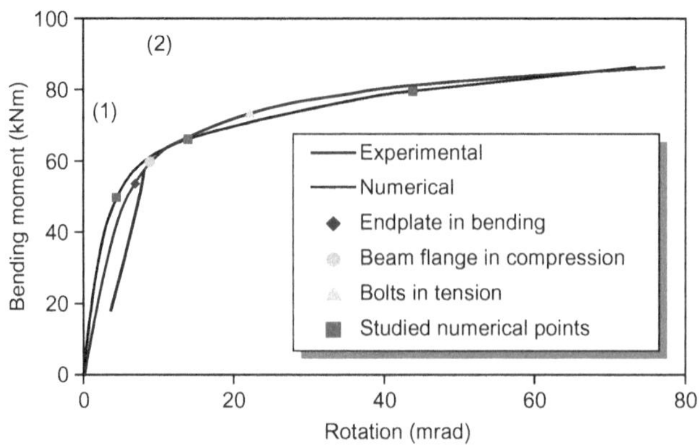

Fig. 3.95 Experimental and numerical moment versus rotation curves—FE01.

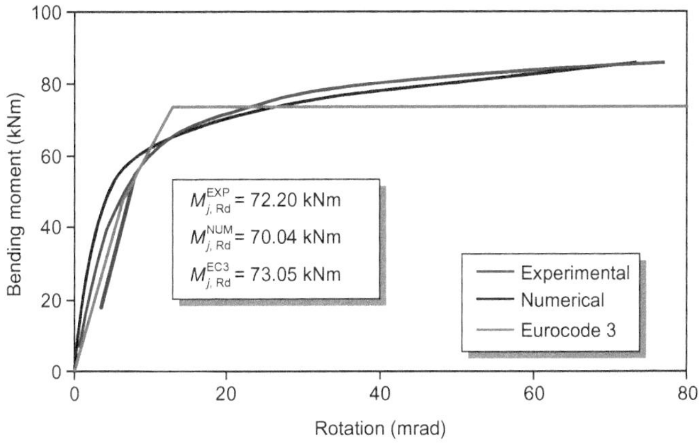

Fig. 3.96 Comparison of the obtained results—FE01.

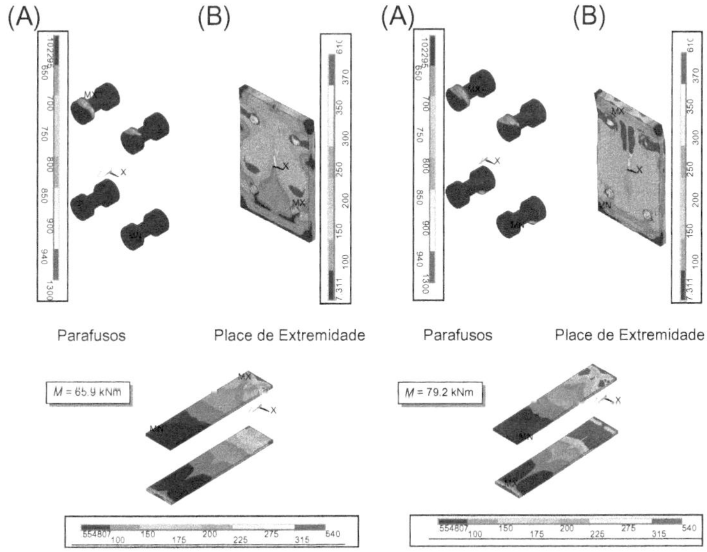

Fig. 3.97 von Mises stress distributions, FE01 Pts. 2 and 3.

Fig. 3.96 presents a comparison of the results for three analyses: experiments, numerical model, and Eurocode 3. The evaluated moment resistant was 73.05 kNm, which is close to both the experimental ($M_{j,Rd}$ = 72.2 kNm), and the numerical model ($M_{j,Rd}$ = 70.0 kNm), indicating that the numerical model led to satisfactory results. It is also possible to note that the initial stiffness calculated through the Eurocode 3 was 11152.2 kNm/rad, higher than the test and the numerical model.

Fig. 3.97 presents the joints von Mises stress distributions, in MPa, in three distinct points of the analysis, Fig. 3.95. In this figure, the components correspond to a moment magnitude of 49.6 kNm, point 1, where the endplate yield starts. Fig. 3.97 depicts the start of the yield of components beam flange in compression, point 2, associated to a 65.9 moment magnitude. Finally, in Fig. 3.97, the components are presented for a bending moment magnitude of 79.2 kNm, where the two components yielded point 3. The comparisons confirmed the similarity between the experiments and numerical models.

Fig. 3.98 presents the comparison of the deformations that occurred in the experiments and the numerical model, confirming, once again, the similarity between the observed displacements.

The behaviour of column flange in bending component was evaluated through the strain gauge 10, located just below the top bolt line, Fig. 3.99. A detailed analysis of this component was not made since it was not one of the most relevant in the investigated joints design.

Moment versus rotation comparative curves will be presented for all the investigated axial force magnitudes. In order to enable a comparative analysis of the obtained results stress distributions for the same bending moment, magnitudes for all the tests will also be depicted. The adopted level was selected, taking into consideration the points obtained in comparison to FE1 model experiments, representing the component yield points. This

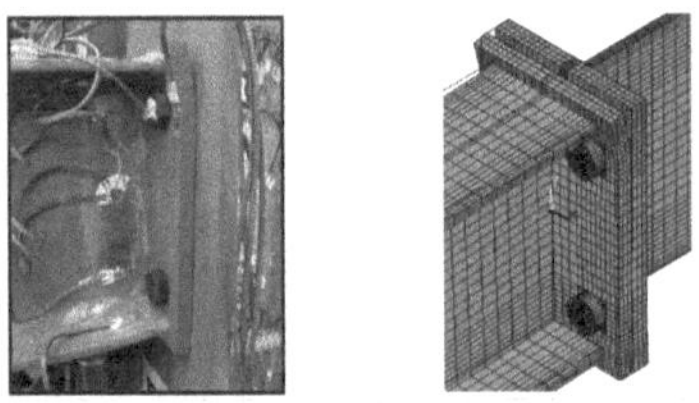

Experiments Numerical

Fig. 3.98 Experimental and numerical strain comparison.

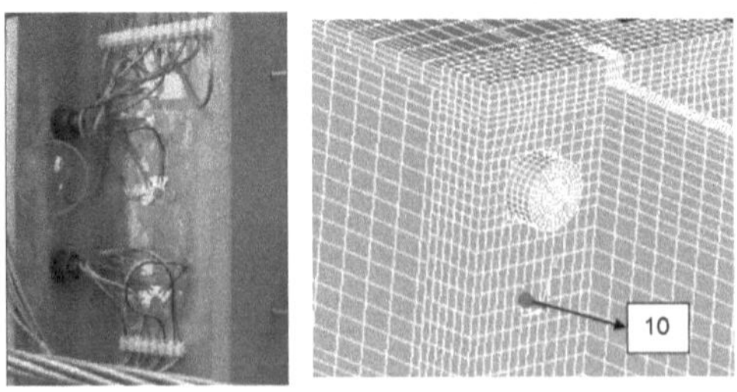

(A) (B)

Fig. 3.99 Location of strain gauge 10.

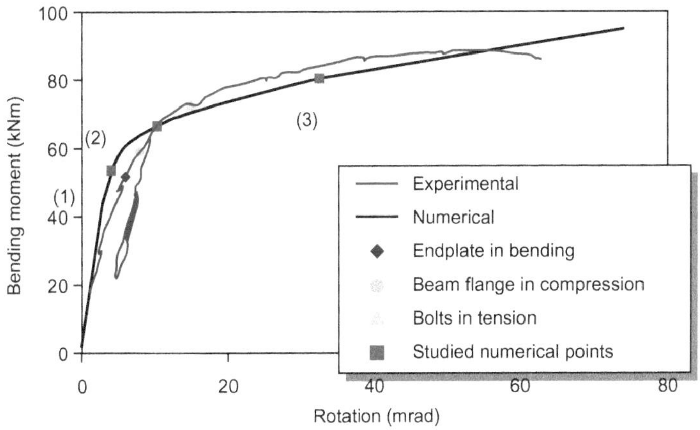

Fig. 3.100 Experimental and numerical moment versus rotation curves—FE03.

strategy could make the assumed point not be the studied model yield point, however, it enables a viable comparative case study.

Test FE03 ($N = -4\%N_{pl}$) is used to illustrate the results comparison. Fig. 3.100 presents a comparison of moment versus rotation curves obtained from the numerical modelling and experiments. It is possible to observe that the numerical curve resembles the experimental curve, but the numerical model presented an initial stiffness 14% higher than the test. These curves also depict the points corresponding to the experiments components yield, which occurred in the following order: endplate in bending, beam bottom flange in compression and bolts in tension. The numbered points correspond to points where the numerical model will be checked.

Fig. 3.101 presents the von Mises stress evolution in MPa, for three different load magnitudes shown in Fig. 3.100 for the three main joint design components. Point 2, associated to a bending moment of 65 kNm, depicts the beginning of the yield at the beam bottom flange, Fig. 3.101 presents a higher bending moment magnitude, 80 kNm, where the endplate and the beam flange are yielded.

Fig. 3.102 presents all the curves obtained in experiments and those are obtained in the numerical modelling. It is possible to note that even for an axial compression, stress magnitude equivalent to 20% of the beam plastic resistance of the moment capacity still is higher than the testing without axial forces (FE01), occurring for both, numerical, and experimental cases. This is due to the fact that the compressed zone components, despite having an increase of the applied force, did not reach their capacities and, consequently, the tension components are reduced by the axial forces. It is worth mentioning that the joint measure rotation was obtained by displacement

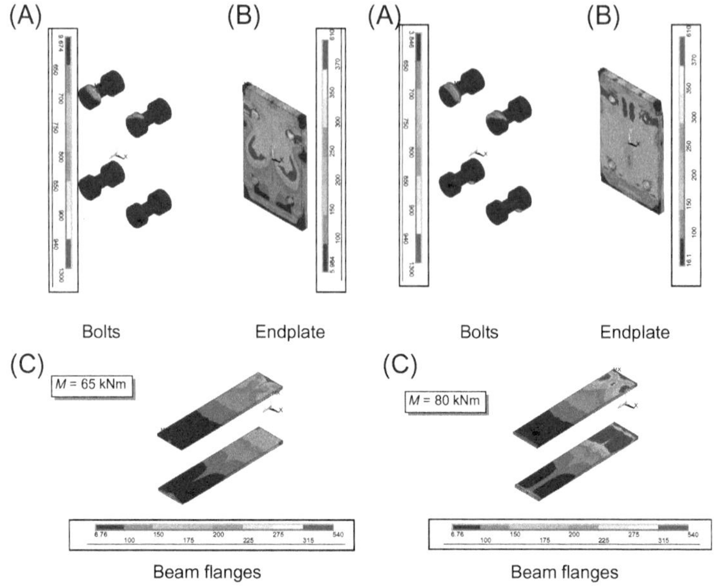

Fig. 3.101 von Mises stress distributions, FE03—Pts. 2 and 3.

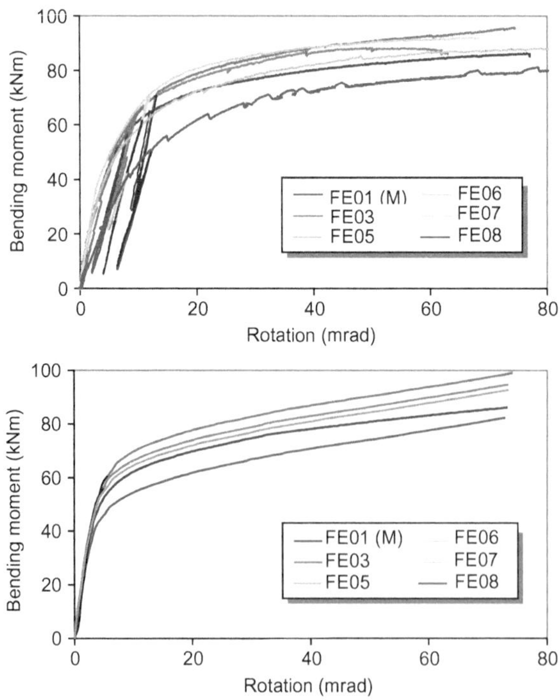

Fig. 3.102 Experimental and numerical moment versus rotation curves.

transducers located on the beam. It can be verified that with the increase of the joint axial compression forces, there is an increase of the joint bending strength, fact that also was noted in the experiments, Fig. 3.102.

In the tests carried out with tension, it is possible to note the occurrence of axial forces opposite effect. There is a decrease in the joint bending strength of the connection. It can be observed that, for the calibration test FE1, the moment versus rotation curve shows a good agreement between numerical and experimental results. On numerical curve, it is observed that from a bending moment equal to 49.60 kNm, the curve becomes nonlinear, due the development of endplate yield zones caused by the component stress increase at the first bolt line. The beam flange in compression starts to yield at a bending moment magnitude of 65.90 kNm. It is observed that from 79.20 kNm, the bolts present stresses at some points along their length that is higher than their tension rupture stress.

In relation to the Eurocode 3 predictions, it can be observed that the standard presented more conservative results when compared to the experiments and the numerical model, although the values obtained are close. For the axial force tests, it could be verified that the numerical model presented higher stiffness than the experiments, but as for the numerical model bending capacities, they were very close to the experimental results. As the joint axial force was increased in the joint, the endplate in bending component ceased to be the controlling component that shifted to the beam bottom flange in compression. These results were well observed experimentally, however, because of boundary conditions imposed to the model, it has not been possible to verify them with such intensity in the numerical model.

The initial stiffness for the models developed with the axial forces led to numerical results stiffer than those obtained experimentally; However, the differences in the order of 15% did not invalidate the results. Numerical and experimental results showed that, when the joints are subject to bending moment and axial forces, the latter may favour the increase in the joint flexural strength and, therefore, allow the use of more economical connections. An increase of the applied compression force leads to the beam bottom flange component to control the joint design.

The values found, both as numerical and experimental models, were higher than those obtained with the Eurocode 3. This means that the values presented in the standard are more conservative, i.e. in favour of safety. Through the moment versus strain curves it was possible to clearly verify the decrease occurred for the FE6 curve associated to the endplate component when compared to other tests with compression forces, indicating that the beam bottom flange started to control the joint design.

The implemented numerical model used contact elements to simulate the various interfaces present between the connections numerous elements. Once more time, care had to be taken, so that the specifications of the support and loading conditions of the model did not generate inaccuracies or even inconsistencies. The material once again had to be modelled using the true stress versus true strain curves due the large deformations that occur as the structural collapse was reached. The mesh had to be very well defined and refined, so as to be able to represent the studied joints with the necessary accuracy. Care on the adopted convergence previously mentioned, once again, was made necessary, as well as the correct specification of support and control of the load through equivalent displacements.

The individualisation of each joint component was extremely helpful to enable a better comparison with the experimental program results. This enabled an assessment of the real contribution of each component to the overall connection response, and what their change in behaviour and importance was, as simultaneous compression or tension loads were applied to the joint.

3.5.11 Tubular T and KT Joint Modelling

Tubular connections can be built with numerous layouts, according to the main global design standards, however, the most adopted solution is welded, presenting advantages associated to their industrial production efficiency. The joints can be produced in various geometries for example: 'T', 'K', 'KT', 'N', 'X' and 'Y', among others. The primary goal of this study was to investigate the behaviour of some of these tubular joints configurations by comparing the results obtained with experiments present in literature and Eurocode 3 [19] recommendations.

All numerical models developed in this study were modelled in the finite element program Ansys [32]. 'T' welded connections represent the structural connection type, involving tubular profiles, associated to the simplest fabrication [132,133]. Seeking to analyse the connection behaviour through a numerical modelling, the literature was assessed to determine the most relevant investigations that have been made on this subject. Lie et al. [133], developed experiments on a 'T' welded connection with square profiles (SHS), the adopted configuration used in the experiments is depicted in Fig. 3.103. These simply supported tests adopted a SHS 350 x 15 chords with 3.0 m in length. The distance between the supports was 2.75 m, while the brace adopted a SHS 200 × 16, both with yield and tension rupture stresses of 380.3 and 529 MPa. The weld had 12 mm thickness with tension rupture stress of 600 MPa. The test considered possible cracks on the weld but this is not the objective of the present study. The numerical model developed by

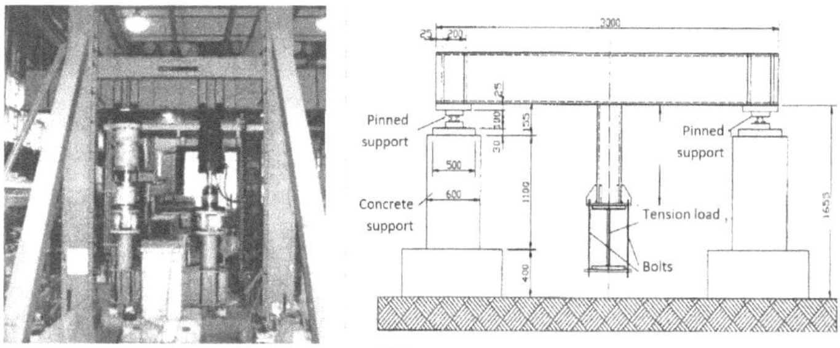

(A) (B)

Fig. 3.103 Experiments and structural model [133]. (A) Experiments. (B) Structural model.

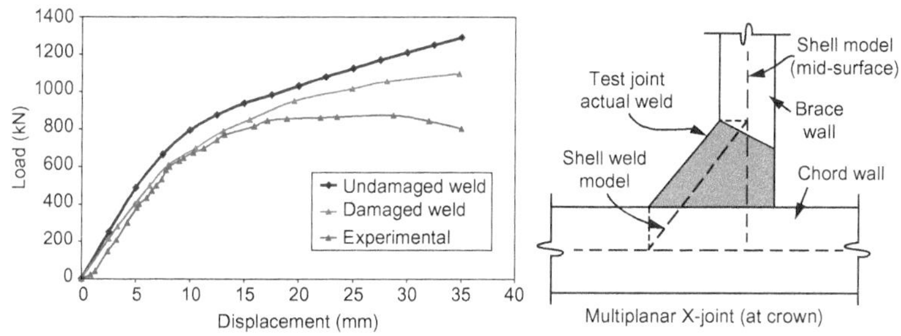

Fig. 3.104 Lie et al. [133] results and weld shell element modelling.

the authors included an analysis of the connections with and without weld cracks, and the results used in this study consider the second case.

Fig. 3.104 presents these results, where it can be observed that the numerical model considering the crack in the weld does not accurately reproduce the experiments, especially in terms of initial stiffness and ultimate load. However, for the segment related to displacements from 7.5 up to 17.5 mm, the curves show a reasonable agreement, and within this range, the deformation limit for the studied joint, corresponding to a 3% displacement of the chord width, i.e. 10 mm, will be used for obtaining the joint load carrying capacity.

Weld analyses were made with SHELL181 element and with SOLID45 element present in the Ansys element library. The load versus displacement response associated to the shell element can be seen in Fig. 3.104 in the average tube surface, as well as the average weld surface representing the geometry to be used in the 'T' connection model of with shell elements. The SHELL181 shell element has four nodes with six degrees of freedom per node, and also considers the suitable membrane effect to deal with for large displacements and rotations in a nonlinear analysis.

Geometric and physical nonlinearities were incorporated into finite element models, in order to fully mobilise the joint resistance to axial and bending forces. Additionally, the use of a geometric nonlinearity enables the prediction of large deformations, considering possible load redistributions in the model after the initial yield. In the performed numerical analysis, the material nonlinearity was considered through the von Mises yield criteria associated to bilinear constitutive law in order to represent an elasto-plastic behaviour with a 10% strain hardening. A 210 GPA a modulus of elasticity and a 380.5 MPa yield stress were assumed. The geometrical nonlinearity was introduced in the model by using the updated Lagrangian formulation. Using the same data as in the model developed by Lie et al. [133], a suitable finite element mesh was developed and used in the model calibration, Fig. 3.105. It is worth noting that support conditions adopted in the numerical study considered the structure as simply supported. In addition, the model adopted a force control strategy.

The weld modelled with SHELL181 shell element proved to be consistent with the numerical results obtained by Lie et al. [133], Fig. 3.106. It is worth noting that modelling performed by Lie et al. [133], was made with solid elements for both the profiles and weld, since the experiments were made considering a weld with a crack that could lead to the failure. Fig. 3.106 depicts a good agreement between developed numerical model and Lie et al. [133] investigations.

This study also modelled the weld with SOLID45 elements that has eight nodes with three degrees of freedom per node. The element is capable of considering plasticity, creep, a stiffness increase, large deformations, and large displacements, also incorporating an option for a reduced integration. The finite element mesh formed considering the weld with SOLID45 element was very similar to the mesh adopted in the analysis with SHELL181 elements, because only the weld region elements have been modified. Initially, it was imagined that the new mesh would be compromised, due to the use of two different elements. This has not occurred since a coupling of the weld element region was made, Fig. 3.107. The results considering

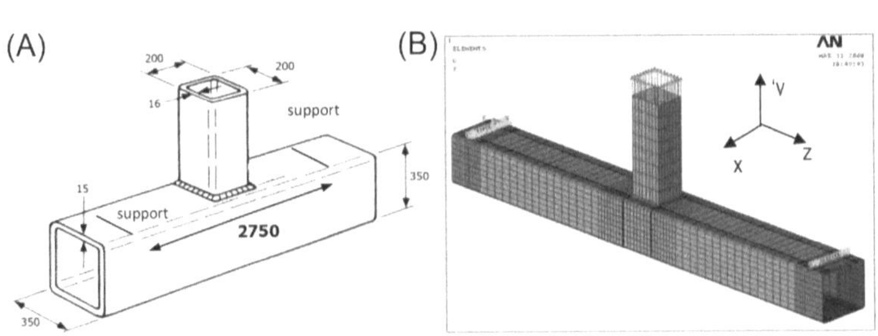

Fig. 3.105 'T' Joint characterisation. (A) Geometrical properties. (B) Finite element mesh.

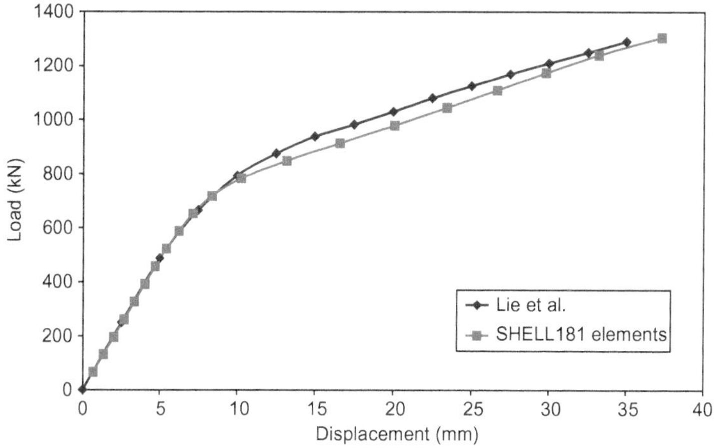

Fig. 3.106 Comparative results between [133] and the proposed numerical models.

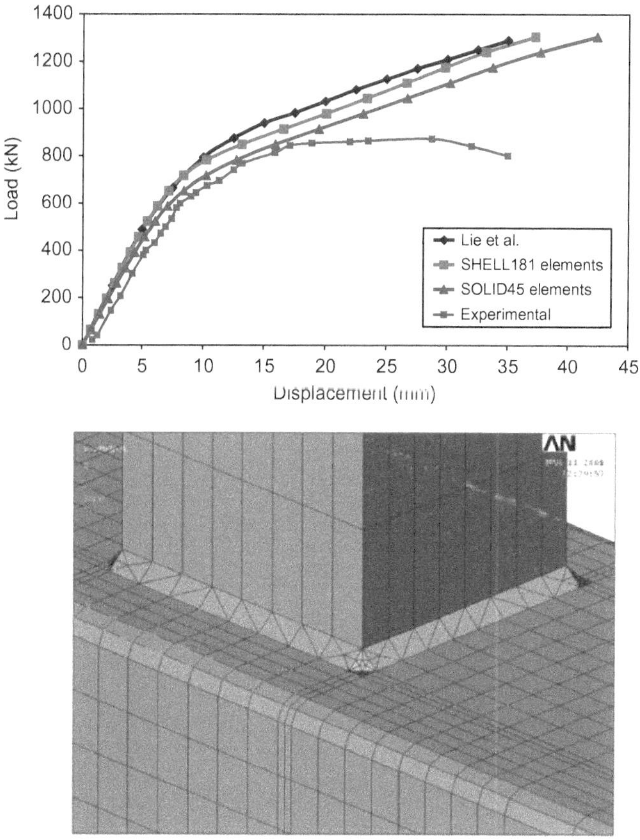

Fig. 3.107 Weld modelling comparison: shell versus solid.

the weld with SOLID45 elements were satisfactory, however a separation from the model curve with solid element to the results of by Lie et al. [133] was noticed, fact that was not observed in the model with shell element, as shown in Fig. 3.107.

On the principle that the shell element could accurately represent the joints behaviour in welded structures, as was observed in the previous model evaluation, a model of a 'K' welded connection between circular tubes was developed, using the welds with the SHELL181 element. Based on the idea that the numerical model developed for the 'T' connection was satisfactory, the same methodology was used for the 'K' joint analysis. The developed numerical model geometry can be viewed in Fig. 3.108.

The numerical model development initially required numerous analyses of structural models that simulated the joint behaviour based on Choo et al. [135], that performed experiments on a structure with circular tubes and welded joints with different support conditions. For the studied joint three structural models were considered for determining a consistency between the numerical model and the actual tubular joint. The first model selected corresponded to the structural model depicted in Fig. 3.109 in which the chord only had vertical restraints while the diagonals had their vertical and horizontal displacements restricted. In respect to the model loads, a horizontal load was imposed to the chords and a bending moment was applied to the member joints. The normal force and bending moment diagrams of the first evaluated structural model can be observed in Fig. 3.109 where it is observed that the bending moment magnitude in bars are small allowing them to be considered as only being submitted to axial forces.

Observing the deformed shape of the first analysed model, Fig. 3.110, it can be noticed that the joint node only moved horizontally, being consistent

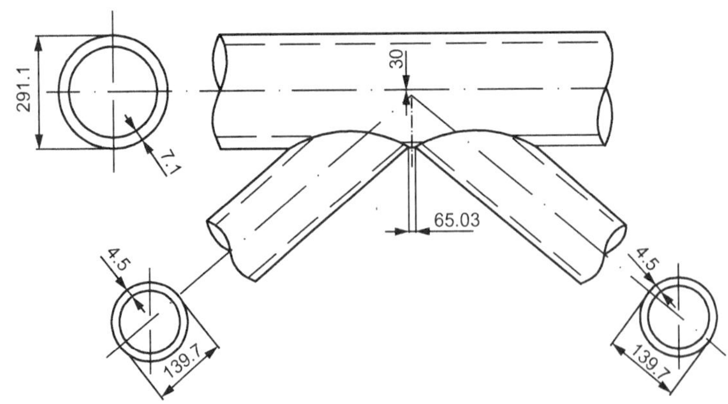

Fig. 3.108 Investigated 'K' joint model.

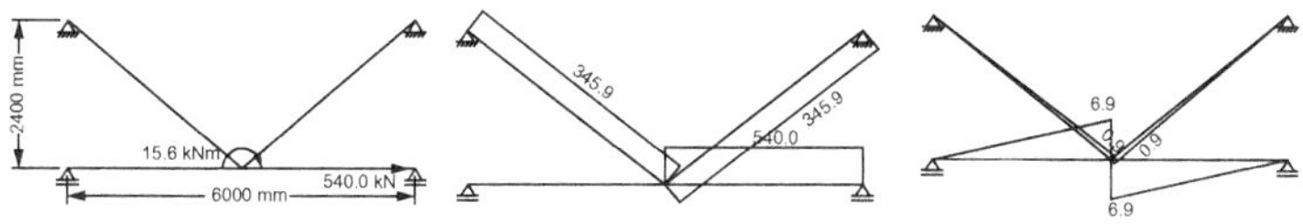

Fig. 3.109 First structural model normal stress and bending moment diagrams.

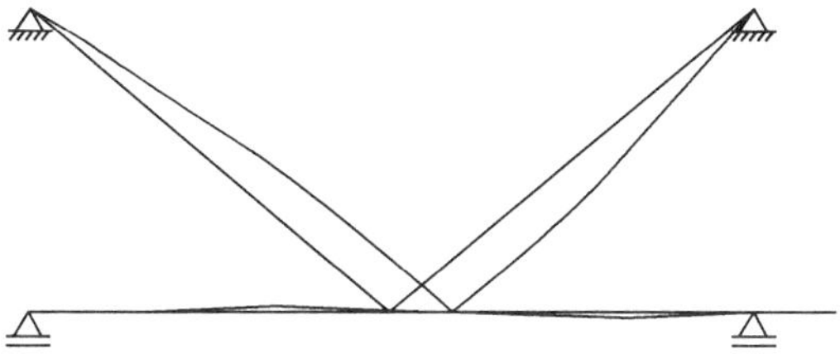

Fig. 3.110 First structural model deformed configuration.

with the expected results. The results obtained for a preliminary analysis of the structural model live up to expectations.

The second evaluated structural model is shown in Fig. 3.111, where the chords were considered only with vertical displacement restrictions, while the diagonal ones had vertical and horizontal displacement restriction. The load corresponded to a single horizontal load applied to the chord. Fig. 3.111 depicts the axial force and bending moment diagrams for the second studied model. Once again, the behaviour of the adopted structural model was consistent with the actual welded joint, whereas the bending moment was still too small and could be neglected for design purposes.

Observing the deformed shape for this structural model it can be concluded that this model, Fig. 3.112, is a representative of the actual welded joint, since it is similar to original joint deformed shape, Fig. 3.110.

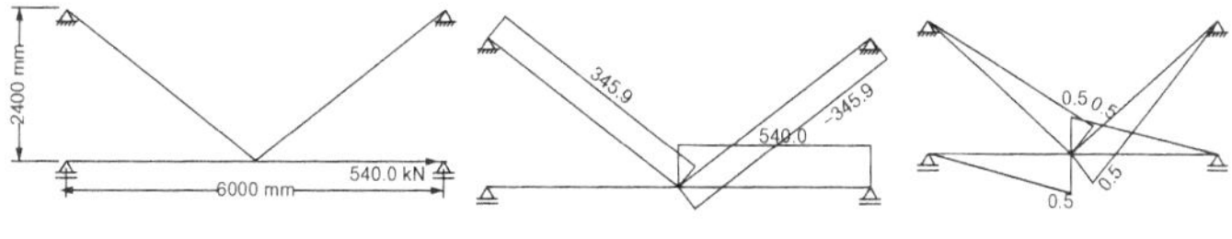

Fig. 3.111 Second structural model normal stress and bending moment diagrams.

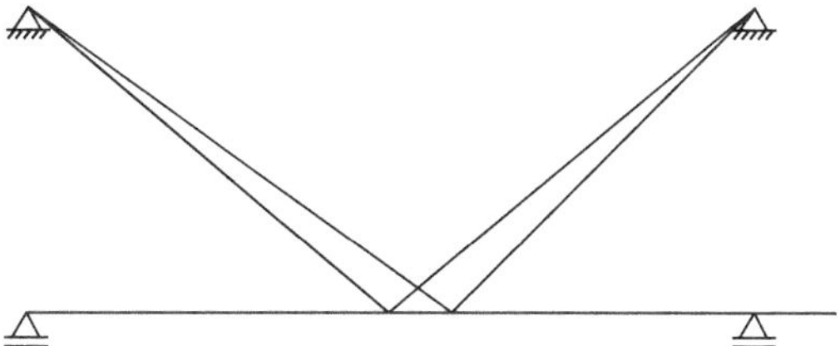

Fig. 3.112 Second structural model deformed configuration.

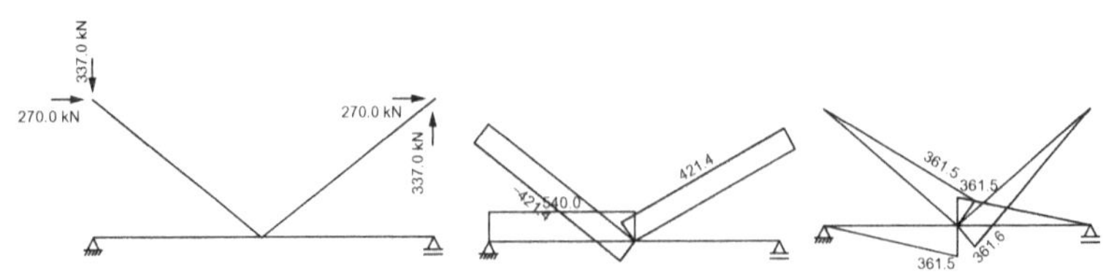

Fig. 3.113 Third structural model normal stress and bending moment diagrams.

The third evaluated model is shown in Fig. 3.113, where horizontal and vertical restrictions were applied to the left support chord, and only vertical restrictions were imposed to the right support chord, while the diagonals had no restrictions whatsoever. The model loads were applied to the diagonal free ends. One more time, the axial force and bending moment diagrams were determined for this model, where the joint presented nonnegligible bending moments. This fact invalidated the model, since its joints would not have the typical response found in truss configurations. In addition, the third deformed structural model, Fig. 3.114, is not consistent with the required response, since only the diagonals were deformed.

The results led to the adoption of the second structural model. Despite the similarity of all models, the second one proved to be simpler, for facilitating the parametric analysis development. It can be observed in Fig. 3.115 that in the second model, tension forces are generated on the chord below the right diagonal and on the left diagonal, while the right diagonal is under compression. It can also be noted that the model loads were applied by using equivalent displacements to facilitate the nonlinear analysis convergence.

The model mesh construction adopted the lines division control system again, so as to obtain more effective mesh. The model final mesh and the monitored displacement points in each element are shown in Fig. 3.116.

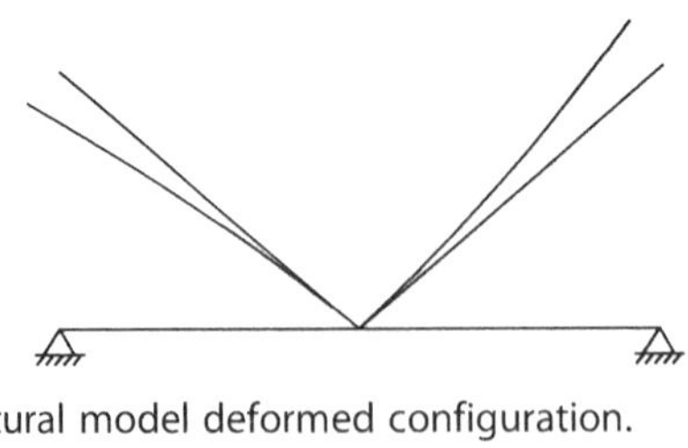

Fig. 3.114 Third structural model deformed configuration.

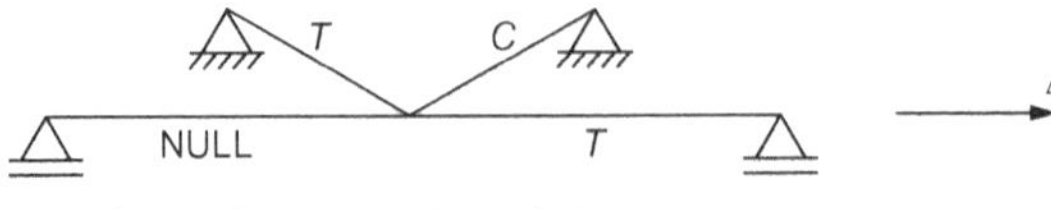

Fig. 3.115 'K' joint adopted structural model.

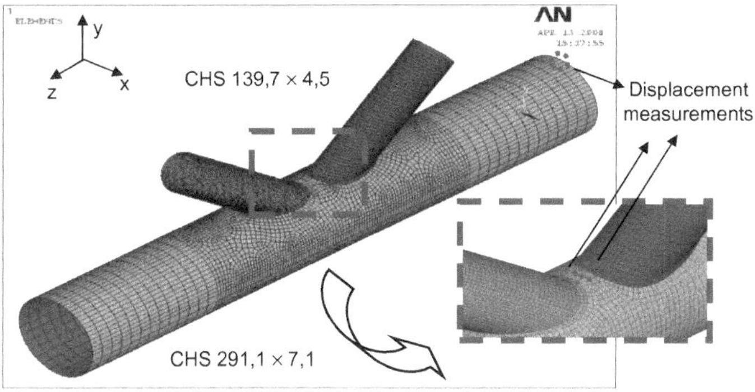

Fig. 3.116 'K' joint model joint mesh.

The 'K' joint numerical analysis used the same considerations of 'T' joint model in terms of the geometric and physical nonlinearities.

The nonlinear analysis was performed to obtain the load versus displacement curves for each of the joint elements, i.e. chord, left and right diagonals. The two diagonals (left and right) curves are presented in Fig. 3.117, where it can be observed that in the right diagonal (compression) curve, the buckling phenomenon reduced the element resistance after the chord yield. In this figure, it can be noted that for the deformation limits proposed by Lu et al. [134], $\Delta_s = 2.9$ mm ($0.01\ d_0$) and $\Delta_u = 8$ mm ($0.03\ d_0$) were obtained from the values of joint strengths of: $N_s = 439.1$ kN and $N_u = 247.9$ kN, respectively. The Eurocode 3 provisions for this joint led to a value of $N_{1,Rd} = 362.4$ kN.

The highlighted points in Fig. 3.117 were obtained from the von Mises stress distribution. It is worth mentioning that the weld was discarded from

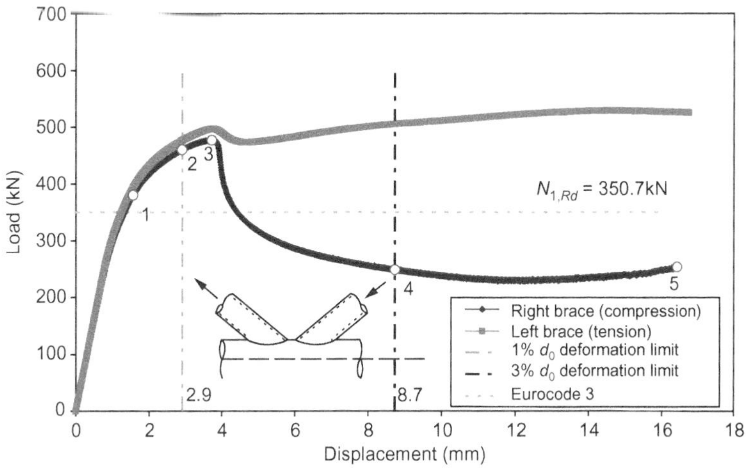

Fig. 3.117 Model diagonals load versus displacement curves.

von Mises stress distribution, once the failure occurred on the top chord face and, if the weld was considered, the used colour scale would hinder the analysis, due to the high difference of materials yield stresses. For this case, the yield stress considered in the colour scale will be associated to the chord, since its yield stress (355 MPa) is higher than the diagonals yield stress (275 MPa).

Fig. 3.118 shows the diagonals von Mises stress distribution associated to point 1 of the load versus displacement curve for the right diagonal

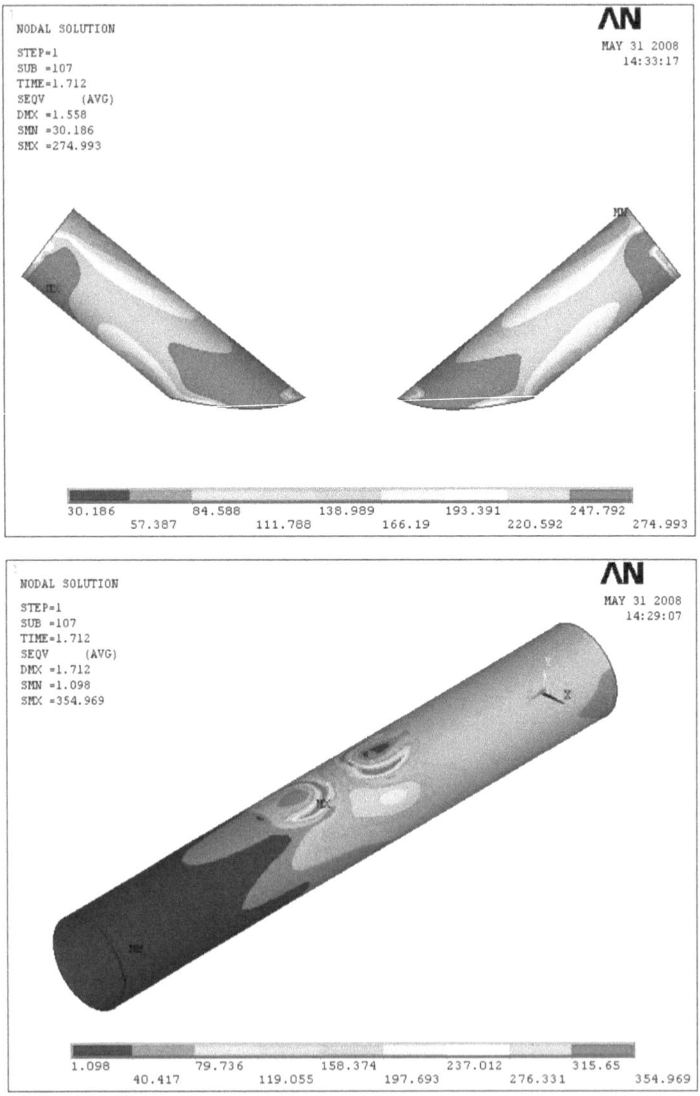

Fig. 3.118 Diagonals and chord von Mises stress distribution, pt. 1, Fig. 3.117.

(compressed) where the start of its yield can be noted. It can also be observed that the chord, associated to the same point 1 of the curve, presents a few yielded points close to the region surrounding the weld.

Fig. 3.118 presents the von Mises stress distribution associated to point 2 of the load versus displacement curve for the compressed diagonal corresponding to the deformation limit equivalent to 1% of the chord diameter as proposed by Lu et al. [134] that corresponds to the serviceability limit state. It can be observed that the compressed diagonal is almost completely yielded unlike the chord, as seen in Fig. 3.118. Fig. 3.120 shows the von Mises stress distribution associated to point 3 of the load versus displacement curve for the diagonals corresponding to the ultimate joint load, while the chord almost remained unaffected. Fig. 3.121 depicts the von Mises stress distribution associated to point 4 corresponding to the ultimate load according to Lu et al. [134], where it can be observed that a reduction on the diagonal forces occurred, fact that explains the sudden load drop. Even with the diagonal failure, Fig. 3.122 shows that the chord was not under high forces even for this ultimate limit state. Fig. 3.122 portrays the von Mises stress distribution associated to point 5 of the load versus displacement curve for the right diagonal (compressed), where it can be noted that this diagonal is associated to a stress concentration at its ends, indicating that the stress loss happened due to the local buckling phenomenon (Fig. 3.119).

The parametric analysis of the 'K' joint was based on the previously developed model only changing the load direction and empha-design. Once again, the load was applied by means of an equivalent displacement. This modification aimed to avoid generating stresses in the chord located directly below the compressed diagonal. The adopted structural model is depicted in Fig. 3.123. The material properties remained unchanged i.e. 355 MPa and 275 MPa for the chord and diagonals yield stresses and 600 MPa for weld ultimate stress. Furthermore, it should be stressed that the adopted weld thickness was in the previously calibrated model i.e. 4.5 mm. Table 3.6 presents the geometry of the elements used in the 'K' joint parametric analysis.

The main evaluated parameter in this parametric analysis was the eccentricity (e), since, as previously seen, when using a negative joint eccentricity, a resistance increase occurs. Such eccentricity also induces a bending moment on the joint that does not significantly affect its behaviour. In this model k_p is always 1 according to Eurocode 3, i.e. the load coming from the diagonals shall be exclusively responsible for the global connection resistance without the influence of the previously mentioned chord normal stresses. The analysis adopted by convention a null eccentricity value with the diagonals convergence point was coincident to the chord axis. Negative and

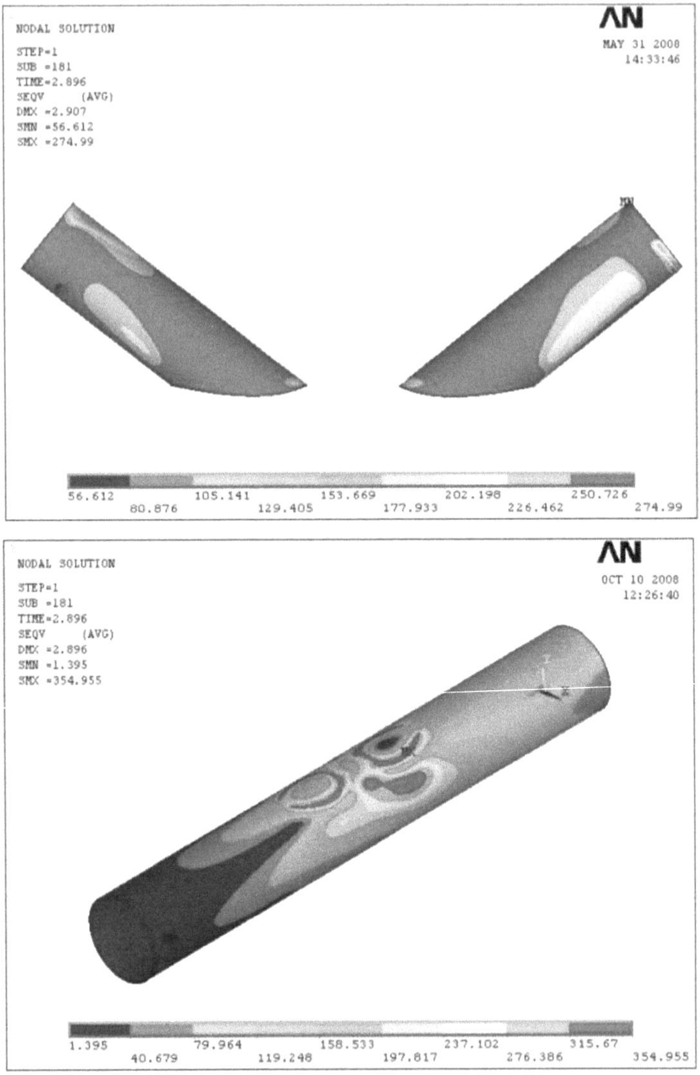

Fig. 3.119 Diagonals and chord von Mises stress distribution, pt. 2, Fig. 3.117.

positive values corresponded to cases where the diagonals convergence point was below and above the chord axis, respectively. Three analyses were made on the proposed model with eccentricities equal to 0, −10, and 10. Fig. 3.124 presents the load versus displacement curve for the 'K' joint chord without eccentricity, where it can be noted that the ultimate chord load was approximately 820 kN.

The equivalent curves for the two diagonals (left and right) are represented in Fig. 3.125 where for the maximum developed load for the left diagonal, corresponding to Fig. 3.125 point 2, showed the start of a local

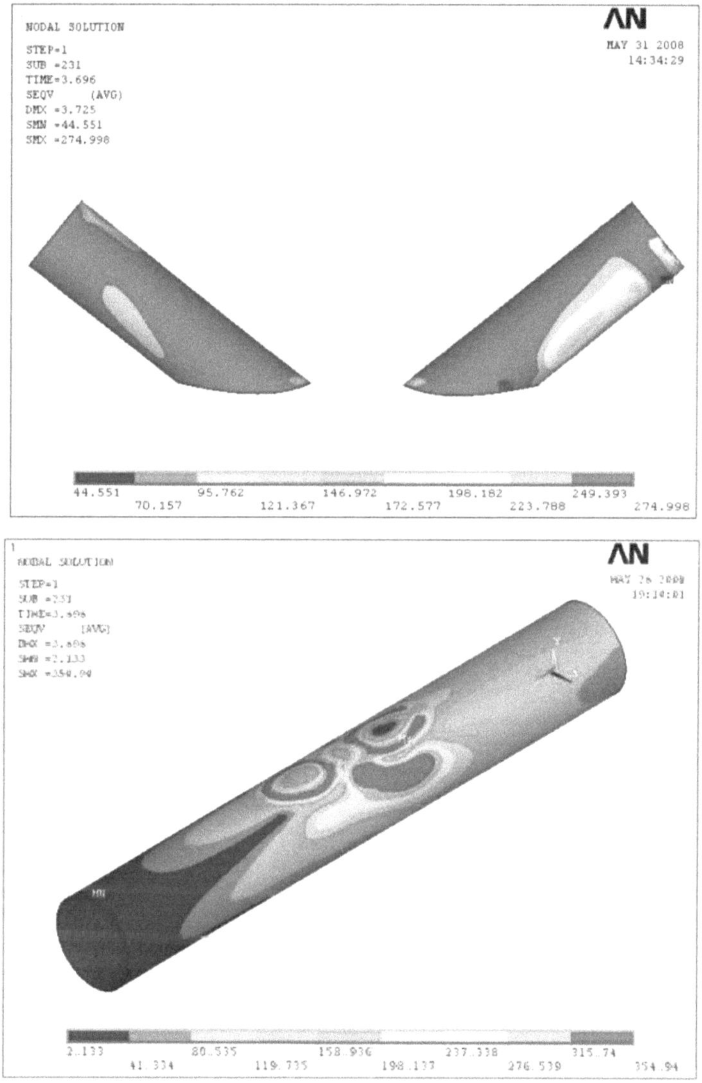

Fig. 3.120 Diagonals and chord von Mises stress distribution, pt. 3, Fig. 3.117.

buckling phenomenon at the upper end of this diagonal. In this figure, it can also be observed that using the deformation limits proposed by Lu et al. [134], i.e. $\Delta_s = 2.19$ mm ($0.01\ d_0$) and $\Delta_u = 6.57$ mm ($0.03\ d_0$), joint serviceability and ultimate load of: $N_s = 501.5$ kN and $N_u = 306.2$ kN, are obtained. Calculating this joint by Eurocode 3 equation a value of $N_{1,Rd} = 437.2$ kN is reached, value lower than the compressed diagonal strength of $N_{b,Rd} = 446.7$ kN, considering that an effective buckling length coefficient, 1.0 or 0.75 would not change the element compressive strength.

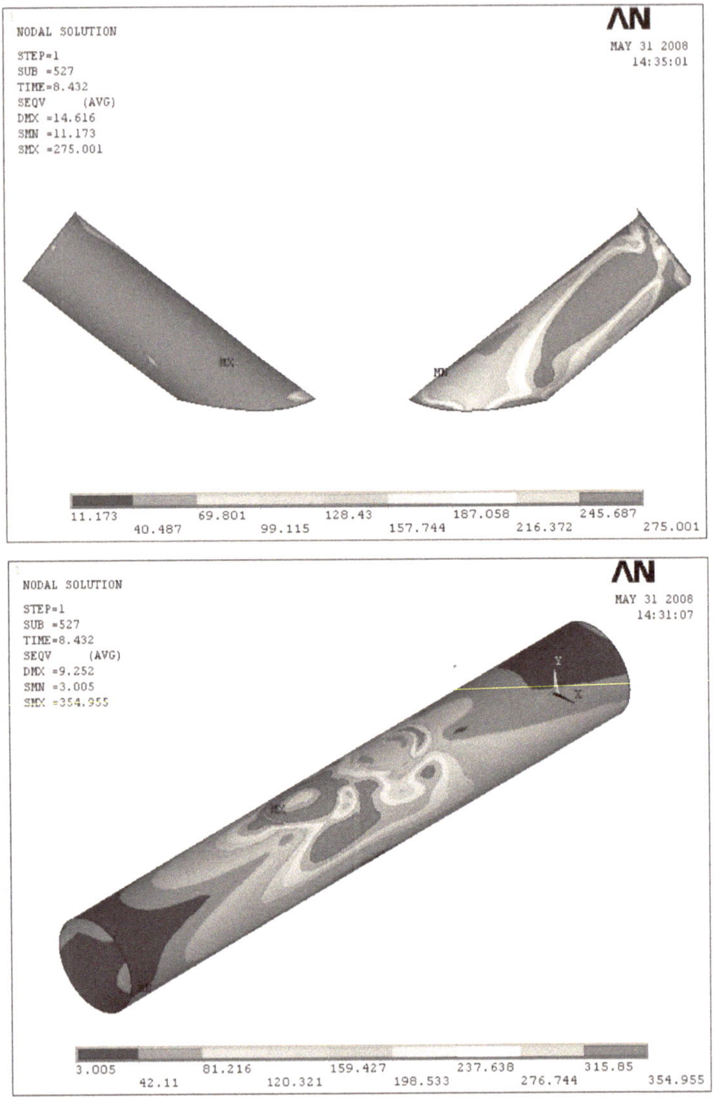

Fig. 3.121 Diagonals and chord von Mises stress distribution, pt. 4, Fig. 3.117.

Von Mises stress distributions were obtained for the highlighted points of the nonlinear analysis curves depicted in Fig. 3.125. It is worth mentioning that the weld was again discarded from von Mises stress distribution, once the failure occurred on the chord and, if the weld was considered, the used colour scale would hinder the analysis, due to the high difference of materials yield stresses. For this case, the yield stress considered in the colour scale was associated to the chord since its yield stress (355 MPa) is higher than the diagonals yield stress (275 MPa).

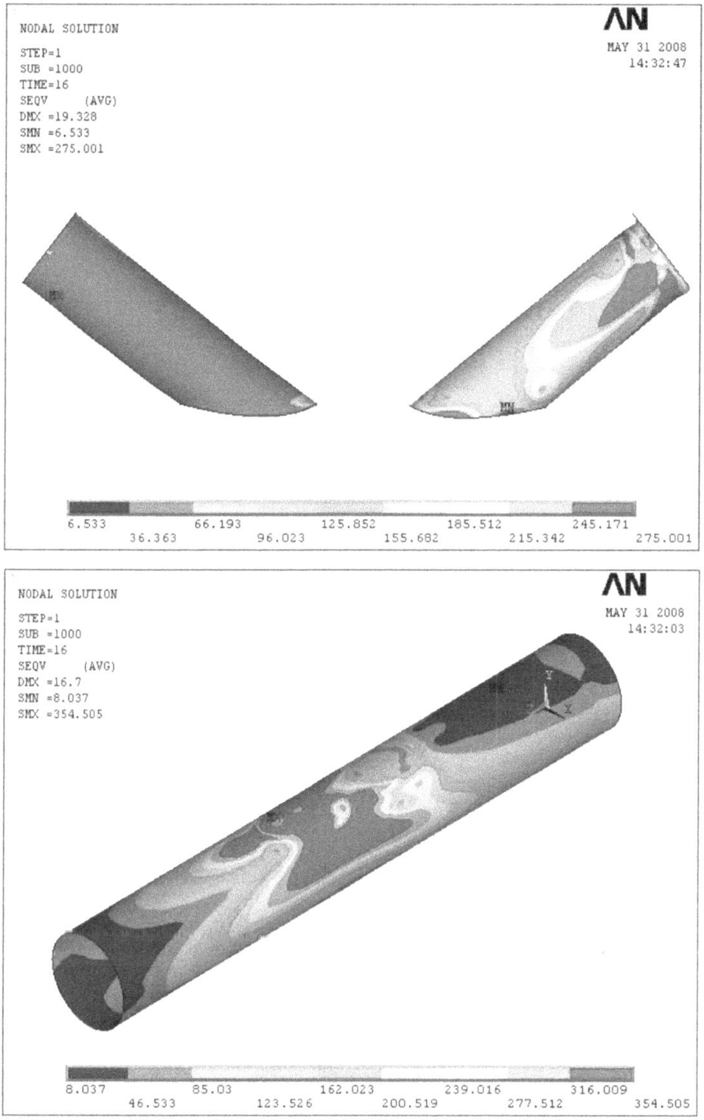

Fig. 3.122 Diagonals and chord von Mises stress distribution, pt. 5, Fig. 3.117.

Fig. 3.126 shows the chord and diagonals von Mises stress distribution associated to point 1 of Fig. 3.125, corresponding to a load acting on the diagonals of 456.1 kN. Fig. 3.126 also shows the compressed diagonals von Mises stress distribution associated to point 1 of the load versus displacement curve, where the start of the diagonals yield can be noted, and the stress distributions are similar for both diagonals. The elements were presented separately, once the stresses of the chord and the diagonals are significantly

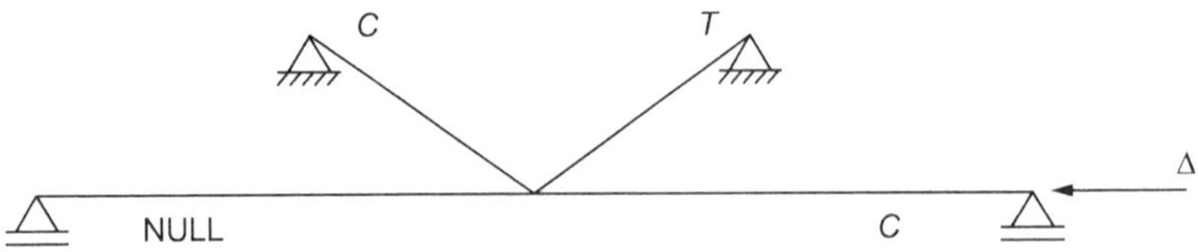

Fig. 3.123 'K' joint structural model adopted in the parametric analysis.

Table 3.6 'K' Joint Model Geometry

Elements	ϕ (mm)	t (mm)
Chord	219.1	7.1
Right diagonal (compression)	139.7	4.5
Left diagonal (tension)	139.7	4.5
Diagonal to chord angle	38.7°	

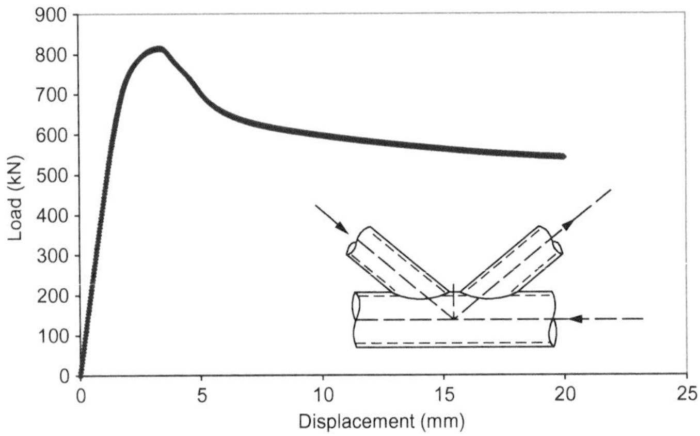

Fig. 3.124 Chord load versus deflection curve without eccentricity.

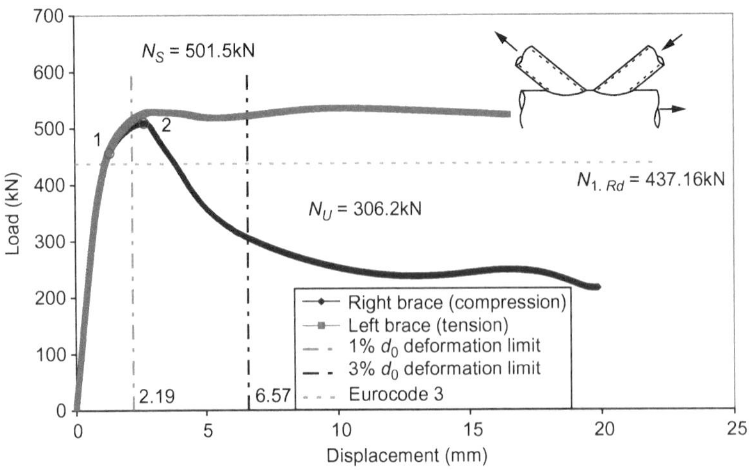

Fig. 3.125 Diagonals load versus deflection curves without eccentricities.

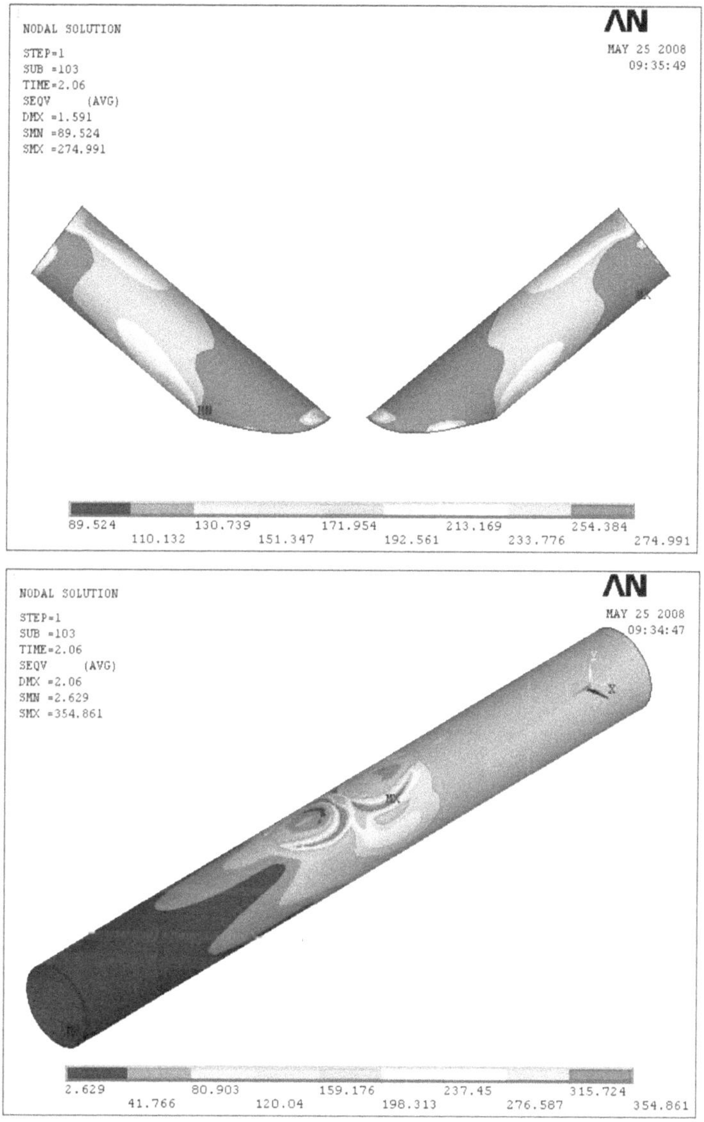

Fig. 3.126 Diagonals and chord von Mises stress distribution, compressed diagonal pt. 1, Fig. 3.125.

different. By observing the chord distributions, the start of the yielding process is detected close to the weld region. At this point, the corresponding diagonals load is 456.1 kN, close to the joint (437.16 kN) and diagonal (446.7 kN) resistances.

Fig. 3.127 presents the diagonal von Mises stress distribution for point 2 of Fig. 3.125. This point represents the maximum load obtained in the

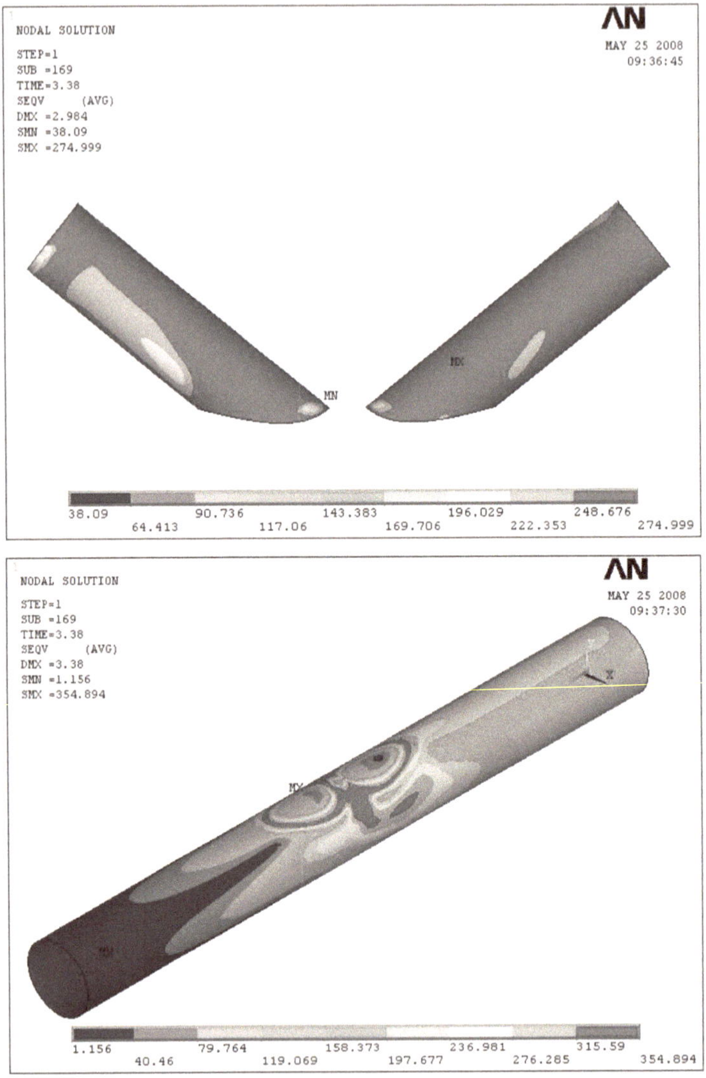

Fig. 3.127 Diagonals and chord von Mises stress distribution, compressed diagonal pt. 2, Fig. 3.125.

analysis for the compressed diagonal i.e. 509.1 kN value. When observing this diagonals stress distributions, it is possible to conclude that both have their sections almost fully yielded.

Fig. 3.128 depicts that the chord von Mises stress distribution for point 2 of Fig. 3.125 yielded regions is not as evident as in the diagonals. The parametric analysis results motivated the development of two additional analyses with intermediate values to obtain a more consistent data curve. Fig. 3.129

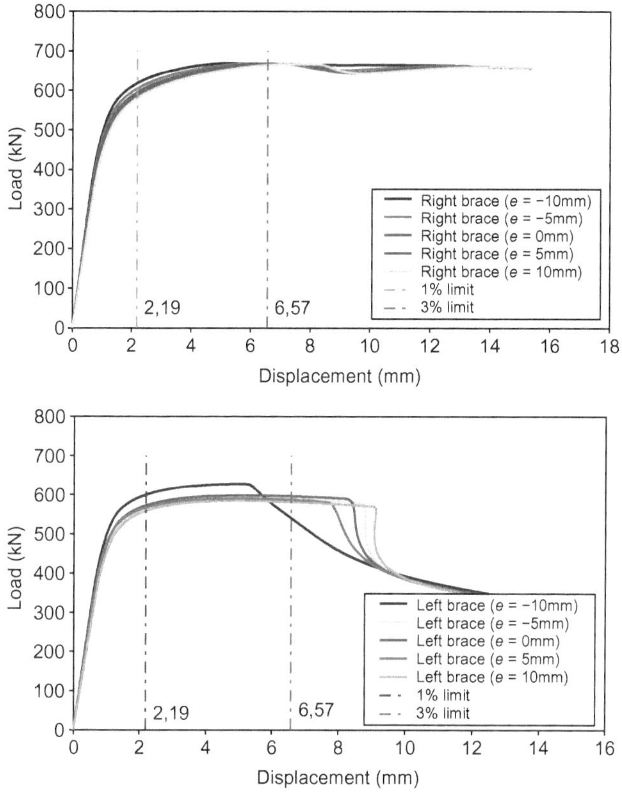

Fig. 3.128 Load versus displacement curves for right (tensioned) and left (compressed) diagonals of adopted parametric analysis 'K' joints.

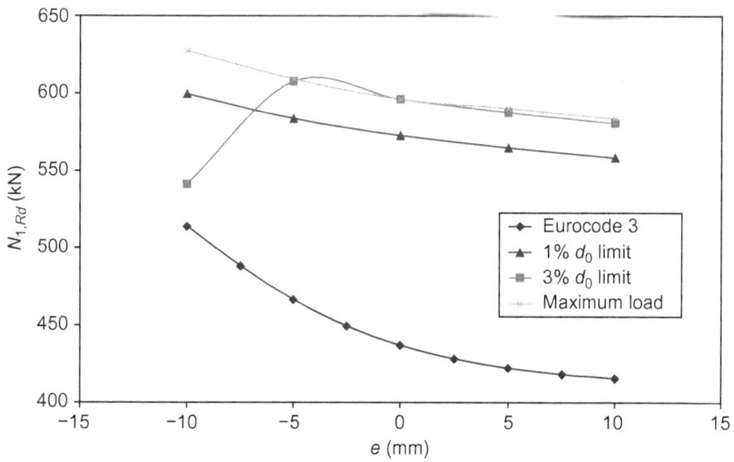

Fig. 3.129 Eccentricity versus joint strength $N_{1,Rd}$.

presents the right diagonal (tension) load curves versus deflection associated to a negative eccentricity, according to the adopted convention.

Fig. 3.129 also portrays the left diagonal von Mises stress distribution from the performed parametric analysis. These curves show that, for negative eccentricities, the joint resistance increases, since the element that controls the design is the compression diagonal failure. Such failure end up causing chord face yield on the joint region that was the objective of the present investigation, since in the Eurocode 3 joint design recommendations are that the most important variables should be associated to the chord geometry or to parameters, where such element is involved.

Fig. 3.129 depicts the eccentricity influence over the joints resistance in the parametric analysis models. These results suggest that the Eurocode 3 recommendations are conservative for the 'K' joint resistance evaluation. However, using Lu et al. [134] deformation limit of 1% of the chord diameter, more consistent results are obtained since they follow the same Eurocode 3 trend, where the joints with negative eccentricities are associated to higher strengths. It should be emphasised that for the ultimate limit state corresponding to 3% of the chord diameter, Lu et al. [134], for a negative eccentricity equal to −10 mm, the local buckling phenomenon occurred before reaching this 3% limit causing the sudden curve drop observed in Fig. 3.129.

The SHELL181 element was again used on the 'K' joint models adopted in the parametric analysis since the results of the previous 'T' joint models were satisfactory. The development of the model considered the previous numerical example with only a few differences. The results obtained in model numerical analysis using the criteria established by Lu et al. [134], and later compared to the results obtained from the Eurocode 3 indicated that the Eurocode 3 results led to safe design predictions. The 'K' joint parametric analysis performed in the present study showed that a negative joint eccentricity i.e. reducing the diagonals distance leads to an increase of the joint resistance also predicted by the Eurocode 3.

The studied tubular joint model demanded an assessment of the support conditions and load application that would enable a more efficient control of the system convergence, especially in load magnitudes close to the joint ultimate limit state. As previously mentioned, in this case, the deflection control proved to be the most appropriate process to be used. On the other hand, if this control is very effective when a single displacement is applied to structure, for cases where more than one displacement is necessary, its correct specification becomes a complex process and often impractical.

The studied tubular 'K' joint model was made of a chord and two diagonals. Usually, the loads are applied on the diagonals and are supported by supports located in the chords. The three imposed loads on the system would generate the need for three equivalent displacements with a difficult correlation. Alternatively, a series of unidimensional models were designed by varying the support position and the applied loads until a similar configuration to the original truss system structure was obtained. An interesting alternative associated to an efficient convergence control is one where the supports are used on the diagonals and an equivalent displacement is imposed at one of the chord ends.

The modelling of the weld region was made with shell and solid elements to determine what the most efficient strategy was. For the cases modelled, the shell proved to be more suitable for generating results that were consistent with the experimental equivalent values adopted in their calibration. A detailed investigation of the collapse modes associated to these joints was also carried out, indicating a complex interaction between the ultimate limits states associated to the global element buckling, local buckling, and squashing.

3.5.12 Modelling of Composite Beams With Full and Partial Interaction

The present study focuses on the modelling of composite girders with full and partial shear connection using the ANSYS program [32]. A three-dimensional model was developed, in which all main structural variables and associated nonlinearities are included (concrete slab, steel beam and shear connectors). Test and numerical data available in the literature are utilised to validate the model, which was able to deal with simply supported systems with I-beams and solid flat slabs.

Based on the calibrated model, an extensive parametric analysis of composite girders was made, specifically aiming at: studying the effect of the continuation of shear connection beyond the supports of simply-supported composite girders; studying the overall structural system response when different concrete compressive strengths were utilised in the slab, and in the related push-out tests. This investigation can be utilised to assess the influence of the effects of partial shear connection, and partial interaction not only on the overall flexural behaviour of composite girders, but also on the related collapse modes for either concrete slab crushing or stud failure, and on the distribution of stud shear forces along the girder length.

The finite element types considered in the model were: elastic-plastic shell (SHELL43) and solid (SOLID65) elements for the steel section and

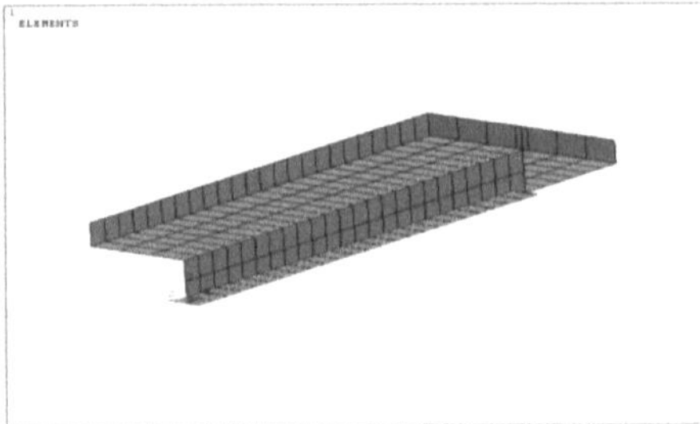

Fig. 3.130 Typical composite girder FE mesh.

the concrete slab, respectively, and nonlinear springs (COMBIN39) to represent the shear connectors. Both longitudinal and transverse reinforcing bars were modelled as smeared throughout the solid finite elements. Symmetry of the composite girders was considered by modelling only half of the girder span. A typical mesh for a composite girder is portrayed in Fig. 3.130.

The von Mises yield criterion with isotropic hardening rule was utilised to represent the steel girder (flanges and web) response. The stress–strain relationship was linear and elastic up to yielding, perfectly plastic between the elastic limit and the start of strain hardening, and followed the constitutive law utilised by Gattesco [139] for the strain hardening branch:

$$\sigma = f_y + E_h(\varepsilon - \varepsilon_h)\left(1 - E_h \frac{\varepsilon - \varepsilon_h}{4(f_u - f_y)}\right) \tag{3.9}$$

where f_y and f_u are the yield and ultimate tensile stresses of the steel component; E_h and ε_h are the strain-hardening modulus (i.e. 3500 N/mm^2), and the strain at strain hardening of the steel component. The von Mises yield criterion with isotropic hardening rule was also utilised for the reinforcing steel. An elastic-linear-work-hardening material was considered, with tangent modulus being equal to 1/10,000 of the elastic modulus, to avoid numerical problems. The values measured in the experiments for the material properties of the steel components (steel beam and reinforcing bars) were utilised in the finite element analyses.

The concrete slab response was modelled by a multilinear isotropic hardening relationship, utilising the von Mises yield criterion coupled with isotropic work hardening. The uniaxial response was described by a piece-wise linear total stress–total strain curve, starting at the origin, with positive stress and strain values, taking into account the concrete compressive strength (f_c) corresponding to a compressive strain of 0.2%. The stress–strain curve also assumed a total increase of 0.05 N/mm^2 in the compressive strength up to the concrete strain of 0.35% to avoid numerical problems due to an unrestricted yielding flow. The concrete element shear transfer coefficients considered were: 0.2 (open crack) and 0.6 (closed crack). Typical values ranged from 0 to 1, where 0 represented a smooth crack (complete loss of shear transfer) and 1, a rough crack (no loss of shear transfer). The default value of 0.6 was utilised as the stress relaxation coefficient. The crushing capability of the concrete element was also disabled to improve convergence. The concrete slab compressive strength was assumed as the actual cylinder strength test values. The concrete tensile strength and the Poisson's ratio were taken as 1/10 of its compressive strength and 0.2, respectively. The concrete elastic modulus was calculated according to Eurocode 4 [64], i.e.:

$$E_c = 9500(f_c + 8)^{1/3}\left(\frac{\gamma_c}{24}\right)^{1/2} \tag{3.10}$$

where γ_c is equal to 24 kN/m^3. The model allowed for any pattern of stud distribution to be considered, for instance: the conventional uniform arrangement and a triangular spacing scheme, where the stud distribution followed the nominal elastic shear diagram. In all analyses, the number/spacing of studs used in the experiments was used. As far as the shear connector response is concerned, the load-slip curves for the studs were utilised (obtained from available push-out tests) by defining a table of force values and relative displacements (slip) as input data for the nonlinear springs.

Concentrated loads were incrementally applied to the model by means of an equivalent displacement to surpass convergence problems (displacement control). For the convergence criterion, the L2-norm (square root sum of the squares) of displacements was considered. Uniformly distributed loads were represented by means of point loads applied at all mid-section concrete nodes. These concentrated loads were also applied to the model incrementally using the load control strategy and the L2-norm. The tolerance associated to this convergence criterion and the load step increment was varied to solve potential numerical problems.

The load control strategy was adopted due to the fact that, in structural problems in which significant nonlinear effects occur, it is difficult to derive a relationship between loads and related displacements for the case of distributed loads, mainly for the plastic range of response. For the case in which only one point load was applied to the system, there was a direct relationship between force and displacement, making the DCM easier to be used. The load control method is, however, less efficient than the DCM in nonlinear analyses. This fact was noted especially when the applied load approached the system ultimate load, as an incremental increase in the load leads to a significant increase in the corresponding displacements, causing difficulties in terms of numerical convergence.

Preliminary attempts to surpass the convergence problems arising from the use of the load control method included the specification of different types of equation solvers. The best approach in terms of numerical performance was the option in which the software ANSYS selects a solver based on the physics of the problem. The arc-length method was also tested, but proved not to be the best option for this particular type of analysis. Two limits were established to define the ultimate load for each finite element study: a lower and an upper bound, corresponding to concrete compressive strains of 0.2% and 0.35%, respectively. These two limits define an interval in which the composite girder failure load is located. A third limit condition, hereinafter referred to as the stud failure point, can also be reached when the composite girder's most heavily loaded stud reaches its ultimate load, as defined from the appropriate push-out tests.

If the stud failure point was located before the lower bound of concrete (i.e. the corresponding load of the stud failure point is smaller than the lower bound load), then the mode of collapse of the composite girder was considered as being stud failure. Alternatively, if the stud failure point was located after the upper bound of concrete, the failure mode was assumed as being concrete crushing. For the intermediate case, where the stud failure point lies between the lower and upper bounds of concrete, than the collapse mode could be either of them. Therefore, the developed finite element model was able to forecast the collapse modes related to either slab crushing or stud failure.

The model was calibrated by comparisons against Chapman and Balakrishnan tests [140]. The tests made by Chapman and Balakrishnan successfully portrayed the response of the composite system which is being studied. The girders spanned 5490 mm with an I-shaped steel member 305 mm deep ($12'' \times 6'' \times 44$ lb/ft BSB) and a concrete slab 152 mm thick × 1220 mm wide. The number and type of studs, as well as the steel and concrete strengths, varied according to the tested composite girder. The slab was longitudinally reinforced with four top and four bottom 8 mm bars. The transverse reinforcement incorporated top and bottom bars of 12.7 mm @ 152 mm centres and 12.7 mm @ 305 mm centres, respectively. The tensile strength, the Young's modulus, and the Poisson's ratio of the reinforcing steel bars were 320 N/mm^2, 205,000 N/mm^2, and 0.3. The girder's material properties is given in [140] and a full description of these girders is depicted in Fig. 3.131.

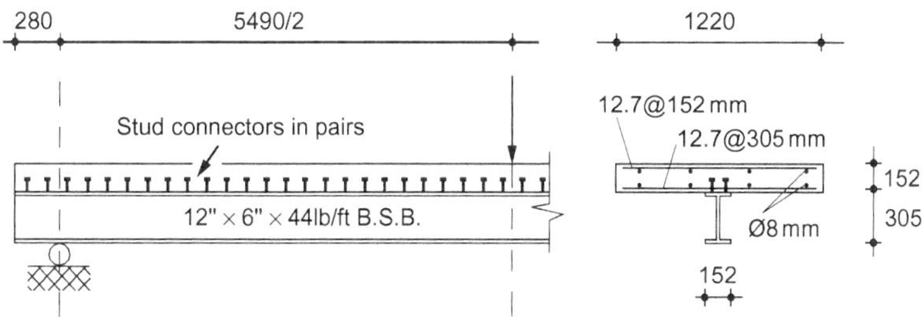

Fig. 3.131 Simply supported girders layout (dimensions in mm).

Based on the composite section strength of the concrete slab, steel components, and shear connectors, the level of shear connection could be evaluated. This value is defined as the ratio between the shear connection capacity and the weakest element capacity (concrete slab or steel beam). Table 3.7 summarises the level of shear connection for all the composite girders, considering two different approaches. The first utilises the nominal values depicted by [140] for the stud strength and steel yield stress. In the second one, the material properties are taken as the actual measured values. Considerable differences among the levels of shear connection according to these two approaches were observed, leading to the conclusion that, to evaluate the level of shear connection of composite systems, the actual material properties of the components related to each experimental programme should be utilised.

For all composite girders depicted in the previous table, load-midspan deflection curves were compared to the tests. Figs 3.132–3.135 (girders A2–E1) and Figs 3.132–3.135 (beams U1–U4) present comparisons between the FE model results and the experimental data for the midspan concentrated and uniformly distributed loaded composite girders, respectively. The limit points for the concrete were represented by a full triangle (lower bound) and by a full square (upper bound), and the stud failure point was represented by a full circle. Good agreement was reached between test and numerical results.

The numerical and test values regarding the slip at the steel-concrete interface along the beam axis for the cases E1 and U4 are plotted in

Table 3.7 Level of Shear Connection of the Composite Beams (%)

Beam	AA2	AA3	AA4	AA5	AA6	BB1	CC1	DD1	EE1	UU1	UU3	UU4
Nominal values	238	213	175	138	101	138	138	313	313	175	175	101
Measured values	231	186	137	123	95	116	114	136	148	155	177	90

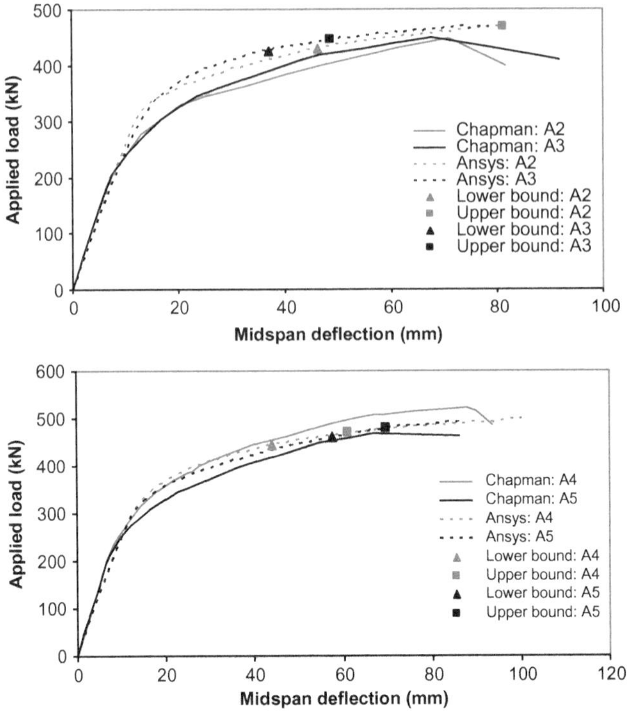

Fig. 3.132 Load (kN) versus midspan deflection (mm)—Girders A2 and A3 and A4 and A5.

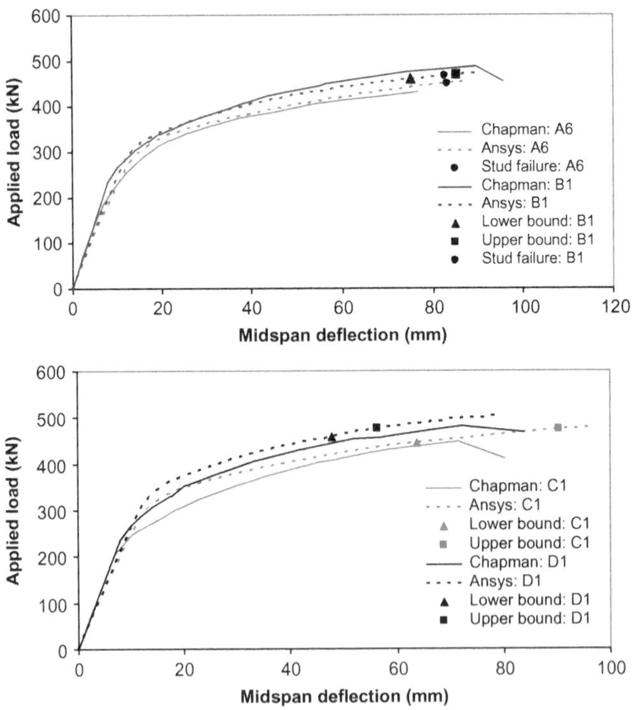

Fig. 3.133 Load (kN) versus midspan deflection (mm)—Girders A6 and B1 and C1 and D1.

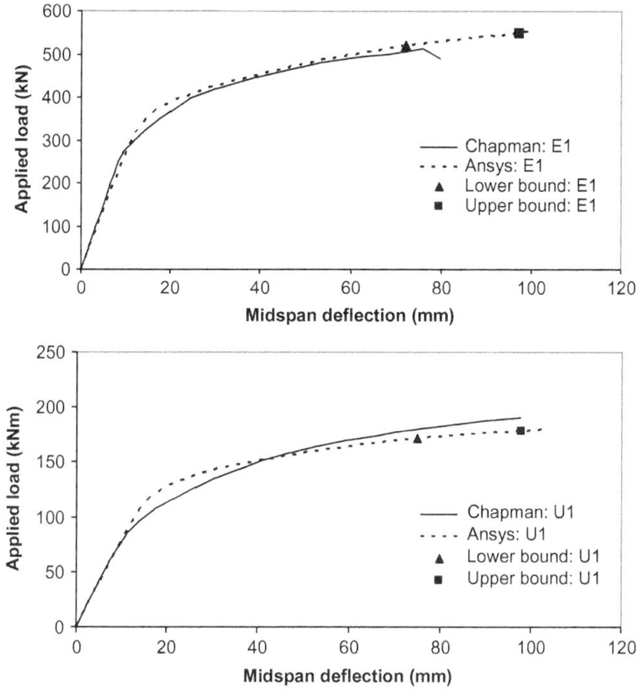

Fig. 3.134 Load (kN) versus midspan deflection (mm)—Girders E1 and U1.

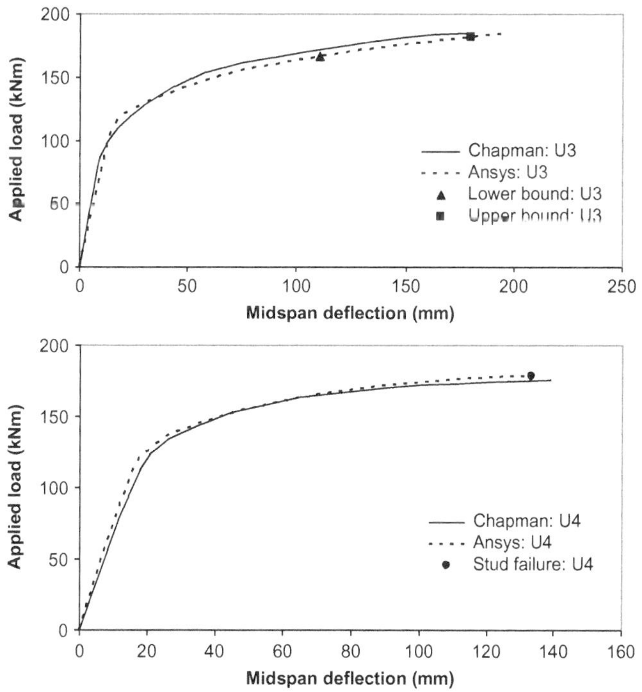

Fig. 3.135 10 Load (kN) versus midspan deflection (mm)—Girders U3 and U4.

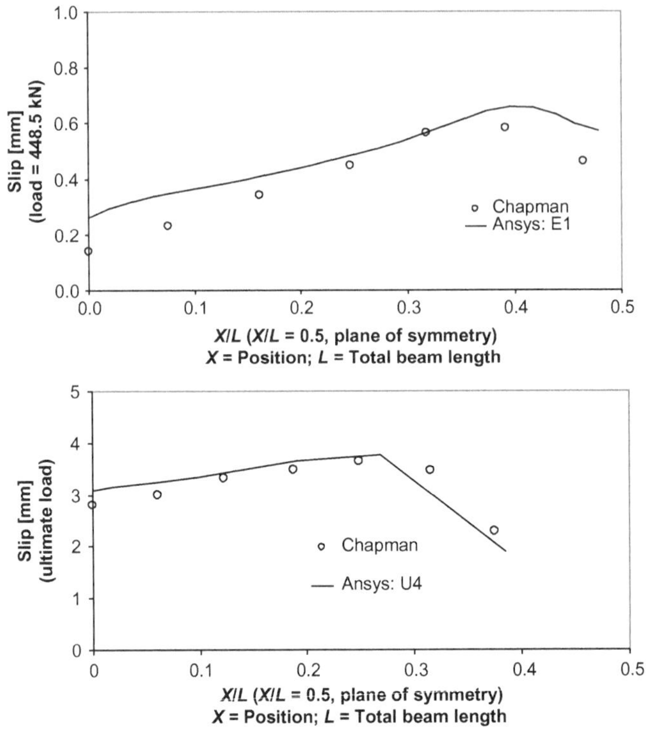

Fig. 3.136 Slip distribution along span—Girders E1 and U4.

Fig. 3.136. The graphs show that the developed model can predict the slip distribution with good precision. It can be observed that the slip is not uniform along the beam length, even where the externally applied shear force was uniform (midspan concentrated load case).

The maximum slip value tended to occur when X/L equals 0.4 (point load case), and X/L equals 0.3 (uniformly distributed load case UDL). According to [140], the large slip which occurred near midspan was due to the high interface shear in the plastic region.

The complete results for all standard girders were presented and discussed in Queiroz et al. [137], including a detailed investigation in terms of the absolute force carried by the shear studs for three different stud overall lengths (for a fixed diameter and spacing), and the absolute force carried by the shear studs for five different connector spacing values (for a fixed diameter and length).

The developed finite element model was also utilised to assess the influence of the effects of partial shear connection and partial interaction, not only on the overall flexural structural system response (represented by the load-deflection curve), but also on the related collapse modes for either slab crushing or stud failure, and on the distribution of stud shear forces along the girder length.

In order to isolate the effect of the level of shear connection, the material properties (steel, concrete and shear connectors) and dimensions of the girder E1 were utilised for all analyses. Two load cases were considered: midspan point load and uniformly distributed load. The levels of shear connection were evaluated utilising the actual material properties measured during the experiments. In the present investigation, levels varying from 47% to 136% were analysed, utilising the standard finite element model.

The load-midspan deflection curves for all cases (each one having a different level of shear connection) are depicted in Figs 3.137 and 3.138 where only the initial branch of these curves is presented to better illustrate the differences between the curves in terms of initial stiffness. It can be noted that when the level of shear connection was decreased, the system became more flexible, with reduced strength and stiffness. However, the decrease in stiffness did not seem to be as relevant as the decrease in the ultimate load.

The three higher levels (118%, 130%, and 136%) resulted in very similar curves, in terms of both, stiffness and ultimate load, represented by the lower and upper bound values, with the collapse mode being slab crushing at the girder midspan. Regarding level 100%, its collapse mode could be either slab crushing or stud failure, as a stud failure point was reached between the lower and upper bounds. As the three lower levels (47%, 71%, and 89%) resulted in stud failure, the level of 100% was an intermediate level between the two possible collapse modes. The ratio between the ultimate load and the load at the end of the elastic behaviour was not much influenced by the distinct levels of shear connection, being in the range from 1.5 to 2

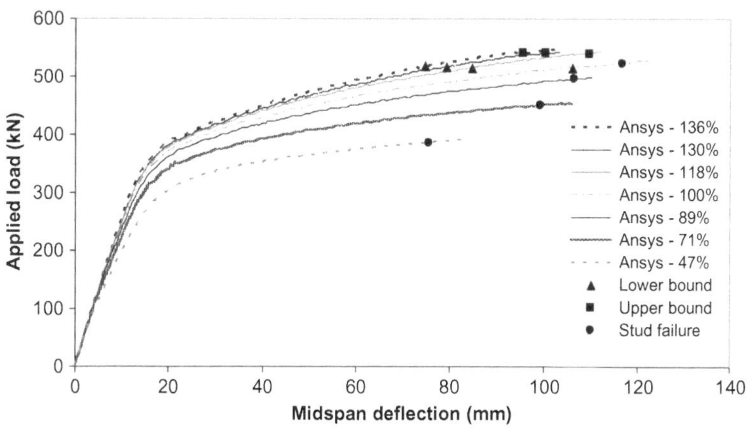

Fig. 3.137 Load-deflection curve—midspan concentrated load.

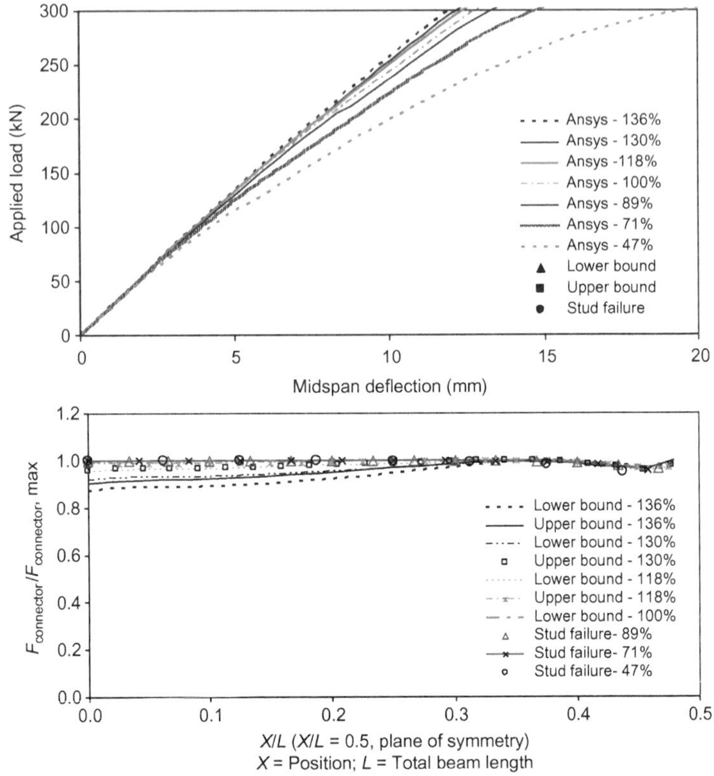

Fig. 3.138 Load-deflection curve—midspan concentrated load—initial stiffness and ratio of stud forces versus relative position of the stud.

for all cases, which is a common value as far as steel-concrete composite girders are concerned.

The stud force distribution related to the composite girder ultimate load is plotted relating force ratio to stud position in Fig. 3.138. It can be observed that the variation of the stud force ratio of all curves is approximately 10%. Moreover, if the 136% level was not considered, this variation is only 5%. This means that, at ultimate load and regardless of the level of shear connection, the connectors were able to redistribute the load almost uniformly among them.

The results demonstrate that, the effects of partial interaction can be neglected for levels of shear connection above 100%, as no significant improvement in terms of either strength or stiffness of the girder was noted. The load-midspan deflection curves for all cases are present in Fig. 3.139. It can be noted that, as occurred for the midspan load case, decreasing the level of shear connection makes the system become more flexible, with reduced strength and stiffness.

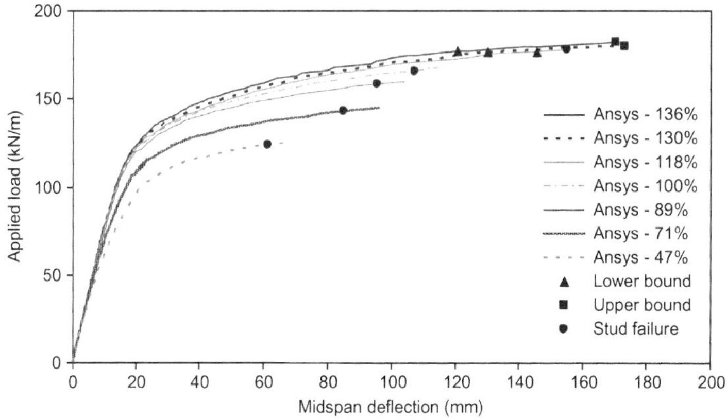

Fig. 3.139 Load-deflection curve—UDL.

Once more, the decrease in stiffness did not seem to be as relevant as the decrease in the ultimate load (Fig. 3.140). The two higher levels (130% and 136%) resulted in very similar curves, in terms of both, stiffness and ultimate load, represented by the lower and upper bound values, with the collapse mode being slab crushing at the beam midspan. Regarding level 118%, its

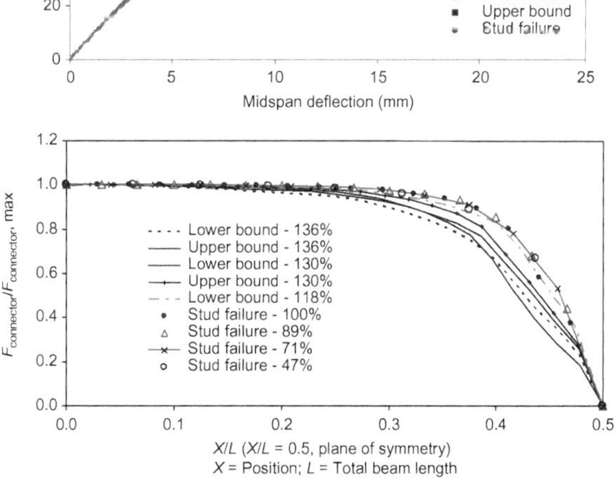

Fig. 3.140 Load-deflection curve—UDL—initial stiffness and ratio of stud forces versus relative position of the stud.

mode of failure could be either slab crushing or stud failure, as a stud failure point was reached between the lower and upper bounds. As the other four levels (47%, 71%, 89%, and 100%) resulted in stud failure, the level 118% was an intermediate level between the two possible collapse modes of failure. The ratio between the ultimate load and the load at the end of the elastic behaviour was similar to the one observed for the midspan point load case (from 1.5 to 2 for all cases).

The stud force distribution graph (Fig. 3.140) is plotted, relating force ratio to stud position. It can be observed that the connectors were able to yield and redistribute the load almost uniformly among them for a length of approximately 30% of the span from each of the supports ($0 \leq X/L \leq 0.3$). In the central region of the girder ($0.3 \leq X/L \leq 0.5$), the studs were less heavily loaded. This response occurred for all levels of shear connection, except the two highest ones (130% and 136%), for which the length of the central region increased by 10%. The results showed that, for the UDL case, the effects of partial shear connection could be important for levels of shear connection above 100%. It was noted that the connectors were not loaded uniformly along the beam length (Fig. 3.140), with the stud forces decreasing towards the beam midspan. Consequently, more shear connectors are needed to obtain a full interaction level, for which the effects of partial interaction may be neglected.

The results show that, for the case with uniformly distributed loads, the effects of partial interaction may be important for higher levels to 100%. It was observed that the connectors were not uniformly loaded along the length of the beam (Fig. 3.140), decreasing pins forces toward the Centre beam. This requires more shear connectors to obtain a total level of interaction, for which partial interaction effects can be neglected. Despite the fact that current design codes assume the same strength for all shear connectors, as well as a uniform distribution of load among them for the ultimate limit state, an 18%-difference between the total-interaction levels for the concentrated load (100%) and UDL (118%) cases was reached. Nevertheless, this difference is not sufficiently relevant to propose modifications to well accepted design methods.

The composite beams modelling strategy used spring interface elements calibrated with shear connectors push-out test results, and made it possible to assess the effects like their spacing and distribution along the composite girder. In addition, the model allowed an assessment of the shear connectors force distribution to evaluate their actual performance along the girder.

The stop control adopted was based on comparison with the connector ultimate loads, whether this controlled the composite girder performance, or a range of minimum and maximum variation of concrete compressive strains. This was performed since the concrete crushing and cracking capabilities of the element had to be suppressed to overcome serious convergence problems in regions close to the investigated structural system ultimate load. It should be stressed at this point, that many investigations have misused ultimate loads, based on the point where the system no longer converges. When some kind of control and calibration can be performed during the experiment, future parametric analysis using this stop criterion must not be trusted, since they are unreliable.

Concentrated and distributed loads were simulated where, in the first case, a displacement control was used, and in the second case, a painstaking and costly load control was adopted. This has given rise to numerous restarts, changes of tolerances, and/or convergence criteria to generate reliable results, varying from case to case and being extremely dependent on the user's experience in dealing with this particular type of convergence problem.

3.5.13 Structural Response of Perfobond and T-Perfobond Shear Connectors

The present investigation is centred on the results of a numerical modelling of alternative shear connectors, aiming to find the best solution to increase the shear connector ductility [141–143]. The first experimental series was held at the Department of Civil Engineering, University of Coimbra, while the second occurred at the Pontifical Catholic University of Rio de Janeiro. In the first series, the T- Perfobond connector was made from S275 IPN 340 profiles, Fig. 3.141, with a 12.2 mm web thickness and an 18.3 mm flange thickness). This specific connector was associated to a rigid block like response during the performed push-out tests. An alternative geometry was studied, reducing the flange thickness to 11.3 mm, while keeping the original web thickness to enable a comparison with the first set of tests that was made to enhance the connector ductility. The present section also depicts a numerical study of the deformation capacity of the connector flange that was performed before the second experimental series. The alternative T- Perfobond connector geometry utilised in the second experimental series is presented in Figs 3.141 and 3.142. The connector was made from

IPN 340 – Coimbra Univ.

HP 200x53 – PUC-Rio

Fig. 3.141 Adopted *T-Perfobond* connectors.

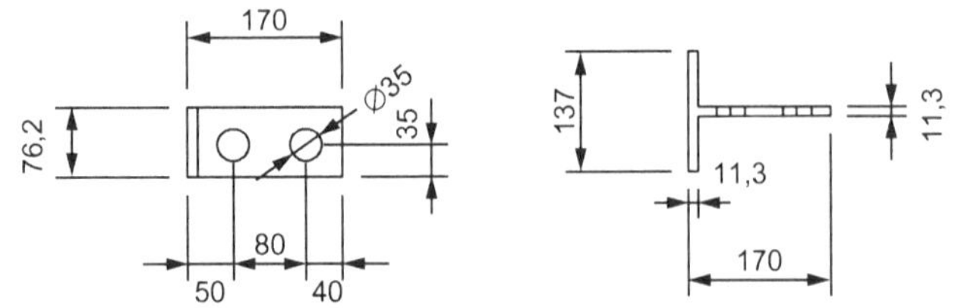

Fig. 3.142 Geometry of *T-Perfobond* connectors the second stage.

a HP200x53 profile equivalent to American Profile HP8x36, using an ASTM A572 Grade 50, equivalent to a S355 steel grade. The T-Perfobond connectors were utilised in 120 mm thick concrete slabs designed to reach a class C30/37 compressive strength.

The investigated numerical models were made using Shell 63 elements, available in the ANSYS program library, to model the T- Perfobond connector flange. This is a plane finite element with four nodes and six degrees of freedom per node, capable of considering the material nonlinearity. A simplified model layout considering the symmetry conditions was used to assess the connector flange deformation capacity. The support nodes representing the part of the connector welded to the beam flange, and the nodes along the symmetry line of the flange had all their degrees of freedom restricted.

The load was applied by means of pressure (around 48.85 MPa), applied at the entire flange area to emulate the push-out test transfer of forces that happen from the concrete slab to the shear connector flange. The model layout is portrayed in Fig. 3.143. A linear elastic analysis was made assuming an isotropic behaviour with a 205 GPa Young's Modulus and a 0.3 Poisson's ratio.

Table 3.8 presents the elastic deformation capacity results and node 9 related displacements illustrated in Fig. 3.144. The 18.3 mm thickness is related to the flange used in the T-Perfobond connectors utilised in the first push-out tests. The thickness of 12 mm was selected, as an initial tentative to increase in almost 3 times the 18.3 thick connector ductility. The 11.3 mm flange thickness corresponds to the flange and web thickness of the HP200x53 profile used in the second set of push-out tests. Fig. 3.144 graphically depicts strain and displacement distributions along the 11.3 mm thick flange.

This simple model of T-Perfobond connector flanges showed that if a reduction of the flange connector thickness from 18.3 to 12 mm was performed a significant gain was achieved in the shear connector ductility. This was the main reason for selecting the HP200x53 profile with a 11.3 mm flange and web thickness in the second push-out tests series. The study also confirmed that T-Perfobond connectors have high load and stiffness capacities. Since this shear connector can be made by rolled sections leftovers a

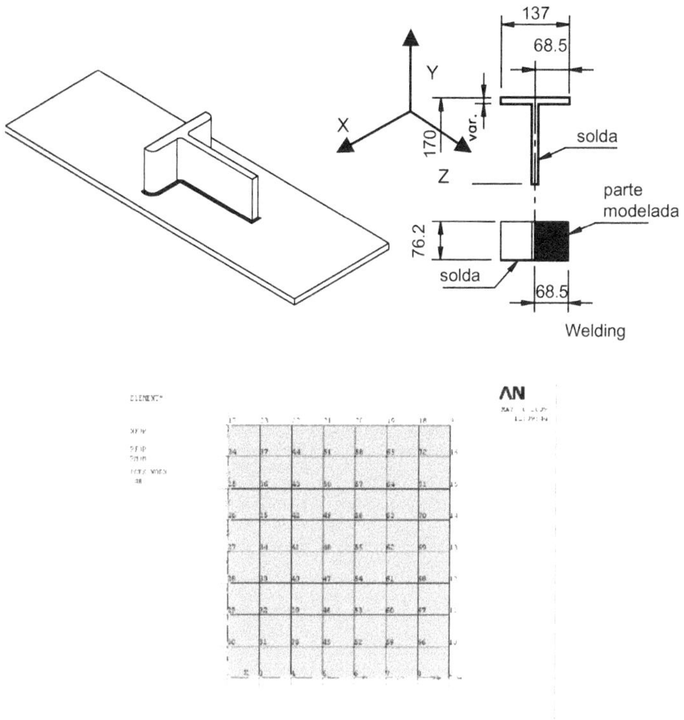

Fig. 3.143 Perfobond connector and Connector flange: mesh and boundary conditions.

Table 3.8 Numerical Modelling Results

Plate Thickness (mm)	*Z*—Displacement (mm)	*Z* Elastic Deformation ($\mu\varepsilon$)
18.3	0.49	203
12.0	1.74	574
11.3	2.09	657

significant economy can be reached when compared to other commonly adopted shear connectors like the studs.

The T-Perfobond connectors made from the IPN340 profile did not present a ductile behaviour in 120 mm thick slabs. These results led to the adoption of an elastic distribution of shear along the beam length for composite beam design. On the other hand, the T-Perfobond connectors, made in Brazil, from a HP200x53profile, with proper reinforcing bars in 120 mm thick slabs depicted a ductile response, thus allowing a plastic design to be performed. The tests performed with the 11.3 mm thick profile were able to meet the minimum slip 6 mm limit required by Eurocode 4 [64], ensuring the connector ductile response. This type of response could be confirmed with the aid of the simple numerical model results.

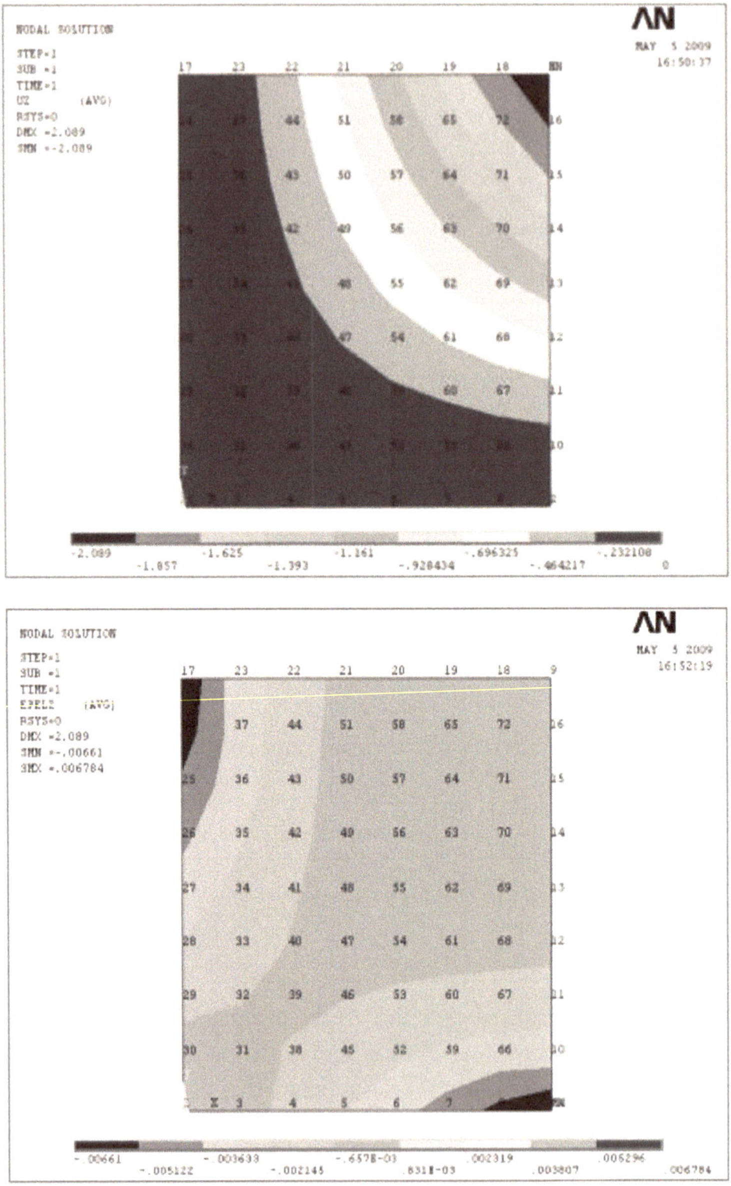

Fig. 3.144 HP200x53 connector results (11.3 mm thick).

The study proceeded with the development of a composite beam structural system model with the aid of the ANSYS program [32]. The used finite elements were: elastic-plastic shell (SHELL43) and solid (SOLID65) elements for the steel section and the concrete slab, respectively, and nonlinear springs (COMBIN39) to model the shear connectors.

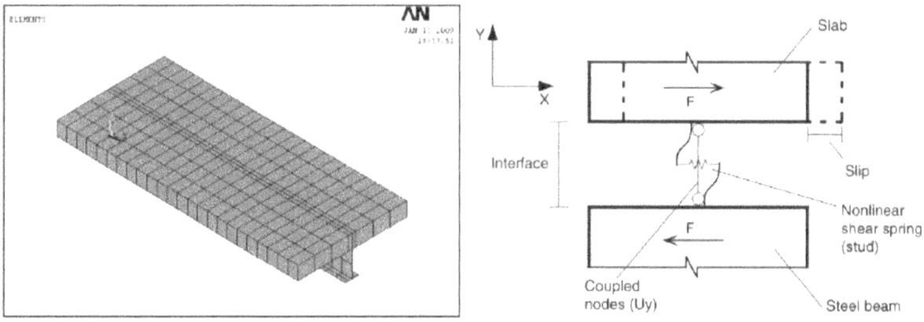

Fig. 3.145 Typical composite beam FE mesh and shear connectors modelling, Queiroz [137].

Both longitudinal and transverse reinforcing bars were assumed as smeared throughout the solid elements. Symmetry of the composite beams is considered by modelling only one half of the beam span. A typical FE mesh for a composite beam is shown in Fig. 3.145.

The von Mises yield criterion with isotropic hardening rule (multilinear work-hardening material) was utilised to represent the steel beam. The stress–strain relationship was linear elastic up to yielding, perfectly plastic between the elastic limit (ε_y) and the start of strain hardening and follows the constitutive law used by Gattesco and presented by Queiroz et al. [137] for the strain-hardening branch. The von Mises yield criterion with isotropic hardening rule is also utilised for the reinforcing steel. An elastic-linear work hardening material is considered, with tangent modulus being equal to 1/10,000 of the elastic modulus, to avoid numerical problems. The values measured in the experiments made by Chapman and Balakrishnan [140] were utilised to model the steel components material properties.

The concrete slab response was modelled by a multilinear isotropic hardening relationship, utilising the von Mises yield criterion coupled with an isotropic work hardening response. The uniaxial behaviour was modelled by a multilinear linear total stress–total strain curve, starting at the origin, with positive stress and strain values, taking into account the concrete compressive strength (f_c) corresponding to a compressive strain of 0.2%. The stress–strain curve also assumes a total increase of 0.05 N/mm_2 in the compressive strength up to the concrete strain of 0.35% to avoid numerical problems due to an unrestricted yielding flow. The concrete element shear transfer coefficients considered were: 0.2 (open crack) and 0.6 (closed crack). Typical values range from 0 to 1, where 0 represents a smooth crack (complete loss of shear transfer) and 1 a rough crack (no loss of shear transfer). The crushing capability of the concrete element was disabled to improve convergence. The concrete slab compressive strength was considered as the actual cylinder strength test value. The concrete tensile strength and the Poisson's ratio were assumed as 1/10 of its compressive strength and 0.2, respectively.

The concrete elastic modulus was calculated according to Eurocode 4 EN 1994 [64]. The model allows for any pattern of stud distribution to be taken into account. In all analyses, the number/spacing of connectors used in the experimental programmes is utilised. The load–slip curves for the connectors were utilised (obtained from available push-out tests) by defining a flange of force values and relative displacements (slip) as inputs for the nonlinear springs. These springs were modelled at the steel–concrete interface.

Concentrated loads were incrementally applied to the model by means of a displacement control. For the convergence criterion, the L2-norm (square root sum of the squares) of displacements was considered. The tolerance related to this convergence criterion (CNVTOL command of ANSYS) and the load step increment were varied to solve potential numerical problems. Whenever the solution did not converge for the set of parameters considered, as far as load step size and convergence criterion were concerned, the RESTART command was utilised in conjunction with the CNVTOL option, Queiroz et al. [137].

Two limits were used to define the ultimate load for each finite element study: a lower and an upper bound, corresponding to concrete compressive strains of 0.2% and 0.35%, respectively. These two limits define an interval in which the composite beam failure load is located. For the shear connector, if the failure point is located below the lower bound of concrete then the mode of collapse of the composite beam is considered as being a connector failure. Alternatively, if the connector failure point is located above the upper bound of concrete, the collapse mode is assumed as being concrete crushing. For the intermediate case, where the connector failure point lies between the lower and upper bounds of concrete, the mode of failure could be either of them. Therefore, the studied finite element model is able to predict the failure modes related to either slab crushing or connector failure.

The model was validated by comparisons against Queiroz et al. [137]. The numerical tests made by Queiroz et al. [137] portray the studied composite system response. The beams spanned 5490 mm with an I-shaped steel member 305 mm deep (12″ × 6″ × 44 lb/ft BSB) and a concrete slab 120 thick × 1220 mm wide.

The number and type of studs, as well as the steel and concrete strengths, varied according to the tested composite beam. The slab was longitudinally reinforced with four top and four bottom 8 mm bars. The transverse reinforcement used top and bottom bars of 12.7 mm @ 152 mm centres and 12.7 mm @ 305 mm centres, respectively. The tensile strength, the Young's

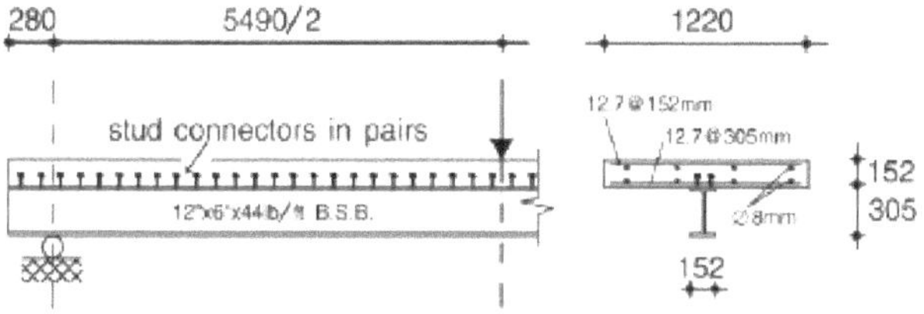

Fig. 3.146 Simply supported beam layout, Queiroz et al. [137].

modulus, and the Poisson's ratio of the reinforcing steel bars were 320 N/mm^2, 205,000 N/mm^2, and 0.3, respectively. A full description of these beams is depicted in Fig. 3.146.

The model initially uses 25 pairs of stud shear connector with a 19 mm diameter, spaced every 114.4 mm, totalising 50 shear connectors. Table 3.9 lists the model geometric and material characteristics in this investigation and as well as their respective ultimate loads corresponding to the concrete failure.

Fig. 3.147 depicts a typical load versus vertical deflection curve of the studied configurations as well as the ultimate load limits related to the concrete slab total strains equal to 0.2% and 0.35%. Fig. 3.147 presents an enlargement of the results comparing the different types of adopted shear connectors. From all the investigated configurations the largest initial stiffness was related to the one with nine Perfobond connectors, simulating a full composite interaction. The other configurations with 5 and 3 Perfobond connectors were utilised to represent the partial interaction design. Fig. 3.147 portrays another set of the results comparing the different types of T-Perfobond connectors. It was noted that the simulations using five T-Perfobond (T-Perf-120-IN-12) shear connectors and three T-Perfobond (T-Perf-120-IN-12-16) presented almost the same initial stiffness values despite using different numbers of shear connectors.

The results also indicate that the proposed Perfobond and the T-Perfobond connectors had similar ultimate loads and structural behaviour,

Table 3.9 Test Settings and Results

			Load (kN)	
Connector Type	**Connector Number**	***s* (mm)**	**0.20%**	**0.35%**
Perfobond	9	343.2	100.01	121.27
	5	686.25	56.88	64.42
	3	1372.5	35.43	39.92
T-Perfobond (IPN 340)	5	686.25	63.14	66.90
	3	1372.5	39.97	44.51
T-Perfobond (W200x46.1)	3	1372.5	34.11	39.04
Stud	50	114.4	67.29	98.39

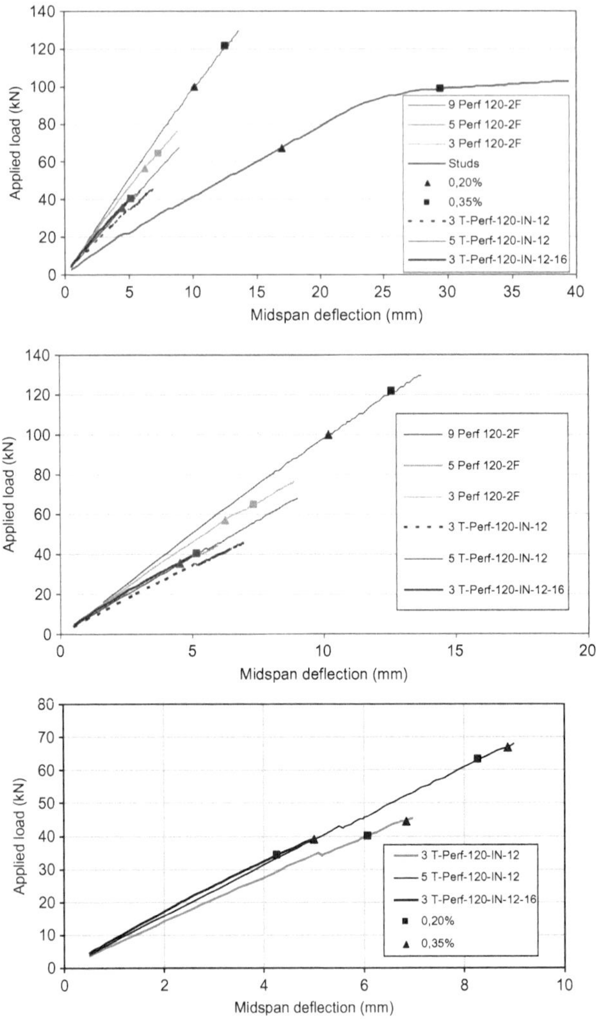

Fig. 3.147 Load versus midspan deflection.

although their related push-out test curves were very different, as it can be observed in Fig. 3.148. A simple inspection of these curves shows that the T-Perfobond connector ultimate load is twice the related collapse load of the equivalent Perfobond shear connector. A possible reason for the similar structural response of these two different connectors can be associated to the fact that in the composite beams the bending response prevails instead of the push-out tests direct shear. Another reason for this similar trend can be understood from the fact that in all the simulations the concrete rupture was directly responsible for the composite beam failure, preventing the

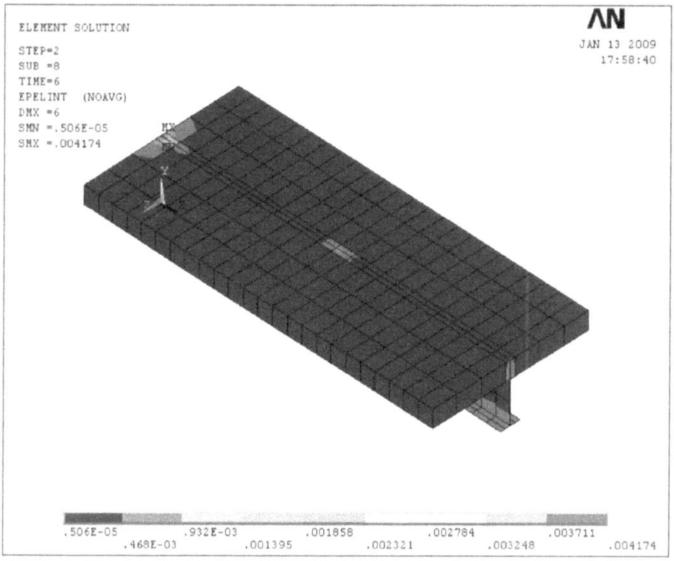

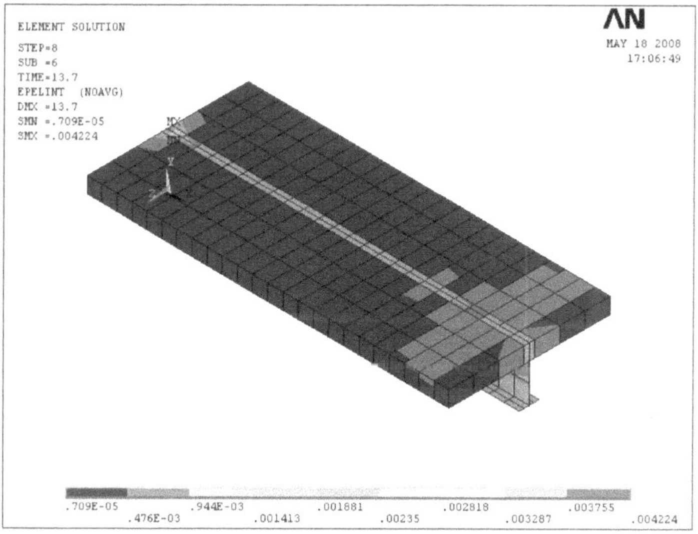

Fig. 3.148 Concrete slab strain distribution—3 and 9 Perfobonds.

shear connector to reach their ultimate load capacities. Figs 3.148–3.151 depict the composite slab total strain distribution for the various studied models close to the ultimate loading stage.

A three-dimensional finite element model of composite beams has proved to be effective in terms of forecasting the load–deflection response for beams under concentrated or uniformly distributed loads, longitudinal slip at the steel–concrete interface, shear force carried by the connector,

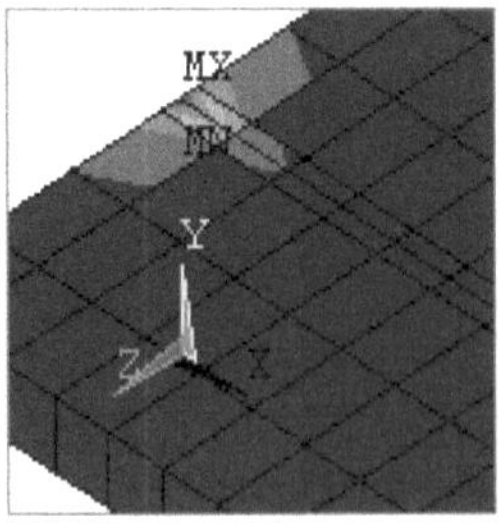

Fig. 3.149 Concrete slab total strain distribution—most stressed regions of Perfobond and T-perfobond models.

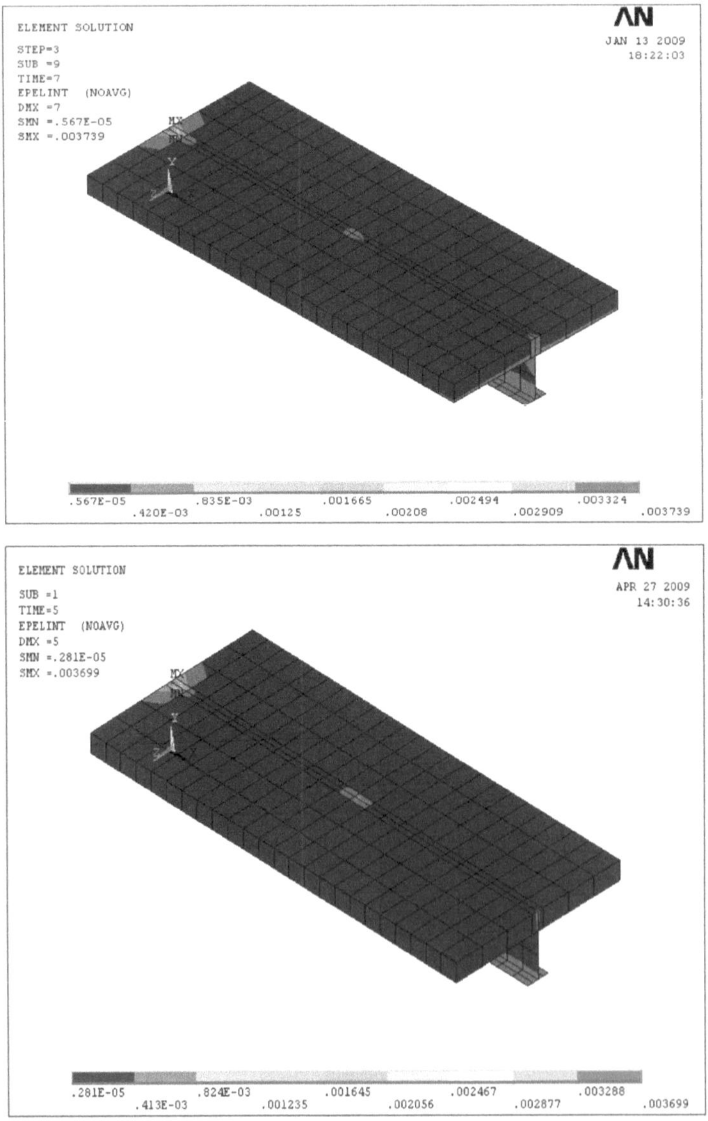

Fig. 3.150 Concrete slab strain distribution—3 T-Perfobonds (IPN 340 and W200x46,1).

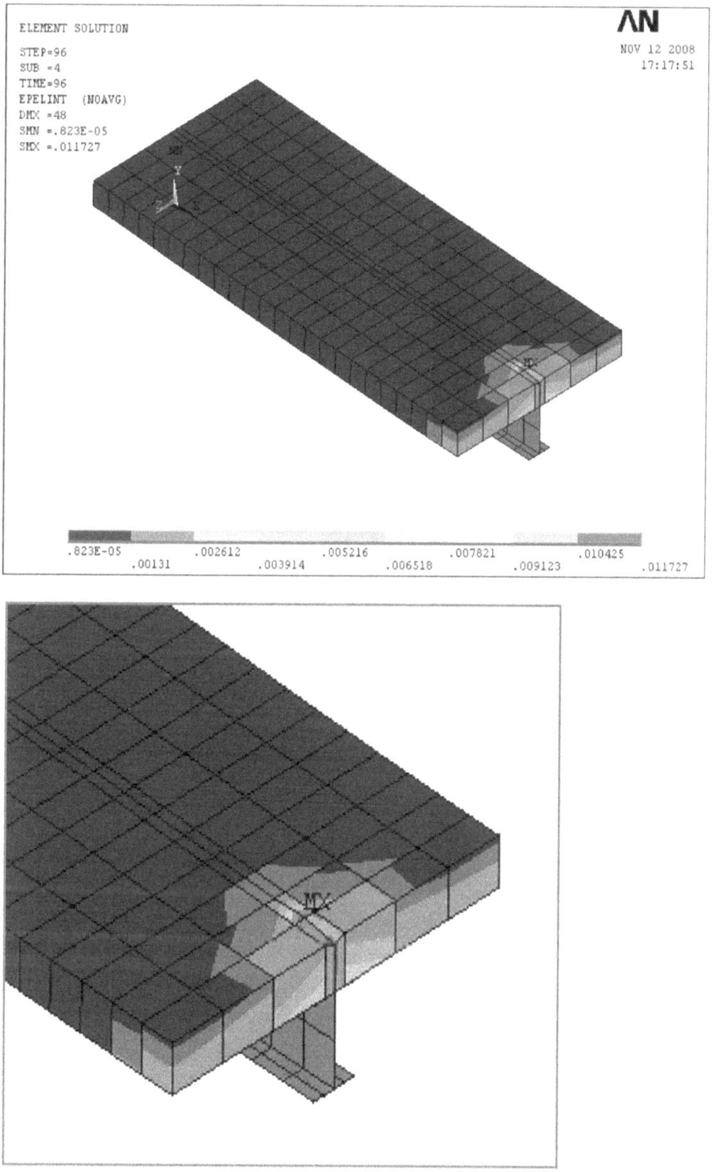

Fig. 3.151 Concrete slab strain distribution—50 Studs.

and the mode of collapse (connector failure or concrete crushing). It is also capable of studying beams with either full or partial shear interaction.

The proposed three-dimensional model provided the opportunity to develop insights that would be virtually impossible using experiments, due to costs and, especially, the dispersion of material properties that inevitably occurs in laboratory work. The advantage of Perfobond and T-Perfobond connectors is closely related to their fatigue response, since their detailing is much more convenient than the stud detailing, leading to advantages in their use for structures subjected to traffic or seismic loads.

The modelling composite beams strategy used, as well as in modelling previously presented, spring interface elements calibrated with shear connectors push-out test results. This made it possible to assess the effects of their spacing and distribution along the composite beam. In addition, the model allowed an assessment of the distribution of forces on the connectors so as to evaluate their real performance along the beam.

The adopted stop control was once more based on comparison to the ultimate loads, whether this controlled the composite beam performance, or according to a range of minimum and maximum variation of concrete compressive deformations. This was made because the concrete solid element crushing and cracking capabilities had to be suppressed to overcome serious problems of convergence in regions near the investigated structural system ultimate load. Concentrated loads were simulated with a displacement control and demanded changes in tolerances and convergence criteria to generate reliable results, varying from case to case. For the first time Perfobond connectors and TPerfobond were investigated to determine whether concentrations of very high stress do not occur in composite beams due to their reduced number along the span when compared to the stud connectors.

T-Perfobond connectors presented higher load capacity and stiffness than Perfobond connectors, and are produced from laminated sections, rather than from specifically fabricated components. In spite of these advantages, their response is still not ductile, and therefore the design of composite girders incorporating this connector's geometry has to be elastic. The work also included very simple elastic modelling of the TPerfobond connector to increase their flexibility without reducing their strength. This modelling has effectively contributed to the specification of the TPerfobond connector flange thickness used in the second phase of experimental series leading to performance compatible with connectors that meet the Eurocode 4ductility criterion and enabling the possibility of their design in the plastic range and their use uniformly spaced along the beam.

3.5.14 Structural Modelling of Push-Out Tests Made With Studs and Perfobond Shear Connectors

Shear connector responses have been extensively investigated due to the fact that its strength is influenced by many variables as for example: concrete resistance, connector geometry, concrete region geometry near the connector, etc. Considering these factors, push-out tests are recommended to assess the shear connector ultimate load and ductility.

Additionally as push-out tests remain a costly and time consuming option, analytical procedures that can forecast the nonlinear response and the ultimate capacity of the push-off test were made to replace a vast number of experiments once the calibration of the numerical method has been performed, fine-tuned experiments. The main aim of the present investigation is to present a nonlinear three-dimensional finite element model developed with the aid of Ansys program [32] to evaluate the ultimate resistance of headed shear stud connectors. The numerical results were calibrated against experimental and numerical results by Lam [145], Vianna [71], and analytical formulations provided in Eurocode 4 [64].

Push-out tests are made utilising a short steel beam length connected to two concrete slabs by means of shear connectors, Fig. 3.152. The present investigation first utilised push-out tests conducted by Lam [145].

These tests were made of a UC 73 (254 × 254 equivalent to a W10 × 49), two concrete slabs attached to the steel beam flanges with dimensions of 619 mm (width), 469 mm (height), and 150 mm (thickness) for each slab. The distance between the connector to the slab edge was 200 mm. The 19 mm stud connector had 100 mm height, Fig. 3.152. The test procedure adopted the Eurocode 4 [64] recommendations.

The work begins with a model with the full connector geometry, using the SOLID45 element, Figs 3.152 and 3.153.

Fig. 3.154 presents the boundary conditions adopted in finite element model: at the concrete slab base of the push-out test and at the concrete slab model to reaction, slab interface restrictions were applied on all nodes in the load direction, y axis, to prevent the structure displacement. Furthermore, in some nodes with distinct coordinates, constraints were applied on the X and Z direction, concrete slab length and width respectively, to avoid the base rotation. The restriction on the X and Y directions was not applied in all nodes of the base having been only adopted to prevent rigid body displacements; above the steel profile top, a joint element was inserted which had its displacements coupled to the profile top of the push-out test, which allowed the application of a displacement on a single point facilitating the results interpretation.

Restrictions were applied on the region between the flange and the web of the steel profile due to the presence of the web since it was not included in the model. The web, which is located in the same plane of the connector, modelling would require the use of very small elements, which would increase the computational force. A coupling in the X and Z axis directions of the nodes present in the region between the steel profile and the concrete slab was also

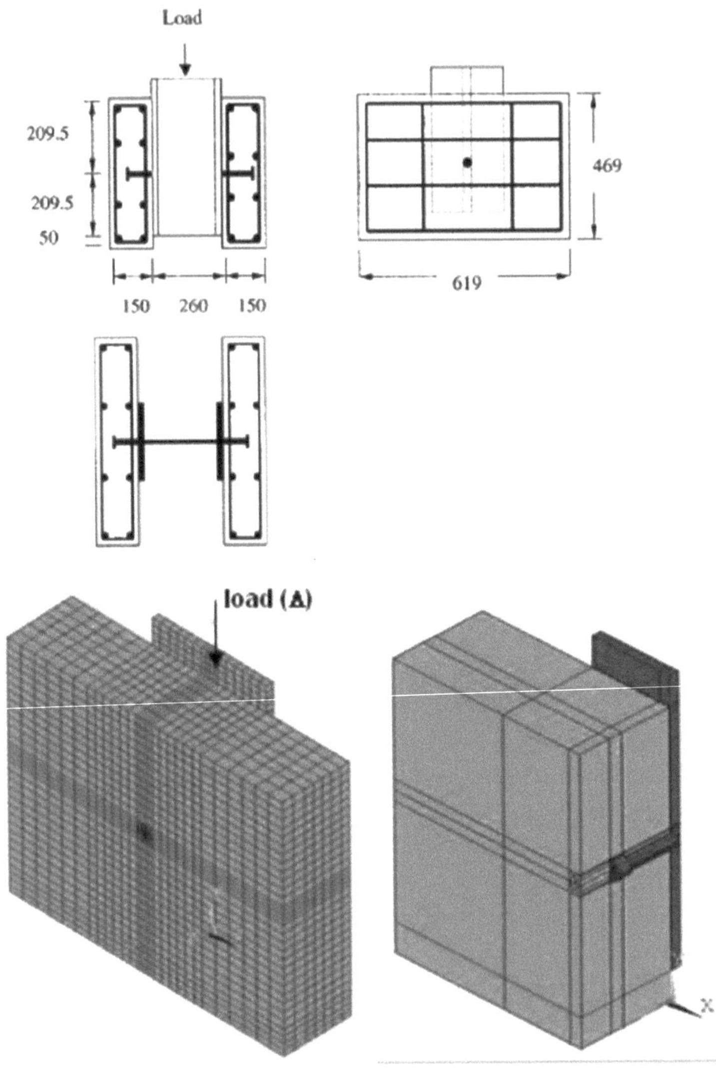

Fig. 3.152 Push-out test, Lam [145], complete and half model.

made. The coupling on the *X* and *Z* axis, different from the load application axis *Y*, were made to prevent the concrete slab and steel profile to, occupy the same physical space. The Eurocode 4 [64], push-out test procedures start with a series of load and unload cycles to induce a slip in the connector and break the steel to concrete chemical bond. In this modelling such effect was simulated by a coupling on the lower interface between the connector and the concrete, Fig. 3.155. On the interface between the steel profile and the connector, a coupling was also made to simulate the weld present in this region.

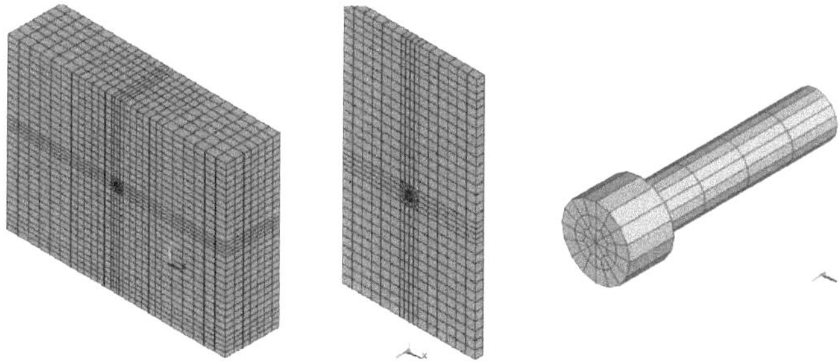

Fig. 3.153 Concrete slab, steel profile, and stud connector.

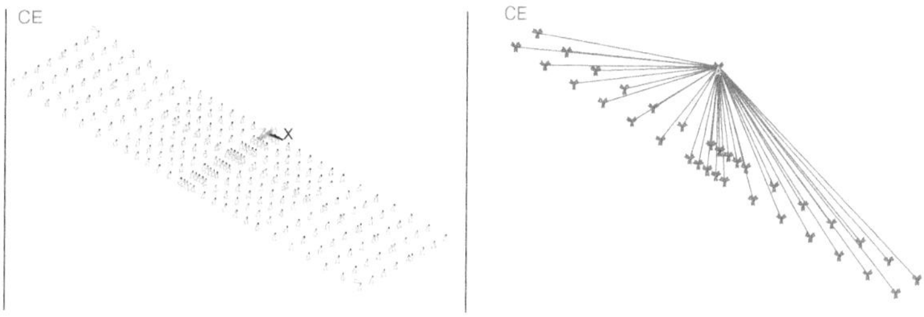

Fig. 3.154 Adopted boundary conditions: restrictions on concrete slab bottom; steel profile top.

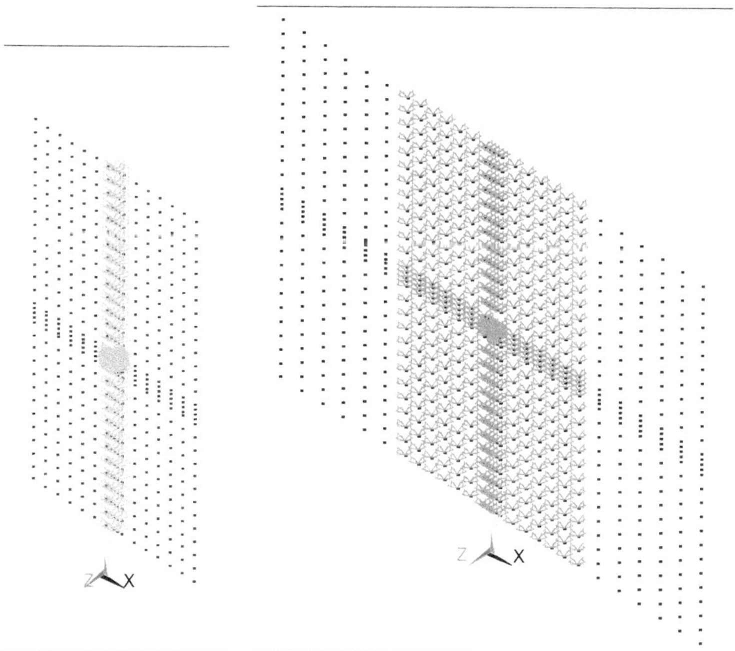

Fig. 3.155 Adopted boundary conditions: restrictions between flange and web; profile to concrete slab coupling.

The contact surface was only used in the upper region of the connector. This approach is justified by the impossibility of coupling this region because such a measure would result in tension stresses nonexistent in concrete. The adopted materials properties were the same present in Lam tests [145], using the following equations for determining the concrete properties:

$$\varepsilon_{yc} = 0.00024\sqrt{f_{cu}} \quad f_{yc} = 0.8 \cdot f_{cu} \quad E_c = f_{yc}/\varepsilon_{yc} \tag{3.11}$$

S275 steel was adopted on the profile's flange while the connector utilised a 200 GPa Young modulus and a 470.8 MPa yield stress, Lam [145]. The steel bars were introduced in the model as smeared in the concrete, as reinforces bars percentage rates with angles and data allowed by the SOLID65 element. They were used as described by Lam [145], i.e. 10 mm diameter bars: six in the horizontal and six in the vertical direction per layer. It could be observed that Lam [145], subdivided each concrete slab in two layers. The model calibration assessed four tests, SP1, SP2, SP3, and SP4. All the tests used a 19 mm diameter stud connector 100 mm high. The adopted concrete compressive strength was: 50, 20, 30, and 35 MPa, respectively. Table 3.10 portrays the results obtained by Lam [145], experimental and numerical models, the Eurocode 4 [64], predictions and those obtained with the proposed finite element model.

The results were considered satisfactory due to the little difference between them and those obtained by Lam [145], experimental and numerical analyses and by the Eurocode 4 [64], Fig. 3.156.

Figs 3.157–3.160 presents the finite element model results following an f_{ck} magnitude ascending order, in terms of comparative illustrations of the connector and steel profile von Mises stress distributions and concrete plastic deformation at its rupture. It can be noted that in addition to the connector von Mises stress distributions its deformed and original configuration are also illustrated where the latter is represented by black lines. This representation seeks to facilitate the visualisation of the connector deformations.

In all cases, the concrete regions exceeded the 0.2%–0.35% limit strain range. A, particular attention should be made to test SP1 in which this region is small and the collapse mode is related to the connection failure.

Table 3.10 Comparison of Eurocode 4 [64], Developed Model, and Lam [145] Experimental and Numerical Results

Test	f_{ck} (MPa)	Lam Experiments (kN)	FE (kN) Lam	EC4 (kN)	FE Ansys (kN)
SP1	50	130.4	116.6	106.8	114.43
SP2	20	71.6	74.4	60.4	73.22
SP3	30	93	91.8	81.9	87.47
SP4	35	102	97.3	91.9	90.02

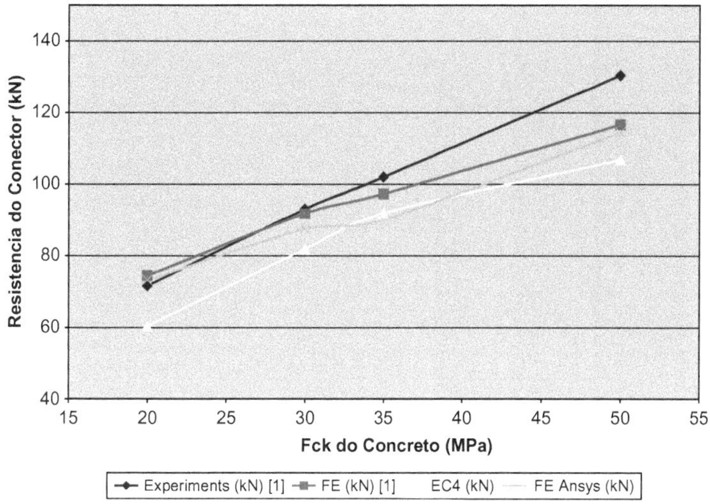

Fig. 3.156 Model comparisons: Eurocode 4 [64], Lam [145] experimental and numerical results.

Similarly to Lam [145] tests the models for SP3 and SP4 test is presented a concrete failure associated to the yield of the connector. The SP2 test presented a large region with excessive concrete deformations indicating that the failure was associated to the concrete rupture.

The collapse modes and load capacities associated to the Eurocode 4 formulae [64], were close to the proposed models since the load supported by the connector was 106.8 kN. Despite the fact the Eurocode 4 [64], only predicted the connector failure for a 50 MPa f_{ck}, Lam [145] notes in his tests results that the connector has yielded. Thus, the model calibration was considered appropriate since all collapse modes were consistent with those found by Lam [145], in his numerical and experimental tests.

With the calibrated results a subsequent study considering the existence of an additional reinforcing bar below the connector, apart from the initially adopted reinforced bars smeared in the concrete, was made. This study used the results of the model with 16 mm diameter stud, 75 mm height and a concrete with a 25 MPa f_{ck}. Initially, the new bar was simulations made by increasing the reinforcing bars rate smeared in the concrete elements close and bellow the connector as indicated in Fig. 3.161.

This strategy had little effect in terms of resistance since only an increase of just 4 kN, 2.24% of the resistance, was reached i.e. from a load capacity of 63.49–65.73 kN. Fig. 3.162 presents these load versus displacement curves where it is even more evident for the little change between the simulations. Fig. 3.163 presents shear connectors at collapse for the two investigated models

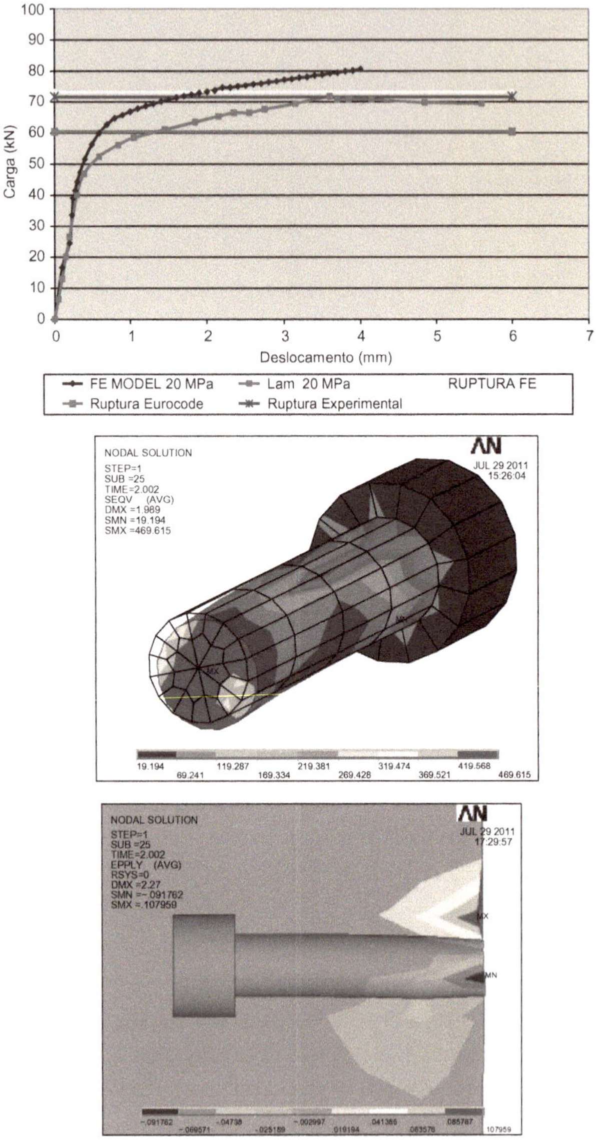

Fig. 3.157 Load versus displacement curve, connector von Mises stress distribution, concrete plastic strain distribution, SP2 test, $f_{ck} = 20$ MPa.

where distinct von Mises stress distributions of the connector are observed. Fig. 3.164 illustrated the collapsed slab plastic deformation where the smaller increase of strains in the models with the additional bars can be observed.

A second strategy was used including a discrete 10 mm reinforcing bar with the aid of the PIPE59 element, Fig. 3.165.

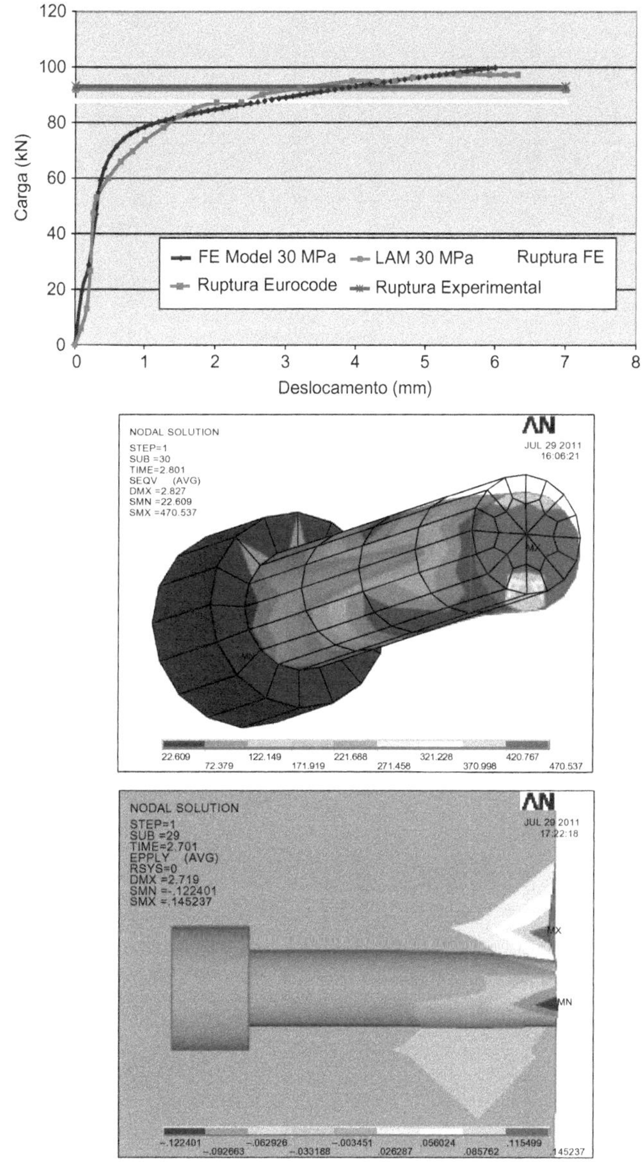

Fig. 3.158 Load versus displacement curve, connector von Mises stress distribution, concrete plastic strain distribution, SP3 test, $f_{ck} = 30$ MPa.

Unlike the previous case, the use of a discrete bar in this region altered the resistance increasing it by 6.78 kN or 10% above the initial connector model i.e. from a load capacity of 63.49 kN to 70.27 kN. Fig. 3.166 presents the load chart 3166 versus where displacement is seen gaining strength in this

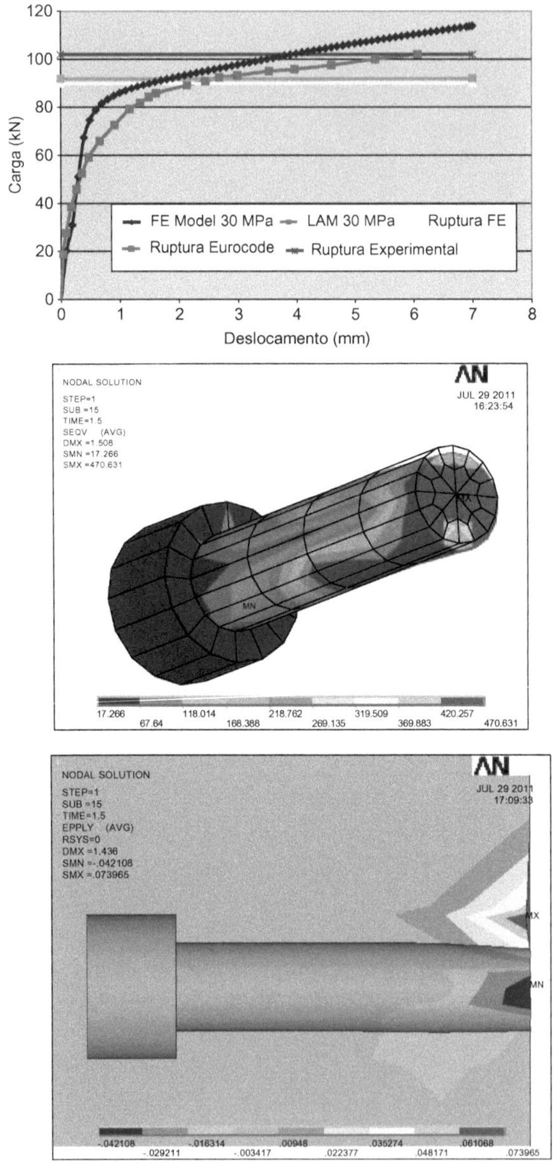

Fig. 3.159 Load versus displacement curve, connector von Mises stress distribution, concrete plastic strain distribution, SP4 test, $f_{ck}=35$ MPa.

simulation. Fig. 3.166 presents these load versus displacement curves where the connection resistance gain is observed.

The shear connector modelling has three crucial elements for its suitable performance, the profile, the connector, and the concrete slab, where the

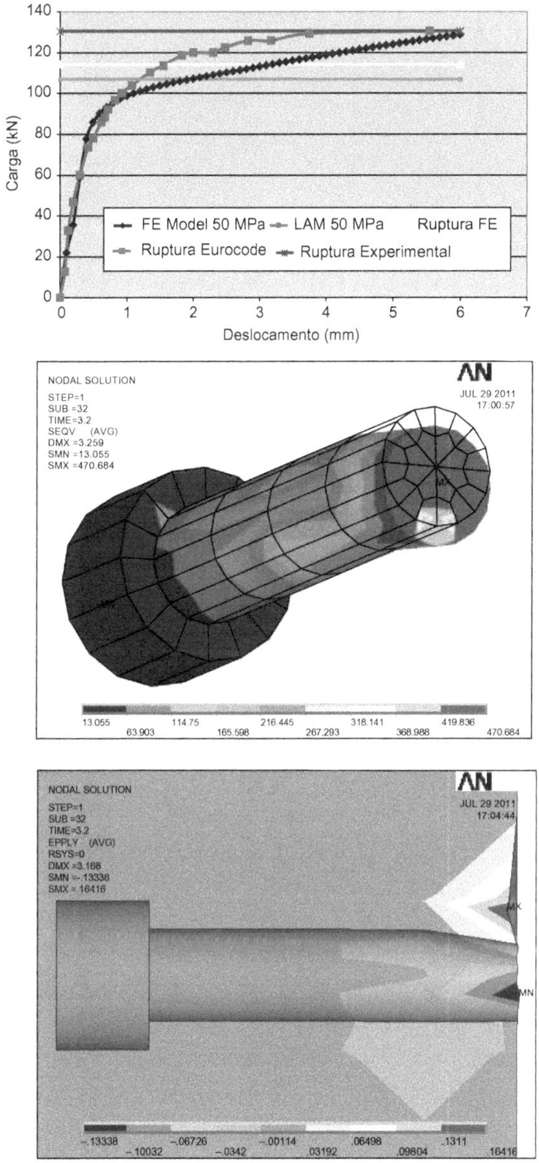

Fig. 3.160 Load versus displacement curve, connector von Mises stress distribution, concrete plastic strain distribution, SP1 test, $f_{ck} = 50$ MPa.

latter becomes even more critical when associated to smaller strengths. It can be observed that in this case the concrete plastic strains distribution was similar to the initial model for the 16 mm stud connector 75 mm height and a concrete slab compressive strength of 25 MPa.

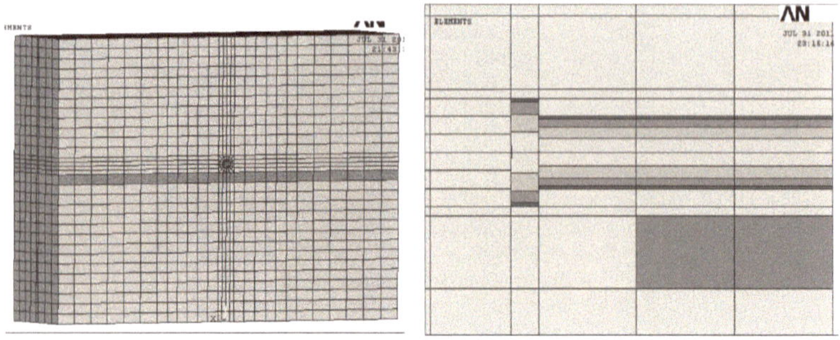

Fig. 3.161 Model with smeared reinforcements simulating the extra reinforcing bars.

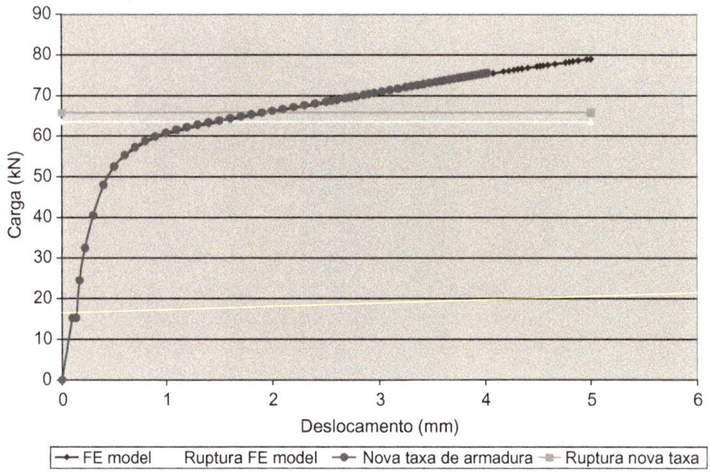

Fig. 3.162 Load versus displacement curve for tests with various reinforcing bars distributions.

A second stage of this investigation contemplated the modelling of other types of connectors. Based on experimental results presented by Vianna [71], push-out tests were modelled with Perfobond shear connectors. The model in question was made with the S275 HEA 200 profile ($f_y = 275$ MPa; $f_u = 430$ MPa), while the Perfobond connectors were used 13 mm thick S355 ($f_y = 355$ MPa; $f_u = 510$ MPa) plates, Fig. 3.167. The reinforcing bars were made of 10 mm diameter S500 steel ($f_y = 500$ MPa), Fig. 3.167. This figure also presents the concrete slab geometry: 120 mm thick, and with height and width of 650 mm and 600 mm, respectively.

The concrete properties required a sharper analysis because, from the adopted 28 MPa f_{ck} and from past experience, it was recognised that the smaller the concrete resistance is, more complex the model would be. This led to the adoption in the proposed model of the concrete characteristic curve present in the Eurocode 2 [146], Fig. 3.168.

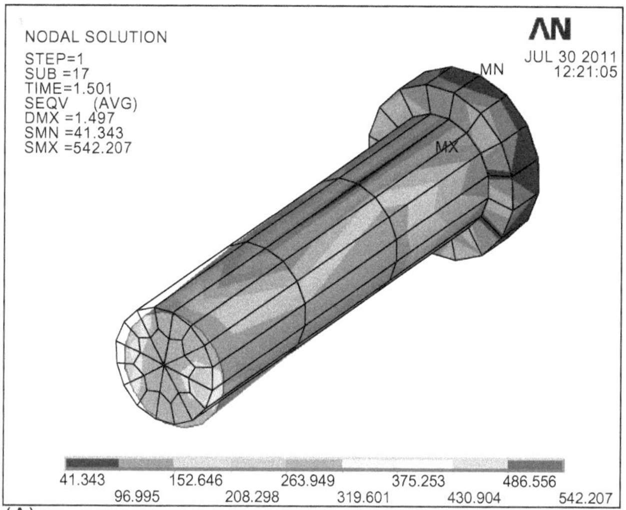

(A)

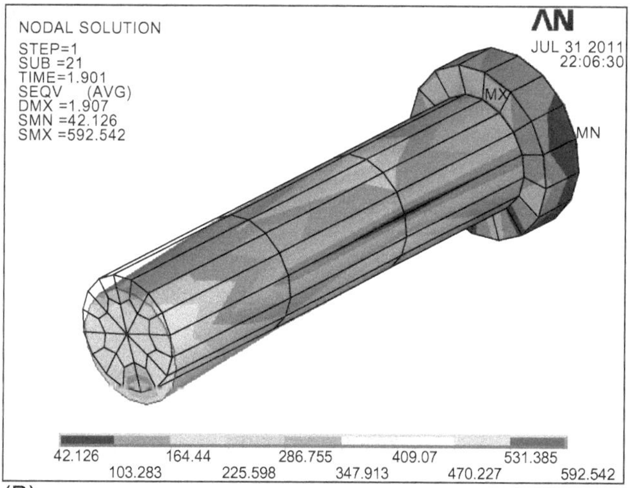

(B)

Fig. 3.163 Tension of von Mises stress distribution: (A) Connector original model (16 mm × 75 mm Stud, $f_{ck}=25$ MPa); and (B) model with altered reinforcing bars.

Fig. 3.169 presents the volumes present in the finite element model while Fig. 3.170 illustrates the finite elements of each part of the model: slab, connector, and profile.

The adopted boundary conditions in the present analysis were based on the conditions present in the experiments and their respective symmetry conditions, being in part similar to the model with stud connectors. At the concrete slab base and the push-out test, at the interface concrete slab to the reaction slab restrictions in the load direction, i.e. y axis, were applied on all nodes to prevent the structure displacement. Furthermore, in some

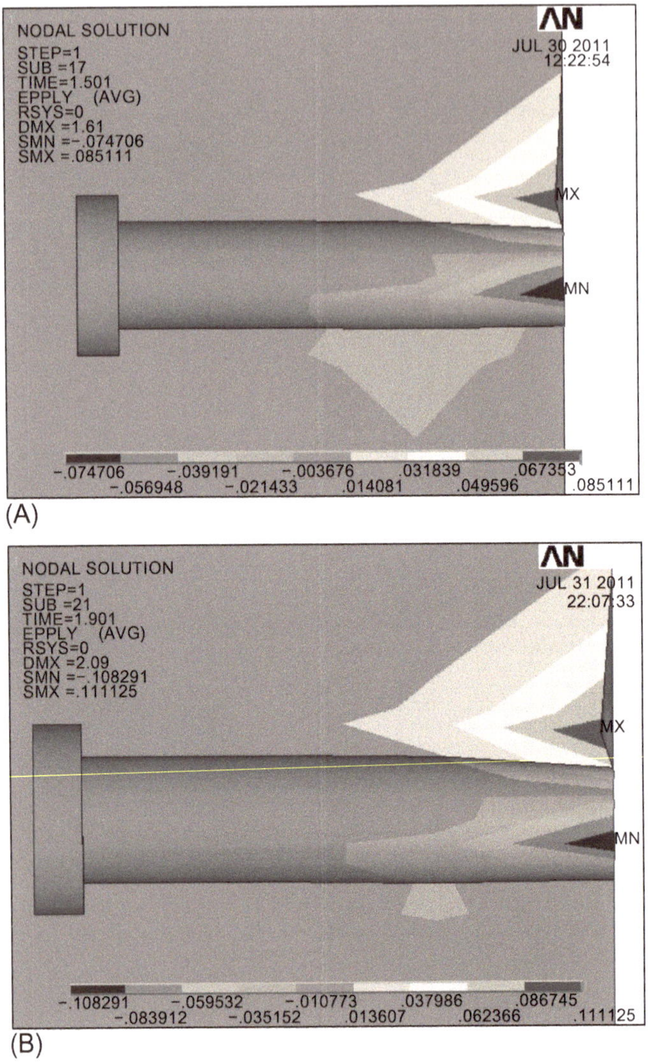

Fig. 3.164 (A) Plastic deformation at concrete region around the connector, original model (16 mm × 75 mm Stud, $f_{ck} = 25$ MPa); and (B) model with altered reinforcing bars.

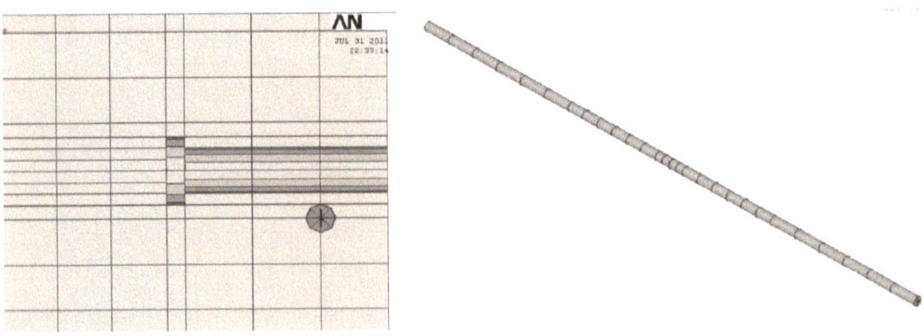

Fig. 3.165 Concrete slab elements, extra reinforcing bar and reinforcing bar mesh.

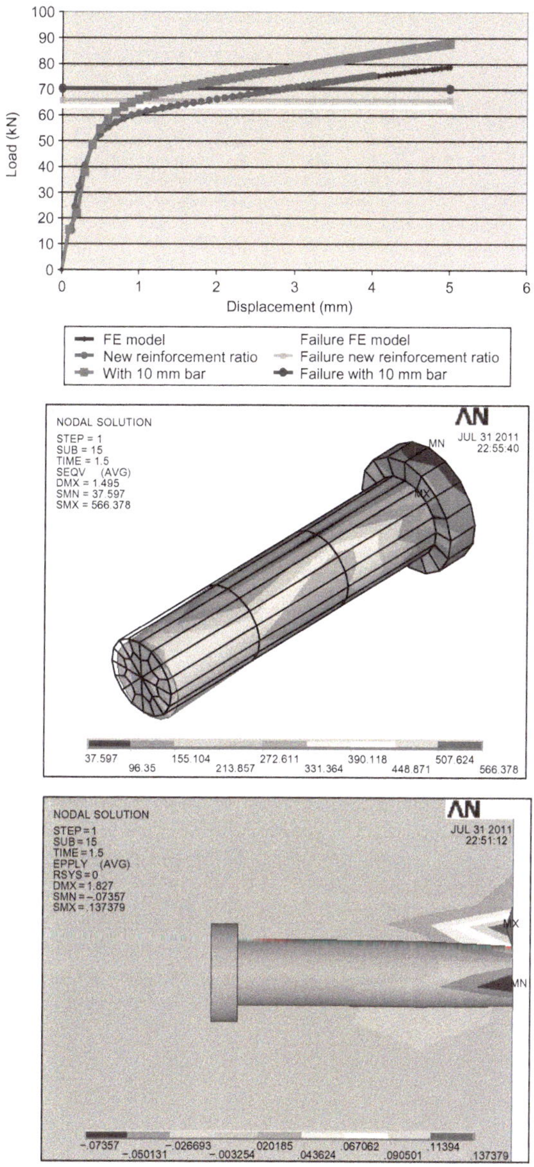

Fig. 3.166 Load versus displacement curve, connector's von Mises stress distribution and concrete plastic strain distribution, SP1 test, $f_{ck} = 50$ MPa.

nodes with distinct coordinates constraints were applied on the X and Z direction, concrete slab length and width of respectively, to avoid the base rotation. The restriction on the X and Y directions was not applied in all nodes of the base, having been only adopted to prevent rigid body displacements; above the steel profile top a joint element was inserted which had

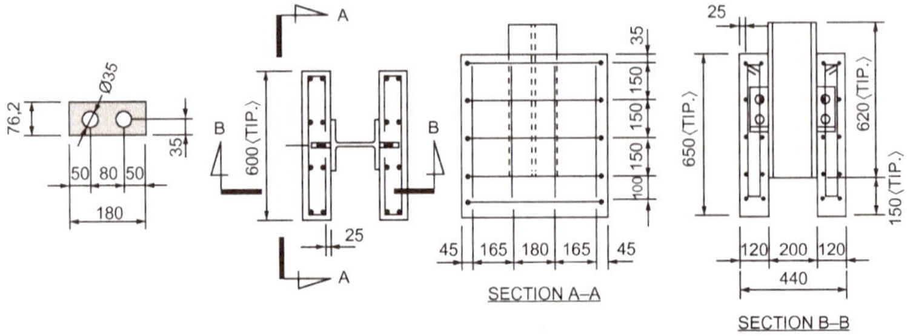

Fig. 3.167 Perfobond connector geometry, push-out test details and reinforcing bars.

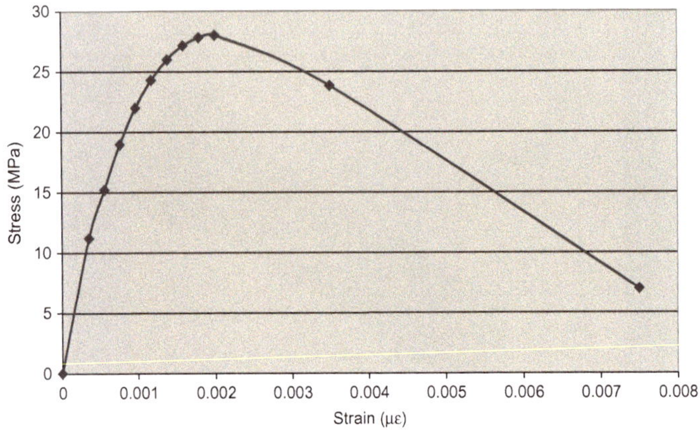

Fig. 3.168 Concrete characteristic curve, $f_{ck}=28$ MPa.

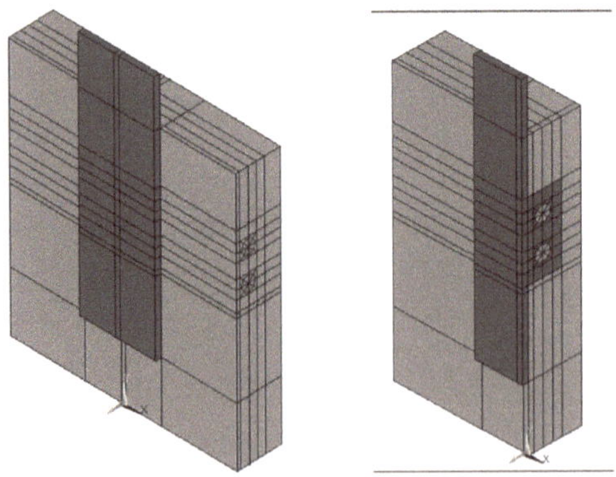

Fig. 3.169 Complete model, connector detail inside the model concrete slab.

their displacements coupled to the profile top of the push-out test, which allowed the application of a displacement on a single point facilitating the results interpretation. Restrictions were applied on the region between the flange and the web of the steel profile due to the presence of the web

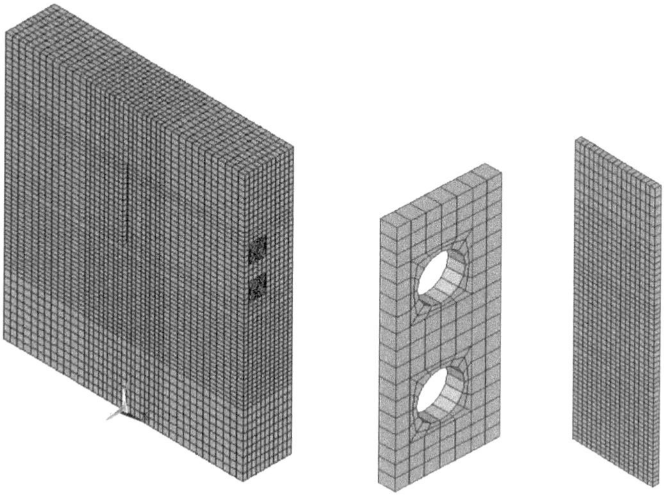

Fig. 3.170 Concrete slab, Perfobond connector details and steel profile.

since it was not included in the model. A coupling in the *X* and *Z* axis directions of the nodes present in the region between the steel profile and the concrete slab was also made. The coupling on the *X* and *Z* axis, different from the load application axis *Y*, were made to prevent the concrete slab and steel profile to, occupy the same physical space. On the interface between the steel profile and the connector a coupling was also made to simulate the weld present in this region.

It can be observed in Table 3.11 that the variation between the values obtained in the experiments and those obtained with the finite element model is rather relevant. A 17.34 kN difference representing a small value of the connector's capacity, 5.35%, was obtained but the rupture was associated to a displacement value smaller than the value recommended in Eurocode 4 [64]. The adopted nomenclature follows the same utilised by Vianna [71], as can be seen in Fig. 3.171, which also features the stress versus deformation curves for the model and for Vianna [71], experiments. It is interesting to note in this curve the reduction of the connector load capacity of the model after the joint failure.

The following Figs 3.172 and 3.173 show the tension of von Mises stresses in the connector and profile at the time of the failure and the associated concrete plastic strains.

Table 3.11 Load Value at Structural Collapse

f_{ck} (MPa)	Type	Vianna [71]	Finite Element Model
28	P-2 F-120	324.10 KN	341.44

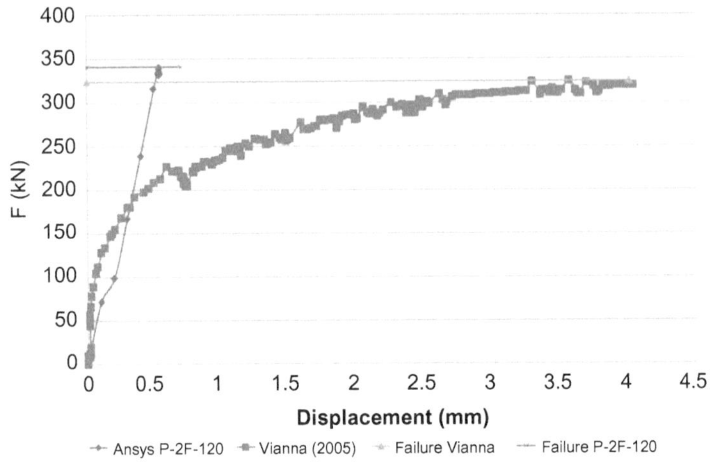

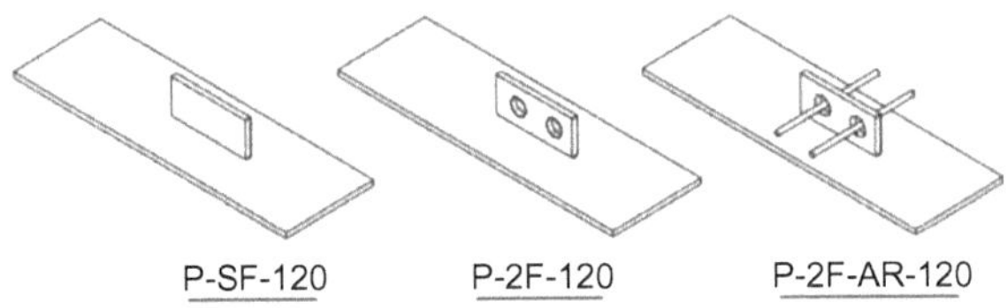

Fig. 3.171 Second series Perfobond connectors load versus displacement curves, Vianna [71].

The work presented by Vianna [71], introduces changes in the use of the Perfobond connectors, motivating the development of numerical models with these characteristics. The following figures show changes in connector that were necessary to adapt the model named by Vianna [71], as P-SF-120. It should be noted that such changes were only geometric keeping the characteristics and boundary conditions presented previously. Fig. 3.174 shows a curve with some interesting aspects, since initially the model with the connector without holes presented a higher strength and results consistency despite its load capacity before failure being less than 326.66 kN when compared to the model with two holes, 341.44 kN. The explanation for this result may be related to the difficulty in modelling circular shapes with nodes suitable coordinated to rectangular shapes, which in terms of finite element models can result in an energy loss in these regions especially if any element is being stressed above their capacities.

Figs 3.175 and 3.176 present the von Mises stress distributions at the profile and in the connector, at the time of the failure and the associated concrete plastic strains in the applied force direction.

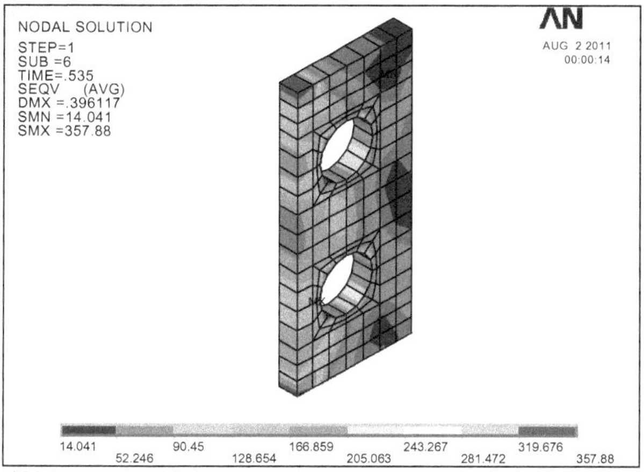

(A)

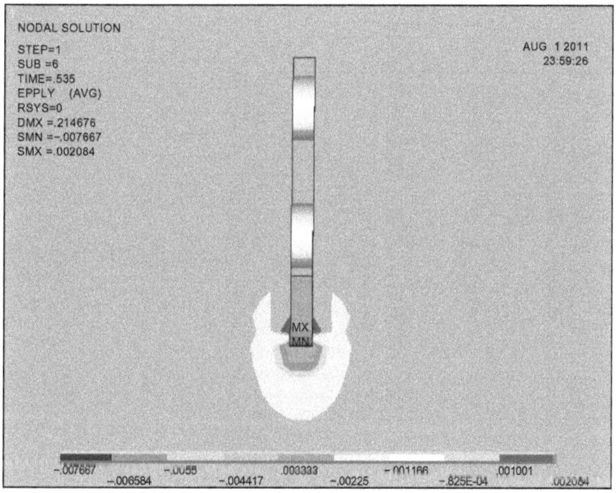

(B)

Fig. 3.172 Profile von Mises stresses at: (A) Perfobond connector; (B) concrete plastic deformation at collapse, $f_{ck} = 28$ MPa.

The next studied model included a 10 mm reinforcing bar in each Perfobond hole as shown in Fig. 3.177.

Fig. 3.178 compares the investigated models where it can be highlighted that the gain with the addition of reinforcing bars was only noticed in terms of load capacity at failure, i.e. 400.26 kN, while the result on the model without the bars was 341.44 kN.

Fig. 3.179 depicts the connector von Mises stress distributions at the time of the failure and the associated concrete plastic strains in the direction of the applied force.

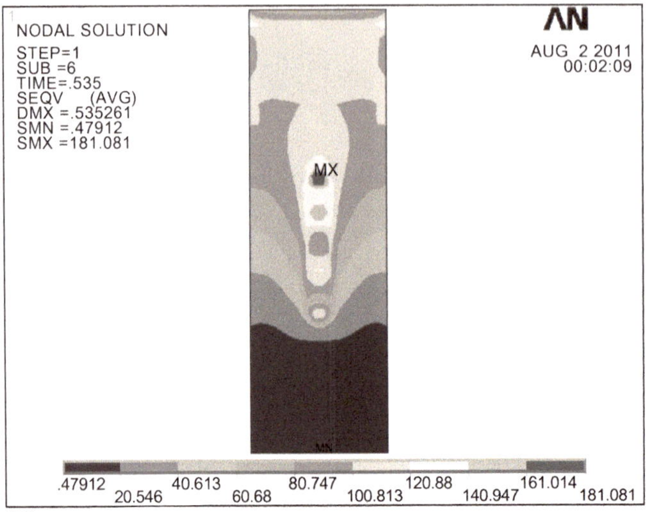

(A)

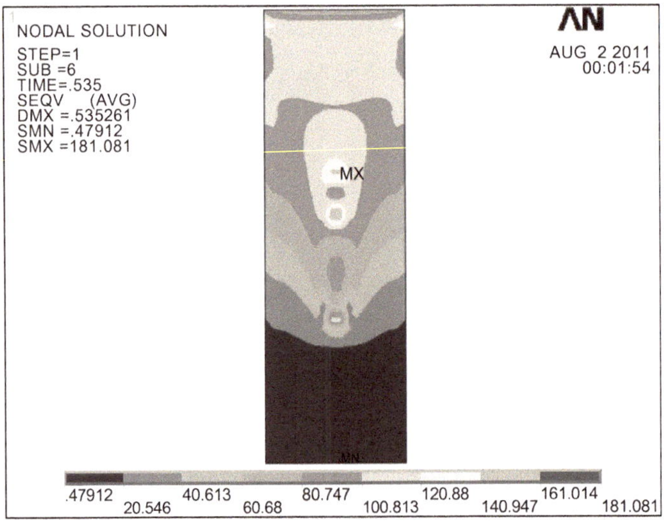

(B)

Fig. 3.173 Profile von Mises stresses at: (A) Perfobond connector interface; (B) interface posterior region at collapse, $f_{ck}=28$ MPa.

Table 3.12 shows the results obtained in finite elements for the P-SF-120 model that is related to the Perfobond connector without holes. The model P-2F-120 corresponds to a Perfobond connector with two holes used for the results validation, and the model P-2F-AIR-120 presented the extra reinforcing bars in the connector holes. The variation, presented in percentage

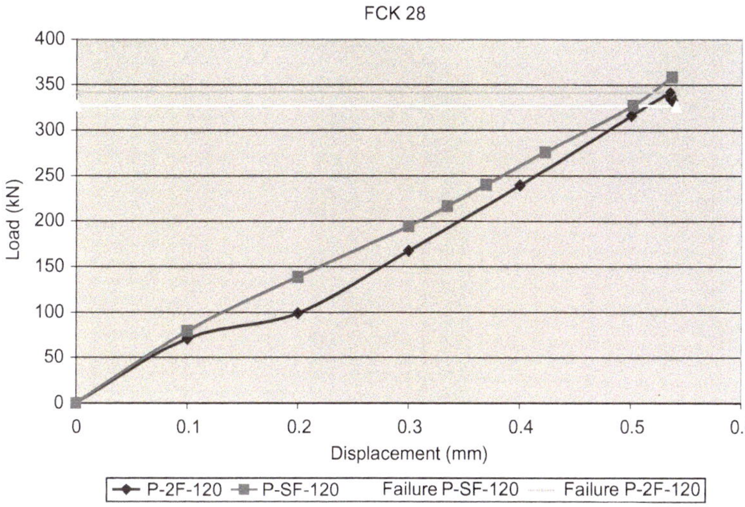

Fig. 3.174 Load versus displacement, Perfobond with two holes and without holes.

terms in the last column, was obtained by the model to the reference model (P-2F-120) ratio.

Despite not having available experiments for all the characteristics adopted in the models, the results indicated a proportional interesting relationship, since, on the basis of the results present in Fig. 3.180, the average variation between models with and without holes is 92.38%/(lowest and highest values of 89.35% and 96.12%). When the variation of the two connector with and without reinforcing bars in the holes is compared an average value of 114.07% is obtained (lowest and highest values of 99.98% and 128.47%).

Push-out test models were developed for stud and Perfobond shear connectors. The modelling adopted a displacement control and support conditions necessary to simulate the effect of the steel web profile, the existing symmetries as well as restricted possible rigid body movements. The head of stud connector was initially not included contributing to a substantial reduction of the concrete confinement in this region by changing their real performance when compared to the experiments used for its calibration.

The contact of the regions between the concrete and the connector and between the profile and the concrete had to be carefully established. The modelling of the additional reinforcing bars was made according to two strategies. The first considered them smeared in a concrete area while the second, which obtained the best results, considered the additional reinforcing bars as discrete bar elements inserted into the reinforced concrete element mesh.

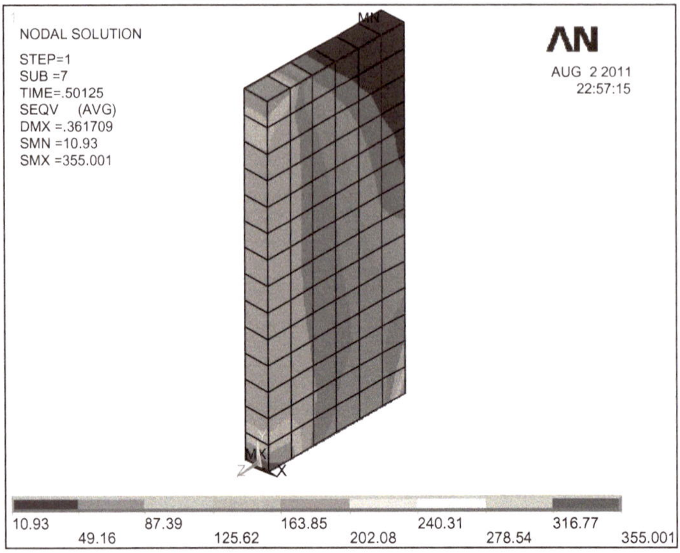

(A)

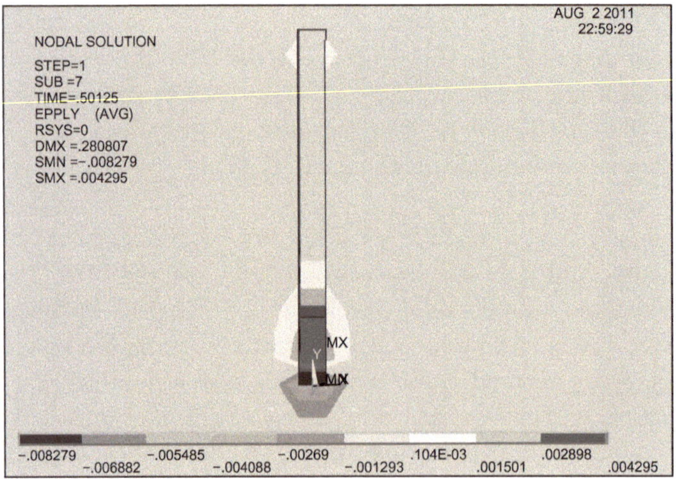

(B)

Fig. 3.175 Profile von Mises stresses at: (A) Perfobond connector without holes; (B) concrete plastic deformation at collapse, $f_{ck} = 28$ MPa.

Once again, convergence was time-consuming and complex, varying from case to case, and demanding several restarts and changes in the criteria and tolerances to achieve the investigated tests ultimate loads. The stop adopted criteria was based, once again, in a comparison with a minimum and maximum range of variation of concrete compressive strain limits. This

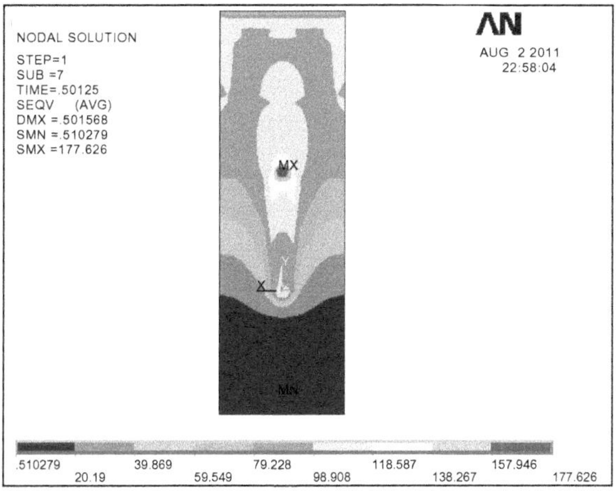

(A)

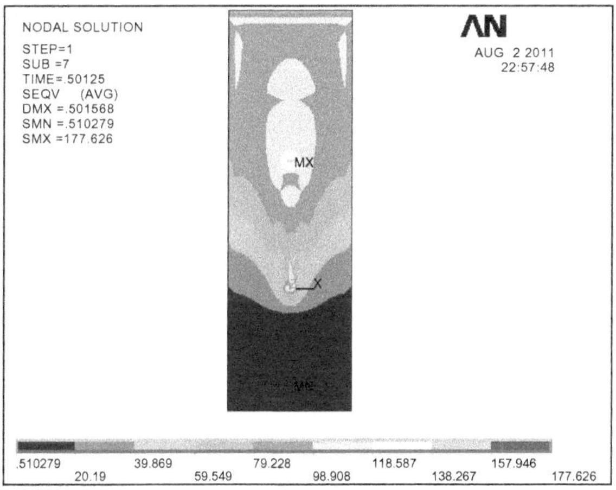

(B)

Fig. 3.176 Profile von Mises stresses at: (A) Perfobond connector without holes interface; (B) interface posterior region at collapse, $f_{ck} = 28$ MPa.

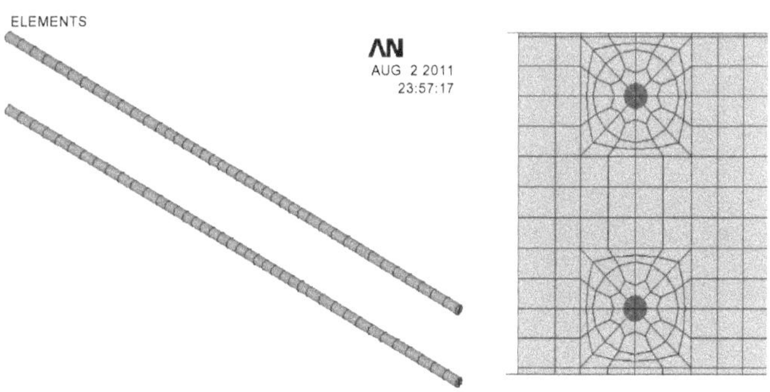

Fig. 3.177 Model with reinforcing bars and Perfobond with reinforcing bars in the connectors' holes detail.

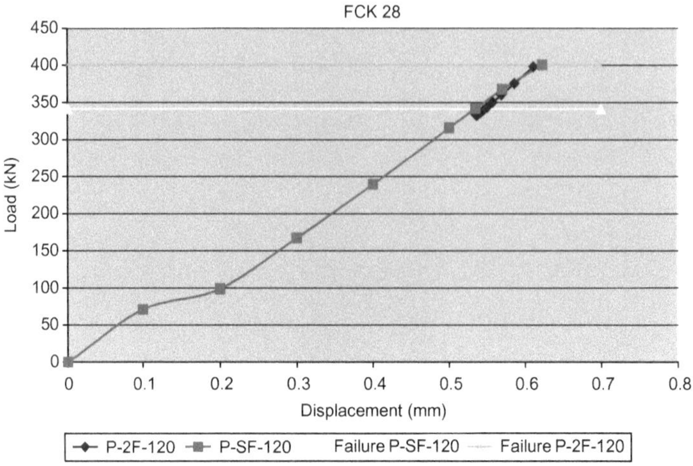

Fig. 3.178 Load versus displacement, Perfobond with reinforcing bars at the connectors' holes.

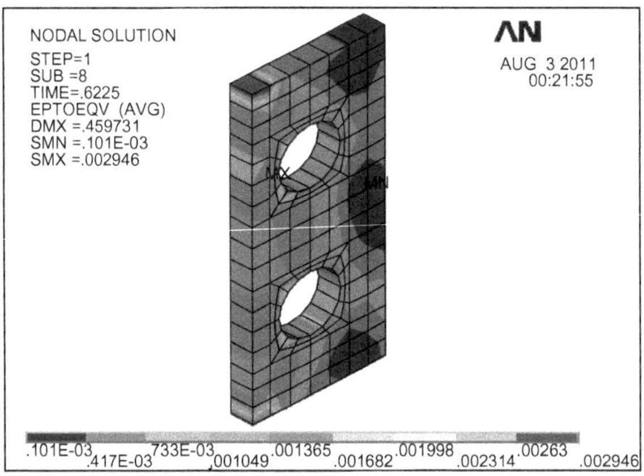

(A)

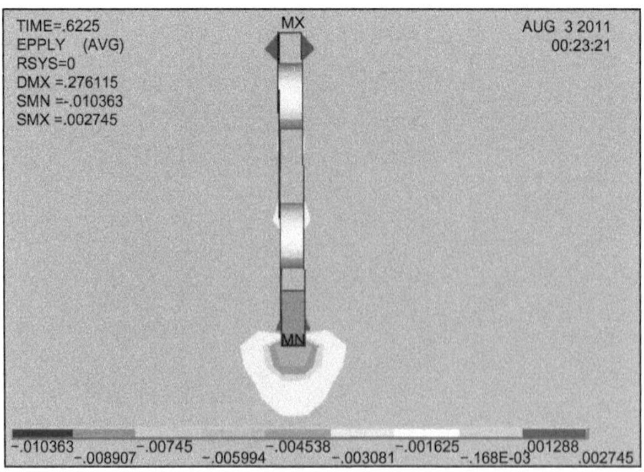

(B)

Fig. 3.179 Profile von Mises stresses at: (A) Perfobond connector; (B) concrete plastic deformation at collapse, $f_{ck} = 28$ MPa.

Table 3.12 Failure Load Values Predicted by the Model

Model	f_{ck} (MPa)	Load (kN)	Variation (%)
P-SF-120	28	326.66	95.67
P-2 F-120	28	341.44	100
P-2 F-AR-120	28	400.26	117.23

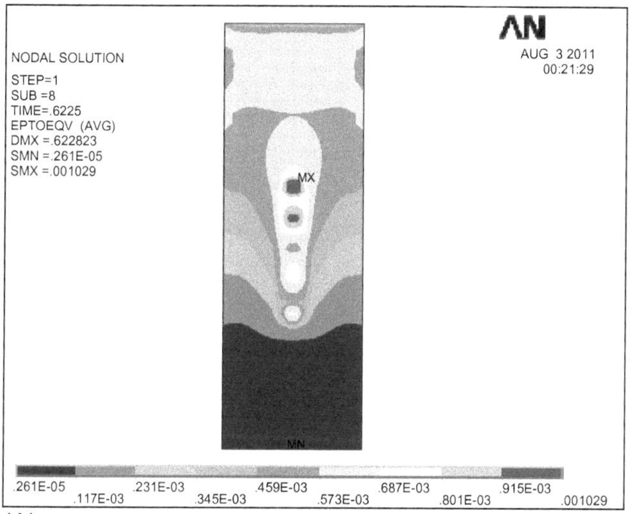

(A)

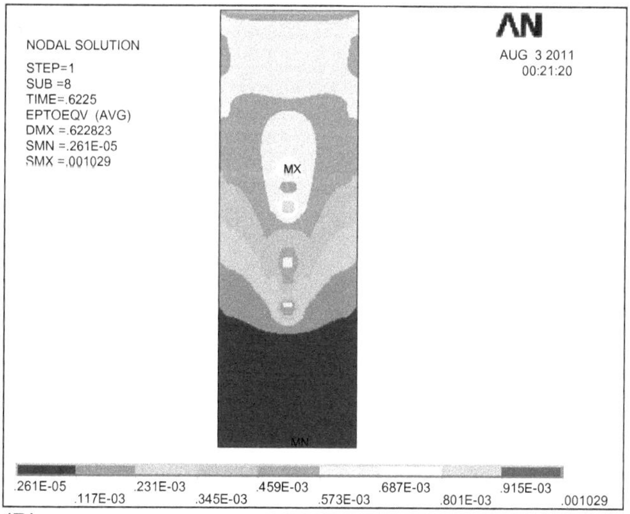

(B)

Fig. 3.180 Profile von Mises stresses at: (A) Perfobond connector interface; (B) interface posterior region at collapse, $f_{ck}=28$ MPa.

was made since the crushing and cracking capabilities of the adopted concrete solid element had to be suppressed to overcome serious convergence problems in regions close to the investigated structural systems ultimate loads.

3.6 FINAL CONSIDERATIONS

This chapter aimed to introduce the necessary steps for the correct design, implementation, and interpretation of numerical model results of using the finite element method for the simulation of the response of steel and composite structures. Other types of modelling including, for example, nonlinear dynamic modal analysis and transient soil structure interaction, and even coupled thermal and mechanical analysis can be and already are objects of much research and development but due to scope limitation issues, they were left out of this chapter.

Once again it must be emphasised that nothing replaces the experience of engineers and researchers who have been dealing with these types of modelling for a long time and, even with limitations of equipment and financial resources, have produced excellent numerical models that have contributed to a better understanding of complex structural engineering problems.

The next chapter contains a similar description to what was done in this chapter for models involving computational intelligence techniques. Thus the main characteristics of these techniques, scope, and the correct treatment and interpretation of the obtained results will presented and discussed in detail.

CHAPTER FOUR

Computational Intelligence Modelling

4.1 INITIAL CONSIDERATIONS

Structural engineers have long been searching for the conception of structures that could express their ability to withstand their imposed loads and usage, not only with lightness and economy, but also with the objective of becoming aesthetic landmarks and being in line with the current sustainability demands. Numerous examples can be seen worldwide such as the Eiffel Tower, Tower Bridge, Sidney Opera House, and Bilbao Guggenheim Museum.

The development of new materials and faster computing processes opened new frontier for the conception and development of new and audacious designs that will set the trend for the future 21th century structures. Various methods, techniques, and procedures have been, and still are being, used to improve and design these structures like optimisation processes, numerical modelling systems involving nonlinear finite element analysis, etc.

Concurrently, the last decade of the 20th century has been related to a large improvement and development of the so-called Computational Intelligent techniques. These techniques are computational systems that try to mimic human behaviour, such as perception, reasoning, learning, evolution, and adaptation. They involve *Neural Networks*, *Genetic Algorithm*, *Fuzzy Logic*, and *Hybrid Intelligent Systems*, such as neuro-fuzzy, neuro-genetic, and fuzzy-genetic models.

Intelligent systems have been successfully applied in forecasting, optimisation, risk analysis, control, inference, and modelling and fraud detection, in a wide range of knowledge fields. These systems offer comprehensive solution for managers and decision makers to a large number of complex and extensive applications that are considered difficult, totally limited, or even impossible by companies in different fields like Trade and Commerce, Engineering, Energy, Logistic, Environment, etc. Therefore, numerical

Modelling Steel and Composite Structures
http://dx.doi.org/10.1016/B978-0-12-813526-6.00004-0

tools based on these ideas have been widely developed and were successfully applied not only to engineering but also to wide range of problems.

This chapter highlights some of the initial attempts to use Computational Intelligent methods to forecast, design, and optimise the structural behaviour. The authors do not aim to exhaust all the attempts to use Computational Intelligence techniques to solve Structural Engineering problems. Alternatively, the idea behind this work was to concentrate on some of these methods to enable a deeper insight of a wide range of structural engineering applications that could be aided by their proper use.

This chapter is divided in sections where initially a brief description of the adopted Computational Intelligence method is presented, followed by some applications. First two basic artificial neural network models (i.e. MultiLayer Perceptron with Back Propagation algorithm and Bayesian Neural Networks) are introduced. This is followed by the Genetic Algorithms (GA) and Genetic Programming (GP). Finally, a brief overview of neuro-fuzzy systems was provided, as well as some of its applications to structural engineering.

4.2 NEURAL NETWORKS

Neural networks are nonlinear computational models, inspired in the human brain structure and operation that aim to reproduce human features, such as learning, association, generalisation, and abstention. Neural networks are made up of various processing elements (PEs) (artificial neurons), highly interconnected, that perform simple operations, transmitting their results to the neighbouring processors. The neural networks ability to perform nonlinear mappings between its inputs and outputs have made then prosperous in pattern recognition and complex systems modelling. Due to its structure, neural networks are very effective for learning patterns from nonlinear data, incomplete with noise or containing contradictory examples. Three basic concepts characterise the various types of neural networks: the artificial neuron model, their interconnection structure (topology), and the learning algorithms.

The neural network PE is a simplified mathematical representation of the biological neuron, which executes the sum of its inputs s_i (dendrites) modified by the associated weights w_{ji} (synaptic weights). Each PE then applies an activation function to that result in order to generate its output s_j (axon). The applied activation function is usually a nonlinear function f (e.g. sigmoid or hyperbolic tangent functions), a feature that enables the ANN to represent more complex problems. Therefore, the output of a PE is calculated as in Eq. (4.1):

$$s_j = f\left(\sum_{i=1}^{N} s_i \cdot \omega_{ji} + \theta_j\right) \tag{4.1}$$

where N is the total number of inputs and θ_j is a bias term that has the effect of increasing or reducing the net input of the activation function. The processor elements can be divided into three types: *input*, *output*, and *hidden*. The input neurons are those that are connected to the global network inputs. Output neurons present the outputs of the neural network to the external environment. The third group is composed of neurons whose inputs and outputs are only connected to other neurons, and where there is no external environment interaction.

The neural network topology defines how the various neurons are interconnected. These neurons are usually arranged in layers, and can be classified according to the neuron type, input layer, output layer, and hidden layer. Two basic topologies are generally considered: Feed-Forward Networks and Recurrent Networks.

Feed-Forward networks consist of an input layer, an output layer, and one or more hidden layers. In this type of network the data flow is always in only one direction; that is, there is no feedback (Fig. 4.1). Recurrent networks have connections between same layer processors and/or previous layers processors (feedback). Recurrent networks, due to its feedback structure, are more suitable for dynamic patterns processing.

The Neural Network processing can be divided into two phases: Information Recall (recall) and learning (learning). Information recall is the process for calculating the network output, for a certain input pattern. Learning is the process for updating the synaptic weights to the knowledge acquisition. There are three basic types of learning: supervised training, nonsupervised training, and 'Batch' training.

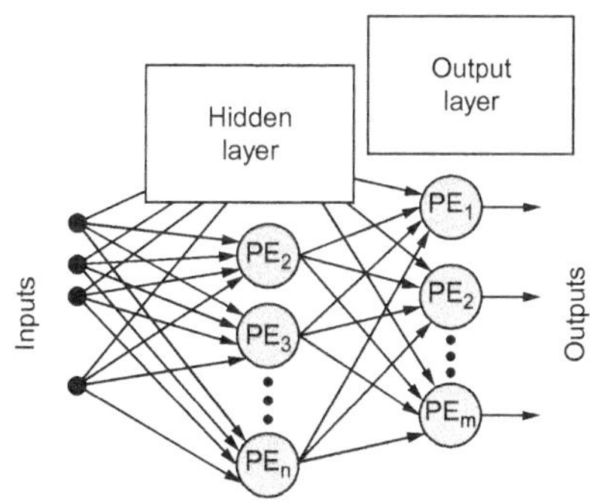

Fig. 4.1 Feed-forward network.

In the supervised training the network is trained through the provision of input values and their corresponding aimed output values on 'training pair'. Alternatively, the nonsupervised training ('self-organisation') does not require the aimed output values. The system extracts the characteristics of the pattern set by grouping them into classes inherent to data (clustering). On the 'Batch' training system the synaptic weights' values are specified a priori, in a single step, also 'Recording'. This book focused on the feed-forward networks, for a static input–output mapping with supervised training.

4.2.1 Back Propagation Training Algorithm

Back Propagation (BP) Neural Networks [148,149] are feed-forward networks of one or more hidden layers. This algorithm defines a systematic way for updating the weights of the various layers based on the idea that the hidden layers' neurons errors are determined by the feedback of the output layer neurons errors.

The BP supervised learning process is based on the Gradient Descent method that usually minimises the sum of squared errors between the target value and the neural network output. Therefore, the weight update (Δw_{ij}) is proportional to the negative of the error partial derivative with respect to its own weight:

$$\Delta\omega_{ji} = -\eta \frac{\partial E_{\mathrm{SSE}}}{\partial \omega_{ji}} \tag{4.2}$$

where η is the learning rate and E_{SSE} is the Sum of Squared Errors function, defined by:

$$E_{\mathrm{SSE}} = \frac{1}{2}\sum_{p=1}^{N_p}\sum_{i=1}^{N_o}\left(t_j^p - s_j^p\right)^2 \tag{4.3}$$

where N_p is the total number of patterns in the training set, N_o is the total number of PEs in the output layer, and t_j^p is the target value of pattern p at the neuron j. By deriving Eq. (4.2), the synaptic weights are updated by:

$$\begin{aligned} &\Delta w_{ji} = \eta \cdot s_i \cdot e_j \\ &e_j = \begin{cases} (t_j - s_j) f'(net_j) & \text{if } j \in \text{output layer} \\ f'(net_j)\displaystyle\sum_{k=1}^{N} \omega_{jk} e_k & \text{if } j \in \text{hidden layer} \end{cases} \end{aligned} \tag{4.4}$$

where η is the learning rate, s_i is the input associated with the weights w_{ij}, e_j is the jth processor error, t_j is the aimed value of the processor output j, s_j is its activation state, net_j is its inner potential, $f'(net_j)$ is the derivative of the activation function, and N is the number of PEs in the following layer of processor j.

As can be seen from the previous equation, the learning process is performed in two stages: feed-forward and feed-backward. In the first phase the inputs are propagated through the layers of PEs, generating an output pattern in response to the input pattern presented. In the second phase the errors calculated in the output layer are then back propagated to the hidden layers where the synaptic weights are updated to reduce the total error. This learning process is repeated until the output error value, for all patterns in the training set, is below a specified value or the total number of cycles specified is reached.

It has been demonstrated that the BP algorithm is a universal approximation method [147], being able to learn any input–output mapping. However, despite the great success of BP in different applications, there are some basic problems: the definition of the network size, the slow training process, and phenomena such as an early network stop (bypassed by decreasing the value of η) and local minimum (which can be solved using statistical methods).

The network layout definition (i.e. the number of hidden layers and number of processors in each of these layers) is a compromise between convergence and generalisation. Convergence is the neural network ability to learn all the patterns of the training set. Generalisation is the capacity to correctly respond to unprecedented patterns (test set). The goal is to use the smallest possible network, in order to obtain a good generalisation, capable to learn all the data patterns.

The learning rate is an important parameter to be tuned in learning. It should not be too small, causing a very slow training, not too large, resulting in oscillations. When the learning rate is small, and depending on the weights initialisation (randomly made), the neural network can get stuck in a local minimum. When the learning rate is large, the neural network could never reach the global minimum due to the large weights' values. The solution to this problem is to use an adaptive learning rate. In addition to this parameter, it is also possible to use a α moment term [148], proportional to the synaptic weight value variation in the previous step. Thus the synaptic weight equation updates w_{ij} as follows:

$$\Delta\omega_{ij}(t+1) = \eta \cdot s_i \cdot e_j + \alpha\Delta\omega_{ij}(t) \quad (4.5)$$

The use of the term moment has the function to speed up the network convergence without causing oscillations.

4.2.2 Bayesian Neural Networks

Bayesian Neural networks are inspired by Bayesian statistics (in contrast with classical statistics), which deal with probability densities instead of frequencies. It assumes a particular model for the probability densities of the data and network weights, and makes use of the Bayes' theorem to infer the optimum set of weights, given the data available. One of the most common learning algorithms in Bayesian neural network is the Gaussian approximation method [150], explained later.

Take into consideration a Multilayer Perceptron Neural Network to be trained, within a total of adjustable W parameters (including weights and biases) gathered together in a vector $w = \{w_1, w_2, \ldots, w_W\}$. And that the training data D comprises N_y pairs (x_j, y_j), where $y_j = y(t_j)$ and x_j is a k-dimensional vector corresponding to a line of the input matrix X_n:

$$D = \left\{ \left(\boldsymbol{x}^1, y_1\right), \left(\boldsymbol{x}^2, y_2\right), \ldots, \left(\boldsymbol{x}^{Ny}, y_{Ny}\right) \right\} \tag{4.6}$$

The Bayesian method aims to infer the weights a posterior probability density, given the data, by employing Bayes' rule, that is:

$$p(\boldsymbol{w}|D) = \frac{p(D|\boldsymbol{w})p(\boldsymbol{w})}{p(D)} \tag{4.7}$$

where $p(w)$ is the prior probability density for the network parameters, $p(D|w)$ is the likelihood of the data given the network parameters, and $p(D)$ is a normalising factor. Both probability densities are assumed to follow the Gaussian distribution:

$$\begin{aligned} p(\boldsymbol{w}) &= \frac{1}{Z_W(\alpha)} \exp(-\alpha E_W) \\ p(D|\boldsymbol{w}) &= \frac{1}{Z_D(\beta)} \exp(-\beta E_D) \end{aligned} \tag{4.8}$$

where α and β are called hyperparameters that, respectively, define the spread of the parameters and the network outputs noise. Z_W and Z_D are normalisation factors that ensure the integrals of the earlier densities are equal to one. From the definitions of sum-squared error and weight decay of conventional BP algorithms:

$$\begin{aligned} E_D &= \frac{1}{2}\sum_{j=1}^{Ny} \left[o_j - y_j\right]^2 \\ E_W &= \frac{1}{2}\|\boldsymbol{w}\|^2 = \frac{1}{2}\sum_{j=1}^{W} w_j \end{aligned} \tag{4.9}$$

It can be shown that [10], in this case, the normalisation factors are given by:

$$Z_W(\alpha) = \left(\frac{2\pi}{\alpha}\right)^{W/2} \; ; \; Z_D(\beta) = \left(\frac{2\pi}{\beta}\right)^{N\gamma/2} \tag{4.10}$$

Now, substituting Eq. (4.9) in Eq. (4.8) yields:

$$p(\boldsymbol{w}|D) = \frac{1}{p(D)} \frac{1}{Z_W(\alpha) Z_D(\beta)} \exp\left(-\alpha E_W - \beta E_D\right) \tag{4.11}$$

This is the value to be maximised during the training process. However, maximising $p(w|D)$ is equivalent to minimising the negative of its logarithm, that is:

$$E = -\ln p(\boldsymbol{w}|D) = \alpha E_W + \beta E_D + \ln p(D) Z_W(\alpha) Z_D(\beta) \tag{4.12}$$

Since the third term in the equation earlier does not depend on the network parameters, given α and β, the Neural Network is trained by minimising:

$$E = \alpha E_W + \beta E_D \tag{4.13}$$

This is equivalent to the BP algorithm with weight decay. Once the optimised posterior probability $p(w|D)$ is found, the weight vector is selected to be, for instance, the probability distribution mode, corresponding to the most probable vector w_{MP}. The Bayesian framework considers a second level of optimisation for adjusting the hyperparameters α and β [151]. Thus the network is trained iteratively in optimisation cycles calculating the networks weights w_{MP}, for fixed values of the hyperparameters α and β, followed by re-estimation of these parameters.

4.2.3 Application Examples

4.2.3.1 Beams Under Patch Loads

Patch load actions on steel beams are frequently found in structural engineering practice. This load type can be encountered in crane girders, secondary beams' reactions acting over the primary girder system and other structural problems. Numerous studies have been performed with the available experiments; however, a 20% error is still found in the current design formulae. A unique design equation for this civil engineering phenomenon proved difficult to be reached, due to the influence of several interdependent variables. On the other hand, creating new experiments in laboratory is time consuming and expensive.

As an alternative solution, numerical methods, like the finite elements, have also been utilised to model the phenomenon but they still present significant differences when compared with experiments. These setbacks and the ability to generalise the knowledge acquired from experiments led to the use of neural networks to predict the ultimate strength of steel beams under patch loads [152–160]. The use of neural networks made possible the development of a comprehensive parametric analysis of girders subjected to patch loads [161].

The ultimate capacity of a steel girder under patch loads depends on many geometrical and material parameters. Fig. 4.2 shows an example of a steel girder under patch loads and its fundamental geometric parameters. The patch load problem has been investigations dating back to the first experiments performed by Lyse and Godfrey [157]. A comprehensive review on this subject can be encountered in Vellasco [160].

The vast majority of studies pointed out that the ultimate load is proportional to the square of the web thickness, whilst all the other parameters were found to be of less significance. Previous investigations [161] demonstrated that new data can be obtained from neural network algorithms. In these studies a neural network system that predicts the ultimate resistance of steel girders under patch loads was established. The neural network system was trained with a set of experiments by Roberts [159] and Kennedy [162]. In these test, each training data sample was made of the eight geometrical and material parameters and by the experimental ultimate load as output. Artificial Neural Networks usually have their outputs limited to values ranging from 0 to 1.

In these investigations a neural network system which provides resistance of steel beams subjected to concentrated loads was developed. The neural network system has been trained with a set of experimental results obtained by Roberts [159] and Kennedy [162]. In these experiments, the training data included eight geometrical and material parameters and had as output, the ultimate load. Usually neural networks have their outputs limited in the 0 to

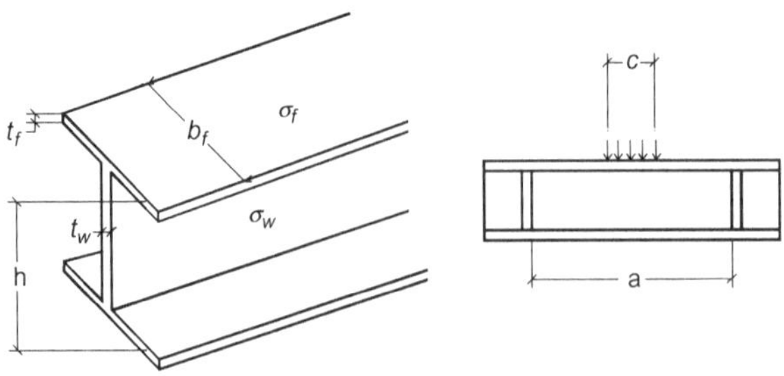

Fig. 4.2 Geometrical and material parameters related to the patch load evaluation.

1 range but the possible ultimate loads range was very large to be normalised into the 0 to 1 interval and still produce accurate results. An initial strategy dividing the available experiments in three classes according to their ultimate load capacity was implemented. The adopted model was made of three predictions and one classifying neural networks, as shown in Fig. 4.3.

This division enables a better normalisation in the training process. The BP algorithm was utilised to train all neural networks. The prediction networks were trained with three ranges of ultimate load. The classifying network architecture was made of 14 inputs and 3 outputs parameters. The first group ranged from 30 to 120 kN, the second ranged from 80 to 250 kN, and the third from 150 to 4010 kN. The classifying neural network splits the data set from 30 to 100 kN, from 100 to 200 kN, and from 200 to 4010 kN. The data superposition presents in the forecasting networks was utilised to expand the data set of each group and to guarantee that the full range of experiments could be used and tested without any discontinuity.

The Predict Software [163] was used to perform a parametric study. This study utilised the trained Neural Networks to forecast the patch load capacity of numerous cases in which no previous experiments was available. The classifying Neural Network was then used to calculate what was the most suitable network to be used in that specific case.

Despite the achieved results, the method used to split the three groups did not properly consider the difference in response of slender, intermediate, and compact girders. The classes' division was purely based on the load level basis and not on the girder's structural response. The new strategy utilises a single neural network with all the 155 experiments using an alternative ultimate load normalisation technique. In this process the ultimate load value is divided by a modified girder's web yield capacity based on Lyse

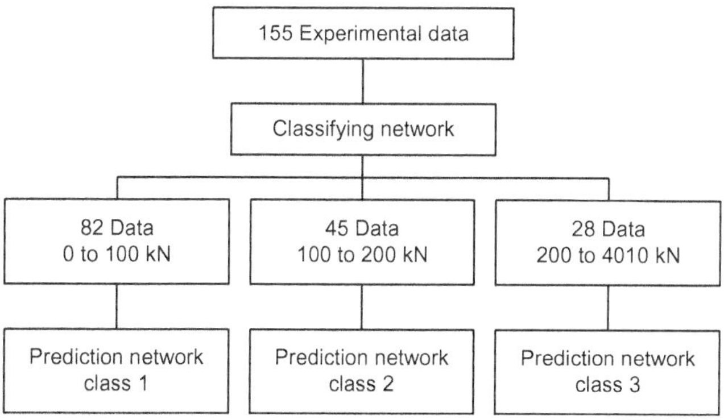

Fig. 4.3 Classification and prediction model.

and Godfrey [157]. The BP algorithm was used to train the neural net using 15 parameters as inputs and the normalised beam resistance as output.

The experimental data were again split into three distinct groups: 70% for training, 20% for testing, and 10% for validating the neural networks. The validation data can be utilised to compare different neural networks to access the best solution in an independent procedure. Several neural network configurations were trained and compared, varying the numbers of hidden layer processors whilst best results were obtained with 15 processors.

A further comparison, in terms of percentile error values, of the neural network predictions (training, testing, and validation) is shown in Fig. 4.4. From this figure it is possible to observe that the unified neural network performance was as similar to the three class neural network system performance and significantly more accurate than Bergfelt [158], Roberts [159], and Eurocode 3 [20]. The neural network training data split into three load classes provided better training and suitability to the problem distinct characteristics.

The networks' errors vis-à-vis the experiments proved to be lower than the errors associated with existing formulae, thus qualifying the neural networks to generate new trustworthy data. The main reason for the difference found in the neural networks and the design formulae is related to the inclusion of the web panel shape factor versus beam's height ratio, a/h, in the neural network training data, that is, part of the errors found in that formulae are due to the absence of this parameter.

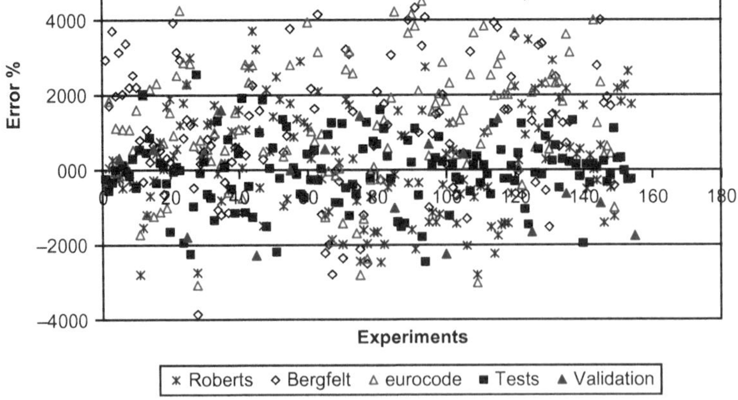

Fig. 4.4 Patch load capacity neural network model results.

4.2.3.2 Block Shear Effect

High strength bolts are extensively used to join steel structural elements in a great variety of applications. The joint resistance is normally reached, due to the bolt high strength, with a reduced number of bolts, located in a small plate area. This fact induces the bolted connections to be associated with a failure mode known as block shear where a block area of the jointed element is separated from the remaining parts. The failure is characterised by the development of two rupture planes. The first is a shear rupture plane, parallel to the load direction, passing through the bolt line. The second is a tension rupture plane, perpendicular to the load direction, passing through the bolt located farthest from the plate edge (Fig. 4.5). Numerous models have been proposed to avoid shear block failure in steel design. These models incorporate a combination of the various hypotheses that control the structural failure. Various design equations for this civil engineering problem were proposed over the years but a definitive solution has not yet been found, due to the influence of several independent variables.

The main aim of the investigation was to present a neural network modelling of the block shear phenomenon [164,165]. The main neural network input parameters were the geometrical and material parameters that control the block shear phenomenon whilst the neural network output parameter was the block shear capacity. The neural network training was

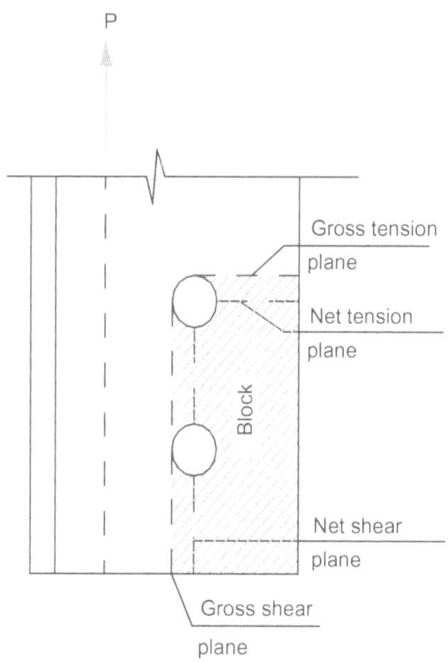

Fig. 4.5 Block shear.

based on experiments found in literature and adopted cross-validation techniques to avoid the neural network overfitting. The present study was made with the aid of the MATLAB software [166]. The investigation was centred on determining of best performance for the neural network with 10 inputs, 5 neurons on the hidden layer, and 1 output. Fig. 4.6 shows the results in terms of MAPE and RMS errors for the neural network test set.

Although the errors with the BP Neural Networks were found to be within the expecting Structural Engineering acceptable limits, the reduced database pointed out for the use of Bayesian Neural Networks. The first strategy for training Bayesian Neural Networks was performed without using noisy data. The database was split in training (75% of the database) and testing (25% of the database) sets. Numerous network architectures were evaluated, varying the number of hidden layer neurons (from 1 to 5). MAPE and RMS error metrics were utilised to evaluate the Neural Network performance (Fig. 4.7). The best performance was reached with a network presenting five hidden layer neurons (MAPE = 1.62 and RMS = 2.15).

Noisy input data were then introduced in a second strategy, to increase the dataset and improve the Neural Network performance. The database

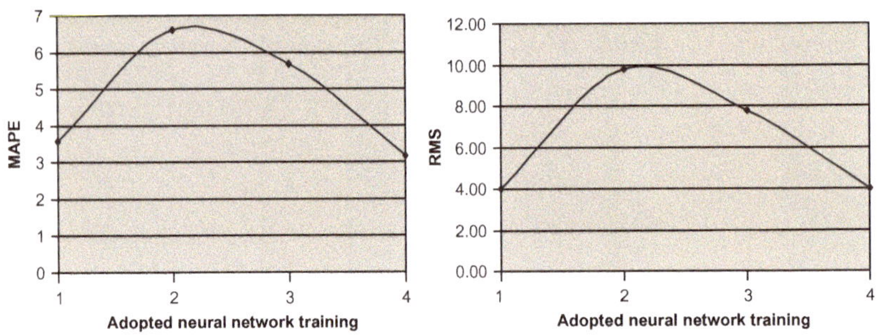

Fig. 4.6 Neural network performance.

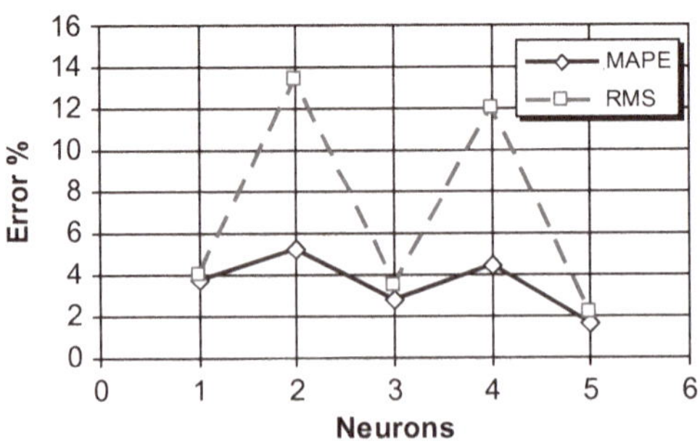

Fig. 4.7 Noiseless Bayesian neural network performance.

was increased 2, 3, and 4 times its original size. Again, 75% and 25% of the database were used in the training and test phases. Numerous network architectures were investigated, also varying the number of hidden layer neurons from 1 to 5. MAPE and RMS error metrics were used to access the network performance. Fig. 4.8 shows the results of a Bayesian Neural Network with a database increased four times its original size. The best performance was reached with a network with four hidden layer neurons (MAPE = 0.74 and RMS = 1.38) and with the database increased four times its original size. The training and test performance of this particular network is shown in Fig. 4.9. The noisy data significantly improved the Bayesian Network performance.

The best performance was reached with a network presenting three hidden layer neurons (MAPE = 1.34 and RMS = 1.95) and with the database increased four times its original size. The third strategy adopted Bayesian neural networks with input and output noisy data. The database was increased 2, 3, and 4 times its original size and, again, 75% and 25% of the database were used in the training and test phases. The number of hidden layer neurons also varied from 1 to 5. MAPE and RMS error metrics were used to evaluate the neural network performance. The used of noisy data

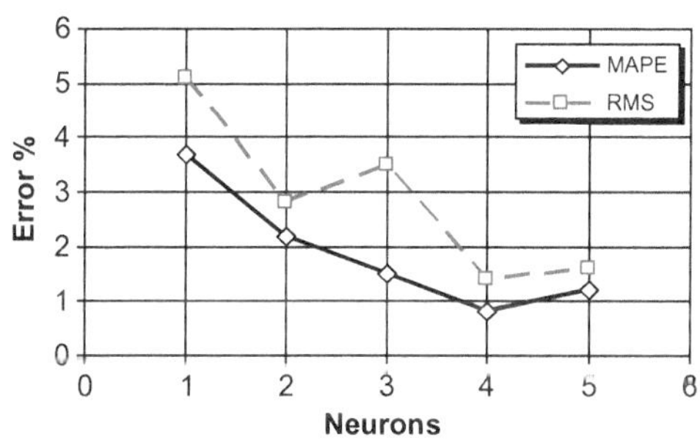

Fig. 4.8 Bayesian neural network with input noisy data.

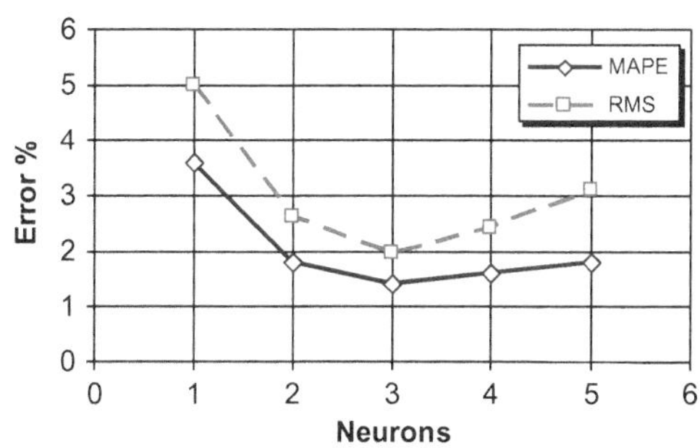

Fig. 4.9 Bayesian neural network with input and output noisy data.

significantly improved the Bayesian Network performance when compared with the noiseless database but the best performance was found with the Bayesian network with only input noisy data.

The Bayesian neural networks adopted to investigate the block shear capacity of coped beams proved to be more accurate and efficient than the BP Neural Networks. The Bayesian Neural Networks' errors were even smaller when input noisy data were adopted to increase the original database to four times its original size. The neural network errors proved to be small (compared with the usual structural engineering safety margins) confirming the possibility of using this method to create new data. It is also important stress that these new data could be easily created by the trained network, reducing the cost and time spent with countless laboratory tests or numerical simulations.

4.2.3.3 Perfobond Shear Connectors

Shear connectors are utilised in composite structures to enable the combined action of the concrete slab and the steel profile. The Perfobond shear connector was developed by Leonhardt to improve the fatigue strength of a Venezuelan composite bridge. The Perfobond shear connector is made from a rectangular steel plate, with holes, welded to the steel beam top flange. Subsequently the Perfobond holes are filled with concrete forming dowels that improve the connector shear strength and inhibits the vertical split at the concrete/steel interface. The Perfobond shear capacity can be further enhanced using reinforcement bars passing through the Perfobond holes (Fig. 4.10). Numerous push-out tests have been performed to investigate the Perfobond structural behaviour (Fig. 4.11). The test layout comprises the steel profile supported by two adjacent concrete slabs [168].

Fig. 4.10 Steel bars used in the Perfobond.

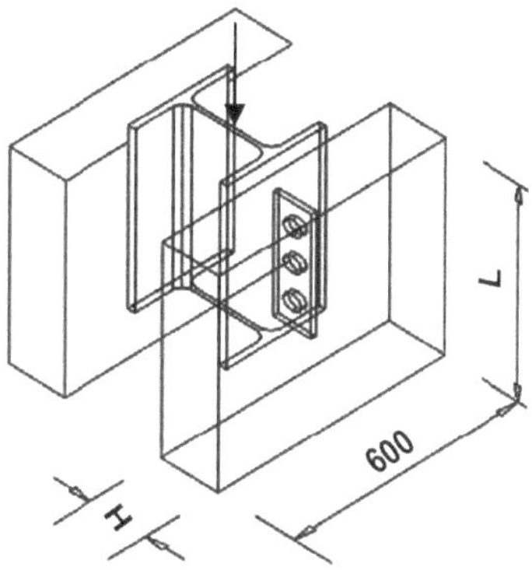

Fig. 4.11 Push-out test layout.

The composite action is assured by the use of shear connectors welded to both beam flanges. The structure is under a load that introduces shear in both sides of the concrete slab to beam interfaces.

This specific type of shear connector has been formerly studied by Oguejiofor and Hosain [169,170], Cruz and Valente [171], amongst others. Various parameters considerably affect the test. However, due to the large number of variables involved the use of neural networks, to predict the Perfobond shear capacity, calibrated against the existing experiments, proved to be one of the most feasible alternatives. This was the main motivation for utilising the neural networks to predict the Perfobond shear resistance. The network input data used current experiments present in the literature.

A Bayesian training algorithm (Bayesian networks) was used since this hybrid model has the competence to improve convergence and overfitting of small input dataset networks. Basically the Bayesian networks utilise the Bayesian inference mechanism to access the neural model variables (weights). Noisy input data were used to increase the dataset and enhance the neural network performance. The original performance was compromised due to the small experimental dataset available [167]. The neural network input and output data were normalised by the maximum/minimum range of values.

The network did not produce trustable results with the original 41 experimental data where the data utilised in the test phase is depicted. When the various trained neural network generalisation errors were compared, the best solution was archived by the network using 123 inputs (41 actual experimental and 82 noisy data) with the following topological characteristics: nine inputs, one hidden layer with five processors, and one output [168]. The mean absolute error of the analytical to experimental predictions was

1.27. This important difference can be explained by the natural difficulty involved in measuring the shear connector structural variables that directly influence the experiments.

The mean absolute error of the experimental to neural networks was 1.07 significantly better than the analytical experimental error of 1.27. Figs 4.12 and 4.13 show the smallest MAPE and RMSE errors for the studied network configurations: without noisy data and second configuration with noisy data in the network. From these graphs, it is clear that the best results in terms of MAPE and RMSE errors were reached for the second configuration (with noisy data in the network) for the network with 41 input and 82 noisy data, where 80 and 43 were utilised for training and testing ($123 \times 80 \times 43$) with three hidden layer processors.

A significant improvement of the network performance is shown in Figs 4.14 and 4.15, for the training and testing results of the second network configuration (with noisy data in the network inputs and outputs and normalisation strategy) with 41 input and 82 noisy data, where 80 and 43 were used for training and testing with three hidden layer processors.

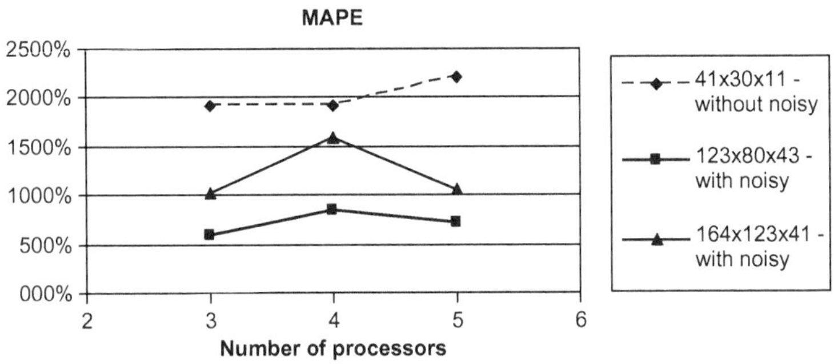

Fig. 4.12 Smaller MAPE error for each evaluated configuration strategy.

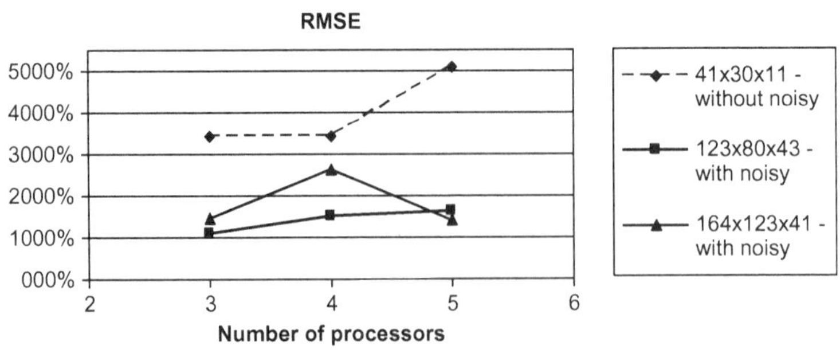

Fig. 4.13 Smaller RMSE error for each evaluated configuration strategy.

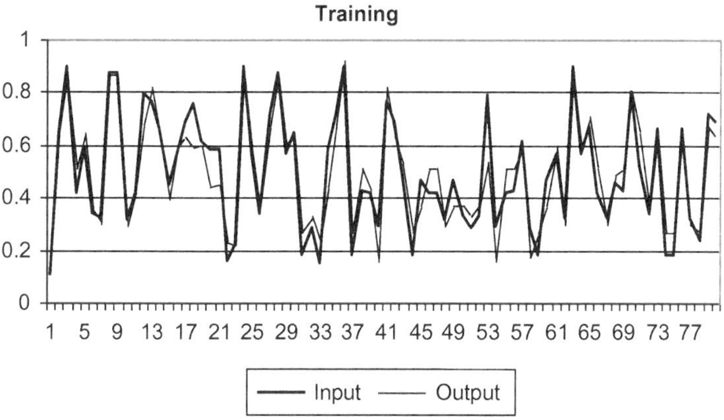

Fig. 4.14 Bayesian neural network with 3 hidden layer neurons and 82 extra noisy data (input and output) training results.

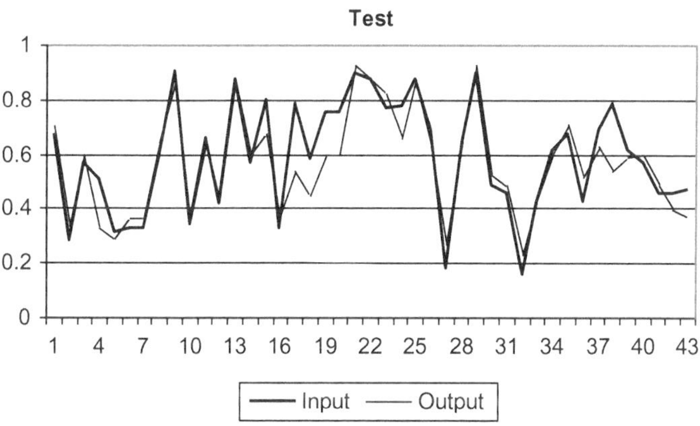

Fig. 4.15 Bayesian neural network with 3 hidden layer neurons and 82 extra noisy data (input and output) test results.

The present investigation highlighted some of the advantages of the Bayesian learning algorithms: minimising overfitting problems in training, controlling the model complexity, achieving a good generalisation, enabling an efficient use of the training data since the parameters and hyperparameters are evaluated from the training data, not requiring a separated set of values for hyperparameters cross-validation, and finally the use of the weight decay technique that penalises parameters with high associated values and enables the neural network to produce smooth values for the approximated output functions. The neural network errors proved to be small (compared with the usual structural engineering safety margins) confirming the use of this method to create new data.

4.2.3.4 Prestressed Stayed Steel Columns

Slender steel columns normally have their load-bearing capacity controlled by the global structural buckling ultimate limit state. These structural elements can be optimised with the use of a prestressed cable or bar system. They become very efficient structural solutions for various challenging structural engineering problems [173]. The use of the prestressed system reduces the column buckling length to at least half of its original size and considerably improves its load-bearing capacity. A specific case of these structural systems is composed of a slender central column, four cross bars perpendicularly positioned in relation to the central column main axis, and four steel cables (or steel bars), as shown in Fig. 4.16. This structure was experimental and numerically studied by Araujo [174,175], with a reduced set of parameters.

The usual method of investigating prestressed stayed steel columns structural behaviour is by means of a nonlinear finite element analysis. A preceding parametric analysis was made, Araujo [175], studying the influence of the prestress force level, cable (steel bar) axial stiffness, and main tubular column initial imperfection on the load-bearing capacity. The parametric analysis proved to be time consuming due to the nonlinear characteristics of the finite element analysis.

However, the large number of required analysis to fully represent their effects has led to the use of Neural Networks in order to reduce the required number of finite element simulations. Neural Networks were then employed to extend the prestressed stayed steel columns parametric analysis and predict their load-bearing capacity. The training and test data utilised by the neural networks were obtained from nonlinear finite element. Due to the small number of available training data, Bayesian Neural Networks were employed [173].

Fig. 4.16 Prestressed stayed steel column full-scale test.

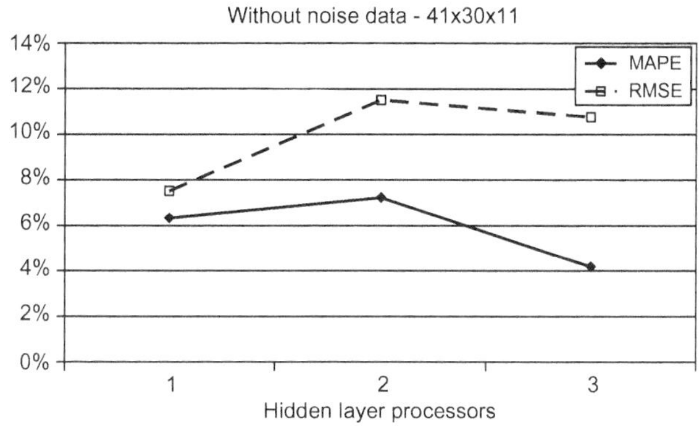

Fig. 4.17 Bayesian neural net. MAP and RMS errors without noisy data and 5 hidden layer neurons.

The network performance without noisy data, with 41 input data, where 30 and 11 were used for training and testing ($41 \times 30 \times 11$), respectively, it is shown in Fig. 4.17 in terms of MAP and RMS errors, varying the number of hidden layer processors. In terms of MAP errors, the best performance was found with three hidden layer processors, whilst in terms of RMS the best performance was obtained with one hidden layer processor [172,173]. The network performance is also shown in Fig. 4.18 for the second network configuration with noisy data in the network inputs and outputs and first normalisation strategy. The results indicated that the best performance was obtained with 41 input and 41 noisy data, where 61 and 21 were used for training and testing ($82 \times 61 \times 21$) with five hidden layer processors.

The Bayesian Neural Networks were very useful for extending the parametric analysis of the load carrying capacity of the prestressed stayed steel

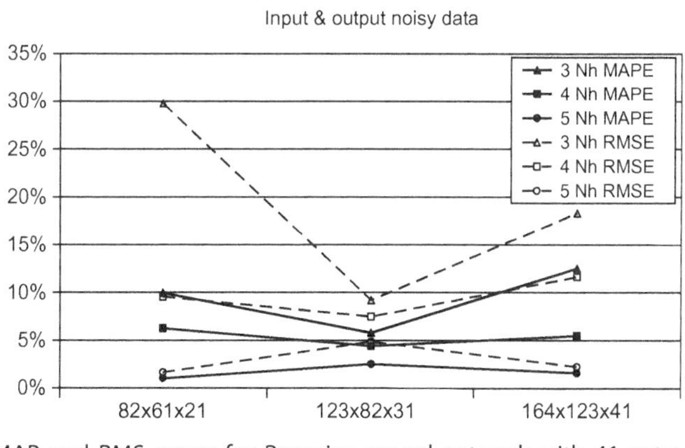

Fig. 4.18 MAP and RMS errors for Bayesian neural network with 41 extra noisy data (input and output).

columns. Despite the small initial database, it was possible to determine the best Neural Network topological configuration that presents errors lower than originally expected. It is important to stress that the use of noisy data was essential to obtain these error levels. When networks with noisy data were used, a significant improvement in the performance was clearly detected.

The best performance was achieved with a configuration containing the number of noisy data (in input and output) equal to the double of the original input data, with five hidden layer processors, and the standard normalisation strategy. Larger numbers of processors made the network unstable. The error levels achieved are well below the acceptable structural engineering errors commonly used in design standards and daily practice.

Finally, the Bayesian Neural Networks proved to be a very useful computational intelligence tool for enhancing this type of parametric analysis without requiring hundreds of time-consuming nonlinear finite element analysis that not always converge. Properly trained networks proved to be very efficient for improving the parametric analysis of prestressed stayed steel columns load-bearing capacity in an easy and simple form.

4.2.3.5 Beam-to-Column Joints

Few studies using artificial neural networks to forecast joints behaviour were found in literature. Abdala and Stavroulakis [177] and Stavroulakis and Abdala [178] have utilised neural networks to forecast the global moment versus rotation curve of single web angle beam-to-column joints. Anderson et al. [179] described the use of neural networks to forecast a bilinear approximation of the moment versus rotation curves of minor axis beam-to-column endplate joints.

This work used of artificial neural networks to forecast the flexural resistance and initial stiffness of semirigid beam-to-column connections. This civil engineering problem is characterised by the influence of several physical and geometric variables and for the great difficulty to generate new laboratory data based on experiments.

In the current study three different types of connections were investigated by the neural networks [180–182]. From these investigated joints it was possible to identify the bolted end-plate joints as one of the most adopted connections (Fig. 4.19). They are widely utilised in constructional steel design because they can cover a wide range of structural solutions, from pinned to rigid joints, by performing minor geometrical changes. The other connections studied in this analysis were welded joints and bolted joint with angles as shown in Figs 4.20 and 4.21.

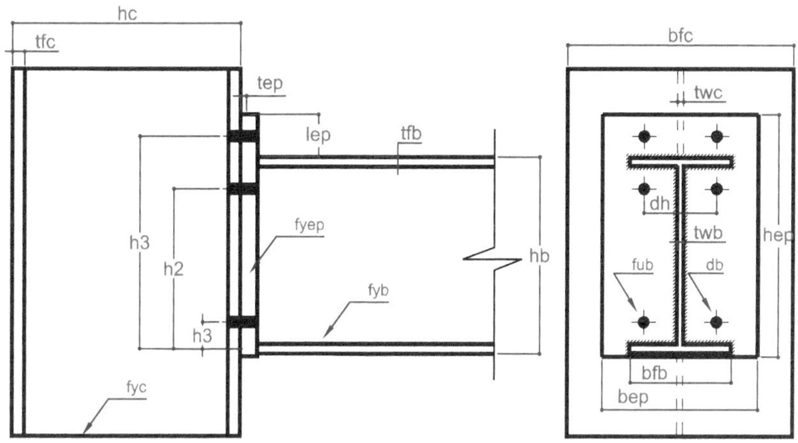

Fig. 4.19 Extended joints.

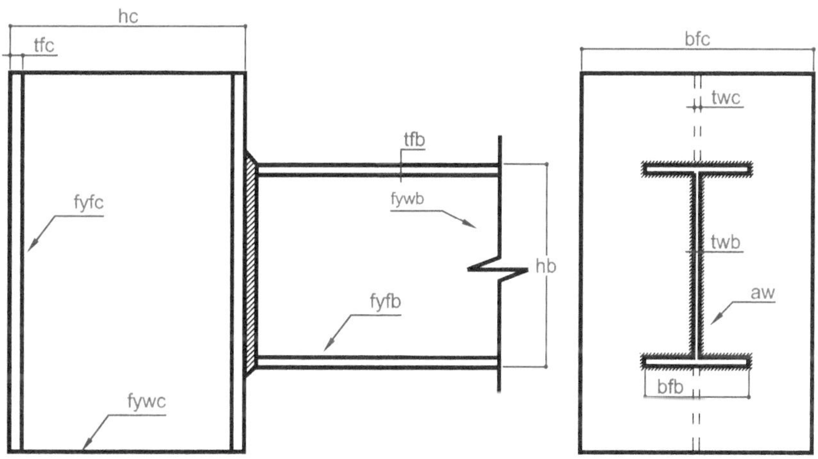

Fig. 4.20 Welded joints.

The response of beam-to-column connections can also be investigated with the aid of the component method, largely used in research studies and incorporated in European Committee for Standardization—Eurocode 3 [19]. Joint components have been introduced to a simple mechanical model to enable the forecasting of beam-to-column moment versus rotation curves. The mechanical model is made of rigid links and springs, created to represent each relevant joint component. A comprehensive description of the joint components is present in Silva and Coelho [183].

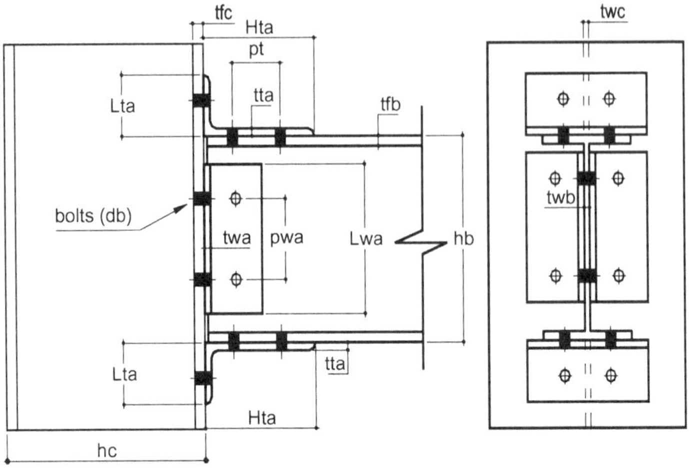

Fig. 4.21 Bolted joints.

The mechanical model can be simplified by substituting each series of springs by an equivalent elasto-plastic spring, which retains all the relevant characteristics. Using this procedure, a general nonlinear equivalent model for the analysis of beam-to-column connections can also be obtained Simões [184]. When the equivalent elastic model is defined, the design process continues with a post-buckling stability analysis using an energy-based formulation [183]. The BP learning algorithm was utilised, where the network is presented with a set of input vectors and their respective desired output vectors. Two neural networks were utilised for each connection type. The first was utilised to forecast the bending moment resistance whilst the second was used to predict the connection initial stiffness.

The input parameters represented by the geometric and mechanical characteristics of all the experiments. The Neural Works Modelling Software (QNet for Windows [185]) was utilised to train and test all neural networks. Several network configurations were investigated varying the hidden processors number (2, 3, 4, 5, and 6). The learning rate was considered adaptive, with an initial value equal to 0.2 whilst the momentum factor varied between 0.4 and 0.8. The optimum neural network was reached through a comparison of the generalisation error obtained with the validation set. For the bolted endplate connections, the results of 26 experiments (21 for training and 5 for testing) were used, producing satisfactory results for the prediction of the flexural resistance where the mean absolute percent error was 8.4%. The minimum and maximum percentile errors were −15.5% and 18%.

Fig. 4.22 shows a comparison of the ratio between the values predicted by the neural networks, the Eurocode 3 formulae, and the experiments. Fig. 4.23

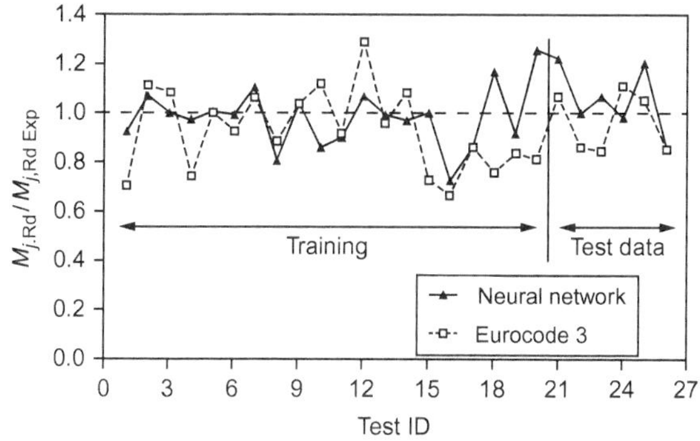

Fig. 4.22 Endplate joints moment capacity.

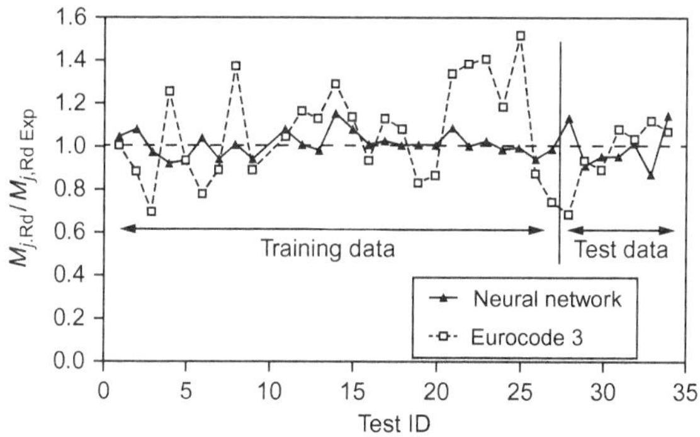

Fig. 4.23 Welded joints moment capacity.

shows the comparison of the results obtained with NN and Eurocode 3 formulae for the welded connections flexural capacities. From this graph it is possible to conclude that the neural network results reached a good agreement with the connections experiments, whilst the Eurocode 3 prediction values led to misleading results. The mean neural network error was 8.9% with −5.4% and 12.4% of minimum and maximum percentile error values, respectively. Fig. 4.24 shows the bolted joints with angles' results. For flexural resistance results, the mean error was 11.6% with −18.3% and 4.7% of minimum and maximum percentile error values, respectively.

The neural networks potential to forecast the connections ultimate moment and initial stiffness was confirmed by these graphs. Although some of the neural network initial stiffness values differ up to 59% from the experiments, this can be explained by the difficulties in the laboratory measuring devices and associated experimental errors, which can produce misleading values for the experiments. The mean errors obtained in this study were

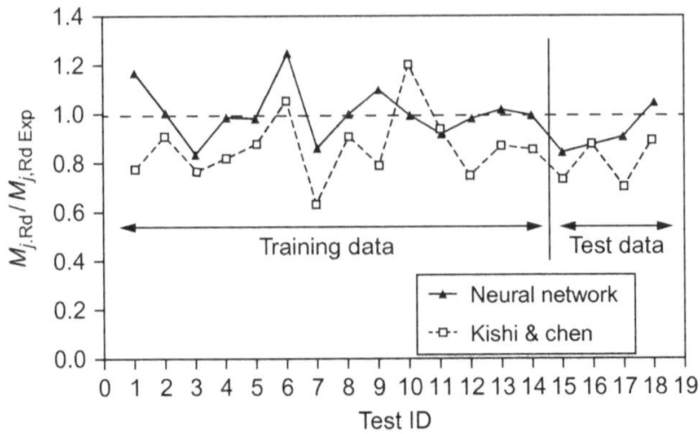

Fig. 4.24 Bolted angle joints moment capacity.

8.4%, 8.9%, and 11.6% for the prediction of the flexural resistance, demonstrating a reasonable agreement between the neural networks and the experiment. These errors are acceptable when compared with the structural engineering safety margins utilised in and the intrinsic errors associated with the adopted experiments.

4.3 EVOLUTIONARY COMPUTATION

4.3.1 Genetic Algorithms

Essentially, GA are search and optimisation methods [186–190] that have their inspiration on the concepts of the theory of natural selection of the species proposed by Darwin. The systems developed from this principle are used to search for solutions to complex problems or with a solution space (search space) very large. When conventional optimisation methods are used on these particular problems they often do not lead to simple solutions.

These algorithms are based on biological organism's genetic processes to determine optimal or suboptimal solutions. The process begins with the encoding of every possible solution for the problem in a structure called the 'chromosome', which is composed of a chain of bits or characters. These chromosomes represent individuals who are evolved over several generations, in a manner similar to living beings, in accordance to the natural selection principles and survival of the fittest, first described by Charles Darwin in his book *The origin of species*. Emulating these processes, GA are able to 'evolve' solutions to real-world problems.

The chromosomes are then subjected to an evolutionary process involving evaluation, selection, crossover, and mutation. After several cycles of evolution the population should contain most able individuals. The GA uses a direct analogy of this nature evolution phenomenon, where each individual represents a possible solution for a specific problem. For every individual an adaptation value is assigned: its fitness, which indicates how much the solution represented by this individual is good when compared with other solutions. In this way the term 'Population' refers to the set of all solutions in which the system works. The most adapted individuals are given the opportunity to reproduce by crossbreeding with other individuals of the population, producing decedents with characteristics from both parts. The mutation has also a significant role, by introducing in the population new individuals randomly generated.

The evolution process begins with the random creation of individuals who will form the initial population. From a selection process based on the each individual fitness, individuals are chosen for reproduction, which creates new solutions using for this a set of genetic operators. Thus the individual ability determines the degree of survival and, therefore, the possibility that the chromosome will be part of the next generations. The genetic algorithm basic procedure is summarised in Fig. 4.25.

The determination of the end of evolution can be set in terms of the number of generations, the number of created individuals, or in a condition where the algorithm should reach a satisfactory solution, that is, an optimum point. Other stop conditions include the processing time and the degree

```
Begin

t←1

initialize population P(t)

evaluate population P(t)

while (not end_condition) do

t←t+1

select population P(t) from P(t-1)

apply genetic operators

evaluate population P(t)

end while

end
```

Fig. 4.25 Genetic algorithm basic procedure.

of similarity between the elements in a population (convergence). The following sections present in detail each of the genetic algorithm components.

4.3.1.1 Representation

The solution of a problem can be represented by a set of parameters (genes), linked to form a value array (chromosome); this process is called encoding. The solutions (chromosomes) are encoded by a sequence consisting of an alphabetic system. Originally, the binary alphabet {0, 1} was used. However, new GA models encode the solutions with other alphabets, such as real numbers [191]. Thus the representation is a key aspect in the GA modelling defining the chromosome structure with the respective composing genes, so that it is able to encompass the entire relevant search space of the problem. The chromosome decoding consists in the construction of the problem real solution from the chromosome. That is, the decoding process builds the solution so that it is evaluated by the problem.

4.3.1.2 Evaluation

The evaluation is the link between the GA and the problem to be solved. It is made through a function that best represents the problem and aims to offer a fitness measure for each individual in the current population, which will direct the search process. The evaluation function for a certain chromosome consists in the associating a numeric value, which is assumed proportional to 'utility' or 'ability' of the individual represented to solve the problem.

4.3.1.3 Genetic Operators

The best-known GA operators in AG's are the reproduction, crossover, and mutation.

Reproduction refers to the process of selecting and copying a specific chromosome to the next population according to its fitness. This means that the fittest chromosomes (larger fitness value) are most likely to contribute to the formation of one or more individuals of the next population. There are basically the following methods: change the entire population, change the entire population with elitism, where all the chromosomes are replaced where the current population fittest chromosome is copied to next population, and partial change of the population (steady state), where the M fittest individuals of the current population are copied to next population [187,188].

Crossover is an operator based on the exchange of chromosomes parts (parents), forming two new solutions (decedents). This process can be observed in Fig. 4.26, where the solution is encoded with a binary alphabet.

To perform the crossover, first it is necessary to select by lot of chromosomes 'parents'. Then there is the crossover implementation or not according

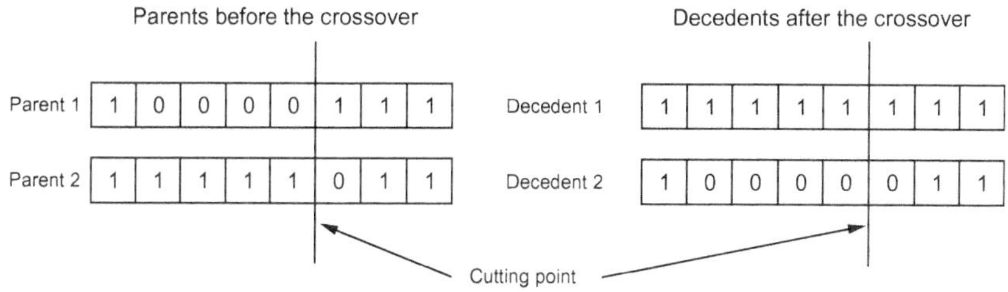

Fig. 4.26 Single point crossover.

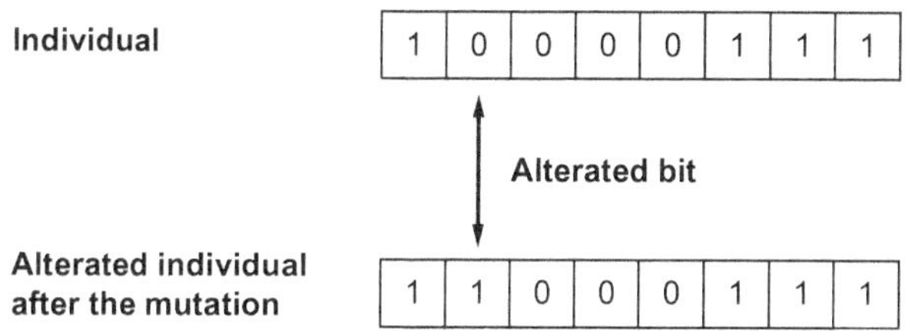

Fig. 4.27 Mutation.

to a parameter, called crossover rate. Thus according to the crossover rate the chromosomes 'parents' could be transferred without modification to the next generation, creating 'decedents' identical to the 'parents'. The point where the crossover cut is random selected; in the example of Fig. 4.27, a single point is used, but more than one point cuts can be performed, featuring a multipoint crossover [188,191,192]. The crossover operator idea is to take advantage (exploit) the genetic material present in the population.

Mutation is a random change of the values contained in the genes of a chromosome by others valid in the alphabet. In the binary alphabet case a change from 0 to 1 and vice versa. Similarly to the crossover, a mutation rate is used that, for each bit of a character sequence, the mutation is performed or not based on a random chance; when the mutation is performed, the bit will be replaced by another valid value belonging to the alphabet (Fig. 4.27).

The mutation ensures a diversity of the individual's characteristics of a population and allows the introduction of information that was not present in any of the individuals. In addition, it provides a random search (*exploration*) in GA space, providing an opportunity for more points to be evaluated.

4.3.1.4 Evolution Parameters

The parameters that most influence the GA performance are follows:

Population size: The size of the population affects the GA overall performance and efficiency. A very small population offers a little coverage of the search space, causing a drop in performance. A sufficiently large population provides a better coverage of the problem domain and prevents the premature convergence to local solutions. However, a large population demands more computer resources, or a longer processing time. Thereafter an equilibrium point in respect to the selected population size should be aimed.

Crossover rate: Probability of an individual being recombined with another. The higher the rate, the faster new structures will be introduced into the population. However, this can generate an undesirable effect as most of the population will be replaced, so loss of genetic variety may occur, a possible loss of high fitness structures and convergence could happen leading to a population with extremely similar individuals, where these individuals could either represent a good solution or not. Low values lead the algorithm to very slow on finding an acceptable answer.

Mutation rate: The probability of a chromosome gene content to be changed. The mutation rate prevents a given population to stay stagnant at a value, in addition to allowing that the search at any point in the space domain. However, very high mutation rate should be avoided, since it can make the search essentially random, strongly inhibiting the convergence to an optimal solution.

Generation interval: Generation interval controls the percentage of the population that will be replaced during the next generation (total replacement, replacement with elitism, replacing the current population worst individuals, and partial replacement of the population without duplicates). This number of replaced individuals is also known as gap.

Number of generations: The number of generations represents the total number of cycles of the genetic algorithm evolution, that is, one to the GA stop criteria. A number of generations too small cause a decline in performance, because the entire search space could not be covered. A large value makes necessary longer processing times, but provides a better coverage of the problem domain avoiding convergence to local solutions.

4.3.2 Genetic Programming

Another evolutionary computation important paradigm is the GP. Similarly to the evolution of living things, the GP aims to develop computer programs to solve real-world problems. These programs are subjected to an evolutionary process involving evaluation, selection, and genetic operations as crossover and mutation. After several evolution cycles the population should contain fittest individuals. For each individual an adaptation value, known as fitness, is assigned indicating how much the solution represented by this individual is good when compared with other solutions within the population.

1. GP structures are computer programs that represent individuals in a population. These individuals are tree-shaped structures that composed

of nodes and leaves. The nodes represent functions or mathematical operations linking one or more leaves representing the terminals. The terminals can be variables or constants.

2. In general, the GP produces computer programs to solve problems by performing the following steps:
   - An initial population is randomly generated by an assembly of the problem functions and terminals (computer programs).
   - The next substeps are processed iteratively until the stop criterion is satisfied:
3. The following paragraphs contain a brief description of the GP individual representation. GP structures are essential to the evolutionary process knowledge and understanding of genetic operations on individuals (c).

Each programme is executed on the population and it associates an aptitude according to an assessment of the problem solution satisfaction.

1. A new population of computer programs is created using the genetic operators. The operations are probabilistically applied to the selected computer programs according to their fitness.
2. The best computer programme that appeared in any generation (the 'best so far') is designated as a GP result. This outcome may be the solution (or the approximate solution) for the problem.

Representation

In GP, the individual represents a computer programme, which consists of functions and terminals. The functions are composed of parameters for input values and returning one or more outputs. The terminals are parameters for input value returning one or more outputs. The combination of all the possible solutions defined by these functions and terminals is called search space.

The structure of each individual in GP is represented in the form of a tree. The terminals are tree leaves located at the extremities. The functions connect one or more terminals situated at the trees vertices (nodes) as shown in Fig. 4.28.

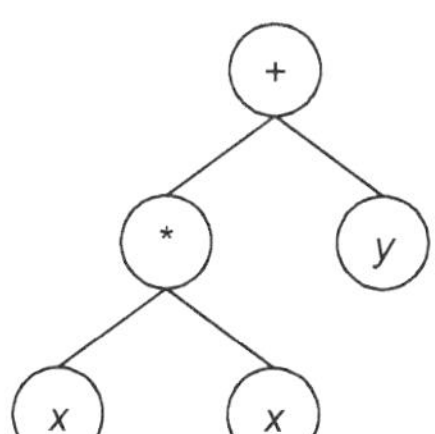

Fig. 4.28 Individual genetic programming tree representation.

Koza [187,193] uses the S-expression term (Symbolic Expression) to designate an individual GP representation in the form of a linear expression. This form is the interpretation of a structure represented by a tree. As an example, the representation of an individual identified by the expression $x+y$ is represented by its linear form $(+y\ (*\ x\ x))$ and by its tree structure in Fig. 4.28.

According to [187,193], the elements of tree structure can be separated in:

Functions: Appear in the tree vertices:

1. arithmetic operators (+, −, *, /, %, etc.);
2. math operator functions (log, exp, sin, cos, etc.);
3. 'logical' operators (and, not, or);
4. conditionals operators (if, then, else);
5. iterative operators (whilst (condition) do);
6. recursive functions; and
7. specific functions of the problem domain.

Terminal: Appear in the tree leafs:

1. Variables.
2. Constants.

The individual represented in Fig. 4.29 has its functions (addition and multiplication) and their terminals (x and y) represented by numerically ordered points from right to left. The ordering of these points is shown in Fig. 4.29. The interpretation of the individual takes place according to this ordering, from the root or top to the leftmost element. From this element a point-to-point trajectory is performed up to the rightmost element.

Having the terminals and functions defined, the PG evolution already has the raw material necessary to create an individual. In order to be valid the individuals must meet the property 'functions should accept as an argument any value or data type that is returned by the functions set and any value and data type that can be assumed by any terminal' [187,193].

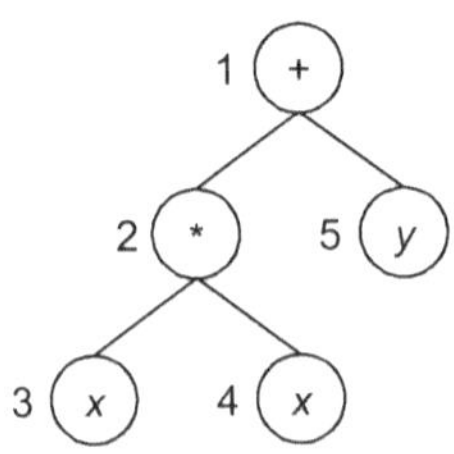

Fig. 4.29 Numbering of points of an individual.

4.3.3 Application Examples

4.3.3.1 Beams Under Patch Loads

This investigation used GA to forecast the ultimate resistance of steel girders under patch loads. The genetic algorithm main objective was to optimise the coefficients used in the patch load equation. Bergfelt [158] and Roberts [159] studies led to some of the most accurate patch load prediction formulae, whilst previous studies [152] involving neural networks led to significant results. The maximum and mean neural network errors were less than 15% and 4%, respectively. Using Roberts equation and the most relevant parameters of Bergfelt equation calculated trough a parametric analysis [152], the production of an equation structure was made together with its associated coefficients to be calibrated, that is:

$$P_{cr} = \left(\frac{t_W}{d_w}\right)^{K1}\left[t_w^{K2}(E\sigma_w)^{K3}\left(\frac{t_f}{t_w}\right)^{K4}\left(K5 + K6c^{k7}t_W^{K8}t_f^{K9}d_w^{k10}b_w^{K11}\right)\right] + K12\left(\frac{t_W}{d_w}\right)^{K13} t_W\sigma_w\left(c + K14t_f\right) \quad (4.14)$$

The main objective of this study [197] was the conception of a single equation that could be used for any steel girder. The equation was split into two terms: the first one is related to the web panel instability, whilst the second is related the web yield collapse. This calibration was made using the $K1$ and $K13$ coefficients. The Evolver software was utilised to perform the GA evolution. The coefficients calibration was made by the minimising of the maximum percentile error between the all 155 experiments and value evaluated using the prediction formula. An additional comparison between the current design formulae and the developed GA, equation for all the 155 data was also made. Although the GA method reduced the equation error, another strategy was developed only using, in the optimisation process, the data from the slender steel beams. The database was reduced from 155 to 127 leading to a performance improvement (Table 4.1) of the proposed equation:

$$P_{cr} = \left(\frac{t_W}{d_w}\right)^{-0.015}\left[t_w^2(E\sigma_w)^{0.5}\left(\frac{t_f}{t_w}\right)^{0.39}\left(0.78 + 0.57c_e^{4.62}t_W^{-2.21}t_f^{2.20}d_w^{1.19}b_w^{-5.8}\right)\right] + 1.72\left(\frac{t_W}{d_w}\right)^{0.56} t_W\sigma_w\left(c + 3.05t_f\right) \quad (4.15)$$

Table 4.1 Calibration of the Genetic Algorithms Equation for the Beam Patch Load Strength

	GA Equation	Roberts [159]	Bergfelt [198]	Lyse and Godfrey [157]
Maximum error	28.49	44.19	125.04	117.01
Mean error	11.43	12.20	23.86	44.05

Despite the formal restrictions, the GA equation has led to satisfactory results confirming the advantages of using a single design equation. Although the GA equation errors were larger than the neural network results, they were less than the traditional design formulas. This fact proves to be more relevant when a single equation approach is considered. The maximum error related to the GA equation was less than all the other formulas, enabling in future steps, a reduction of the structural engineering safety factors related to this ultimate limit state.

4.3.3.2 Composite and Steel Semirigid Endplate Joints' Structural Design

One of the most desirable objectives of the Engineering Science is related to the development of a cost-effective production of structural members attained with the implementation of new design philosophies and procedures. When the search for optimal structural connections is considered within this context, the GA prove to be a valuable tool for optimising steel and composite semirigid connections. Focusing on the aim of minimising the force spent in the structural design, the study presents an automatic optimisation procedure implemented to determine the optimum joint [200,201]. This optimum connection has the required flexural strength and initial stiffness aiming at the minimum connection fabrication cost.

The developed software starts the design process executing an elastic analysis to find out the first estimative of the structural response. This preliminary data are used in the genetic algorithm system with its basic operations to determine a new parameter value that is subsequently compared with the desired objectives. This iterative procedure is repeated until the optimum values are reached (i.e. minimising the connection cost for a required flexural capacity). This process is based on the use of GA software, Evolver [199], producing reliable results in terms of economy and efficiency, with a reduced computational force. This study utilised the procedures for the endplate beam-to-column steel and composite connection design according to the component method proposed in the Eurocodes 3 [19] and 4 [64] as well as recommendations proposed by Ahmed and Nethercot [202] and Ramires et al. [203].

The evaluation of the joint global response requires the consideration of their full geometrical and mechanical properties. With these results in hand, the connection mechanical model can be characterised according to the connection type. This followed by the component resistance and the connection moment versus rotation curve evaluations. The variables used in the optimisation process are geometric properties like distances, thickness, etc., and are manipulated by the GA.

The authors described a process to reduce the cost of composite and steel semirigid endplate joints. The process is based on a suitable selection of the most significant parameters that control the connections structural performance to obtain cost-effective solutions with the initial stiffness and flexion capacity required in structural design. This process is centred on the search for an optimum connection within reliable ultimate and serviceability limit state constraints.

All the examples presented were limited to two types of structural joints, that is, flush and extended endplate joints and started from an initial basic joint configuration (Fig. 4.30). The first example involves flush and extended steel endplate joints aiming to transmit bending moment values close to 120 and 423 kNm. The optimisation results depicted a large dispersion of the initial stiffness, since no restriction was applied to these variables. A different trend was noticed with the flush endplate connections' results that depicted a significant stiffness increase when compared with the initial steel joint model. Composite flush and extended endplate joints were also optimised to transmit bending moment values close to 120 and 423 kNm, where in both cases a sudden increase of the initial stiffness was observed when compared with their equivalent steel joint models, confirming the composite joints advantages.

The final example considered composite flush endplate connections and steel extended endplate connections able to transmit bending moment levels corresponding to 60%, 80%, and 100% of the plastic bending moment capacity of the used beam IPE steel profiles, aiming to minimise their global fabrication costs. Fig. 4.31 contains a comparison between the composite and steel connections in terms of a ratio between their bending capacities to their global costs. The composite connection was associated to higher bending capacities and to reduced costs when compared with their steel counterparts.

The main aim of the optimisation was to reduce the global connection cost ensuring that the associated connection flexural capacity and initial stiffness did not surpassed the required ultimate and serviceability limit states. It could also be noticed that the composite connections' performance was improved when heavier steel sections were utilised. The performed examples validated and proved the viability of using GA to optimise composite and steel semirigid joints.

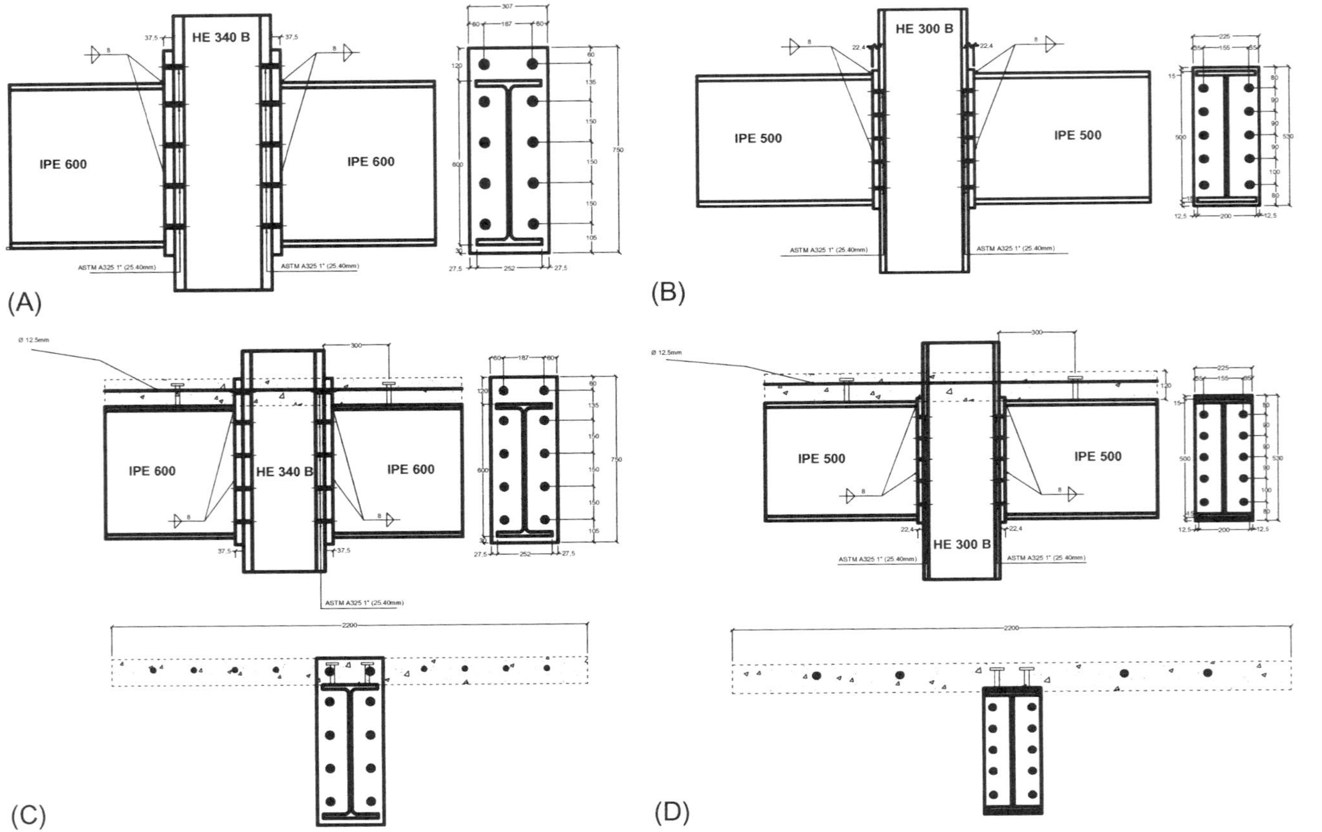

Fig. 4.30 Adopted joint models: (A) extended endplate, (B) flush endplate set, (C) composite extended endplate, and (D) composite flush endplate.

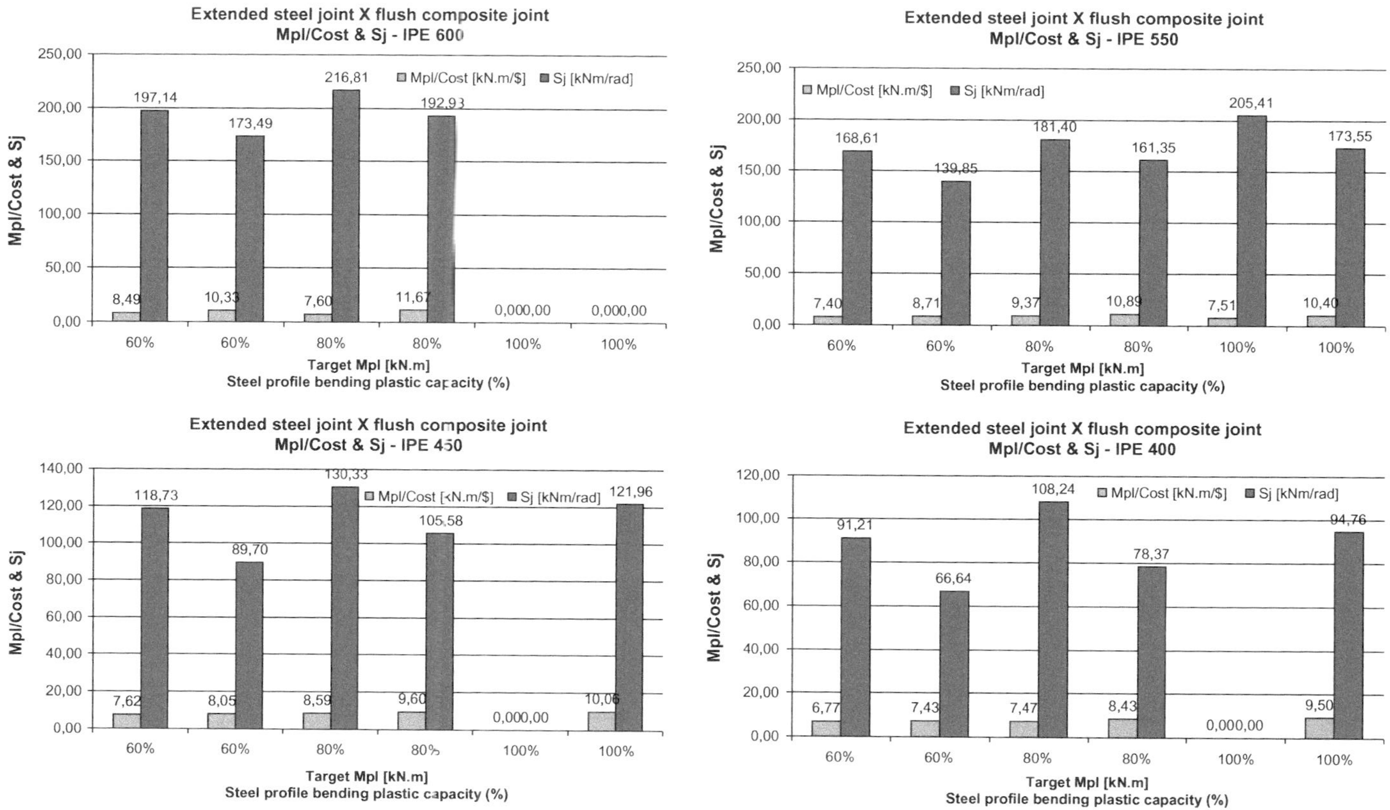

Fig. 4.31 Cost and structural response of composite and steel models (IPE profiles).

4.3.3.3 Semirigid Joints' Post-Stiffness Evaluation

This investigation proposes the use of GA to calibrate the post-limit stiffness of the various connection components proposed in the Eurocode 3 [19] component models. NASCon, specialised computer software purposely developed for the analysis of steel joints, was used to calibrate the post-limit stiffness of each individual connection component. An automatic iterative procedure was implemented combining the component method solver (Nonlinear Analysis of Steel Connections, NASCon) with the genetic algorithm software, Evolver, to yield reliable predictions of the post-limit stiffness of each component from available experiments for the global connection response.

The endplate beam-to-column connections studied were designed according to the component method present in the Eurocode 3 [19]. The connection overall response can be found based on the force–deformation properties of its components. Subsequently, the components are assembled into a mechanical model, made of extensional springs and rigid links.

Generally, each component is characterised by a nonlinear force versus displacement curve, adequately represented by a bi-linear approximation when only the resistance and the initial stiffness of the joint are required [19]. The post-limit stiffness of the components is not considered since perfect elastic-plastic behaviour is assumed [19].

The NASCon software [205] was conceived to provide a user-friendly tool that can aid researchers and civil engineers to easy implement the component method procedures. This software mimics the connection response by numerically evaluating its generalised force–deformation behaviour from the complete characterisation of the relevant components. It permits comparison to experiments and the corresponding error assessment. NASCon calculates the moment versus rotation curve by successive comparison to the experimental curve.

The solution fitness is measured according to the distance between the idealised and the experimental curves. The procedure continues with the genetic algorithm operating the changes in the chromosome and a new iteration is made. This interactive procedure was implemented with the help of genetic algorithmic software, Evolver [199]. Numerous genetic algorithm parameters configurations were tried such as crossover rate, mutation rate, and population size [204–206]. The best results were achieved with crossover rate: 0.30, mutation rate: 0.008, and population size: 50.

The first test, a flush endplate beam-to-column connection, was tested at the Civil Engineering Department, University of Coimbra [47]. In this test, the column was simply supported at both ends and consisted of an HEB240 section whilst the girders were made of an IPE240 and a 15-mm thick end-plate, all made from an S275 steel. The bolts were M20, class 10.9.

The other beam-to-column connection tests, extended endplate connections with backing plates (Humer, 109.005 and 109.006), and welded connections (Klein, 105.018) were obtained from the SERICON II Database [208]. The component resistance for all tests was calculated according to EUROCODE 3 using the measured material properties.

In test FE1, the slope of the obtained post-limit curve from the mechanical model was close to the experiments (see Fig. 4.32). For the extended endplate connections, the genetic algorithm optimisation results reached a good agreement with the experiments as shown in Fig. 4.33. The results for the welded connections, Fig. 4.34, also depicted a good agreement to the experimental curves. In this connection, the reduced number of components obviously made the convergence process faster.

The computational evolutionary techniques potential for the assessment of structural engineering problems has been fulfilled by the trustworthy results obtained with the semirigid connection. The use of the GA procedure simplified the components calibration against the experiments. The main contribution of this investigation was to present the use of genetic algorithm as a design tool to forecast the full response of the beam-to-column connections taking into account when appropriate, the components post-limit stiffness.

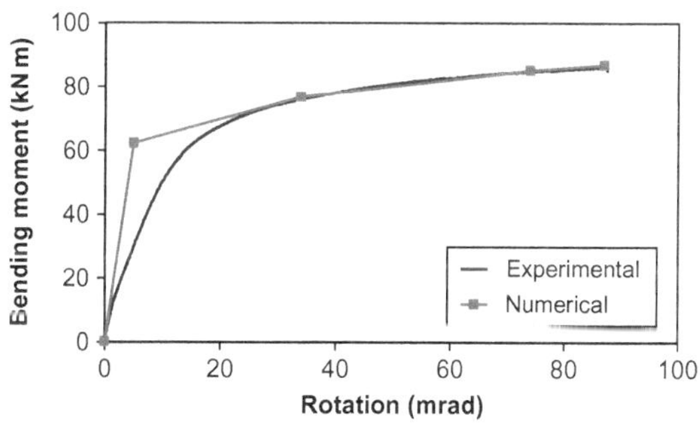

Fig. 4.32 Moment versus rotation, FE1 [47].

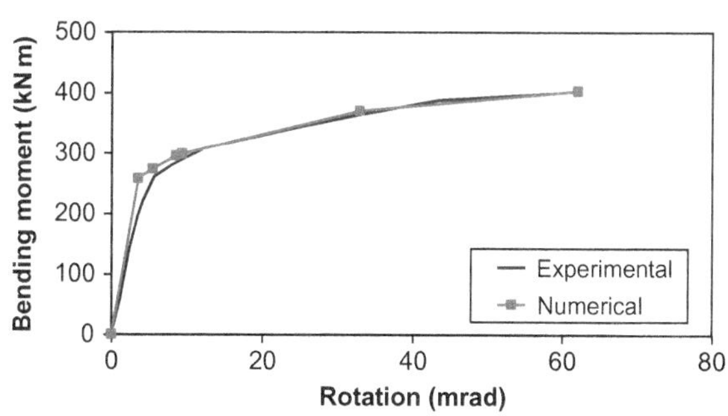

Fig. 4.33 Moment versus rotation, 109.005 [208].

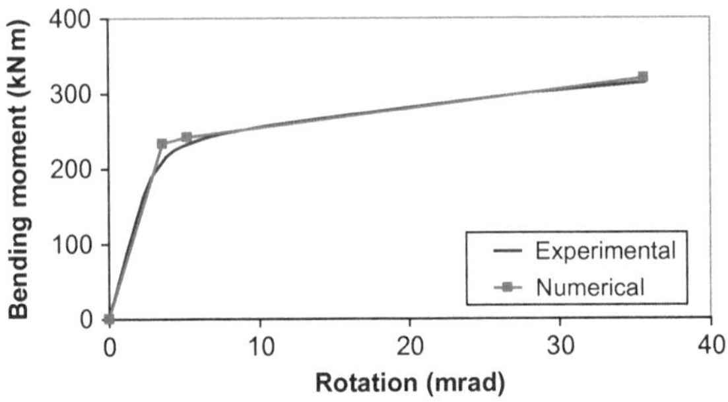

Fig. 4.34 Moment versus rotation, 105.018 [208].

It is also important to observe that before implementing the GA procedure the components' post-limit stiffness determination was only possible through a user-dependent time-consuming manual process. The best results were reached for the welded connections although all the results proved to be reliable when compared with the available tests. Despite this fact the results showed that the GA can be safely utilised to predict the required connection components post-limit stiffness.

4.3.3.4 Minor Axis Semirigid Joints' Structural Design

The joints studied were made of double web angles, a support angle or a transverse web stiffener welded to the column web. A simple mechanical model [19] was utilised to assess the joint's structural response. In this model, each joint component contributes to the global joint stiffness. When minor axis behaviour is considered, an extra component, the web column out of plane stiffness, is not taken into account producing misleading results. The adopted chromosome modelling was made in a tree format. At the evolution end, the web column stiffness values (when compared with finite element simulations) presented a 9% mean error [211].

When beam-to-column connections to the column minor axis are considered, the used design process generally assumes such connections as pinned. This is not true for the great majority of structural joints and motivated the development of a mechanical model, in accordance to the Eurocode 3 [19] to forecast the joint's structural response. The connections studied were made of web angles, a seat angle or a transverse web stiffener welded to the column web with the exception of the *k*14 component (column web panel subjected to out of plane loads) that still did not existed in any design code.

Although each component is characterised by a nonlinear force–displacement curve, their response can be approximated by multilinear or simple polynomial models [49]. A validation study [77] of the proposed model was accomplished through comparisons with the acquired experimental results and finite element simulations using the ANSYS computer programme [78]. Whilst a nonlinear load versus deflection curve characterises each component, simple multilinear or polynomial models can approximate their response. A validation [209] of the proposed model was performed through comparisons to experiments and finite element simulations using the ANSYS programme [32].

The experimental investigation [37] comprised a set of three tests made in column's minor axis using a cantilever-loading girder. The beams and angles were made of rolled steel sections [S shape 10″ × 37.7, L 76 × 76 × 9.5 (first test), L 127 × 76 × 9.5 (other tests)]. The adopted column was a three-plate welded steel section, W 310 × 60. In the first test, due to the web plate flexibility, large deformations occurred and the maximum joint moment was 38 kNm.

The GA were used to identify, through a symbolic regression, the equation that correlates the joint geometric properties and the column web out of plane resistance. The input data consisted of the column web thickness (t_w) and the column depth (d), and the target was the portion of stiffness of the web column under out of plane loads ($k14$). The adopted programme was Evolver 4.0 [199] and the configuration parameters were crossover rate: 0.65 and mutation rate: 0.008. The input data, the $k14$ theoretical value, and GA results as their respective errors are shown in Table 4.2 whilst the final evolution equation of $k14$ is:

$$k_{14} = \frac{\left(t_w^{2,6} - f_y^{4,2}\right)}{\left(E - d\sqrt{t_w^3}\right)\left(2.f_y^{\sqrt{t_w}} + E\sqrt{d^3}\right)} \tag{4.16}$$

An inspection of the results showed that the potential use of GP for dealing with structural and civil engineering problems was fulfilled with the studied semirigid connections.

Table 4.2 Genetic Algorithms Input Data and Results

t_w (mm)	d (mm)	$k14$ Theory (m)	$k14$ GA (m)	Error (%)
6.3	300	0.01769	0.01769	0.00
8.0	300	0.03109	0.03290	5.83
10.0	300	0.05425	0.05419	0.12
6.3	600	0.00810	0.00642	20.76
8.0	600	0.01286	0.01218	5.27
10.0	600	0.02810	0.02166	22.89
6.3	800	0.00381	0.00424	11.17
8.0	800	0.00667	0.00812	21.74
10.0	800	0.01476	0.01477	0.07
MAPE				9.76

4.3.3.5 Structural Design Optimisation of Beam-to-Column Semirigid Joints

The main objective of this study was to optimise the flexural capacity of steel structures by varying the stiffness and resistance of its associate semirigid connections. This optimisation process used a genetic algorithm developed and implemented in a structural analysis programme, FTOOL/SRC [211] This algorithm was inspired in the optimum-use algorithm, incorporating the problem domains, the examples adopted: ten experiments with 50 generations each, a population of 50 individuals, the selection technique included an accumulated relative evaluation, steady-state reproduction with a GAP value of 50, and the operator roulette technique. The search domain limits for the problem variable were, for the Vierendel beam example, zero as the lower limit and two upper limits of $1.0e^{+16}$ and $1.0e^{+08}$.

A 12-m span Vierendel structural system, 1 m height was optimised [211]. The columns (vertical member) and beams (horizontal members) were, respectively, HEB240 and IPE240 steel sections. Fig. 4.35 shows (A) the rigid connection initial modelling used as GA seeds and (B) the bending moment distribution of the associate frame.

Fig. 4.36 shows the GA performance in terms of an average curve of the best evaluations in ten experiments and of the best evaluations in 100 individuals associated with one experiment. When the initially rigid modelling is compared with the GA final results, it is evident that the GA method led to semirigid connection configurations with associate stiffness values significantly smaller than the rigid connection values.

The new connection stiffness configuration led to a substantial improvement of the bending moment distribution. A close inspection of Fig. 4.36 shows that the search for the best average evaluation is influenced by the

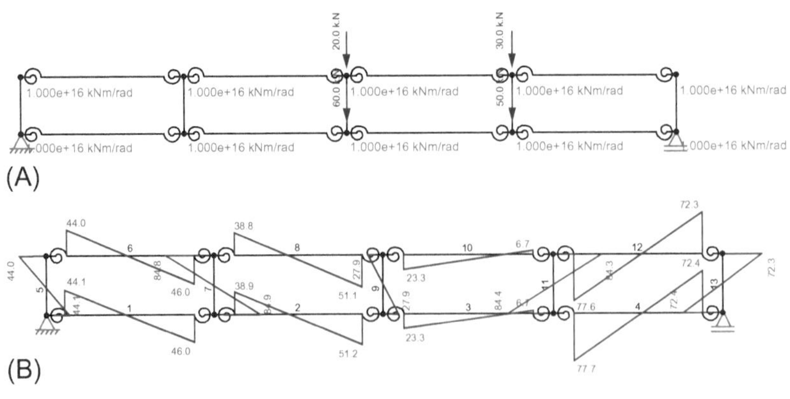

Fig. 4.35 (A) Initial joint stiffness values. (B) Bending moment distribution.

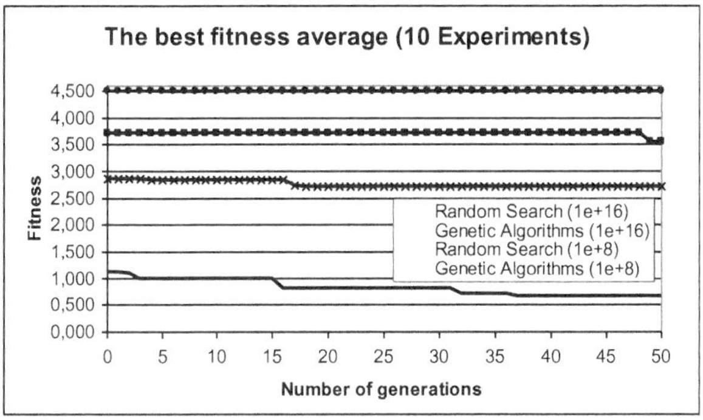

(Ten experiments)

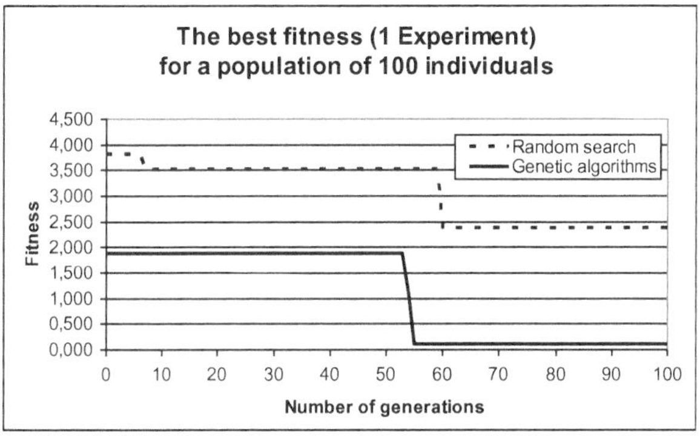

(One experiment, for 100 individuals population

Fig. 4.36 Comparison between genetic algorithm and random search.

variable search space limits. When the search space was limited within zero and $1.0E^{+16}$, the GA method led to an average evaluation of 3.539. Alternatively, when the search space was limited within zero and $1.0E^{+8}$, the GA method led to an average evaluation of 0.668.

In the comparisons performed in Fig. 4.36 the GA demonstrated a superior performance than the random search procedure. When the best evaluation achieved with the use of the random search procedure (2.364) was compared with the GA method (0.113), a difference of 2.251 was found. If this process was repeated for the results of Fig. 4.36, differences of 2.042 and 0.969 were determined for the search space with superior limits of $1.0E^{+08}$ and $1.0E^{+16}$. These results validated the GA implementation and the methodology adopted in the problem evaluation function.

The present investigation was centred in a nonconventional semirigid connection stiffness optimisation procedure where individuals were evaluated to generate an optimum bending moment distribution for a structure.

This search methodology, varying the joint stiffness values, was used in two different examples. The best solution was always associated with the genetic algorithm when compared with the random search procedure.

The difference between these two methods is amplified when the variable search space is increased. This difference was also magnified by increasing the structural problem complexity. A significantly better bending moment distribution was reached when the adopted objective function was globally applied to the studied problems.

The Vierendel structural system example showed that the semirigid solution performance, found with the GA method, was better than the usually adopted rigid solution. The semirigid connections created a balanced moment distribution with a smooth transition between the sagging and the hogging moment regions. An optimum bending moment distribution is a direct consequence of an adequate choice of the structural connection stiffness. This is usually a complex procedure when manually made due to the tedious and significant amount of work required to evaluate the connection stiffness, combination that could lead to a balanced bending moment distribution.

The use of GA considerably simplified this process, leading to bending moment distributions that could produce efficient and economic steel structures. The results shown that the implemented GA procedure and associated methodology are efficient, representing an important contribution to the development of steel frame structural design with semirigid connections.

4.4 NEURO-FUZZY NETWORKS

Neuro-Fuzzy systems [212] are some most widely studied hybrid systems due to the fact that they combine associate the advantages of Neural Networks [148] and Fuzzy Logic [213]. These systems merge the learning capability of Artificial Neural Nets with the power of linguistic interpretation offered by Fuzzy Systems.

One of the main restrictions of conventional neuro-fuzzy systems, however, is their limitation to work with a small number of inputs. This restriction is associated with the 'curse of dimensionality' generated by the large number of rules that comes from the splitting of the input space in a grid form. A neuro-fuzzy system with 5 input variables, each with its universe of discourse divided into 4 fuzzy sets, may obtain a total of 1024 rules (4^5). If the number of inputs is increased to 20, and using the same division of the universes of discourse the result is an unmanageable total of 1.099.511.627.776 rules (4^{20}). Most existing neuro-fuzzy systems, for example, NEFPROX [214], NEFCLASS [215], and ANFIS [216], make use of grid partitioning, which significantly limits the total number of possible variables to be dealt within the model. Recursive partitioning can be used to minimise this problem, such as Quadtree [217] and BSP (Binary Space Partitioning) [218], which employ recursive processes in their generation. By preserving the input characteristics of independence, these types of partitioning maintain the model interpretability in a fuzzy rule format.

The BSP partitioning was then used to create a new class of Neuro-Fuzzy systems and the Hierarchical Neuro-Fuzzy Binary Space Partitioning models (HNFB) [219]. This class of models overcomes two of the main weaknesses of current Neuro-Fuzzy Systems: their fixed structure and limited number of inputs. One particular version of these models is the HNFB-Class—Hierarchical Neuro-Fuzzy BSP Model for Classification [220], which will be described in the following section.

4.4.1 Hierarchic Neuro-Fuzzy BSP Models for Classification

The Hierarchical Neuro-Fuzzy BSP model dedicated to pattern classification and rule extension, called HNFB-Class, was developed to automatically extract classification rules such as *If x is A and y is B then input-pattern belongs to class Z*. The HNFB-Class model is composed of one or several standard cells called NFB-Class cells. These cells are arranged in a hierarchical structure in a binary tree form. The cell outputs in the lower levels are the consequents of the higher-level cells. The following paragraphs will define the basic cell, the hierarchical structure, and the learning algorithm.

4.4.1.1 Basic NFB-Class Cell

A NFB-Class cell is a neuro-fuzzy mini-system that performs fuzzy binary partitioning of a particular input space according to the complementary membership functions $\mu(x)$ and $\rho(x)$ described by Eqs (4.17) and (4.18).

$$\mu(x) = \frac{1}{1 + \exp[-(ax - b)]} \tag{4.17}$$

$$\rho(x) = 1 - \mu(x) \tag{4.18}$$

where b is the sigmoid inflexion point and a is the sigmoid inclination at $x = b$. Fig. 4.37A shows the NFB-Class cell basic representation whilst Fig. 4.37B shows its interior. In this cell, x_i represents the input variable, whilst $\rho(x)$ and $\mu(x)$ are the membership functions low and high, respectively.

After a defuzzification process, the NFB-Class cell generates two crisp outputs, y_1 and y_2, given by Eqs (4.17) and (4.18), where the idea that

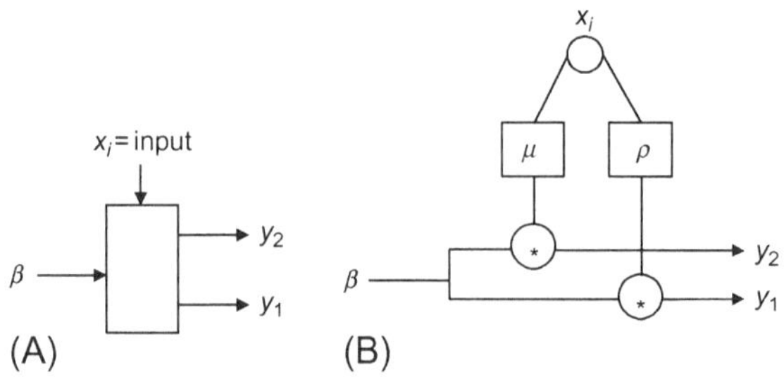

Fig. 4.37 (A) NFB-Class cell simplified representation. (B) NFB-Class cell Interior.

membership functions μ (*high*) and ρ (*low*) are complementary has already been considered.

$$y_1 = \beta^* \rho(x) \tag{4.19}$$

$$y_2 = \beta^* \mu(x) \tag{4.20}$$

The variable β in the previous equations corresponds to one of the two possible cases:

- β *is* the first cell input: Case in which $\beta = 1$, where the value '1' represents the entire input space, that is, the entire universe of discourse of the variable x_i that is being used as the cell's input.
- β *is* the stage of a previous level output: Case in which $\beta = y_j$, where y_j represents one of the two outputs of a generic '*j*' cell, whose value is also calculated by Eq. (4.19) or (4.20).

4.4.1.2 HNFB-Class Architecture

The HNFB-Class architecture is formed by the interconnection of basic NFB-Class cells in a binary tree structure. The architecture is divided in two basic modules: the *decision-tree module* and the *classification module* (see Fig. 4.37).

The decision-tree module contains the input partitioning structure automatically created by a learning algorithm based on the gradient descent optimisation method. The classification module is made of n T-conorm (OR fuzzy operator) neurons that represent the n classes present in the database. The T-conorm cell with the highest output value defines the class which the input pattern belongs.

Fig. 4.38A shows a decision-tree module made of three levels, obtained during the training phase of a certain database with three classes whilst Fig. 4.38B shows its respective input space partitioning. This specific partitioning structure contains five leaf cells, named d_i, where $i \in \{1, \ldots, 5\}$. The output value of each leaf cell d_i is evaluated with the following equations [using the complementary membership functions ($\rho + \mu = 1$)]:

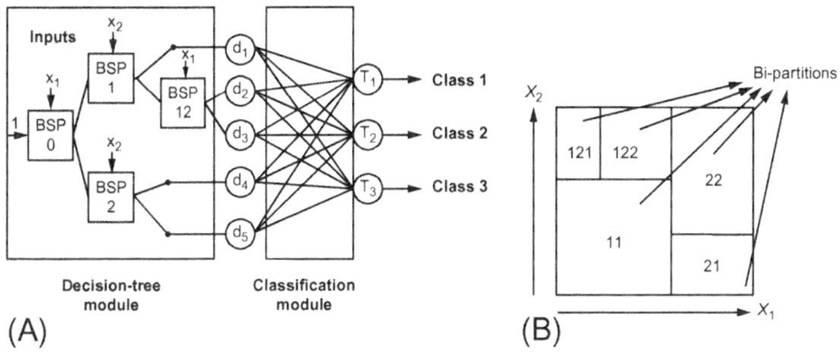

Fig. 4.38 (A) HNFB-Class architecture. (B) Decision-tree module input space partitioning.

$$d_1 = \rho_0 \cdot \rho_1 \tag{4.21}$$

$$d_2 = \rho_0 \cdot \mu_1 \cdot \rho_{12} \tag{4.22}$$

$$d_3 = \rho_0 \cdot \mu_1 \cdot \mu_{12} \tag{4.23}$$

$$d_4 = \mu_0 \cdot \rho_2 \tag{4.24}$$

$$d_5 = \mu_0 \cdot \mu_2 \tag{4.25}$$

where

- d_i are the outputs of the input partitioning structure;
- ρ_0, ρ_1, ρ_2, and ρ_{12} are the membership degrees of the input pattern in the *low* membership function of partitions 0, 1, 2, and 12, respectively; and
- μ_0, μ_1, μ_2, and μ_{12} are the membership degrees of the input pattern in the *high* membership function of partitions 0, 1, 2, and 12, respectively.

Each d_i is connected to all T-conorms neurons (T_i) of the Classification Module, with a weight value associated with every joint. These weights are assigned by a learning method based on the Least Mean Squares [148], as detailed in the next section. Each T-conorm neuron output is then calculated using the Limited Sum T-conorm operator [220,213] over the weighted d_i inputs. In the example of Fig. 4.38A, the outputs of the three T-conorm neurons are evaluated through:

$$\text{T} - \text{conorm class } 1 = y_1{}^*w_{11} \oplus y_2{}^*w_{21} \oplus y_3{}^*w_{31} \oplus y_4{}^*w_{41} \oplus y_5{}^*w_{51} \tag{4.26}$$

$$\text{T} - \text{conorm class } 2 = y_1{}^*w_{12} \oplus y_2{}^*w_{22} \oplus y_3{}^*w_{32} \oplus y_4{}^*w_{42} \oplus y_5{}^*w_{52} \tag{4.27}$$

$$\text{T} - \text{conorm class } 3 = y_1{}^*w_{13} \oplus y_2{}^*w_{23} \oplus y_3{}^*w_{33} \oplus y_4{}^*w_{43} \oplus y_5{}^*w_{53} \tag{4.28}$$

where

- y_i, $i \in \{1,2,3,4,5\}$, is the leaf cell outputs;
- W_{ij}, $i \in \{1,2,3,4,5\}$ and $j \in \{1,2,3\}$, is the joint weight between the leaf cell i and the T-conorm neuron j; and
- $\oplus$ is the Limited Sum T-conorm operator.

The final HNFB-Class system output is evaluated when the highest output obtained amongst all the T-conorm neurons is utilised to determine the class to which the input pattern belongs. The training algorithm is based on the gradient descent method for learning the model structure and the linguistic rules. The parameters that define the antecedents membership functions profiles are considered as fuzzy weights of the neuro-fuzzy system.

A tuning parameter δ, that is, the decomposition rate prevents the structure from growing indefinitely. A detailed description of the Hierarchical Neuro-Fuzzy BSP models and their learning algorithm can be found in [220–223].

4.4.2 Application Examples

4.4.2.1 Beams Under Patch Loads

Bergfelt [158] described the collapse of a beam under patch loads in three phases. The first step lasts until some yielding is detected in the web surface. The second phase extends from this point until some small folds can be found in the web. The last phase goes up to the collapse. The structural failure was then classified in three different cases: yielding, overall web buckling, and a localised zone under the load where bigger folds and high stress concentrations are presented (commonly known as web crippling). Previous studies [152–156,161] demonstrated that new data could be obtained from neural network models.

The girders ultimate load was forecasted using a data set composed of 155 experiments with ultimate loads varying from 30 to 4010 kN. The results have been compared with experiments and existing design formulae, indicating a good agreement. Subsequent investigations split the experiments available into three classes according to the ultimate load magnitude. Further studies utilising a single neural network with all the 155 experiments, used a different ultimate load normalisation technique. In this process the ultimate load value is divided by a modified girder web yield capacity. The BP algorithm was utilised to train the neural net made of 15 beam parameters as inputs and the normalised beam resistance as output.

The experimental data were split in three distinct groups: 70% for training, 20% for testing, and 10% for validating the neural networks. Numerous neural network configurations were trained and compared, varying the number of hidden layer processors. However, despite the results accuracy, the system architecture did not explicitly consider the main difference in the beam ultimate limit state associated with the structural failure (web and flange yield, web buckling, and web crippling).

This was the motivation for the study that presented the practical use of a neuro-fuzzy system that can consider the patch load response according to the related physical phenomena (web yielding, web buckling, and web crippling) [224–228]. A previously trained neuro-fuzzy network was utilised to classify the evaluated beams according to their pertinence to a specific structural response. With the pertinence values established by the neuro-fuzzy classification network, a neural network was utilised to predict the girder ultimate patch load capacity (see Fig. 4.39).

The data used in the Neuro-Fuzzy classification system training and validation phases used geometric material properties as inputs and the associated physical phenomenon as the output. This approach required that all available data must be correctly labelled, according to each path load case physical phenomenon. The Neuro-Fuzzy system was trained with the data presented in Fonseca et al. [224–228] with 50 additional data from a rolled section data set (data 162–211). These additional data were necessary to expand the number of cases in the database associated with the web yielding collapse, third class. The database was split into three groups: 70% for training, 10% for validation, and 20% for testing the model.

After several trainings were made using the above-mentioned data, the classification errors for training and validation were 5.75% and 7.41%, respectively. The number of rules generated was 377 with a decomposition rate equals to 0.0015. All the testing data were properly classified. In spite of the results accuracy there were still too many rules when compared with the data set size, providing more than one rule for each experimental data. Thus the decomposition rate was increased to reduce the number of rules. Adopting a 0.01 decomposition rate, only 85% of data were correctly classified. The first training

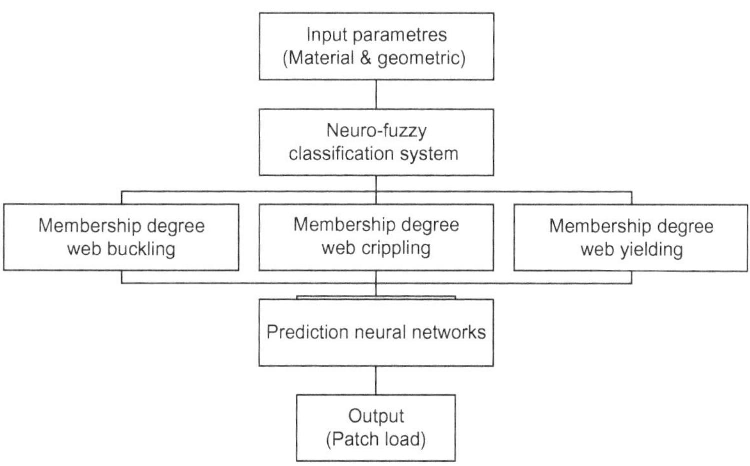

Fig. 4.39 Neuro-fuzzy system.

sets adopted the activation levels provided by the initial neuro-fuzzy system with 377 rules leading to maximum errors of 70%.

The neuro-fuzzy classification network results led to some stimulating conclusions. As the flange thickness is reduced, crippling collapse becomes more frequent, whilst beams with thicker flanges presented a web buckling failure, as shown in Fig. 4.40. Another interesting conclusion is associated with the web thickness parameter. As expected, the web yielding class degree for thin web beams was close to zero whilst the web buckling and particularly web crippling class degrees were very relevant (Fig. 4.41). The opposite case occurred for thicker web beam that was related to higher web yielding class degrees. The neural network results were compatible to the experiments acceptable error margin.

The present investigation is focused on steel sections similar to the commercial WWF (welded wide flange). The flange thicknesses varied from 6.3 to 31.5 mm, whilst the web thicknesses varied from 4 to 12.5 mm according to the usual values present in commercial WWF sections. The girder yield stress was equal to 345 MPa, whilst the web panel aspect ratio was equal to 3 when not mentioned otherwise. The adopted methodology was based on an evaluation of the physical phenomena activation levels in terms of the girder most relevant physical and geometrical parameters. These activation levels indicated the importance of each structural collapse type in the beam behaviour.

The web panel aspect ratio was generally not significant on the physical phenomena type. However, this influence tended to increase in girders with thinner flanges and smaller loading lengths (Fig. 4.42). In these cases the yielding activation levels are reduced with an increase of the transverse stiffener

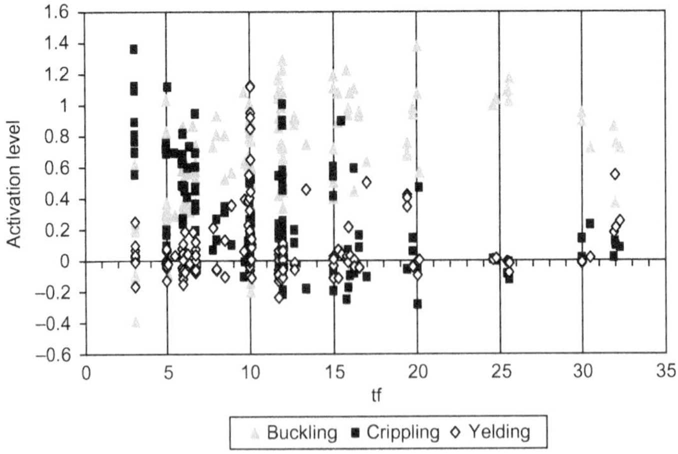

Fig. 4.40 Flange thickness influence over the patch load collapse.

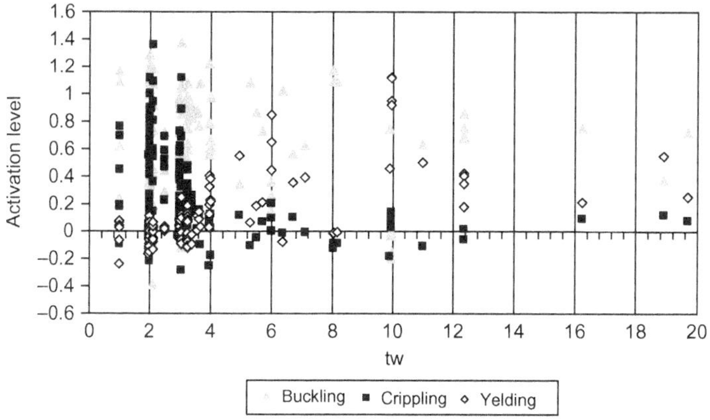

Fig. 4.41 Web thickness influence over the patch load collapse.

spacing. Similarly, the crippling and web buckling levels increase linearly due to the support efficiency provided by the transverse stiffeners.

Some of the most interesting conclusions regarding the parametric study will be reproduced here. The crippling ultimate limit state is more frequent in beams with thinner flanges, whilst the flange thickness has a significant influence on the collapse type, particularly when smaller loading lengths are considered. The flange thickness influence over the patch load capacity was more significant in thicker web beams and the patch load resistance of beams with slender and thinner webs was influenced by the flange thickness. Several of these parameters are interconnected. It was easy to observe that an increase of the beam height also implies an increase of the web slenderness, and reductions of the web panel aspect ratio and the beam loading length to height ratio. The beam height increase proved to generate substantial nonlinear consequences to the controlling physical phenomena also decreasing the patch load capacity.

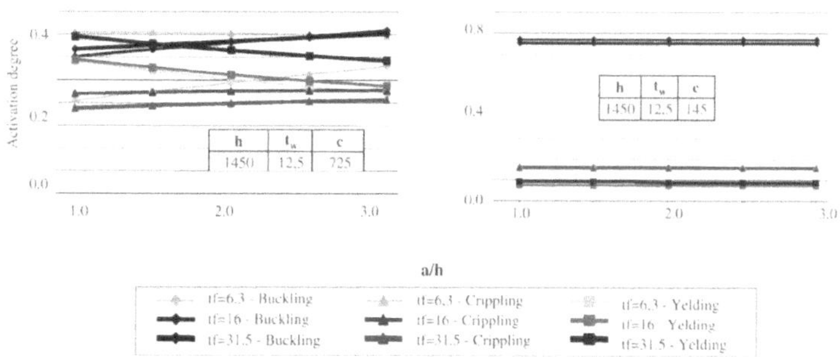

Fig. 4.42 Activation degree variation, for the three classes studied, in terms of web panel aspect ratios and flange thickness.

This study proved the efficacy of the use of computational intelligence technique to solve a complex structural design problem. It is important to stress that the results reliability can be improved as the ratio of experimental evidence to the number of involved variables is increased. If a larger number of experiments could not be incorporated in the neuro-fuzzy system, the generation of new data should be made respecting the limits imposed by the range of parameters used to train the system.

4.5 FINAL CONSIDERATIONS

The constant development of new materials, faster computing processes, and methods have opened a new frontier for the conception and development of new and bolder designs that will set the trend for the future 21th century structures. Various methods, techniques, and tools have been, and still are being, used to improve and design these structures like optimisation processes, numerical modelling systems involving nonlinear finite element analysis, etc.

At the same time, the last decade of the 20th century has been associated with a substantial improvement and development of Computational Intelligent methods. Decision support systems based on these Computational Intelligence models have been widely developed and used successfully in a wide range of applications.

This chapter presented a brief introduction of neural networks, genetic algorithm, and neuro-fuzzy systems, which were used to forecast, design, and optimise the structural behaviour. As mentioned earlier, it is not intended to provide a list running out all applications of Computational Intelligence techniques to solve Structural Engineering problems. The main idea was to focus on some of these methods to provide a deeper insight of structural engineering problems that could be modelled with their proper use.

After presenting the basic concepts of neural networks, GA, and a neuro-fuzzy model based on hierarchical partitioning (HNFB-Class—Hierarchical Neuro-Fuzzy BSP model for Classification problems), this chapter detailed some Structural Engineering case studies. Examples of these case studies were patch load prediction, block shear design, Perfobond shear connectors, stayed steel columns, steel and composite joint design and response, etc. All these case studies corroborate the great potential of Computational Intelligence techniques to solve problems that were considered difficult, limited, or even impossible by many researchers in different fields. Some case studies had a performance, in some cases, beyond expected, suggesting that these techniques might be a good solution in many other structural engineering applications.

CHAPTER FIVE

Final Considerations

5.1 SYNTHESIS

This book aimed to describe, through a series of examples, the methods and techniques used for modelling of steel and composite structures. Initially, the main models used to represent the structural behaviour of these elements were presented and it was followed by a detailed description and discussion of experimental models, as well as the main requirements and precautions that should be taken for a steel structure or composite test in the laboratory.

The focus was then changed to numerical models, developed based on the finite element method. Various simulations depicting nonlinear responses were presented as well as their main details and restrictions in terms of boundary conditions, major difficulties, solution strategies, and methods to overcome convergence difficulties.

Finally, examples of the application of computational intelligence techniques in the simulation of steel and composite structures behaviour were presented. Neural networks, neuro-fuzzy networks, and genetic algorithms were described and applied focusing on their main advantages, scope, and limitations.

In summary, the main contribution of this book was to bring together in a single source by means of a clear, objective, and synthetic form, the main techniques for steel and composite structures modelling. Initially, the laboratory experiments were introduced as well as their main steps aiming to understand their design, procedures development, instrumentation, and results interpretation.

Contributions for understanding the various steps related to the numerical models based on the finite element method were also made, as well as their most relevant aspects related to their design development, how to overcome convergence problems and an effective result interpretation.

Additional contributions were also made towards the understanding of the correct use of computational intelligence models in the simulation of

Modelling Steel and Composite Structures
http://dx.doi.org/10.1016/B978-0-12-813526-6.00005-2

structural response focused on significant steps to their complete understanding and interpretation, scope, and application to problems of steel and composite structures.

5.2 ADDITIONAL CONSIDERATIONS

Experimental modelling is constantly evolving and improving with the development and enhancement of new systems for acquisition and real-time monitoring like fibre optics, laser measurements, and piezoelectric materials amongst many others. Despite this enormous evolution and innovation, it should be stressed that nothing can replace the experience of technicians and researchers who have been working for a long time with these types of modelling and, even with limitations of space, time, equipment, and financial resources, have been producing over the last few years, many excellent experimental models.

Similarly in finite element method numerical modelling, the systems have evolved significantly to enable increasingly complex solutions involving modal and transient nonlinear dynamic analysis, soil structure interaction, and even thermal and mechanical coupled analyses. However, it should be one mere time stressed that nothing replaces the experience of engineers and researchers who have been dealing with these types of modelling for a long time and, even with limitations of equipment and financial resources, have produced numerical models with excellent performance and accuracy.

In addition, it can be noted that the last decade of the 20th century has been associated with a substantial improvement and development of computational intelligent methods. Decision support systems based on these computational intelligence models have been widely developed and successfully used in a wide range of applications.

Their use in structural engineering problems corroborated the great potential of computational intelligence techniques for solving problems that were considered difficult, limited, or even impossible for many researchers in numerous science fields. All the case studies had a performance, in some cases, beyond the expectance suggesting that these techniques can be a good solution for many other structural engineering applications.

REFERENCES

[1] Chan SL. Vibration and modal analysis of steel frames with semi-rigid connections. Eng Struct 1994;16(1):25–31.

[2] Richard RM, Abbott BJ. Versatile elastic–plastic stress–strain formula. J Eng Mech Div ASCE 1975;101(4):511–5.

[3] Lui EM, Chen WF. Analysis and behavior of flexible-jointed frames. Eng Struct 1986;8:107–18.

[4] Lui EM, Chen WF. Behavior of braced and unbraced semi-rigid frames. Int J Solids Struct 1988;24(9):893–913.

[5] Al-Bermani FGA, et al. Cyclic and seismic response of flexibly jointed frames. Eng Struct 1994;16(4):249–55.

[6] Frye MJ, Morris GA. Analysis of flexibly connected steel frames. Can J Civ Eng 1975; 2(3):280–91.

[7] Ang KM, Morris GA. Analysis of three-dimensional frames with flexible beam-column connections. Can J Civ Eng 1984;11:245–54.

[8] Simões RAD. Eurocódigo 3: Projecto de Estruturas Metálicas, Parte 1-1: Regras Gerais e Regras para Edifícios. Manual de Dimensionamento de Estruturas Metálicas. Coimbra: CMM—Ass. Port. de Construção Metálica e Mista; 2005 [in portuguese].

[9] Lochner RH, Matar JE. Designing for quality: an introduction to the best of Taguchi and western methods of statistical experimental design. White Plains, NY: Quality Resources; 1990.

[10] Vellasco PCGdaS, Hobbs REA. Design model for local alma buckling in tapered composite beams. Adv Struct Eng 2001;4:125–35.

[11] Vellasco PCGdaS, Hobbs RE. Local alma buckling in tapered composite beams. Struct Eng (Lond) 1996;74:41–6.

[12] Vellasco PCGdaS. Local web buckling in tapered composite beams [Tese de Doutorado]. London, UK: Imperial College of Science Technology, University of London; 1992.

[13] D'este AV. Comportamento de Estruturas Tubulares Espaciais Padronizadas. [M.Sc. dissertation (Civil Eng.)]. Pontifícia Universidade Católica do Rio de Janeiro; 1998 [in portuguese].

[14] Andrade SALde, Vellasco PCGdaS, Silva JGSda, Lima LROda, D'este AV. Tubular space trusses with simple and reinforced end flattened nodes—an overview and experiments. J Constr Steel Res 2005;61:1025–50.

[15] D'este AV, Andrade SALde, Vellasco PCGdaS. Structural behaviour of standardized tubular space trusses. In: 2nd European conference on steel structures, Eurosteel, Praga; 1999. p. 1–12. República Tcheca: Editora da Universidade Técnica de Praga.

[16] Rosa YRS. Comportamento de Estruturas Espaciais Metálicas com Nós da ligação do Tipo Ponta Amassada. [M.Sc. dissertation (Civil Eng.)]. Pontifícia Universidade Católica do Rio de Janeiro; 2001 [in portuguese].

[17] Rosa YRS, Vellasco PCGdaS, Andrade SALde, Silva JGSda, D'este AV. An experimental investigation of tubular space trusses with simple and reinforced end flattened nodes. In: International symposium on lightweight structures in civil engineering, Varsóvia; 2002. p. 645–50.

[18] Cochrane VH. Rules for rivet hole deduction in tension members. Eng News Rec 1922;80.

[19] Eurocode 3. EN 1993-1-8, design of steel structures—part 1.8: design of joints. Brussels: CEN, European Committee for Standardisation; 2005.

[20] Eurocode 3. ENV—1993-1-1: design of steel structures—part 1-1: general rules and rules for buildings. Brussels: CEN, European Committee for Standardisation; 2003.

[21] Eurocode 3. EN 1993-1-4, design of steel structures—part 1.4: general rules—stainless steel. Brussels: CEN, European Committee for Standardisation; 2003.
[22] Santos JdeJ. Comportamento Estrutural de Elementos em Aço Inoxidável [M.Sc. dissertation (Civil Eng.)] Univ. Estado do Rio de Janeiro; 2008 [in portuguese].
[23] Santos JdeJ, Vellasco PCGdaS, Lima LROde, Andrade SALde. Comportamento de Elementos Formados com Aço Carbono e Aço Inoxidável com Furos Defasados sob Tração. Metálica; 2011. p. 8–10 [in portuguese].
[24] Santos JdeJ, Silva ATda, Lima LROde, Vellasco PCGdaS, Andrade SALde, Silva JGSda. Avaliação da Resistência a Tração de Ligações Aparafusadas em Aço Inoxidável. In: CILAMCE'2008—XXVII Congresso Ibero Latino Americano de Métodos Computacionais em Eng; 2008 [in portuguese].
[25] Araújo RRde. Estudo Teórico-Experimental de Colunas de Aço Estaiadas e Protendidas. Rio de Janeiro [M.Sc. dissertation (Civil Eng.)] PUC-Rio; 2005 [in portuguese].
[26] Araújo RRde. Comportamento Estrutural de Colunas de Aço Estaiadas e Protendidas. [Ph.D. thesis (Civil Eng.)]. Pontifícia Universidade Católica do Rio de Janeiro; 2009 [in portuguese].
[27] Araújo RR, Andrade SALde, Silva LAPSda, Vellasco PCGdaS, Silva JGSda, Lima LROde. Experimental analysis of prestressed stayed steel columns. In: Fifth European conference on steel and composite structures. Brussels. ECCS European convention for constructional steelwork. Eurosteel; 2008. p. 1605–10.
[28] de Araújo RR, de Andrade SAL, Vellasco PCGdaS, Silva JGSda, Lima LROde. Experimental and numerical assessment of stayed steel columns. J Constr Steel Res 2008;64:1020–9.
[29] Araújo RRde, Silva LAPSda, Andrade SALde, Vellasco PCGdaS, Silva JGSda, Lima LROde. Structural evaluation of steel columns with reinforced bar stays. In: 7th EUROMECH solid mech conf, Lisboa; 2009. p. 1–10.
[30] Araújo RRde, Vellasco PCGdaS, Andrade SALde, Silva JGSda, Lima LROde. Structural response of pre-stressed stayed steel columns. In: International colloquia on stability and ductility of steel structures—SDSS'06, Lisboa; 2006. p. 241–8.
[31] Associação Brasileira de Normas Técnicas. NBR 8800, Projeto e Execução de Estruturas de Aço em Edifícios; 1986 [in portuguese].
[32] ANSYS ®. Release 10.0, ANSYS Inc. Theory Reference; 2005.
[33] Carvalho LCV. Avaliação Estrutural de Ligações Aparafusadas Semi-rígidas [M.Sc. dissertation (Civil Eng.)] Pontifícia Universidade Católica do Rio de Janeiro; 1997 [in portuguese].
[34] Carvalho LCV, Andrade SALde, Vellasco PCGdaS. Experimental analysis of bolted semi-rigid connections. J Constr Steel Res 1998;46:1–20.
[35] Lima LROde, Silva LAPSda, Vellasco PCGdaS, Andrade SALde. Experimental evaluation of extended endplate beam-to-column joints subjected to bending and axial force. Eng Struct 2004;26:1333–47.
[36] Lima LROde, Vellasco PCGdaS, Andrade SALde, Silva LAPSda. Experimental and mechanical model for predicting the behaviour of minor axis beam-to-column semi-rigid joints. Int J Mech Sci 2002;44(6):1047–65.
[37] Lima LROde. Avaliação de Ligações Viga-Coluna em Estruturas de Aço Submetidas a Flexão no Eixo de Menor Inércia [M.Sc. dissertation (Civil Eng.)]. Pontifícia Universidade Católica do Rio de Janeiro; 1999 [in portuguese].
[38] Lima LROde, Vellasco PCGdaS, Andrade SALde. Ultimate capacity of bolted semi-rigid connections to the column minor axis. In: Structural failure and plasticity IMPLAST2000, Melbourne; 2000. p. 513–8.
[39] Lima LROde, Vellasco PCGdaS, Andrade SALde. Bolted semi-rigid connections in the Column's minor axis. In: 2nd European conference on steel structures, Eurosteel, Praga. Eurosteel, 2; 1999. p. 1–14.

[40] Dally JW, Rilley WF. Experimental stress analysis. 4rd ed. New York: College House Interprises; 2005.
[41] Dally JW, Mulc A. Polycarbonate as a model material for three-dimensional photoelasticity. J Appl Mech 1973;E95:600–5.
[42] Whitfield JK. Characterization of polycarbonate as a photoelastic materia. [Ph.D. thesis].Faculty of Virginia Polytechnic Institute; 1969.
[43] Neves LFdaC. Comportamento Monotónico E Cíclico de Ligações de Menor Inércia e Tubulares em Estruturas Metálicas e Mistas Aço-Concreto [Ph.D. thesis (Civil Eng.)].Universidade de Coimbra; 2004 [in portuguese].
[44] Neves LFdaC, Silva LAPSda, Vellasco PCGdaS. Experimental behaviour of end plate minor-axis steel and composite joints under static monotonic loading. In: 3rd European conference on steel structures, Eurosteel 2002, vol. 1; 2002. p. 1101–10.
[45] Neves LFdaC, Silva LAPSda, Vellasco PCGdaS. Characterisation of the behaviour of the column Web loaded in out-of-plane bending in the framework of the component method. In: 9th International conference on civil and structural engineering computing, Egmond aan Zee; 2003. p. 1–17.
[46] Silva LAPSda, Vellasco PCGdaS. Model for predicting the stiffness of beam to concrete filled column and minor axis joints under static monotonic loading. In: 4th European conference on steel and composite structures, Eurosteel 2005, Maastricht; 2005. p. 10131–8.
[47] Lima LROde. Comportamento de Ligações com Placa de Extremidade em Estruturas de Aço Submetidas a Momento Fletor e Força Axial [Ph.D. thesis (Civil Eng.)]. Pontifícia Universidade Católica do Rio de Janeiro; 2003 [in portuguese].
[48] da Silva LAPS, de Lima LRO, Vellasco PCGdaS, Andrade SALde. Behaviour of flush end-plate beam-to-column joints under bending and axial force. Steel Compos Struct 2004;4:77–94.
[49] de Lima LRO, da Silva LAPS, Vellasco PCGdaS, Andrade SALde. Experimental evaluation of extended endplate beam-to-column joints subjected to bending and axial force. Eng Struct 2004;26:1333–47.
[50] Silva LAPSda, Lima LROde, Vellasco PCGdaS, Andrade SALde. Experimental and numerical assessment of beam-to-column joints under bending and axial force. In: 1st international conference on steel & composite structures, ICSCS; 2001. p. 715–22.
[51] Lima LROde, Silva LAPSda, Vellasco PCGdaS, Andrade SALde. Experimental analysis of extended End-plate beam-to-column joints under bending and axial force. In: 3rd European conference on steel structures, Eurosteel 2002, coimbra; 2002. p. 1121–30.
[52] Ferreira LTS. Sistemas Construtivos Semi-Rígidos Mistos para Edfificações, [Ph.D. thesis (Civil Eng.)]. Pontifícia Universidade Católica do Rio de Janeiro; 2000 [in portuguese].
[53] Vellasco PCGdaS, Andrade SALde, Ferreira LTS, Lima LROde. Semi-rigid composite frames with perfobond and T-rib connectors part 1: full scale tests. J Constr Steel Res 2007;63:263–79.
[54] Andrade SALde, Vellasco PCGdaS, Ferreira LTS, Lima LROde. Semi-rigid composite frames with perfobond and T-rib connectors part 2: dimensionamento models assessment. J Constr Steel Res 2007;63:280–92.
[55] Ferreira LTS, Andrade SALde, Vellasco PCGdaS. Connections—a design model for bolted composite semi-rigid connections. Stability and ductility of steel structures. Kidlington, Oxford: Elsevier Science Ltd; 1998. p. 293–306.
[56] Oguejiofor EC, Hosain MU. A parametric study of perfobond rib shear connectors. Can J Civ Eng 1994;21:614–25.
[57] Taguchi G. Introduction to quality engineering: designing quality into products and processes. Tokyo: Asian Productivity Organization; 1986.

[58] Ramires FB. Contribuição das Componentes Mistas em Ligações Semi-rigid [Ph.D. thesis (Civil Eng.)]. Pontifícia Universidade Católica do Rio de Janeiro; 2010 [in portuguese].
[59] Ramires FB, Andrade SALde, Vellasco PCGdaS. Avaliação da Contribuição da Armadura em Ligações Semi-r[igidas Mistas. In: Cilamce 2005, Iberian Latin American congress on computation methods eng., Guarapari; 2005. p. 1–8.
[60] Ramires FB, Andrade SALde, Vellasco PCGdaS. Experimental analysis of composite semi-rigid beam to colum joints. In: Eurosteel 2008—fifth European conference on steel and composite structures, Graz; 2008. p. 471–6.
[61] Mergulhão AJR. Comportamento de Vigas Mistas Constituídas por Perfis de Aco Preenchidos com Concreto Armado [M.Sc. dissertation (Civil Eng.)]. Pontifícia Universidade Católica do Rio de Janeiro; 1994 [in portuguese].
[62] Andrade SALde, Vellasco PCGdaS, Mergulhão AJR. Structural asessment of cold-formed composite structures. Steel Compos Struct 2002;2:397–410.
[63] Vianna JdaC, Andrade SALde, Vellasco PCGdaS, Lima LROde, Silva JGSda. Um Sistema de Laje-Mista para Edificações Residenciais Usando Perfis de Chapa Dobrada com Corrugações. Revista da Escola de Minas 2007;60:325–31 [in portuguese].
[64] Eurocode 4. EN 1994. design of composite steel and concrete structures part 1.1 general rules and rules for buildings Document CEN/TC 250/SC 4. Brussels: CEN, European Committee for Standardisation; 2001. p. 135.
[65] Takey TH. Sistemas de Laje Mista para Edificações com uso de Perfis de Chapa Metálica [M.Sc. dissertation (Civil Eng.)]. Pontifícia Universidade Católica do Rio de Janeiro; 2001 [in portuguese].
[66] Takey TH, Andrade SALd, Vellasco PCGdaS, Silva JGSda. A composite slab system for low-cost building construction. In: EASEC8—The Eighth East Asia Pacific conf structural eng & construction, Singapore, (CD-ROM); 2001.
[67] Andrade SALd, Vellasco PCGdaS, Silva JGSda, Takey TH. Standardized composite slab systems for building constructions. J Constr Steel Res 2003;60:493–524.
[68] Takey TH, Andrade SALde, Vellasco PCGdaS, Silva JGSda. An innovative composite slab system for residential buildings. In: 3rd European conference on steel structures, Eurosteel 2002, Coimbra; 2002. p. 451–60.
[69] Beltrão AJdoN. Comportamento Estrutural de Laje Mista com Corrugações de Alma de Perfis de Chapa Dobrada [M.Sc. dissertation (Civil Eng.)]. Pontifícia Universidade Católica do Rio de Janeiro; 2003 [in portuguese].
[70] Vianna JC. Sistema de Laje Mista para Edificações Residenciais com Uso de Perfis Embossados de Chapa Dobrada. Rio de Janeiro [M.Sc. dissertation (Civil Eng.)]. PUC-Rio; 2005 [in portuguese].
[71] Vianna JdaC. Comportamento Estrutural de Conectores de Cisalhamento Tipo Perfobond e T-rib [Ph.D. thesis (Civil Eng.)]. Pontifícia Universidade Católica do Rio de Janeiro; 2009 [in portuguese].
[72] Vianna JC, Costa-Neves LF, Vellasco PCGS, Andrade SAL. Experimental assessment of perfobond and TPerfobond shear connectors. J Constr Steel Res 2009;65:408–21.
[73] Vianna JdaC, Neves LFdaC, Vellasco PCGdaS, Andrade SALde. Structural behaviour of TPerfobond shear connectors in composite girders: an experimental approach. Eng Struct 2008;30:2381–91.
[74] Vianna JdaC, Neves LFdaC, Vellasco PCGdaS, Andrade SALde. Estudo Comparativo de Conectores de Corte para Estruturas Mistas de Aço e Concreto. 23:Porto: Construção Magazine; 2008. p. 23–30 [in portuguese].
[75] Vianna JdaC, Andrade SALde, Vellasco PCGdaS, Neves LFdaC. A parametric study of shear connectors in steel and concrete composite girders. In: 4th international conference on steel & composite structures (ICSCS'10), Sydney; 2010. p. 341–6.

[76] Vianna JdaC, Andrade SALde, Vellasco PCGdaS, Neves LFdaC. Parametric analysis of composite beams with T- perfobond shear connectors. In: International colloquium on stability and ductility of steel structures—SDSS2010, 2010, Rio de Janeiro, 2; 2010. p. 879–86.
[77] Vianna JdaC, Neves LFdaC, Vellasco PCGdaS, Andrade SALde. An experimental evaluation of composite structures perforated shear connectors. In: 7th international conference on steel bridges (ICSB08), Guimarães; 2008. p. 1–8.
[78] Vianna JdaC, Vellasco PCGdaS, Neves LFdaC, Andrade SALde. Experimental analysis of TPerfobond shear connectors for steel and composite structures. In: Eurosteel 2008—fifth European conference on steel and composite structures, Graz; 2008. p. 2 49–254.
[79] Paulino GH, Menezes IFM, Lages EN. A unified approach for solving nonlinear finite element systems—implementation and applications. Personal notes; 2003.
[80] Krenk S. An orthogonal residual procedure for nonlinear finite element equations. Int J Numer Methods Eng 1995;823–39.
[81] Yang YB, Shieh MS. Solution method for nonlinear problems with multiple critical points. AIAA J 1990;2110–6.
[82] Wempner GA. Discrete approximations related to nonlinear theory of solids. Int J Solids Struct 1971;1581–99.
[83] Ricks E. The application of Newton's method to the problem of elastic stability. J Appl Mech 1972;1060–6.
[84] Ramm E. Strategies for tracing nonlinear responses near limit points. Nonlinear finite element analysis in struc. mechanics. Berlin: Spring-Verlag; 1981. p. 68–89.
[85] Crisfield MA. A fast incremental/iterative solution procedure that handles snap-through. Comput Struct 1981;55–62.
[86] Batoz JL, Dhatt G. Incremental displacement algorithms for nonlinear problems. Int J Numer Methods Eng 1979;1262–7.
[87] Bathe KJ, Dvorkin EN. On the automatic solution of nonlinear finite element equations. Comput Struct 1983;871–9.
[88] Yang YB. Linear and nonlinear analysis of space frames with nonuniform torsion using interactive computer graphics [Ph.D. thesis]. School of Civil and Environmental Engineering—Cornell University, Ithaca, NY, USA; 1984.
[89] Crisfield MA. Non-linear finite element analysis of solid and structures Vol. 1: essentials. England: John Wiley and Sons; 1987.
[90] Vellasco PCGdaS, Hobbs RE. Local Web buckling in composite tapered beams—a parametric study. J Braz Soc Mech Sci 1998;23:411–25.
[91] Vellasco PCGdaS, Andrade SALde, Silva JGSda, Lima LROde, Brito Júnior OF. A parametric analysis of steel and composite portal frames with semi-rigid connections. Eng Struct 2006;28:543–56.
[92] Brito Jr. OF. Desenvolvimento de Sistemas Estruturais Semi-Rígidos em Aço e Mistos para Edificações Residenciais Multi-Familiares [M.Sc. dissertation (Civil Eng.)]. Pontifícia Universidade Católica do Rio de Janeiro; 2001 [in portuguese].
[93] Usiminas. Usiteto—Solução Usiminas Para Habitação Popular. 2000.
[94] Steenhuis M, Weynand K, Gresnigt AM. Strategies for economic dimensionamento of unbraced steel frames. J Constr Steel Res 1998;46(1–3).
[95] Lima LROde. Avaliação de Ligações Viga-Coluna Em Estruturas de Aço Submetidas A Flexão No Eixo de Menor Inércia [M.Sc. dissertation (Civil Eng.)]. Pontifícia Universidade Católica do Rio de Janeiro; 1999 [in portuguese].
[96] Lima LROde, Freire JLF, Vellasco PCGdaS, Andrade SALde, Silva JGSda. Structural assessment of minor axis steel joints using photoelasticity and finite elements. J Constr Steel Res 2009;65:466–78.
[97] Del Savio AA. A component-based model for semi-rigid steel joints including bending moment-axial force interaction [Ph.D. thesis (Civil Eng.)]. Pontifícia Universidade Católica do Rio de Janeiro; 2008.

[98] Del Savio AA, Andrade SAde, Vellasco PCGS, Martha LF. Structural evaluation of semi-rigid steel portal frames. In: Eurosteel 2005, forth European conference on steel and composite structures, Maastricht; 2005. p. 4.49–56.

[99] Savio AA, Del Nethercot DA, Vellasco PCGdaS, Lima LRODE, Andrade SALde, Martha LFCR. An assessment of beam-to-column endplate and baseplate joints including the axial-moment interaction. Adv Steel Constr 2010;65:548–66.

[100] Savio AA, Del Nethercot DA, Vellasco PCGdaS, Andrade SALde, Martha LFCR. Generalised component-based model for beam-to-column connections including axial vs. Moment interaction. J Constr Steel Res 2009;65:1876–95.

[101] Del Savio AA, de Andrade SAL, Martha LFCR, Vellasco PCGdaS. Um Sistema Não-Linear para Análise de Pórticos Planos Semi-Rígidos. Revista Sul-Americana de Engenharia Estrutural 2005;2:97–125 [in portuguese].

[102] Chan SL. Vibration and modal analysis of steel frames with semi-rigid connections. Eng Struct 1994;16(1):25–31.

[103] van Keulen DC, Nethercot DA, Snijder HH, Bakker MCM. Frame analysis incorporating semi-rigid joint action: applicability of THE half initial secant stiffness approach. J Constr Steel Res 2003;59:1083–100.

[104] Steenhuis M, Gresnigt N, Weynand K. Design of semi-rigid joints in steel frames. In: Wald F, editor. Proceedings of the second state of the art workshop on semi-rigid behaviour of civil engineering structural connections, COST C1, Prague, Czech Republic; 1994. p. 131–40.

[105] Souza RS. Flambagem Local de Vigas de Aco Sujeitas a Cargas Concentradas [M.Sc. dissertation (Civil Eng.)]. Pontifícia Universidade Católica do Rio de Janeiro; 1995 [in portuguese].

[106] Souza RS, Vellasco PCGdaS, Andrade SALde. Plates and plated structures—a design study of the patch load instability phenomenon, stability and ductility of steel structures. Elsevier; 1998. p. 233–246, ISBN:0080433200.

[107] Souza RS, Andrade SALde, Vellasco PCGdaS. Pacth load instability by the finite element method. In: 5th International colloquium on structural stability, SSRCIC/96, Rio de Janeiro. SSRC IC/96. Rio de Janeiro; 1996. p. 561–71.

[108] Andrade S.A.L.de. Non-linear behaviour of haunched steel beams by the finite element method [Ph.D. thesis]. Victoria University of Manchester; 1983.

[109] Bergfelt A. Patch loading on a slender web. Influence of horizontal and vertical web stiffeners on the load carrying capacity. In: Goteborg: Chalmers University of Technology, Division of Steel and Timber Structures; 1979. p. 1–143. Publ. s79:1.

[110] Araújo RRde, Andrade SALde, Vellasco PCGdaS, Silva JGSda, Lima LROde. Structural response of pre-stressed stayed steel columns. In: International colloquia on stability & ductility of steel structures; 2006. p. 241–8.

[111] Araújo RRde, Vellasco PCGdaS, Silva JGSda, Andrade SALde. Dynamic response of pre-stressed steel columns. In: 11th int conference on civil, structural and environmental eng computing—CC2007, Saint Julians; 2007. p. 1–14.

[112] Belenya E. Prestressed load-bearing metal structures. Moscow: MIR Publishers; 1977.

[113] Sirqueira AdaS. Comportamento Estrutural de Torres de Aço para Apoio de Turbinas Eólicas [M.Sc. dissertation (Civil Eng.)]. Universidade do Estado do Rio de Janeiro; 2008 [in portuguese].

[114] Sirqueira AdaS, Vellasco PCGdaS, Lima LROde, Silva JGSda, Andrade SALde. Computational modelling of the static and dynamic behaviour of wind turbine tower structures. In: CST2008—the ninth international conference on computational structures technology, Atenas; 2008.

[115] Sirqueira AdaS, Vellasco PCGdaS, Silva JGSda, Lima LROde, Andrade SALde. Finite element modelling of wind turbine tower structures. In: Eurosteel 2008—fifth European conference on steel and composite structures, Graz; 2008. p. 747–52.

[116] Kim TS, Kuwamura H. Finite element modeling of bolted connections in thin-walled stainless steel plates under static shear. Thin-Walled Struct 2007;45:407–21.
[117] Silva ATda. Modelagem Numérica de Elementos Tracionados com Parafusos Defasados em Aço Inoxidável [M.Sc. dissertation (Civil Eng.)]. Universidade do Estado do Rio de Janeiro; 2009.
[118] Silva ATda, Santos J de J, Lima, LROde, Vellasco PCGda, Silva JGSda, Andrade SALde. A numerical modelling evaluation of the tension capacity of stainless steel plates with staggered bolts. In: 4th international conf on steel & composite structures (ICSCS'10), Sydney; 2010. p. 432–7.
[119] Silva ATda, Santos JdeJ, Lima LROde, Vellasco PCGda, Andrade SALde, Silva JGSda. Numerical evaluation of stainless steel joints subjected to tension. In: CC2009—12th international conference on civil, structural and environmental engineering computing, Funchal, Madeira; 2009. p. 1–13.
[120] Rodrigues F. Comportamento Estrutural de Vigas de Aço com Aberturas na Alma [M. Sc. dissertation (Civil Eng.)]. Universidade do Estado do Rio de Janeiro; 2006 [in portuguese].
[121] Rodrigues F, Vellasco PCGdaS, Lima LROde, Andrade SALde, Silva JGSda. Finite element modelling of steel beams with web openings. In: 11th international conference on civil, structural and environmental engineering computing—CC2007, Saint Julians; 2007. p. 1–13.
[122] Chung KF, Liu TCHACH. Investigation on vierendeel mechanism in steel beams with circular web openings. J Constr Steel Res 2001;57:467–90.
[123] Chung KF, Lawson RM. Simplified design of composite beams with large web openings to eurocode 4. J Constr Steel Res 2001;57:135–63.
[124] Chung KF, Liu TCH, Ko ACH. Steel beams with large Web openings of various shapes and sizes: an empirical design method using a generalised moment-shear interaction curve. J Constr Steel Res 2003;59:1177–200.
[125] Redwood RG, Mccutcheon JO. Beam tests with un-reinforced web openings. J Struct Div 1968;94(1):1–17.
[126] Rodrigues MC. Modelagem Numérica de Ligações Viga-Coluna em Aço Sob Momento Fletor e Força Normal [M.Sc. dissertation (Civil Eng.)]. Universidade do Estado do Rio de Janeiro; 2009 [in portuguese].
[127] Rodrigues MC, Lima LROde, Andrade SALde, Vellasco PCGdaS, Silva JGSda. Numerical analysis of endplate beam-to-column joints under bending and axial force. In: 4th international conference on steel & composite structures (ICSCS'10), Sydney; 2010. p. 561–6.
[128] Rodrigues MC, Lima LROde, Andrade SALde, Vellasco PCGdaS, Silva JGSda. Numerical analysis of endplate beam-to-column joints under bending and axial force. In: International colloquium on stability and ductility of steel structures—SDSS2010, Rio de Janeiro; 2010. p. 279–86.
[129] Nunes PCdaL, Lima LROde, Silva JGSda, Vellasco PCGdaS, Andrade SALde. Parametrical analysis of extended semi-rigid endplate joints subjected to bending moment and axial forces. Int J Solids Struct 2007;4:39–59.
[130] Lima LROde, Vellasco PCGdaS, Neves LFdaC, Silva JGSda, Andrade SALde. Structural response of K and T tubular joints under static loading. In: Connections VI—sixth international workshop on connections in steel structures, ChicagoChicago: AISC; 2008.
[131] Lima LROde, Neves LFdaC, Silva JGSda, Vellasco PCGdaS, Andrade SALde. Parametric analysis of RHS "T" joints under static loading. In: 11th int conf civil, structural & environmental eng comp—CC2007; 2007.
[132] Bittencourt MC. Avaliação de Ligações Soldadas de Perfis Tubulares em Estruturas de Aço Através do Método dos Elementos Finitos [M.Sc. dissertation (Civil Eng.)]. Universidade do Estado do Rio de Janeiro; 2008.

[133] Lie ST, Chiew SP, Lee CK, Yang ZM. Static strength of cracked square hollow section T joints under axial loads. I: experimental. J Struct Eng 2006;132(3):368–77.
[134] Lu LH, de Winkel GD, Yu Y, Wardenier J. Deformation limit for the ultimate strength of hollow section joints. In: 6th international symposium on tubular structures, Melbourne, Australia; 1994. p. 341–7.
[135] Choo YS, Qian XD, Liew JYR, Wardenier J. Static strength of thick-walled CHS X-joints—part I. New approach in strength definition. J Constr Steel Res 2003;59:1201–28.
[136] Lee MMK. Strength, stress and fracture analyses of offshore tubular joints using finite elements. J Constr Steel Res 1999;51:265–86.
[137] Queiroz FD, Vellasco PCGS, Nethercot DA. Structural assessment of composite beams using the finite element method. In: Eurosteel conf steel & composite structures; 2005. p. 4.3-49–58.
[138] Queiroz FD, Vellasco PCGS, Nethercot DA. Finite element modelling of composite beams with full & partial shear connect. J Constr Steel Res 2007;63:505–21.
[139] Gattesco N. Analytical modelling of nonlinear behaviour of composite beams with deformable connection. J Constr Steel Res 1999;52:195–218.
[140] Chapman JC, Balakrishnan S. Experiments on composite beams. Struct Eng 1964;42 (11):369–83.
[141] Vianna JC. Comportamento Estrutural de Conectores de Cisalhamento Tipo Perfobond e T-rib [Ph.D. thesis (Civil Eng.)]. Pontifícia Universidade Católica do Rio de Janeiro; 2009 [in portuguese].
[142] Vianna JC, Neves LFC, Vellasco PCGS, Andrade SAL. Experimental assessment of perfobond and TPerfobond shear connectors structural response. J Constr Steel Res 2009;65:408–21.
[143] Vianna JdaC, Neves LFdaC, Vellasco PCGdaS, Andrade SALde. A parametric analysis of composite beams with TPerfobond shear connectors. In: CC2009—twelfth international conference on civil, structural and environmental engineering computing, Funchal, Madeira; 2009. p. 1–13.
[144] Cruz BCda. Comportamento Estrutural de Conectores de Cisalhamento Através de uma Análise Numérica/Experimental [M.Sc. dissertation (Civil Eng.)]. Universidade do Estado do Rio de Janeiro; 2010 [in portuguese].
[145] Lam D, El-Lobody E. Behavior of headed stud shear connectors in composite beam. J Struct Eng 2005;131:96–107.
[146] Env 1992-1-1. Eurocode-2: design of concrete structures, part 1: general rules and rules for building. Brussels: European Standard, CEN; 1992.
[147] Hornik K, Stinchcombe M, White H. Multilayer feedforward networks are universal approximators. Neural Netw 1989;2:359–66.
[148] Haykin S. Neural networks—a comprehensive foundation. New York: Macmillan College Publishing Company; 1999.
[149] Wasserman PD. Advanced methods in neural computing. New York: Van Nostrand Reinhold; 1993.
[150] Bishop CM. Neural networks for pattern recognition. Oxford: Oxford University Press; 1995.
[151] MacKay DJC. Probable networks and plausible predictions—a review of practical bayesian methods for supervised neural networks. Netw Comput Neural Syst 1995;6:469–505.
[152] Fonseca ET, Vellasco MMBR, Vellasco PCGS, Andrade SAL, Pacheco MAC. Class-based neural network system for patch load prediction. In: Eann99, Int conf app eng neural net, Warsaw; 1999.
[153] Fonseca ET, Vellasco MMBR, Andrade SALde, Vellasco PCGdaS. A unified neural network approach for steel beams patch load capacity. In: 6th conf app art int civil & structural eng, Vienna; 2001.

[154] Fonseca ET, Vellasco MMBR, Vellasco PCG, Andrade SALde, Pacheco MACdaS. A neural network system for patch load prediction. J Intell Robot Syst 2001; 31(1–3):185–200.
[155] Fonseca ET, Vellasco PCGdaS, Andrade SALde, Vellasco MMBR. Neural network evaluation of steel beam patch load capacity. Adv Eng Softw 2003;34(11–12):763–72.
[156] Fonseca ET, Vellasco PCGdaS, Andrade SALde, Vellasco MMBR. A patch load parametric analysis using neural networks. J Constr Steel Res 2003;59(2):251–67.
[157] Lyse I, Godfrey HJ. Investigation of web buckling in steel beams. 100. ASCE Transactions; 1935. p. 675–95. paper 1907.
[158] Bergfelt A. Girder web stiffening for patch loading. In: Goteborg: Chalmers University of Technology, Div. Steel & Timber Struct.; 1983. p. 1–78. Pub. s83:1.
[159] Roberts TM, Newark ACB. Strength of webs subjected to compressive edge loading. J Struct Eng ASCE 1997;123(2):176–83.
[160] Vellasco PCGdaS. Local web buckling in tapered composite beams [Tese de doutorado]. London: Imperial College; 1992.
[161] Fonseca ET, Vellasco MMBR, Vellasco PCGdaS, Andrade SALde. Neural network evaluation of the patch load phenomenon. In: International conference on enhancement and promotion of computational methods in eng and science, Macao; 1999. p. 797–806.
[162] Kennedy SJ, Kennedy DJL, Prabha KR, Prowse D. The bearing resistance of webs: a plate post-buckling strength problem. In: Structural stability research council, annual tech. session & meeting, Toronto; 1997.
[163] NeuralWare. Neuralworks predict for Excel 7.0, manual. Pittsburgh, USA: NeuralWare; 1994.
[164] Rodrigues F, Biondi Neto L, Vellasco PCGdaS, Lima LROde, Vellasco MMBR. A neural network modelling of steel joints block shear capacity. In: 9th int conf application of artificial intelligence to civil and structural engineering, Malta; 2007.
[165] Rodrigues F, Vellasco PCGdaS, Lima LROde, Biondi Neto L, Vellasco MMBR. Steel joints block shear strength forecast using bayesian neural networks. In: 3rd Asian-Pacific congress on comp mechanics & EPMESCXI, 11th I. Conf Enhanc & promotion of comp methods in eng & science, Kyoto; 2007.
[166] MATLAB © The Mathworks Inc, version 7.
[167] Vianna, JdaC, Andrade SALde, Vellasco PCGdaS, Vellasco MMBR. Neural network modelling of perfobond shear connector resistance. In: 3rd int conf bridge maint, safety & management, Porto; 2006.
[168] Vianna JdaC, Andrade SALde, Vellasco PCGdaS, Vellasco MMBR. An evaluation of perfobond shear connector capacity by bayesian neural networks. In: CILAMCE 2006 Iberian Latin American congress computation methods in engineering, Belém; 2006.
[169] Oguejifor EC, Hosain MU. A parametric study of perfobond rib shear connectors. Can J Civ Eng 1994;21:614–25.
[170] Oguejifor EC, Hosain MU. Numerical analysis of push-Out specimens with perfobond rib connectors. Comput Struct 1997;62(4):617–24.
[171] Valente I, Cruz PJS. Experimental analysis of perfobond shear connection between steel & lightweight concrete. J Constr Steel Res 2004;60:465–79.
[172] Araújo RRde, Andrade SALde, Vellasco PCGdaS, Vellasco MMBR. An evaluation of prestressed stayed steel columns capacity using bayesian neural networks. In: CILAMCE 2006 Iberian Latin American congress on computation methods in engineering, Belém; 2006.
[173] Araújo RRde, Andrade, SALde, Vellasco PCGdaS, Vellasco MMBR, Fonseca ET. Utilização de Projeto de Experimentos e Redes Neurais Bayesianas para Previsão de Cargas Críticas em Colunas de Aço Estaiadas e Protendidas. In: CILAMCE 2007 Iberian Latin American cong on comp methods in eng, Porto; 2007 [in portuguese].

[174] Vellasco PCGdaS, Andrade SALde, Silva JGSda. Sistema Construtivo e Montagem do Palco Principal do Rock in Rio III, vol. 7. Porto, Portugal: Construção Magazine; 2003. p. 30–55 [in portuguese].
[175] Araujo R. Estudo Teórico-Experimental de Colunas de Aço Estaiadas e Protendidas [M.Sc. dissertation (Civil Eng.)]. PUC-RIO; 2005 [in portuguese].
[176] Owens GW, Cheal BD. Structural steelwork connections. England: Butterworks; 1989.
[177] Abdala KM, Stavroulakis GEA. Backpropagation neural network model for semi-rigid joints. Comput Aided Civ Inf Eng 1995;10:77–87.
[178] Stavroulakis GE, Abdala KM. A systematic neural network classificator in mechanics: application in semi-rigid steel joints. Int J Eng Anal Des 1994;1:279–92.
[179] Anderson D, et al. Application of artificial neural networks to the prediction of minor axis joints. Comput Struct 1997;63:685–92.
[180] Lima LROde, Vellasco PCGdaS, Andrade, SALde, Vellasco MMBR, Silva JGSda. Neural networks assessment of beam-to-column joints. J Braz Soc Mech Sci 2005;27(3):314–24.
[181] Lima LROde, Vellasco PCGdaS, Andrade SALde, Vellasco MMBR. Neural networks forecasting of endplate steel connections capacity. In: AICIVIL-COMP2001 6th int conf applications of artificial intelligence to civil & structural engineering, Vienna; 2001.
[182] Lima LROde, Vellasco MMBR, Vellasco PCGdaS, Andrade SALde. Avaliação de Ligações Viga-Coluna em Estruturas de Aço Utilizando-se Redes Neurais. In: CILAMCE 2000 Iberian Latin Amer cong comp methods in eng, Rio de Janeiro; 2000 [in portuguese].
[183] da Silva LS, Coelho AMGA. Ductility model for steel joints. J Constr Steel Res 2001;57:45–70.
[184] Simões R, Simões da Silva L, Cruz P. Behaviour of end-plate beam-to-column composite joints under cyclic loading. Steel Compos Struct 2001;1:355–76.
[185] Qnet for Windows. Neural network modelling, V2K build 721. "user manual". Winnetka, IL: Vesta Services, Inc; 2000.
[186] Mitchell M. An introduction to genetic algorithms. Cambridge, MA: The MIT Press; 1994.
[187] Koza JR. Genetic programming: on the programming of computer by means of natural selection. Cambridge, MA: MIT Press; 1992.
[188] Goldberg DE. Genetic algorithms in search, optimization, and machine learning. Boston, MA: Addison-Wesley Publishing Company; 1989.
[189] Back T. Evolutionary algorithms in theory and practice. Oxford, UK: Oxford University Press; 1996.
[190] Fogel LJ, Owens AJ, Walsh MJ. Artificial intelligence through simulated evolution. New York: John Wiley & Sons; 1966.
[191] Michalewicz Z. Genetic algorithms + data structures = evolution programs. 3rd ed. (Revised and Extended). Berlin Heidelberg: Springer-Verlag; 1996.
[192] Holland JH. Adaptation in natural and artificial systems. Boston, MA: MIT Press; 1992.
[193] Koza JR. Genetic programming II: automatic discovery of reusable programs. Cambridge, MA: MIT Press; 1994.
[194] Barbosa HJC. Algoritmos Genéticos para Otimização em Engenharia: Uma Introdução aos Algoritmos Genéticos. In: Lecture Notes, 2ª Escola de Verão em Computação Científica, Laboratório Nacional de Computação Científica—LNCC and UFJF, Rio de Janeiro; 1997 [in portuguese].
[195] De Jong KA. Genetic algorithms are not function optimizers. In: FOGA-92, foundations of genetic algorithms, Vail; 1992. p. 24–9.

[196] Michielssen E, Weile DS. Electromagnetic system design using genetic algorithms. In: Genetic algorithms in engineering and computer science. England: John Wiley & Sons; 1995. p. 345–69.

[197] Fonseca ET, Pacheco MAC, Vellasco PCGdaS, Andrade SALde. Estudo do Fenômeno de Cargas Concentradas Utilizando Algoritmos Genéticos. In: CILAMCE 2000 Iberian Latin American cong. on computation methods in engineering, Rio de Janeiro; 2000 [in portuguese].

[198] Bergfelt A. Patch Loading on a Slender Web. Influence of Horizontal and Vertical Alma Stiffeners on the Load Carrying Capacity. Göteborg, Sweden: Chalmers University of Technology; 1979. s79:1.

[199] Evolver 4.0. Genetic optimization for Microsoft Excel. Palisade Decision Tools; 1999.

[200] Ramires FB, Lima LROde, Andrade SALde, Vellasco PCGdaS, Silva JGSda. Genetic algorithms structural optimisation of beam-to-column semi-rigid joints. In: 7th int conf comp struct tech, Lisbon; 2004.

[201] Ramires FB, Andrade SALde, Vellasco PCGdaS. A Otimização Estrutural de Ligações Semi-rigid Mistas Através de Algoritmos Genéticos. In: CILAMCE 2006 Iberian Latin American congress on computation methods in engineering, Belém; 2006 [in portuguese].

[202] Ahmed B, Nethercot DA. Prediction of initial stiffness and available rotation capacity of major axis composite flush End plate connections. Nottingham: University of Nottingham; 1996.

[203] Ramires SFB, Andrade SALde, Vellasco PCGdaS. Experimental analysis of composite semi-rigid beam to colum joints. In: 5th European conf. steel & composite structures. Brussels, ECCS, v. A; 2008. p. 471–6.

[204] Lima LROde, Vellasco PCGdaS, Silva, JGSda, Borges LAC. Post-limit stiffness prediction of semi-rigid joints using genetic algorithms. Int J Solids Struct 2005; 2(4):305–20.

[205] Borges LAC, Lima LROde, Silva LAPSda, Vellasco PCGdaS. Evaluation of the post-limit stiffness of beam-to-column semi-rigid joints using genetic algorithms. In: 9th international conf on civil and structural engineering computing, Egmond aan Zee; 2003.

[206] Lima LRO, Borges LAC, Silva JGda, Vellasco PCGdaSS, Silva LAPSda. Calibraçâo da Rigidez Pós-Limite de Ligações Viga-Coluna Utilizando-se Algoritmos Genéticos. In: CILAMCE 2004 Iberian Latin American cong comp methods eng, Recife; 2004 [in portuguese].

[207] Gervásio H, da Silva LS, Borges LAC. Reliability assessment of the post limit stiffness and ductility of steel joints. 3rd European conf on steel structures, Coimbra. 2002; 2: 1027–38.

[208] Cruz P, Silva LSda, Rodrigues D, Simões RD. Database for the semi-rigid behaviour of beam to column connections in seismic regions. J Constr Steel Res 1998; 46(1–3):233–4.

[209] Lima LROde, Pacheco MAC, Andrade SALde, Vellasco PCGdaS. Genetic algorithms assessment of minor axis semi-rigid connections. In: EASEC8, Eighth East Asia Pacific conference on structural engineering and construction, Singapore; 2001.

[210] Faella C, Piluso V, Rizzano G. Structural steel semi-rigid connections: theory, design and software. Boca Raton, FL: CRC Press LLC; 2000.

[211] Del Sávio AA, Andrade SALde, Vellasco PCGdaS, Martha LFCR. Genetic algorithm optimization of semi-rigid steel structures. In: 8th int conf appl art intel civil and structural engineering, Roma; 2005.

[212] Jang JR, Sun CT, Mizutani E. Neuro-fuzzy and soft computing: computational app to learning & machine intel. Upper Saddle River, NJ: Prentice-Hall; 1997.

[213] Mendel JM. Fuzzy logic systems for engineering: a tutorial. Proc IEEE 1995;83 (3):345–77.

[214] Nauck D, Kruse R. A neuro-fuzzy approach to obtain interpreflange fuzzy systems for function approximation. In: IEEE int conf on Fuzzy Systems (FUZZ-IEEE'98), Anchorage; 1998. p. 1106–11.
[215] Kruse R, Nauck D. NEFCLASS—a neuro-fuzzy approach for the classification of data. In: ACM symp applied computing, Nashville; 1995.
[216] Jang JSR. ANFIS: adaptive-network-based fuzzy inference system. IEEE Trans Syst Man Cybern 1993;23(3):665–85.
[217] Finkel RA, Bentley JL. Quad trees, a data structure for retrieval on composite keys. Acta Inform 1974;4:1–9.
[218] Chrysanthou Y, Slater M. Computing dynamic changes to BSP trees. Comput Graphics Forum 1992;11(3):321–32 (EUROGRAPHICS '92).
[219] Vellasco MMBR, Pacheco MAC, Neto LSR, de Souza FJ. Electric load forecasting: evaluating the novel hierarchical neuro-fuzzy BSP model. Int J Electr Power Energy Syst 2004;26(2):131–42.
[220] Gonçalves LB, Vellasco MMBR, Pacheco MAC, de Souza FJ. Inverted hierarchical neuro-fuzzy BSP system: a novel neuro-fuzzy model for pattern classification and rule extension in databases. IEEE Trans Syst Man Cybern Part C Appl Rev 2006;36 (2):236–48.
[221] de Souza FJ, Vellasco MMBR, Pacheco MAC. Load forecasting with the hierarchical neuro-fuzzy binary space partitioning model. Int J Comput Syst Signals 2002;3 (2):118–32.
[222] Vellasco MMBR, Pacheco MA, Figueiredo K, Souza FJ. Hierarchical neuro-fuzzy systems—part I. In: Rabuñal JR, Dorado J, Pazos A, editors. Encyclopaedia of artificial intelligence. Hershey PA: Information Science Referente; 2008.
[223] Vellasco MMBR, Pacheco MA, Figueiredo K, Souza FJ. Hierarchical neuro-fuzzy systems—part II. In: Rabuñal JR, Dorado J, Pazos A, editors. Encyclopaedia of artificial intelligence. Hershey PA: Information Science Reference; 2008 (formerly Idea Group Reference).
[224] Fonseca ET, de Andrade SAL, Vellasco PCGdaS, Vellasco MMBR. A parametric analysis of the patch load behaviour using a neuro-fuzzy system. J Constr Steel Res 2007;63(2):194–210.
[225] Fonseca ET, Vellasco PCGdaS, Andrade SALde, Vellasco MMBR. A neuro-fuzzy evaluation of steel beams patch load behaviour. Adv Eng Softw 2008;39(7):558–72.
[226] Fonseca ET, Vellasco PCGdaS, Vellasco MMBR, Andrade SALde. A neuro-fuzzy system for patch load prediction. In: 7th int conf appl artificial intel civil & structural engineering, Egmond aan Zee; 2003.
[227] Fonseca ET, Vellasco PCGdaS, Vellasco MMBR, Andrade SALde. Neuro-fuzzy system for steel beams patch load prediction. In: 5Th int conf hybrid intelligent systems (HIS'05), Rio de Janeiro; 2005. p. 110–5.
[228] Fonseca ET, Vellasco PCGdaS, Vellasco MMBR, Andrade SALde. A parametric analysis of the patch load behaviour using a neuro-fuzzy system. In: 9th int conf on the application of artificial intelligence to civil & structural engineering, Malta; 2007.
[229] Vellasco PCGdaS, Vellasco MMBR. Applications of computational intelligence techniques to steel and composite structures. In: Soft computing in civil and structural engineering. 1st ed. Kippen: Saxe-Coburg; 2009. p. 73–126, ISBN:9781874672432 [Chapter 3].

INDEX

Note: Page numbers followed by *f* indicate figures, and *t* indicate tables.

D

E

R

S

T

FSC
www.fsc.org
MIX
Papier | Fördert gute Waldnutzung
FSC® C083411

Kolibri 360 GmbH
Ferdinand-Jühlke-Straße 7
99095 Erfurt, Deutschland
produktsicherheit@kolibri360.de